Nonlinear Differential Problems with Smooth and Nonsmooth Constraints

Mathematical Analysis and its Applications Series

Nonlinear Differential Problems with Smooth and Nonsmooth Constraints

Dumitru Motreanu
University of Perpignan, Department of Mathematics,
66000 Perpignan, France

Series Editor
Themistocles Rassias

ELSEVIER

ACADEMIC PRESS
An imprint of Elsevier

Academic Press is an imprint of Elsevier
125 London Wall, London EC2Y 5AS, United Kingdom
525 B Street, Suite 1800, San Diego, CA 92101-4495, United States
50 Hampshire Street, 5th Floor, Cambridge, MA 02139, United States
The Boulevard, Langford Lane, Kidlington, Oxford OX5 1GB, United Kingdom

Notices

Knowledge and best practice in this field are constantly changing. As new research and experience broaden our understanding, changes in research methods, professional practices, or medical treatment may become necessary.

Practitioners and researchers must always rely on their own experience and knowledge in evaluating and using any information, methods, compounds, or experiments described herein. In using such information or methods they should be mindful of their own safety and the safety of others, including parties for whom they have a professional responsibility.

To the fullest extent of the law, neither the Publisher nor the authors, contributors, or editors, assume any liability for any injury and/or damage to persons or property as a matter of products liability, negligence or otherwise, or from any use or operation of any methods, products, instructions, or ideas contained in the material herein.

Library of Congress Cataloging-in-Publication Data
A catalog record for this book is available from the Library of Congress

British Library Cataloguing-in-Publication Data
A catalogue record for this book is available from the British Library

ISBN: 978-0-12-813386-6

For information on all Academic Press publications
visit our website at https://www.elsevier.com/books-and-journals

Publisher: Candice Janco
Acquisition Editor: Graham Nisbet
Editorial Project Manager: Jaclyn A. Truesdell
Production Project Manager: Sreejith Viswanathan
Designer: Mark Rogers

Typeset by VTeX

Dedication

This book is dedicated to the memory of my wife Mariana Claudia Motreanu

Dumitru Motreanu

Contents

About the Author

Dumitru Motreanu is Professor of Mathematics at the University of Perpignan, in France. Dr. Motreanu received his PhD degree in 1978 from the University of Iasi, Romania. In 1991 he was a recipient of Simeon Stoilov Award from the Romanian Academy of Sciences. His areas of expertise cover partial differential equations and nonlinear analysis. He obtained original results in smooth and nonsmooth variational methods, nonlinear eigenvalue problems, multiple solutions for elliptic equations and systems. He coauthored seven monographs in mathematics and published more than 190 professional articles in prestigious international journals of mathematics, as well as more than 20 chapters in research volumes. He also edited a handbook of nonconvex analysis and three special issues of international journals. His results are cited and used in more than 1900 refereed papers. He serves in the editorial boards of more than 20 academic journals in mathematics.

Preface

The book comprises a wide range of subjects passing from eigenvalue problems for nonlinear elliptic operators to control problems for evolution equations driven by variational inequalities. In the large panel of topics covered in the book, the reader will discover surprisingly strong links between the underlying components. The core of the whole material is represented by the study, through structural differential properties, of nonlinear smooth or nonsmooth problems subject to constraints of different type. The choice of statements and approach reflects the author's scientific preferences and original contributions during the last five years.

The book is organized in nine chapters arranged gradually and having to a large extent an independent existence. Many of the results included here appear for the first time in the book form. The notes given in the final section of every chapter address the evolution of main ideas within a specific theme attempting to make a bridge from classical mathematics to recent research including open problems. The references are carefully selected listing titles strictly tied to the mentioned results.

The author is deeply indebted to Dr. Lucas Fresse and Dr. Viorica Venera Motreanu for their huge efforts in reading, correcting and improving the text.

The author expresses his gratitude to Prof. Themistocles M. Rassias, the editor of the series "Mathematical Analysis and its Applications" for his kind invitation and generous support.

The author thanks the Editorial Project Manager Jackie Truesdell and the Senior Acquiring Editor Graham Nisbet for their outstanding editorial work. Many thanks are also due to Publisher's Academic Press with Elsevier Inc., specifically, the Project Manager, Reference Content Production, Sreejith Viswanathan for his great assistance.

Dumitru Motreanu
Perpignan

October 2017

Acknowledgment

I thank my beloved daughter Dr. Viorica Venera Motreanu and Dr. Lucas Fresse for thoroughly checking and greatly improving the text. I thank Prof. Themistocles M. Rassias for kindly inviting me to contribute with a volume in the series entitled "Mathematical Analysis and its Applications" for which he is the Series Editor. I thank my Editors and Publishers with Academic Press of Elsevier Inc., and first of all the Editorial Project Manager Jackie Truesdell, the Senior Acquiring Editor Graham Nisbet, and the Project Manager, Reference Content Production, Sreejith Viswanathan for highly professional assistance and constant support.

Dumitru Motreanu

November 2017

Introduction

This book is built on two main ideas regarding the topics under consideration: to exploit the presence of the available differentiability properties (even in a weak sense) of the data in the problem and to figure out the natural (smooth or nonsmooth) constraints that the solutions should satisfy. The interplay between these two essential features is systematically taken into account in every studied subject.

In order to give a hint about the meaning of this mutual relationship, let us focus on a simplified but relevant situation. Consider an inclusion problem

$$0 \in A(x, u, \nabla u)$$

on a bounded domain $\Omega \subset \mathbb{R}^N$, with $A : \Omega \times \mathbb{R} \times \mathbb{R}^N \to 2^{\mathbb{R}}$ satisfying a (global) growth condition

$$\sup |A(x, s, y)| \leq \alpha(s) + \beta(y) \quad \text{on } \Omega \times \mathbb{R} \times \mathbb{R}^N$$

for $\alpha : \mathbb{R} \to \mathbb{R}$ and $\beta : \mathbb{R}^N \to \mathbb{R}$ monotone, and seek for a solution $u : \Omega \to \mathbb{R}$ with a gradient $\nabla u : \Omega \to \mathbb{R}^N$ subject to the pointwise constraints on the pattern of subsolution–supersolution and obstacle bound

$$\underline{u}(x) \leq u(x) \leq \overline{u}(x) \quad \text{and} \quad |\nabla u(x)| \leq \tilde{u}(x),$$

with $\underline{u}, \overline{u}, \tilde{u}$ such that $\alpha(\underline{u}), \alpha(\overline{u}), \beta(\tilde{u})$ are in some $L^p(\Omega)$, $p \geq 1$. Then the needed growth for the nonlinear operator A driving the problem is actually local and pointwise, namely

$$\sup |A(x, s, y)| \leq \max\{\alpha(\underline{u}(x)), \alpha(\overline{u}(x))\} + \beta(\tilde{u}(x)) \equiv w(x)$$

for all $x \in \Omega$, $s \in [\underline{u}(x), \overline{u}(x)]$, $|y| \leq \tilde{u}(x)$, with $w \in L^p(\Omega)$.

The book consists of nine chapters upon which we give a brief overview. Chapter 1 provides a concise presentation of selected facts related to Sobolev spaces, differentiability and generalized monotonicity. Chapter 2 discusses regularity properties of solutions and a general strong maximum principle. Chapter 3 treats eigenvalue problems for important nonlinear operators as p-Laplacian, (p, q)-Laplacian, nonhomogeneous differential operators by admitting various boundary conditions. Chapter 4 focuses on the nonlinear elliptic problems exhibiting full dependence on the gradient of the solution, which prevents the application of variational methods. Chapter 5 presents results ensuring the existence of solutions with precise sign information, namely positive, negative, and nodal (i.e., sign-changing) solutions for quasilinear elliptic problems, possibly with multivalued terms. Chapter 6 reports on systems of quasilinear elliptic equations paying special attention to the general method of subsolution–supersolution giving rise to the prominent concept of a trapping region. Chapter 7 sets forth existence of positive solutions for systems of quasilinear elliptic equations allowing singularities with respect to the solutions and the gradients of the solutions. Chapter 8 contains various theorems on the solvability of variational and quasilinear inequalities incorporating in a unifying way the elliptic and evolutionary cases. Chapter 9 aims to handle general evolutionary inclusion problems involving control parameters and constraints for the solutions subject to variational inequalities. We close every chapter with a section where one finds notes on related developments, open problems with partial answers, and historic comments.

Chapter 1

Elements of Functional Analysis and Operator Theory

1.1 SOBOLEV SPACES

We start by presenting some basic elements of integration theory in a Banach space. Let I be an open interval in $\mathbb{R}$ of Lebesgue measure $|I|$ and let X be a real Banach space with the norm $\|\cdot\|$, dual space X^*, and duality bracket $\langle\cdot,\cdot\rangle$ between X and X^*. We denote by $C_c(I,X)$ the Banach space of continuous functions from I to X with compact support.

Definition 1.1. The function $f : I \to X$ is measurable if there exists a sequence $\{f_n\} \subset C_c(I,X)$ such that $f_n(t) \to f(t)$ in X for a.e. $t \in I$. A measurable function $f : I \to X$ is called integrable if there exists a sequence $\{f_n\} \subset C_c(I,X)$ such that

$$\lim_{n\to\infty} \int_I \|f_n(t) - f(t)\|\, dt = 0. \tag{1.1}$$

Proposition 1.2. *If the function $f : I \to X$ is integrable, then there exists a unique point in X denoted $\int_I f(t)\,dt$ (called the integral of f over I) such that for every sequence $\{f_n\} \subset C_c(I,X)$ satisfying* (1.1) *one has*

$$\lim_{n\to\infty} \int_I f_n(t)\,dt = \int_I f(t)\,dt. \tag{1.2}$$

Proof. It is straightforward to show from (1.1) that $\left\{\int_I f_n(t)\,dt\right\}$ is a Cauchy sequence in X, so the limit in (1.2) exists. Furthermore, again through (1.1), a direct verification ensures that the limit in (1.2) does not depend on the sequence $\{f_n\}$ chosen in (1.1). □

A powerful criterion of integrability is provided in the following result.

Theorem 1.3 (Bochner's theorem). *A measurable function $f : I \to X$ is integrable if and only if $\|f\| : I \to \mathbb{R}$ is integrable.*

Moreover, in the case of integrability, it holds

$$\left\| \int_I f(t)\,dt \right\| \le \int_I \|f(t)\|\,dt. \tag{1.3}$$

Nonlinear Differential Problems with Smooth and Nonsmooth Constraints. DOI: 10.1016/B978-0-12-813386-6.00001-8

Proof. Assume that $f : I \to X$ is integrable. If the sequence $\{f_n\} \subset C_c(I, X)$ fulfills (1.1), then the integrability of $\|f\| : I \to \mathbb{R}$ follows readily by considering the sequence $\{\|f_n\|\} \subset C_c(I, \mathbb{R})$.

Conversely, by the measurability of $f : I \to X$ and the integrability of $\|f\| : I \to \mathbb{R}$, we can find a sequence $\{f_n\} \subset C_c(I, X)$ with $f_n(t) \to f(t)$ in X for a.e. $t \in I$ and a sequence $\{g_n\} \subset C_c(I, \mathbb{R})$ with $g_n \to \|f\|$ in $L^1(I)$. Setting

$$h_n = \frac{\|g_n\| f_n}{\|f_n\| + \frac{1}{n}},$$

by Lebesgue's dominated convergence theorem we infer that

$$\lim_{n\to\infty} \int_I \|h_n(t) - f(t)\| \, dt = 0,$$

so $f : I \to X$ is integrable.

Inequality (1.3) follows because

$$\| \int_I h_n(t) \, dt \| \le \int_I \|h_n(t)\| \, dt$$

and we pass to the limit as $n \to \infty$. □

We recall the Hahn–Banach separation theorem. For its proof we refer to [31, Theorems 1.6, 1.7].

Theorem 1.4. *Let A and B be two nonempty, disjoint, and convex subsets of a normed space X.*

(i) *Assume that one of them is open. Then there exist $f \in X^* \setminus \{0\}$ and $\alpha \in \mathbb{R}$ such that*

$$\langle f, x \rangle \le \alpha, \ \forall x \in A, \ \text{and} \ \langle f, y \rangle \ge \alpha, \forall y \in B.$$

(ii) *Assume that the set A is closed and the set B is compact. Then there exist $f \in X^* \setminus \{0\}$ and real numbers $\alpha < \beta$ such that the strict separation holds*

$$\langle f, x \rangle \le \alpha, \ \forall x \in A, \ \text{and} \ \langle f, y \rangle \ge \beta, \forall y \in B.$$

Two important consequences of Theorem 1.4 (ii) are needed in the sequel.

Corollary 1.5. *A linear subspace S of a normed space X is dense in X if and only if the following holds: $f \in X^*$ with $\langle f, x \rangle = 0$ for all $x \in S$ implies $f = 0$.*

Proof. The necessity part follows from the continuity of f. For the sufficiency, suppose there is $x_0 \in X \setminus \overline{S}$. Then Theorem 1.4 provides $f \in X^* \setminus \{0\}$ with the property

$$\langle f, x \rangle < f(x_0), \ \forall x \in \overline{S}.$$

Replacing x by λx, with $\lambda \in \mathbb{R}$, enables us to deduce that f vanishes on S, hence $f = 0$ in view of our hypothesis. This is impossible because the above inequality implies $f(x_0) > 0$ (take $x = 0$). The reached contradiction achieves the proof. □

Corollary 1.6. *Let $f : I \to X$ be an integrable function on an open interval I and let K be a closed convex subset of a real Banach space X such that $f(t) \in K$ for all $t \in I$. Then it holds*

$$\frac{1}{|I|}\int_I f(t)\,dt \in K. \tag{1.4}$$

Proof. Arguing by contradiction, assume that (1.4) is not true. Then Theorem 1.4 provides $g \in X^* \setminus \{0\}$ and $\varepsilon > 0$ satisfying

$$\langle g, x\rangle + \varepsilon < \langle g, \int_I f(t)\,dt\rangle, \quad \forall x \in K.$$

In particular, for a.a. $t \in I$, we get

$$\langle g, f(t)\rangle + \varepsilon < \langle g, \int_I f(t)\,dt\rangle.$$

Integrating over I leads to a contradiction, which completes the proof. □

At this point we focus on certain spaces of integrable functions.

Definition 1.7. Given an open interval I in $\mathbb{R}$, a real Banach space X, and a number $p \in [1, +\infty]$, it is said that a function $f : I \to X$ (in fact, its equivalence class with respect to equality a.e.) belongs to $L^p(I, X)$ if f is measurable and $\|f\| \in L^p(I)$.

When endowed with the norm

$$\|f\|_{L^p(I,X)} = \left(\int_I \|f(t)\|^p\,dt\right)^{\frac{1}{p}} \quad \text{for } 1 \le p < +\infty,$$
$$\|f\|_{L^\infty(I,X)} = \operatorname*{ess\,sup}_{t\in I} \|f(t)\|,$$

$L^p(I, X)$ (for $1 \le p \le +\infty$) becomes a Banach space. The notation $f \in L^p_{\text{loc}}(I, X)$ stands for the fact that $f \in L^p(J, X)$ for every open subinterval J of I with $\overline{J}$ compact and contained in I.

In order to pass to Sobolev spaces, we need weak differentiability properties. Denote by $C_c^k(I, X)$, with $1 \le k \le \infty$, the Banach space of functions of class C^k from I to X with compact support. If $X = \mathbb{R}$, we just denote $C_c^k(I)$.

Lemma 1.8. *Let $f, g \in L^1_{\text{loc}}(I, X)$ satisfy*

$$\int_I \varphi(t)g(t)\,dt = -\int_I \varphi'(t)f(t)\,dt \ \textit{for all}\ \varphi \in C_c^\infty(I).$$

Then for every $t_0 \in I$ there corresponds $x_0 \in X$ (actually, f is continuous on I and $x_0 = f(t_0)$) such that

$$f(t) = x_0 + \int_{t_0}^t g(s)\,ds \ \textit{for all}\ t \in I.$$

Proof. Without loss of generality we may suppose that $I = \mathbb{R}$ and $g \in L^1(\mathbb{R}, X)$. Then we can approximate $g_n \to g$ in $L^1(\mathbb{R}, X)$ with a sequence $\{g_n\} \subset C_c^\infty(\mathbb{R}, X)$. For a $\varphi \in C_c^\infty(\mathbb{R})$, the integration by parts yields

$$\begin{aligned}\int_{\mathbb{R}} \varphi(t)g(t)\,dt &= \lim_{n\to\infty}\int_{\mathbb{R}} \varphi(t)g_n(t)\,dt \\ &= -\lim_{n\to\infty}\int_{\mathbb{R}} \varphi'(t)\left(\int_{t_0}^t g_n(s)\,ds\right)dt \\ &= -\int_{\mathbb{R}} \varphi'(t)\left(\int_{t_0}^t g(s)\,ds\right)dt.\end{aligned}$$

Through the hypothesis, we find

$$\int_{\mathbb{R}} \varphi'(t)\left(f(t) - \int_{t_0}^t g(s)\,ds\right)dt = 0.$$

Taking into account that $\varphi \in C_c^\infty(\mathbb{R})$ is arbitrary, the conclusion ensues. □

We are ready to discuss the Sobolev spaces of functions from I to X.

Definition 1.9. Given an open interval I in $\mathbb{R}$, a real Banach space X and a number $p \in [1, +\infty]$, it is said that $f \in W^{1,p}(I, X)$ if $f \in L^p(I, X)$ and there exists $g \in L^p(I, X)$ (necessarily unique, see Lemma 1.8) such that

$$\int_I \varphi(t)g(t)\,dt = -\int_I \varphi'(t)f(t)\,dt \ \text{for all}\ \varphi \in C_c^1(I).$$

We denote $g = f' = \frac{df}{dt}$. Endowed with the norm

$$\|f\|_{W^{1,p}(I,X)} = \|f\|_{L^p(I,X)} + \|f'\|_{L^p(I,X)},$$

$W^{1,p}(I, X)$ is a Banach space. If $p = 2$, $H^1(I, X) := W^{1,2}(I, X)$ is a Hilbert space.

We point out a useful characterization of the elements of the space $W^{1,p}(I, X)$.

Theorem 1.10. *For any* $f \in L^p(I, X)$, *the following properties are equivalent:*

(i) $f \in W^{1,p}(I, X)$;

(ii) $f(t) = f(s) + \int_s^t g(\tau)\,d\tau$ *for a.a.* $s, t \in I$,

with some $g \in L^p(I, X)$ *(actually,* $g = f'$*, see Definition 1.9).*

Proof. The equivalence can be easily established on the basis of Lemma 1.8. □

Corollary 1.11. *The continuous embedding* $W^{1,p}(I, X) \hookrightarrow C_{b,u}(\overline{I}, X)$ *holds, where the notation* $C_{b,u}(\overline{I}, X)$ *means the Banach space of bounded and uniformly continuous functions from* $\overline{I}$ *to* X*. In particular, one has* $W^{1,p}(I, X) \hookrightarrow L^\infty(I, X)$*. If* $p > 1$*, the Hölder regularity can be achieved* $W^{1,p}(I, X) \hookrightarrow C^{\frac{p-1}{p}}(\overline{I}, X)$.

Proof. Using Theorems 1.3 and 1.10, for every $f \in W^{1,p}(I, X)$ it turns out that

$$| \|f(t)\| - \|f(s)\| | \le \|f(t) - f(s)\| \le \left| \int_s^t \|f'(\tau)\|\,d\tau \right| \quad \text{for a.a. } s, t \in I. \tag{1.5}$$

The second inequality in (1.5) shows the uniform continuity of f on $\overline{I}$, whereas by means of the first inequality in (1.5) we can infer that $\|f\| \in W^{1,p}(I)$, so f is bounded on $\overline{I}$. If $p > 1$, (1.5) in conjunction with Hölder's inequality implies

$$\|f(t) - f(s)\| \le |t - s|^{\frac{p-1}{p}} \|f'\|_{L^p(I,X)} \quad \text{for a.a. } s, t \in I,$$

whence the conclusion. □

Remark 1.12. If $1 \le p, q < +\infty$ and if $X \hookrightarrow Y$ is a continuous embedding of Banach spaces, then $C_c^\infty(\overline{I}, X)$ is dense in $L^p(I, X) \cap W^{1,q}(I, Y)$ (cf. [32, Proposition A.2.46]).

For higher order Sobolev spaces of functions from I to X we proceed inductively. Given an open interval I in $\mathbb{R}$, a real Banach space X, a number $p \in [1, +\infty]$, and an integer $m > 1$, we set

$$W^{m,p}(I, X) = \{f \in W^{m-1,p}(I, X) : f' \in W^{m-1,p}(I, X)\}. \tag{1.6}$$

Corollary 1.13. *For every integer* $m \ge 1$ *and number* $p \in [1, +\infty]$, $W^{m,p}(I, X)$ *is a Banach space and one has the continuous embedding* $W^{m,1}(I, X) \hookrightarrow C^{m-1}(\overline{I}, X)$, *and* $W^{m,p}(I, X) \hookrightarrow C^{m-1,\frac{p-1}{p}}(\overline{I}, X)$ *provided* $p > 1$.

Proof. The result follows inductively from (1.6) and Corollary 1.11. □

We pass to Sobolev spaces of functions defined on multidimensional domains. Let Ω be an open subset of $\mathbb{R}^N$ with Lebesgue measure $|\Omega|$. For every multiindex $\alpha = (\alpha_1, \dots, \alpha_N)$ of length $|\alpha| = \alpha_1 + \cdots + \alpha_N$, where $\alpha_1, \dots, \alpha_N$ are nonnegative integers, the weak derivative $D^\alpha u$ of a locally integrable function $u \in L^1_{\mathrm{loc}}(\Omega)$ is the distribution defined by

$$\langle D^\alpha u, \varphi \rangle = (-1)^{|\alpha|} \int_\Omega u(x) D^\alpha \varphi(x)\, dx \quad \text{for all } \varphi \in C_c^\infty(\Omega).$$

Important particular cases are the (weak) partial derivatives $\partial_i u = \frac{\partial u}{\partial x_i}$ for $i = 1, \dots, N$, and the (weak) gradient $\nabla u = (\partial_1 u, \dots, \partial_N u)$.

Definition 1.14. Given an open set Ω in $\mathbb{R}^N$, a nonnegative integer m, and a number $p \in [1, +\infty]$, it is said that $u \in W^{m,p}(\Omega)$ if $u \in L^p(\Omega)$ and $D^\alpha u \in L^p(\Omega)$ for every multiindex α with $|\alpha| \le m$.

With respect to the norm

$$\|u\|_{W^{m,p}(\Omega)} = \sum_{|\alpha| \le m} \|D^\alpha u\|_{L^p(\Omega)},$$

$W^{m,p}(\Omega)$ is a Banach space. The closure of $C_c^\infty(\Omega)$ in $W^{m,p}(\Omega)$, with $p < +\infty$, is the Banach space $W_0^{m,p}(\Omega)$. When $p = 2$, $H^m(\Omega) := W^{m,2}(\Omega)$ and $H_0^m(\Omega) := W_0^{m,2}(\Omega)$ are Hilbert spaces.

We mention a significant tool to generate elements of the Sobolev space $W^{1,p}(\Omega)$. For more details we refer to [146,78].

Theorem 1.15 (Marcus–Mizel). *If $F : \mathbb{R} \to \mathbb{R}$ is Lipschitz continuous with $F(0) = 0$ and $u \in W^{1,p}(\Omega)$, then for the composition $F \circ u$ there hold $F \circ u \in W^{1,p}(\Omega)$ and $\nabla(F \circ u)(x) = F'(x)\nabla u(x)$ for a.e. $x \in \Omega$. Moreover, if $p < +\infty$, the Nemytskii operator $N_F : W^{1,p}(\Omega) \to W^{1,p}(\Omega)$ given by $N_F(u) = F \circ u$ is continuous.*

Corollary 1.16. *For every $u \in W^{1,p}(\Omega)$, there hold $u^+ := \max\{u, 0\} \in W^{1,p}(\Omega)$, $u^- := \max\{-u, 0\} \in W^{1,p}(\Omega)$, and*

$$\nabla u^+ = \begin{cases} \nabla u & \text{if } u > 0, \\ 0 & \text{if } u \le 0, \end{cases} \qquad \nabla u^- = \begin{cases} -\nabla u & \text{if } u < 0, \\ 0 & \text{if } u \ge 0. \end{cases}$$

Proof. It suffices to prove the assertion for u^+ because $u^- = u^+ - u$. Theorem 1.15 can be applied with the function $F(t) = t^+$, which gives $u^+ \in W^{1,p}(\Omega)$. In order to compute its gradient, for every $\varepsilon > 0$ we set

$$F_\varepsilon(t) = ((t^+)^2 + \varepsilon^2)^{\frac{1}{2}} \quad \text{for all } t \in \mathbb{R}.$$

Then Theorem 1.15 yields

$$\nabla(F_\varepsilon \circ u)(x) = \frac{(u(x))^+}{F_\varepsilon(u(x))}\nabla u(x) \quad \text{for a.a. } x \in \Omega.$$

Letting $\varepsilon \to 0$ provides the result. □

A characteristic feature of the Banach space $W_0^{1,p}(\Omega)$ is set forth.

Theorem 1.17 (Poincaré's inequality). *If the open set $\Omega \subset \mathbb{R}^N$ is bounded in at least one direction, then there exists a constant $C = C(p, \Omega) > 0$ such that*

$$\|u\|_{L^p(\Omega)} \leq C\|\nabla u\|_{L^p(\Omega)} \ \textit{for all } u \in W_0^{1,p}(\Omega).$$

Proof. On the basis of the assumption, without loss of generality we may admit that $\Omega \subset \mathbb{R}^{N-1} \times (0, a)$, with some $a > 0$. In view of the definition of $W_0^{1,p}(\Omega)$, it is sufficient to argue with $u \in C_c^1(\Omega)$. Writing $x = (x', x_n) \in \mathbb{R}^{N-1} \times (0, a)$, it is seen that

$$\begin{aligned} |u(x)|^p &= p\int_0^{x_N} |u(x', s)|^{p-2}u(x', s)\frac{\partial u}{\partial x_N}(x', s)\,ds \\ &\leq p\int_0^a |u(x', s)|^{p-1}\left|\frac{\partial u}{\partial x_N}(x', s)\right| ds. \end{aligned}$$

The integration over Ω, combined with Hölder's inequality and Fubini's theorem, enables us to find

$$\|u\|^p_{L^p(\Omega)} \leq pa\,\|u\|^{p-1}_{L^p(\Omega)}\left\|\frac{\partial u}{\partial x_N}\right\|_{L^p(\Omega)},$$

which entails the result. □

Remark 1.18. The conclusion of Theorem 1.17 remains valid if the measure $|\Omega|$ is finite (for the proof see, e.g., [116, pp. 34–35]). Poincaré's inequality ensures that $\|\nabla u\|_{L^p(\Omega)}$ is an equivalent norm on $W_0^{1,p}(\Omega)$.

In order to clarify the relationship of the Sobolev spaces with other function spaces we need regularity conditions on Ω.

Definition 1.19. An open set Ω in $\mathbb{R}^N$ is said to be of class C^m, with an integer $m \geq 1$, if for every point $x \in \partial\Omega$ there exist a neighborhood U of x in $\mathbb{R}^N$ and a C^m-diffeomorphism Φ of U onto $\{x = (x', x_N) \in \mathbb{R}^{N-1} \times (-1, 1) : |x'| < 1\}$ such that

$$\begin{aligned} \Phi(U \cap \Omega) &= \{x = (x', x_N) \in \mathbb{R}^{N-1} \times (0, 1) : |x'| < 1\}, \\ \Phi(U \cap \partial\Omega) &= \{x = (x', 0) : |x'| < 1\}. \end{aligned}$$

Theorem 1.20 (Sobolev's and Rellich–Kondrachov embedding theorems). *Assume that the open set $\Omega \subset \mathbb{R}^N$ is bounded and of class C^m, with an integer $m \geq 1$, and let $1 \leq p < +\infty$.*

- **(i)** *If $mp < N$, one has the continuous embedding $W^{m,p}(\Omega) \hookrightarrow L^q(\Omega)$ for each $q \leq \frac{Np}{N-mp}$, which is compact provided $q < \frac{Np}{N-mp}$.*
- **(ii)** *If $mp = N$, one has the compact embedding $W^{m,p}(\Omega) \hookrightarrow L^q(\Omega)$ for each $q < +\infty$.*
- **(iii)** *If $mp > N$, one has the continuous embedding $W^{m,p}(\Omega) \hookrightarrow C^{k,\alpha}(\overline{\Omega})$, with k the biggest integer such that $k \leq m - \frac{N}{p}$ and $0 \leq \alpha \leq m - k - \frac{N}{p}$, which is compact provided $\alpha < m - k - \frac{N}{p}$.*

The above assertions remain valid without assuming that Ω is of class C^m when $W^{m,p}(\Omega)$ is replaced by $W_0^{m,p}(\Omega)$.

Remark 1.21. One can relax the regularity condition on Ω required in Theorem 1.20 (see Adams [1]).

We end this section with the characterization of the dual space $W_0^{1,p}(\Omega)^*$, denoted $W^{-1,p'}(\Omega)$ with $\frac{1}{p} + \frac{1}{p'} = 1$ (see, e.g., [31, p. 291]). If $p = 2$, we denote $H^{-1}(\Omega) := H_0^1(\Omega)^*$.

Theorem 1.22. *Let Ω be an open subset of $\mathbb{R}^N$ and $1 \leq p < +\infty$. Then $\xi \in W^{-1,p'}(\Omega)$ if and only if there exist $f_0, f_1, \ldots, f_N \in L^{p'}(\Omega)$ such that*

$$\langle \xi, v \rangle = \int_\Omega f_0(x) v(x)\, dx + \sum_{i=1}^N \int_\Omega f_i(x) \frac{\partial v}{\partial x_i}(x)\, dx \quad \text{for all } v \in W_0^{1,p}(\Omega).$$

In addition, it holds

$$\|\xi\|_{W^{-1,p'}(\Omega)} = \sum_{i=0}^N \|f_i\|_{L^{p'}(\Omega)}.$$

In the case where Ω is bounded, one can choose $f_0 = 0$.

1.2 OPTIMIZATION, SUBDIFFERENTIALS AND GENERALIZED GRADIENTS

This section deals with functionals on a Banach space X, i.e., real-valued functions on X. The first case that we consider is of a linear and continuous functional $f : X \to \mathbb{R}$, which means that $f \in X^*$. We start with the classical description of X^* when X is a Hilbert space, which is known as Riesz representation theorem.

Theorem 1.23. *Let H be a real Hilbert space endowed with the scalar product $(\cdot,\cdot)_H$ and associated Euclidean norm $\|u\|_H = (u,u)_H^{1/2}$. For every $f \in H^*$*

there corresponds a unique $u_f \in H$ *such that*

$$\langle f, v\rangle = (u_f, v)_H \text{ for all } v \in H. \tag{1.7}$$

Moreover, it holds

$$\|f\|_{H^*} = \|u_f\|_H. \tag{1.8}$$

Proof. If $f = 0$, the proof is achieved by taking $u_f = 0$. Suppose that $f \neq 0$. Then we can find $w \in H$, $w \neq 0$, to have the orthogonal direct sum decomposition

$$H = \ker(f) \oplus \mathbb{R}w. \tag{1.9}$$

Precisely, there is $w \in \ker(f)^{\perp} := \{v \in H : (u, v)_H = 0,\ \forall u \in \ker(f)\}$ with $f(w) \neq 0$ (so, $w \neq 0$) because $f \neq 0$. If $v \in \ker(f)^{\perp}$, we note that

$$v - \frac{f(v)}{f(w)} w \in \ker(f) \cap \ker(f)^{\perp},$$

thus $v - \frac{f(v)}{f(w)} w = 0$, which ensures that $v \in \mathbb{R}w$. Then (1.9) follows from the equality $\ker(f)^{\perp} = \mathbb{R}w$.

Let us check that for every $f \in H^*$ the vector $u_f = \frac{f(w)}{\|w\|_H^2} w$ satisfies the requirements. Indeed, observing that $v - \frac{f(v)}{f(w)} w \in \ker(f)$ whenever $v \in H$, the orthogonality of w to $\ker(f)$ in (1.9) results in

$$\begin{aligned}(u_f, v)_H &= \frac{f(w)}{\|w\|_H^2}(w, \left(v - \frac{f(v)}{f(w)} w\right) + \frac{f(v)}{f(w)} w)_H \\ &= \frac{f(w)}{\|w\|_H^2}(w, \frac{f(v)}{f(w)} w)_H = \langle f, v\rangle\end{aligned}$$

for all $v \in H$, so (1.7) holds true.

Using (1.7) and Cauchy–Schwarz inequality allows obtaining the estimate

$$\|f\|_{H^*} = \sup_{\|v\|_H \le 1} |(u_f, v)_H| \le \|u_f\|_H.$$

On the other hand, the expression of u_f renders

$$\|u_f\|_H = \frac{|f(w)|}{\|w\|_H} \le \|f\|_{H^*}.$$

Altogether, (1.8) ensues.

The uniqueness of u_f follows readily from (1.7), which completes the proof. □

Corollary 1.24. *If H is a Hilbert space, then the dual H^* is a Hilbert space.*

Proof. Applying Theorem 1.23 and with the notation therein, for every $f, g \in H^*$ we set $(f, g)_{H^*} = (u_f, u_g)_H$. Hence we obtain a scalar product on H^*. Moreover, from (1.8) it turns out that

$$\|f\|_{H^*} = (f, f)_{H^*}^{1/2} = (u_f, u_f)_H^{1/2} = \|u_f\|_H, \ \forall f \in H^*.$$

Since the normed space $(H, \|\cdot\|_H)$ is complete, the same follows for the normed space $(H^*, \|\cdot\|_{H^*})$, which completes the proof. □

We indicate how Theorem 1.23 can be directly used to study linear boundary value problems whose prototype is discussed below.

Corollary 1.25. *Let Ω be an open subset of $\mathbb{R}^N$ and let $f \in H^{-1}(\Omega)$. Then the Dirichlet problem*

$$\begin{cases} -\Delta u + u = f \ \text{ in } \Omega, \\ u = 0 \ \text{ on } \partial\Omega \end{cases} \tag{1.10}$$

has a unique weak solution $u \in H_0^1(\Omega)$, that is,

$$\int_\Omega (\nabla u, \nabla v)_{\mathbb{R}^N}\, dx + \int_\Omega uv\, dx = f(v), \ \forall v \in H_0^1(\Omega). \tag{1.11}$$

Proof. Since equality (1.11) reads as

$$(u, v)_{H_0^1(\Omega)} = f(v), \ \forall v \in H_0^1(\Omega),$$

the conclusion regarding the unique solvability of problem (1.10) is readily obtained from Theorem 1.23. □

Remark 1.26. A careful examination of the proof of Theorem 1.23 reveals that the unique solution of problem (1.10) is of the form

$$u = \frac{f(w)}{\|w\|_{H_0^1(\Omega)}^2} w$$

for

$$w \in \ker(f)^\perp = \{v \in H_0^1(\Omega) : (z, v)_{H_0^1(\Omega)} = 0 \quad \text{for all } z \in \ker(f)\},$$

with $f(w) \neq 0$ provided $f \neq 0$.

We review a few basic facts regarding differentiability. For a comprehensive theory and applications, we refer to [13].

Definition 1.27. A function $F : U \to \mathbb{R}$ defined on an open subset U of a normed space X is said to admit the directional derivative (or differential) at $u \in U$ in the direction $v \in X$ if there exists the limit

$$dF(u; v) := \lim_{t \to 0^+} \frac{F(u + tv) - F(u)}{t} \in \mathbb{R}.$$

Example 1.28. Let U be an open subset of $X = \mathbb{R}^n$ endowed with the standard scalar product. The partial derivatives $(\frac{\partial F}{\partial x_i}(u))_{1 \le i \le n}$ of a function $F : U \to \mathbb{R}$ at $u \in U$ are the directional derivatives $(dF(u; e_i))_{1 \le i \le n}$, where $\{e_1, \dots, e_n\}$ stands for the canonical basis of $\mathbb{R}^n$.

Definition 1.29. A function $F : U \to \mathbb{R}$ defined on an open subset U of a normed space X is said to be Gâteaux differentiable (or derivable) at $u \in U$ if there exists $l \in X^*$ such that

$$dF(u; v) := l(v) \quad \text{for all } v \in X.$$

We denote $F'(u) := l$ (called the (Gâteaux) differential or derivative of F at u).

Definition 1.30. A function $F : U \to \mathbb{R}$ defined on an open subset U of a normed space X is said to be Fréchet differentiable (or derivable) at $u \in U$ if there exists $l \in X^*$ such that

$$\lim_{h \to 0} \frac{F(u + h) - F(u) - l(h)}{\|h\|_X} = 0.$$

We denote $F'(u) := l$ (called the (Fréchet) differential or derivative of F at u), which is justified by the next statement.

Proposition 1.31. *If the function $F : U \to \mathbb{R}$ is Fréchet differentiable at u, then it is Gâteaux differentiable at u and the differentials coincide.*

Proof. It suffices to insert $h = tv$ in Definition 1.30, with $v \in X$ and $t > 0$, to get the desired conclusion. □

A partial converse of Proposition 1.31 is available.

Proposition 1.32. *If the function $F : U \to \mathbb{R}$ is Gâteaux differentiable on a neighborhood V of u and the differential $F' : V \to X^*$ is continuous at u, then F is Fréchet differentiable at u.*

Proof. Without loss of generality we may suppose that $V = U$. For every $h \in X$, based on Definition 1.29 we may write

$$\begin{aligned} F(u + h) - F(u) - F'(u)h &= \int_0^1 \frac{d}{dt} F(u + th)\, dt - F'(u)h \\ &= \int_0^1 (F'(u + th) - F'(u))h\, dt. \end{aligned}$$

Now it is straightforward to derive that F is Fréchet differentiable at u. □

In the case of a Hilbert space, the differential is expressed through the gradient.

Proposition 1.33. *Let U be an open subset of a Hilbert space X endowed with the scalar product $(\cdot,\cdot)_X$. If the function $F : U \to \mathbb{R}$ is Gâteaux differentiable at $u \in U$, then there exists a unique vector $\nabla F(u) \in X$ (called the gradient of F at u) such that*

$$\langle F'(u), v\rangle = (\nabla F(u), v)_X \quad \textit{for all } v \in X.$$

Proof. One applies Theorem 1.23 on the Hilbert space X taking $f = F'(u)$. □

Example 1.34. **(i)** If U is an open subset of $X = \mathbb{R}^n$ endowed with the standard scalar product and $F : U \to \mathbb{R}$ is Gâteaux differentiable at $u \in U$, then

$$\nabla F(u) = \left(\frac{\partial F}{\partial x_i}(u)\right)_{1\le i\le n}.$$

(ii) For the function $F(u) = \frac{1}{2}\|u\|_X^2$ on the Hilbert space X, the gradient of F at every point $u \in X$ is $\nabla F(u) = u$.

Remark 1.35. The differentiability notions in Definitions 1.27, 1.29, and 1.30 can similarly be introduced for mappings with values in a Banach space in place of scalar values. Propositions 1.31 and 1.32 are valid for vector-valued mappings (see, e.g., [13]).

The implicit function theorem as stated below will be very useful in the following.

Theorem 1.36. *Let E, X, Y be Banach spaces, let $F : D \to Y$ be a mapping of class C^k with $k \ge 1$ defined on an open subset D of $E \times X$ and let a point $(u_0, x_0) \in D$ such that*

$$F(u_0, x_0) = 0$$

and the partial differential $F_x(u_0, x_0)$ is an isomorphism of X onto Y. Then there exist a neighborhood U of u_0 in E, a neighborhood V of x_0 in X, and a unique mapping $g : U \to V$ of class C^k such that

(i) *$g(u_0) = x_0$;*
(ii) *$F(u, g(u)) = 0$, $\forall u \in U$;*
(iii) *if $F(u, x) = 0$ with $(u, x) \in U \times V$, then $x = g(u)$;*
(iv) *$g'(u) = -F_x(u, g(u))^{-1} \circ F_u(u, g(u))$, $\forall u \in U$.*

Proof. Let us define $G : D \to E \times Y$ by

$$G(u, x) = (u, F(u, x)), \ \forall (u, x) \in D.$$

It is clear that G is of class C^k and its differential at (u_0, x_0) is expressed as follows:

$$G'(u_0, x_0)(u, x) = (u, F'(u_0, x_0)(u, x)), \ \forall (u, x) \in E \times X.$$

Using the hypothesis, it is straightforward to check that $G'(u_0, x_0)$ is an isomorphism, which permits applying the inverse function theorem. Since $G(u_0, x_0) = (u_0, 0)$, we find that there exist a neighborhood U of u_0 in E and a neighborhood V of x_0 in X such that $G|_{U\times V}$ is a C^k-diffeomorphism onto a neighborhood of $(u_0, 0)$ in $E \times Y$. Denote the inverse $Q = (G|_{U\times V})^{-1}$ on that neighborhood and set $Q(v, y) = (h(v, y), \varphi(v, y))$.

In view of the definition of Q, on a neighborhood of $(u_0, 0)$ in $E \times Y$ we have

$$(v, y) = G(Q(v, y)) = (h(v, y), F(h(v, y), \varphi(v, y)),$$

so

$$Q(v, y) = (v, \varphi(v, y)) \tag{1.12}$$

and

$$F(v, \varphi(v, y)) = y. \tag{1.13}$$

Now let us define

$$g(v) = \varphi(v, 0), \ \forall v \in U. \tag{1.14}$$

From (1.13) and (1.14) we infer that $F(v, g(v)) = 0$, thus assertion (ii) holds.

On the basis of (1.12), (1.13) we derive that

$$\begin{aligned}(u_0, x_0) &= Q(G(u_0, x_0)) = Q(u_0, F(u_0, x_0)) = Q(u_0, 0) = (u_0, \varphi(u_0, 0)) \\ &= (u_0, g(u_0)),\end{aligned}$$

whence (i).

In order to show (iii), let $(u, x) \in U \times V$ satisfy $F(u, x) = 0$. Using again (1.13) we note that

$$(u, x) = Q(G(u, x)) = Q(u, F(u, x)) = Q(u, 0) = (u, \varphi(u, 0)) = (u, g(u)),$$

which confirms that (iii) is valid.

Finally, differentiating $F(u, g(u)) = 0$ implies

$$F_u(u, g(u)) + F_x(u, g(u))g'(u) = 0.$$

This enables us to get the formula in (iv) on a possibly smaller neighborhood U of u_0 by making use of the hypothesis that $F_x(u_0, x_0)$ is an isomorphism and the set of isomorphisms of X onto Y is an open subset of $L(X, Y)$. □

Next we focus on the minimization problems with constraints in the form of an equality, that is,

$$\inf_{v \in S} J(v), \tag{1.15}$$

where $J : X \to \mathbb{R}$ is a function of class C^1 on a Banach space X and the set of constraints S is described by

$$S = \{v \in X : F(v) = 0\}, \tag{1.16}$$

with a function $F : X \to \mathbb{R}$ of class C^1. By a solution of problem (1.15) (with S in (1.16)) we mean any point $u_0 \in S$ satisfying

$$J(u_0) = \inf_{v \in S} J(v).$$

We seek necessary conditions of optimality for problem (1.15), (1.16), which gives rise to the first result on the Lagrange multiplier rule. The main tool is the implicit function ensured by Theorem 1.36.

Theorem 1.37. *Assume that* 0 *is a regular value of* F, *that is,* $F'(u) \neq 0$ *for every* $u \in S$, *with* S *in (1.16). Then for every solution* $u_0 \in S$ *of problem (1.15) there exists some* $\lambda \in \mathbb{R}$ *(called Lagrange multiplier associated to* u_0*) such that*

$$J'(u_0) = \lambda F'(u_0). \tag{1.17}$$

Proof. Let $u_0 \in S$ be a solution of problem (1.15), (1.16). Since $F'(u_0) \neq 0$, there is $w \in X$ with $F'(u_0)w \neq 0$. Then, denoting $X_0 = \ker F'(u_0)$, we have the direct sum splitting

$$X = X_0 \oplus \mathbb{R}w. \tag{1.18}$$

Indeed, for any $u \in X$ it is seen that $u - F'(u_0)u \in X_0$, whereas $X_0 \cap \mathbb{R}w = \{0\}$, which proves (1.18). Motivated by (1.18), we introduce the function $G : X_0 \times \mathbb{R} \to \mathbb{R}$ defined by

$$G(v,t) = F(u_0 + v + tw), \ \forall (v,t) \in X_0 \times \mathbb{R}. \tag{1.19}$$

From (1.19) we obtain that G is a C^1 function,

$$G(0,0) = F(u_0) = 0 \quad \text{and} \quad G_t(0,0) = F'(u_0)w\mathrm{id}_{\mathbb{R}}.$$

Therefore Theorem 1.36 can be applied, ensuring the existence of a neighborhood V of 0 in X_0, a neighborhood I of 0 in $\mathbb{R}$, and a unique function $g : V \to I$ of class C^1 with the properties:

$$g(0) = 0, \tag{1.20}$$

$$\forall (v,t) \in V \times I, \quad G(v,t) = 0 \Leftrightarrow t = g(v), \tag{1.21}$$

$$g'(0) = -(G_t(0,0))^{-1} \circ G_v(0,0) = -(F'(u_0)w)^{-1} F'(u_0)|_{X_0} = 0. \tag{1.22}$$

We define $\tilde{J} : V \to \mathbb{R}$ by

$$\tilde{J}(v) = J(u_0 + v + g(v)w), \ \forall v \in V.$$

It turns out that $\tilde{J}$ is of class C^1 on V and attains its minimum at 0 because (1.20), (1.21), (1.16), and (1.19) entail

$$\tilde{J}(v) = J(u_0 + v + g(v)w) \geq J(u_0) = \tilde{J}(0), \ \forall v \in V,$$

thereby $\tilde{J}'(0) = 0$. On the other hand, from (1.22) it is seen that

$$\tilde{J}'(0) = J'(u_0)(\mathrm{id}_{X_0} + g'(0)w) = J'(u_0)|_{X_0},$$

which renders

$$J'(u_0)|_{X_0} = 0.$$

The obtained result reads as

$$\ker F'(u_0) \subset \ker J'(u_0).$$

If $x \in X$, one has

$$x - (F'(u_0)x)w \in \ker F'(u_0) = X_0,$$

so

$$x - (F'(u_0)x)w \in \ker J'(u_0),$$

or

$$J'(u_0)x = J'(u_0)(w)F'(u_0)x, \ \forall x \in X.$$

Consequently, (1.17) holds by choosing $\lambda = J'(u_0)(w)$, thus the proof is complete. □

We look again at the minimization problem with constraints (1.15) for a function $J : X \to \mathbb{R}$ on a normed space X, but this time the constraints are given in the form of a system of inequalities. We provide a second version of Lagrange multiplier rule adapted to this type of constraints.

Theorem 1.38. *Assume that X is a Banach space, $J, F_i : X \to \mathbb{R}$ are Gâteaux differentiable functions ($i = 1, \ldots, m$) and $u \in X$ is locally a solution of the minimization problem*

$$\inf_{F_i(u) \leq 0, \ i=1,\ldots,m} J(u)$$

such that there exists $v_u \in X$ *satisfying*

$$F_i(u) + F_i'(u)v_u < 0 \text{ for all } i = 1, \dots, m.$$

Then there exist $\lambda_1, \dots, \lambda_m \in \mathbb{R}$ *called Lagrange multipliers such that*

(i) $J'(u) + \sum_{i=1}^{m} \lambda_i F_i'(u) = 0;$

(ii) $\sum_{i=1}^{m} \lambda_i F_i(u) = 0;$

(iii) $\lambda_i \geq 0$ *for all* $i = 1, \dots, m.$

Proof. Consider in the product space $\mathbb{R}^m \times \mathbb{R} = \mathbb{R}^{m+1}$ the subsets

$$A = \{(x, s) = (x_1, \dots, x_m, s) \in \mathbb{R}^m \times \mathbb{R} : x_i < 0 \text{ for } i = 1, \dots, m, \ s < 0\},$$
$$B = \{(y, t) = (y_1, \dots, y_m, t) \in \mathbb{R}^m \times \mathbb{R} : \text{there exists } v \in X \text{ with}$$
$$F_i(u) + F_i'(u)v \leq y_i \text{ for all } i = 1, \dots, m, \ J'(u)v \leq t\}.$$

We see that A and B are convex sets, A is open, $(0, 0) \in B$.

Let us show that the sets A and B are disjoint. Arguing by contradiction, suppose that there is $(x, t) \in A \cap B$, which implies the existence of $v \in X$ such that

$$F_i(u) + F_i'(u)v < 0$$

for all $i = 1, \dots, m$, and $J'(u)v < 0$. By the Gâteaux differentiability of the functions F_i $(i = 1, \dots, m)$ and J (see Definition 1.29), we know that

$$F_i(u + \tau v) = (1 - \tau)F_i(u) + \tau(F_i(u) + F_i'(u)v + a_i(\tau))$$
$$\text{with } \ a_i(\tau) \to 0 \text{ as } \ \tau \to 0^+$$

and

$$J(u + \tau v) = (1 - \tau)J(u) + \tau(J(u) + J'(u)v + b(\tau))$$
$$\text{with } \ b(\tau) \to 0 \text{ as } \ \tau \to 0^+.$$

We are thus led to

$$F_i(u + \tau v) < 0 \quad \text{for } \ \tau > 0 \text{ small}, \ i = 1, \dots, m$$

and

$$J(u + \tau v) < J(u) \quad \text{for } \ \tau > 0 \text{ small},$$

contradicting the optimality of the solution u. Hence the sets A and B are disjoint.

Therefore the separation in Theorem 1.4 can be applied, which ensures the existence of $(\xi, r) = (\xi_1, \dots, \xi_m, r) \in \mathbb{R}^{m+1} \setminus \{(0,0)\}$ and $\alpha \in \mathbb{R}$ such that

$$\sum_{i=1}^{m} \xi_i x_i + rs \le \alpha \le \sum_{i=1}^{m} \xi_i y_i + rt$$

whenever $(x, s) = (x_1, \dots, x_m, s) \in \overline{A}$ and $(y, t) = (y_1, \dots, y_m, t) \in B$. If we set $(x, s) = (y, t) = (0, 0)$, it follows that $\alpha = 0$, so we have

$$\sum_{i=1}^{m} \xi_i x_i + rs \le 0 \le \sum_{i=1}^{m} \xi_i y_i + rt \tag{1.23}$$

whenever $(x, s) = (x_1, \dots, x_m, s) \in \overline{A}$ and $(y, t) = (y_1, \dots, y_m, t) \in B$. Inserting $(x, s) = (0, -1) \in \overline{A}$ in (1.23), we find that $r \ge 0$. Moreover, for $x_i = -1$, $x_j = 0$ if $j \ne i$ and $s = 0$, from (1.23) it follows that

$$\xi_i \ge 0 \quad \text{for } i = 1, \dots, m. \tag{1.24}$$

We claim that

$$r > 0. \tag{1.25}$$

Indeed, if $r = 0$ then (1.23) becomes

$$\sum_{i=1}^{m} \xi_i x_i \le 0 \le \sum_{i=1}^{m} \xi_i y_i \tag{1.26}$$

with $\xi \ne 0$ because $(\xi, r) \ne (0, 0)$. On the other hand, the hypothesis regarding $v_u \in X$ leads to

$$\sum_{i=1}^{m} \xi_i (F_i(u) + F_i'(u) v_u) < 0.$$

Since this contradicts (1.26), the claim in (1.25) is proven.

Thanks to (1.25) we may set

$$\lambda_i = \frac{\xi_i}{r} \quad \text{for } i = 1, \dots, m.$$

In view of (1.24), assertion (iii) holds true.

Using (1.25), we can express (1.23) in the form

$$\sum_{i=1}^{m} \lambda_i x_i + s \le 0 \le \sum_{i=1}^{m} \lambda_i y_i + t \tag{1.27}$$

whenever $(x,s) = (x_1, \dots, x_m, s) \in \overline{A}$ and $(y,t) = (y_1, \dots, y_m, t) \in B$. Let us insert in (1.27), for every $w \in X$, $y_i = F_i(u) + F_i'(u)w$, with $i = 1, \dots, m$, and $t = J'(u)w$, which is possible in view of the definition of the set B. This yields

$$J'(u)w + \sum_{i=1}^{m} \lambda_i (F_i(u) + F_i'(u)w) \geq 0 \text{ for all } w \in X. \tag{1.28}$$

Taking $w = 0$ in (1.28) gives

$$\sum_{i=1}^{m} \lambda_i F_i(u) \geq 0.$$

Then by (iii) and since u is admissible for the minimization problem, we infer that assertion (ii) is fulfilled.

If we combine (1.28) and (ii), then we derive

$$J'(u)w + \sum_{i=1}^{m} \lambda_i F_i'(u)w \geq 0 \text{ for all } w \in X.$$

Changing w with $-w$, this is just (i). The proof is thus complete. □

In the following we review some elements of the subdifferentiability and critical point theory for locally Lipschitz functions that will be required in the sequel.

A function $\varphi : X \to \mathbb{R}$ on a Banach space $(X, \|\cdot\|)$ is called locally Lipschitz if for every $u \in X$ there correspond a neighborhood U of u and a constant $L > 0$ such that

$$|\varphi(v) - \varphi(w)| \leq L\|v - w\|, \quad \forall\, v, w \in U.$$

The generalized directional derivative of φ at $u \in X$ along the direction $v \in X$ is defined as

$$\varphi^0(u; v) := \limsup_{\substack{w \to u \\ \tau \to 0^+}} \frac{\varphi(w + \tau v) - \varphi(w)}{\tau}.$$

The generalized gradient of φ at $u \in X$ is defined to be the set

$$\partial\varphi(u) := \left\{ u^* \in X^* \ : \ \langle u^*, v \rangle \leq \varphi^0(u; v) \quad \forall\, v \in X \right\}.$$

The next proposition collects some useful properties of the preceding notions.

Proposition 1.39. *If $\varphi : X \to \mathbb{R}$ is locally Lipschitz, then:*

(i) *$\partial\varphi(u)$ is nonempty, convex, closed, and weak* compact for all $u \in X$;*
(ii) *the multifunction $\partial\varphi : X \to 2^{X^*}$ is upper semicontinuous for the weak* topology on X^*;*

(iii) *for all* $y, z \in X$, *one has* $\varphi^0(y; z) = \max\{\langle y^*, z\rangle : y^* \in \partial\varphi(y)\}$;
(iv) *if* $\varphi \in C^1(X)$, *then* $\partial\varphi(u) = \{\varphi'(u)\}$ *for all* $u \in X$.

A key result in this area is Lebourg's mean value theorem (see [38, Theorem 2.177]).

Proposition 1.40. *If* $\varphi : X \to \mathbb{R}$ *is locally Lipschitz and* $u, v \in X$, *then there exist* $\tau \in (0, 1)$ *and* $w^* \in \partial\varphi(\tau u + (1-\tau)v)$ *such that*

$$\varphi(v) - \varphi(u) = \langle w^*, v - u\rangle.$$

We turn to a brief introduction to nonsmooth critical point theory for locally Lipschitz functions. In particular, it covers the smooth (i.e., C^1) critical point theory. For more details we refer to [51,169].

Definition 1.41. We say that $u \in X$ is a critical point of a locally Lipschitz function $\varphi : X \to \mathbb{R}$ if $0 \in \partial\varphi(u)$.

We denote

$$K(\varphi) := \{u \in X \ : \ u \text{ is a critical point of } \varphi\}$$

and, for any $c \in \mathbb{R}$,

$$K_c(\varphi) := \{u \in K(\varphi) \ : \ \varphi(u) = c\} \quad \text{and} \quad \varphi^c := \{u \in X \ : \ \varphi(u) \le c\}.$$

Definition 1.42. A locally Lipschitz function $\varphi : X \to \mathbb{R}$ is said to satisfy the (nonsmooth) Palais–Smale condition (for short, (PS)) if every sequence $\{u_n\}$ in X such that $\{\varphi(u_n)\}$ is bounded and

$$\varphi^0(u_n; v - u_n) + \varepsilon_n \|v - u_n\| \ge 0, \ \forall\, v \in X, \ n \in \mathbb{N},$$

with a sequence $\{\varepsilon_n\} \subset (0, +\infty)$ such that $\varepsilon_n \to 0$ as $n \to \infty$, has a convergent subsequence in X.

The following nonsmooth version of the deformation theorem given in [51] permits the extension of the minimax principles from the smooth critical point theory to the nonsmooth critical point theory.

Theorem 1.43. *If the locally Lipschitz function* $\varphi : X \to \mathbb{R}$ *on the reflexive Banach space* X *satisfies condition* (PS) *and* $c \in \mathbb{R}$ *verifies* $K_c(\varphi) = \emptyset$, *then for every* $\varepsilon_0 > 0$ *there exist* $\varepsilon \in (0, \varepsilon_0)$ *and a homeomorphism* $\eta : X \to X$ *such that*

(i) $\eta(u) = u$ *for all* $u \notin \varphi^{c+\varepsilon_0} \setminus \varphi^{c-\varepsilon_C}$;
(ii) $\eta(\varphi^{c+\varepsilon_0}) \subset \varphi^{c-\varepsilon_0}$.

The following minimax result is the nonsmooth version of the mountain pass theorem (see [51]).

Theorem 1.44. *If the locally Lipschitz function $\varphi : X \to \mathbb{R}$ on a reflexive Banach space X satisfies condition (PS) and there exist $u_0, u_1 \in X$ and $0 < r < \|u_0 - u_1\|$ such that*

$$\max\{\varphi(u_0), \varphi(u_1)\} < \sigma_r := \inf_{u \in \partial B_r(u_0)} \varphi(u),$$

then for

$$\Gamma := \{\gamma \in C([-1,1], X) \ : \ \gamma(-1) = u_0, \ \gamma(1) = u_1\},$$
$$c := \inf_{\gamma \in \Gamma} \max_{t \in [-1,1]} \varphi(\gamma(t)),$$

one has $c \geq \sigma_r$ and $K_c(\varphi) \neq \emptyset$.

Proof. The inequality $c \geq \eta_r$ is the direct consequence of the definition of c because every path in Γ crosses the sphere $\partial B_r(u_0)$.

Now, arguing by contradiction, suppose that $K_c(\varphi) = \emptyset$. Setting

$$\varepsilon_0 = \sigma_r - \max\{\varphi(u_0), \varphi(u_1)\},$$

we note that the imposed hypothesis ensures $\varepsilon_0 > 0$. Then Theorem 1.43 provides a number $\varepsilon \in (0, \varepsilon_0)$ and a homeomorphism $\eta : X \to X$ satisfying assertions (i) and (ii) in Theorem 1.43.

By the definition of c, we can choose $\gamma \in \Gamma$ such that

$$\varphi(\gamma(t)) \leq c + \varepsilon, \ \forall t \in [-1, 1].$$

Since

$$\varphi(u_0) + \varepsilon_0 \leq \sigma_r \leq c \ \text{ and } \ \varphi(u_1) + \varepsilon_0 \leq \sigma_r \leq c,$$

we find from property (i) in Theorem 1.43 that $\eta(u_0) = u_0$ and $\eta(u_1) = u_1$. Therefore

$$\eta(\gamma(-1)) = u_0 \ \text{ and } \ \eta(\gamma(1)) = u_1,$$

so $\eta \circ \gamma \in \Gamma$. Then property (ii) in Theorem 1.43 asserts

$$c \leq \max_{t \in [-1,1]} \varphi(\eta(\gamma(t))) \leq c - \varepsilon.$$

The obtained contradiction completes the proof. □

We close this section by formulating a nonsmooth version of the second deformation theorem, which is a special case of [94, Theorem 2.1.1].

Theorem 1.45. *If the locally Lipschitz function $\varphi : X \to \mathbb{R}$ satisfies the (PS) condition in Definition 1.42 and there are real numbers $a < b$ such that $K_a(\varphi)$ is a finite set consisting only of local minimizers of φ while $K_c(\varphi) = \emptyset$ for all $c \in \,]a, b]$, then there exists a continuous function $h : [0,1] \times \varphi^b \to \varphi^b$ such that*

(i) $h(0,u) = u$ *and* $h(1,u) \in K_a(\varphi)$ *for all* $u \in \varphi^b$;
(ii) $\varphi(h(t,u)) \le \varphi(u)$ *for all* $(t,u) \in [0,1] \times \varphi^b$.

1.3 MONOTONE AND PSEUDOMONOTONE OPERATORS

First, we discuss significant classes of linear unbounded operators having monotonicity properties.

Definition 1.46. A linear operator $A : D(A) \to X$ defined on a vector subspace $D(A)$ (called the domain of A) of a Banach space X is said to be unbounded if

$$\sup_{u \in D(A),\ \|u\|_X \le 1} \|Au\|_X = +\infty.$$

Example 1.47. Let X be the Banach space of continuous, bounded functions $f : \mathbb{R} \to \mathbb{R}$ endowed with the norm

$$\|f\|_X = \sup_{t \in \mathbb{R}} |f(t)|.$$

Let $D(A)$ be the vector subspace of X formed by the differentiable functions f whose derivative f' belongs to X and introduce $A : D(A) \to X$ by $Af = f'$. Notice that A is unbounded in the sense of Definition 1.46 because, for instance, one has for the functions $f_n(t) = \sin(nt)$ that $\|f_n\|_X = 1$ and $\|Af_n\|_X = n$.

Definition 1.48. A linear (possibly unbounded) operator $A : D(A) \to X$ on a Banach space X is said to be accretive if

$$\|u + \lambda Au\|_X \ge \|u\|_X, \ \forall u \in D(A), \ \forall \lambda > 0. \tag{1.29}$$

An accretive operator $A : D(A) \to X$ is said to be m-accretive if for every $v \in X$ and for every $\lambda > 0$ the equation

$$u + \lambda Au = v \tag{1.30}$$

has a solution $u \in D(A)$ (denoted $J_\lambda v$).

Lemma 1.49. *Assume that the linear operator* $A : D(A) \to X$ *is* m*-accretive. Then the following is true:*

(i) *The solution* $u = J_\lambda v \in D(A)$ *of Eq. (1.30) is unique.*
(ii) *The map* $J_\lambda = (I + \lambda A)^{-1} : X \to X$ *(called the resolvent of* A*) is linear and continuous, with* $\|J_\lambda\|_{L(X)} \le 1$*, where* I *denotes the identity map.*

Proof. The assertions are direct consequences of inequality (1.29). □

Definition 1.50. The Yosida approximation of an m-accretive operator $A : D(A) \to X$ on a Banach space X is the linear operator $A_\lambda := AJ_\lambda : X \to X$.

The following statement lists useful properties of Yosida approximation.

Lemma 1.51. *Assume that the linear operator $A : D(A) \to X$ is m-accretive. Then:*

- **(i)** $A_\lambda = \frac{1}{\lambda}(I - J_\lambda)$.
- **(ii)** *The linear map $A_\lambda : X \to X$ is continuous, with $\|A_\lambda\|_{L(X)} \leq \frac{2}{\lambda}$.*
- **(iii)** $A_\lambda|_{D(A)} = J_\lambda A$.

Proof. (i) Eq. (1.30) reads as

$$J_\lambda u + \lambda A J_\lambda u = u, \ \forall u \in X. \tag{1.31}$$

Taking into account Definition 1.50, we obtain the given expression of A_λ.

(ii) This results easily from part (i) and Lemma 1.49 (ii).

(iii) From (1.31) and the linearity of A we derive

$$A J_\lambda u + \lambda A(A J_\lambda u) = Au, \ \forall u \in D(A).$$

In view of the uniqueness in Lemma 1.49 (i), we can infer that $J_\lambda Au = AJ_\lambda u$ for all $u \in D(A)$, which completes the proof. □

The next result describes the approximation of the identity map by the resolvent J_λ and of the linear unbounded operator A by the Yosida approximation A_λ.

Proposition 1.52. *Assume that the linear operator $A : D(A) \to X$ is m-accretive. Then:*

- **(i)** $\|J_\lambda u - u\|_X \leq \lambda \|Au\|$ *for all* $u \in D(A)$.
- **(ii)** *If $D(A)$ is dense in X, then $J_\lambda u \to u$ in X as $\lambda \to 0^+$ whenever $u \in X$.*
- **(iii)** *If $D(A)$ is dense in X, then $A_\lambda u \to Au$ in X as $\lambda \to 0^+$ whenever $u \in D(A)$.*
- **(iv)** *If $D(A)$ is dense in X, then $J_\lambda u \to u$ in $D(A)$ endowed with the graph norm*

$$\| \cdot \|_{D(A)} = \| \cdot \|_X + \|A \cdot \|_X \tag{1.32}$$

as $\lambda \to 0^+$ whenever $u \in D(A)$.

Proof. (i) By (1.31), Lemma 1.51 (iii), and Lemma 1.49 (ii), for every $u \in D(A)$ we have

$$\|u - J_\lambda u\|_X = \lambda \|J_\lambda Au\|_X \leq \lambda \|J_\lambda\|_{L(X)} \|Au\|_X \leq \lambda \|Au\|_X.$$

(ii) One uses (i) and the density of $D(A)$ in X.

(iii) Through (ii) and Lemma 1.51 (iii), for every $u \in D(A)$ it turns out that

$$\|A_\lambda u - Au\|_X = \|J_\lambda(Au) - Au\|_X \to 0 \text{ as } \lambda \to 0^+.$$

(iv) It suffices to combine (1.32), (ii), and (iii). □

In Hilbert spaces we retrieve the linear monotone operators.

Proposition 1.53. *Assume that X is a Hilbert space with the scalar product $(\cdot,\cdot)_X$. Then:*

(i) *The linear operator $A : D(A) \to X$ is accretive if and only if it is monotone, that is,*

$$(Au, u)_X \geq 0, \ \forall u \in D(A). \tag{1.33}$$

(ii) *If the linear operator $A : D(A) \to X$ is m-accretive, then $D(A)$ is dense in X.*

Proof. (i) Assume that the linear operator $A : D(A) \to X$ is accretive. Then, for every $u \in D(A)$, (1.29) implies

$$\|u\|_X^2 \leq \lambda\|u + \lambda Au\|_X^2 = \|u\|_X^2 + 2\lambda(Au, u)_X + \lambda^2\|Au\|_X^2,$$

which readily leads to (1.33).

Conversely, suppose (1.33). Then, for every $u \in D(A)$ we are able to get

$$\|u\|_X^2 \leq \|u\|_X^2 + \lambda(Au, u)_X = (u + \lambda Au, u)_X \leq \|u\|_X \|u + \lambda Au\|_X.$$

Thus we obtain (1.29).

(ii) Combining Corollary 1.5 and Theorem 1.23, it suffices to show that $(u, v)_X = 0$ for all $v \in D(A)$ ensures $u = 0$. Indeed, setting $v = J_\lambda(u)$, we see that $(u, J_\lambda u)_X = 0$. Then, due to (1.31) and (1.33), it follows that $J_\lambda(u) = 0$. Invoking again (1.31), we obtain $u = 0$ as desired. □

An important example of m-accretive operator in partial differential equations is provided by the following result.

Theorem 1.54. *The negative Laplacian operator $-\Delta : H_0^1(\Omega) \to H^{-1}(\Omega)$ on an open subset $\Omega \subset \mathbb{R}^N$ defined by*

$$\langle -\Delta u, v\rangle = \int_\Omega (\nabla u, \nabla v)_{\mathbb{R}^N}\, dx \quad \text{for all } v \in H_0^1(\Omega) \tag{1.34}$$

is m-accretive with a dense domain $H_0^1(\Omega)$.

Proof. We have to check Definition 1.48 with $X = H^{-1}(\Omega)$, $D(A) = H_0^1(\Omega)$ and $A = -\Delta$. First, we note that an obvious modification of Corollary 1.25 ensures that Eq. (1.30) has a solution $J_\lambda v \in H_0^1(\Omega)$ unique for each $v \in H^{-1}(\Omega)$ (Corollary 1.25 deals with the case $\lambda = 1$).

It remains to verify that the linear operator $-\Delta : H_0^1(\Omega) \to H^{-1}(\Omega)$ is accretive. To this end we make use of Proposition 1.53 (i) taking advantage that

$H^{-1}(\Omega) = H_0^1(\Omega)^*$ is a Hilbert space (see Corollary 1.24). Precisely, we must show that

$$(-\Delta u, u)_{H^{-1}(\Omega)} \geq 0, \ \forall u \in H_0^1(\Omega). \tag{1.35}$$

On the basis of the definition of the scalar product on the dual of a Hilbert space as done in Corollary 1.24 and using (1.11), for every $u \in H_0^1(\Omega)$ we have

$$(-\Delta u, u)_{H^{-1}(\Omega)} = (J_1(-\Delta u), J_1 u)_{H_0^1(\Omega)} = (J_1 u, u)_{H_0^1(\Omega)} - \|J_1 u\|^2_{H_0^1(\Omega)}. \tag{1.36}$$

On the other hand, the definition of the scalar product on $H_0^1(\Omega)$, (1.10) and (1.34) entail that

$$(J_1 u, u)_{H_0^1(\Omega)} = (J_1 u - \Delta(J_1 u), u)_{L^2(\Omega)} = \|u\|^2_{L^2(\Omega)} \tag{1.37}$$

and in addition

$$\|J_1 u\|^2_{H_0^1(\Omega)} = \|u\|^2_{H^{-1}(\Omega)} = \left(\sup_{v \in H_0^1(\Omega),\ \|v\|_{H_0^1(\Omega)} \leq 1} (u, v)_{L^2(\Omega)} \right)^2$$
$$\leq \left(\sup_{v \in H_0^1(\Omega),\ \|v\|_{L^2(\Omega)} \leq 1} (u, v)_{L^2(\Omega)} \right)^2 \leq \|u\|^2_{L^2(\Omega)}. \tag{1.38}$$

From (1.36), (1.37), and (1.38), we see that (1.35) holds true, which ends the proof. ☐

We also mention some facts related to pseudomonotone operators. This is a class of nonlinear operators exhibiting a monotonicity property in a generalized sense. Let X be a Banach space and X^* its topological dual. By $\langle \cdot, \cdot \rangle$ we denote the duality brackets of the pair (X^*, X), by $\rightharpoonup$ the weak convergence, while $\to$ stands for the strong convergence. A map $A : X \to X^*$ is said to be pseudomonotone if for every sequence $\{x_n\} \subset X$ such that $x_n \rightharpoonup x$ in X and

$$\limsup_{n \to +\infty} \langle A(x_n), x_n - x \rangle \leq 0,$$

one has

$$\langle A(x), x - y \rangle \leq \liminf_{n \to +\infty} \langle A(x_n), x_n - y \rangle \text{ for all } y \in X.$$

A map $A : X \to X^*$ is said to satisfy the $(S)_+$-property if for every sequence $\{x_n\} \subset X$ such that $x_n \rightharpoonup x$ in X and

$$\limsup_{n \to +\infty} \langle A(x_n), x_n - x \rangle \leq 0,$$

one has $x_n \to x$ in X. The fundamental theorem for pseudomonotone operators (see [38, Theorem 2.99]) reads as follows.

Theorem 1.55. *Let X be a reflexive Banach space, $A : X \to X^*$ a pseudomonotone, bounded, coercive operator, and $b \in X^*$. Then a solution of the equation $Au = b$ exists.*

Let $C_0^1(\overline{\Omega}) = \{u \in C^1(\overline{\Omega}) : u = 0 \text{ on } \partial\Omega\}$. In what follows it is worth to know that the cone of nonnegative functions $C_0^1(\overline{\Omega})_+ = \{u \in C_0^1(\overline{\Omega}) : u \geq 0$ in $\Omega\}$ has a nonempty interior in the Banach space $C_0^1(\overline{\Omega}) = \{u \in C^1(\overline{\Omega}) : u = 0$ on $\partial\Omega\}$ given by

$$\operatorname{int}(C_0^1(\overline{\Omega})_+) = \{u \in C_0^1(\overline{\Omega}) \,:\, u(x) > 0,\ \forall x \in \Omega,\ \text{and } \frac{\partial u}{\partial \nu}(x) < 0,\ \forall x \in \partial\Omega\}, \tag{1.39}$$

where $\nu = \nu(x)$ is the outer unit normal at $x \in \partial\Omega$ (see [169, p. 220]).

Finally, we cite from [169, Proposition 3.5] a result describing basic properties of the p-Laplacian on the spaces $W^{1,p}(\Omega)$ and $W_0^{1,p}(\Omega)$ on an open set $\Omega \subset \mathbb{R}^N$.

Proposition 1.56. *The nonlinear map $A : W^{1,p}(\Omega) \to W^{1,p}(\Omega)^*$, with $1 < p < +\infty$, defined by*

$$\langle A(u), v\rangle := \int_\Omega (|\nabla u|^{p-2}(\nabla u, \nabla v)_{\mathbb{R}^N} + |u|^{p-2}uv)\,dx \quad \textit{for all } u, v \in W^{1,p}(\Omega)$$

is bounded, continuous, coercive, strictly monotone (so, maximal monotone and pseudomonotone) and satisfies the $(S)_+$-property. The same properties are satisfied by the nonlinear map $-\Delta_p : W_0^{1,p}(\Omega) \to W^{-1,p'}(\Omega)$, with $1 < p < +\infty$, defined by

$$\langle -\Delta_p u, v\rangle := \int_\Omega |\nabla u|^{p-2}(\nabla u, \nabla v)_{\mathbb{R}^N}\,dx \quad \textit{for all } u, v \in W_0^{1,p}(\Omega).$$

We need some facts regarding the spectrum of the operator $-\Delta_p$ on $W_0^{1,p}(\Omega)$. Consider the nonlinear weighted eigenvalue problem driven by the p-Laplacian:

$$\begin{cases} -\Delta_p u = \lambda m(x)|u|^{p-2}u & \text{in } \Omega, \\ u = 0 & \text{on } \partial\Omega, \end{cases} \tag{1.40}$$

where $m \in L^\infty(\Omega)_+$, $m \neq 0$. Problem (1.40) admits a (possibly nonexhaustive if $p \neq 2$) sequence of eigenvalues $0 < \lambda_1(m) < \lambda_2(m) \leq \cdots \to +\infty$ which are known as Lusternik–Schnirelman eigenvalues, where $\lambda_1(m)$ and $\lambda_2(m)$ are the first two eigenvalues of problem (1.40).

Proposition 1.57. *If $m \in L^\infty(\Omega)_+ \setminus \{0\}$, then $\lambda_1(m)$ is simple (in the sense that the associated eigenfunctions together with 0 constitute a one-dimensional vector subspace), isolated, and the corresponding eigenfunctions have constant sign in Ω. The eigenfunctions associated with any eigenvalue $\lambda > \lambda_1(m)$ are nodal (i.e., sign changing). If $m = 1$ then $\lambda_1 := \lambda_1(1)$ admits the following variational characterization:*

$$\lambda_1 = \inf_{u \in W_0^{1,p}(\Omega)\setminus\{0\}} \frac{\|\nabla u\|^p_{L^p(\Omega)}}{\|u\|^p_{L^p(\Omega)}},$$

with the infimum being attained at an eigenfunction which can be chosen such that $\hat{u}_1 \in \operatorname{int}(C_0^1(\overline{\Omega})_+)$ and $\|\hat{u}_1\|_p = 1$. If $m, m' \in L^\infty(\Omega)_+ \setminus \{0\}$, $m \geq m'$, and $m \neq m'$, then $\lambda_1(m) < \lambda_1(m')$.

Denoting by $\lambda_2 := \lambda_2(1)$ the second eigenvalue of the operator $-\Delta_p$ on $W_0^{1,p}(\Omega)$, i.e., of (1.40), we quote from [60] a useful variational characterization of λ_2.

Proposition 1.58. *It holds*

$$\lambda_2 = \inf_{\gamma \in \Gamma_0} \max_{t \in [-1,1]} \|\nabla \gamma(t)\|^p_{L^p(\Omega)}$$

with $\Gamma_0 := \left\{\gamma \in C([-1,1], S) : \gamma(\pm 1) = \pm \hat{u}_1\right\}$, where $S := \left\{u \in W_0^{1,p}(\Omega) : \|u\|_{L^p(\Omega)} = 1\right\}$ is endowed with the induced $W_0^{1,p}(\Omega)$-topology.

1.4 NOTES

Section 1.1 is devoted to fundamental facts regarding the functional setting for the entire monograph. The Sobolev spaces that are considered here consist of functions exhibiting derivatives in the weak (or distributional sense). These spaces are formed either of vector-valued functions defined on a real-line interval, which will be employed for evolutionary partial differential equations, or of real-valued functions defined on a domain in some Euclidean space, which will be employed for elliptic partial differential equations. The integrability in the sense of Bochner and weak differentiability are outlined. For a detailed study we refer to [31,32,36,38,115,204].

From the vast theory of Sobolev spaces we selected the essential elements necessary in the sequel. For a variety of topics in this direction we indicate [1,31, 32,38,103,169,228]. Theorem 1.20 can be proven by transporting the problem on $\mathbb{R}^N$ through the extension operator $P : W^{m,p}(\Omega) \to W^{m,p}(\mathbb{R}^N)$, which exists because Ω is supposed to be of class C^m. The development of this approach can be found in [31].

Section 1.2 deals with elements of differentiability and subdifferentiability theories, as well as optimization and smooth and nonsmooth critical point theories. For us the very beginning lies in Riesz representation theorem (stated as

Theorem 1.23). If in Theorem 1.23 one identifies $f \equiv u_f$ for every $f \in H$, one obtains the identification $H = H^*$ of the Hilbert space H with its dual H^*. This is a relevant example of determination of the dual of a Banach space. Another relevant example is offered by Theorem 1.22 where the dual $W^{-1,p'}(\Omega)$ of the Sobolev space $W_0^{1,p}(\Omega)$ is explicitly described. Other important descriptions of dual spaces are $L^p(\Omega)^* = L^{p'}(\Omega)$, where $p' = \frac{p}{p-1}$ with $1 < p < +\infty$, and $L^1(\Omega)^* = L^\infty(\Omega)$ (see, e.g., [31]).

The differentiability theories that are object of treatment here are those in the sense of Fréchet and Gâteaux, which are both of prime interest. A key result related to differentiability is the implicit function theorem (Theorem 1.36). Important information concerning differentiability is available in [13,116,152]. The differentiability framework is implemented to study optimization. Specifically, we stress the necessary conditions of optimality of Lagrange multiplier rule type. This is set forth in two separate situations, according to description of the set of constraints through equalities or inequalities. Theorem 1.37 formulates the Lagrange multiplier rule for an equality constraint. The extension of Theorem 1.37 allowing to have finitely many equality constraints in the minimization problem (1.15) is as follows (see [116]).

Theorem 1.59. *Assume that X is a Banach space, $F_1, \ldots, F_m \in C^1(X)$ are such that $F_1'(u), \ldots, F_m'(u)$ are linearly independent in X^* whenever*

$$u \in S := \{v \in X : F_1(v) = \ldots = F_m(v) = 0\}$$

and $J : U \to \mathbb{R}$ is a C^1 function on a neighborhood U of S in X. If $u_0 \in S$ satisfies

$$J(u_0) = \inf_{v \in S} J(v),$$

then there exist $\lambda_1, \ldots, \lambda_m \in \mathbb{R}$ (called Lagrange multipliers associated to u_0) such that

$$J'(u_0) = \lambda_1 F_1'(u_0) + \cdots + \lambda_m F_m'(u_0).$$

Our result regarding Lagrange multiplier rule for minimization with constraints described by inequalities is stated in Theorem 1.38. It is inspired by Ponstein [202].

The final part of Section 1.2 is a brief exposition of critical point theory. The setting is that of nonsmooth critical point theory for locally Lipschitz functions, which incorporates the smooth (i.e., C^1) critical point theory. The nonsmooth critical point theory for locally Lipschitz functions is due to Chang [51]. It relies on the notion of generalized gradient, which was introduced by Clarke (see [55]). An essential result for calculus with generalized gradients is given in Proposition 1.40. The nonsmooth deformation result formulated as Theorem 1.43 is Theorem 3.1 in [51]. It represents the core of the construction of the critical point theory for locally Lipschitz functionals. A second deformation result is the object of Theorem 1.45.

Efficient tools of finding critical points (or, equivalently, solutions of variational inclusions, in particular equations) are the minimax theorems. Amongst them, a leading place is occupied by the famous mountain pass theorem, which is shown in a nonsmooth form for locally Lipschitz functionals in Theorem 1.44. A recent linking theorem for locally Lipschitz functions can be found in [56]. Another important nonsmooth critical point theory was constructed by Szulkin [221] for the class of nonsmooth functionals that can be expressed as the sum of a C^1 function and a convex, lower semicontinuous $\not\equiv +\infty$ function. A unification of the nonsmooth critical point theories of Chang and Szulkin was achieved in Motreanu–Panagiotopoulos [174, Chapter 3] covering the class of functions that can be expressed as the sum of a locally Lipschitz function and a convex, lower semicontinuous $\not\equiv +\infty$ function.

The aim of Section 1.3 is twofold: on the one hand, to create the passage to evolution problems, and on the other hand, to link with classes of linear and nonlinear operators. In this respect, the central and far reaching notion is that of monotone operator, and the various generalizations, which is at the basis of the study of evolutionary equations and differential inclusions. In order to cast the principal features, we focus on linear, but unbounded operators, with monotonicity property. Namely, we provide a brief treatment of m-accretive operators, with the fundamental example of the negative Laplacian $-\Delta$ on $H_0^1(\Omega)$. For the general theory of nonlinear maximal monotone operators and applications to evolution problems, we refer to [21,30–32,115,217,228,236,238]. Section 1.3 is also concerned with a powerful generalization of maximal monotone operators defined on the whole space, namely the pseudomonotone operators. The surjectivity result stated as Theorem 1.55 constitutes an efficient tool to prove existence of solutions. For relevant additional information, we address to [38, 169,228,238].

A striking example of pseudomonotone operator is the negative p-Laplacian $-\Delta_p$ for $1 < p < +\infty$. The only linear case is for $p = 2$. The complete description of the spectrum of $-\Delta_p$ when $p \neq 2$ is an open problem. One knows only what is called the beginning of its spectrum, basically the characterization of the first two eigenvalues, which is discussed in Propositions 1.57 and 1.58. For important developments and proofs, we refer to [2,14,38,60,67,69,75,89,131,149, 169,186,211,216].

Chapter 2

Elements of Regularity Theory and Maximum Principle

2.1 PROPERTIES OF THE SOLUTION SET

First we present an abstract result from the transversality theory that will be used in the sequel.

Let U be an open set of a separable reflexive Banach space X, let E be another separable Banach space, and consider a function $I : U \times E \to \mathbb{R}$ with the properties:

(H_1) There exists the partial derivative $I_u : U \times E \to X^*$ of I with respect to $u \in U$ and this is a C^r mapping with $r \geq 1$;

(H_2) There exists the second order partial derivative $I_{uu}(u, a) : X \to X^*$ of I with respect to $u \in U$ at any point $(u, a) \in U \times E$ with

$$I_u(u, a) = 0 \tag{2.1}$$

and it is a Fredholm operator of index zero, that is,

$$\dim \ker I_{uu}(u, a) = \operatorname{codim} \operatorname{Im} I_{uu}(u, a) < +\infty.$$

Lemma 2.1. *Assume that $I : U \times E \to \mathbb{R}$ verifies hypotheses (H_1) and (H_2). Then the next conditions are equivalent:*

(H_3) *For every point $(u, a) \in U \times E$ satisfying (2.1), the kernel $\ker(I_u)'(u, a)$ splits $X \times E$, and the kernels of the linear operators given by the second order partial derivatives $I_{uu}(u, a) : X \to X^*$ and $I_{ua}(u, a) : X \to E^*$ fulfill*

$$\ker I_{uu}(u, a) \cap \ker I_{ua}(u, a) = \{0\}; \tag{2.2}$$

(H_3') *$0 \in X^*$ is a regular value of $I_u : U \times E \to X^*$, i.e., for each $(u, a) \in U \times E$ solving (2.1), the derivative $(I_u)'(u, a) : X \times E \to X^*$ is surjective and its kernel splits.*

Proof. If (u, a) satisfies (2.1), then $(I_u)'(u, a)$ is surjective if and only if for every $v \in X^*$ there is $w \in E$ such that

$$v - I_{ua}(u, a)(\cdot, w) \in \operatorname{Im}(I_{uu}(u, a)). \tag{2.3}$$

Nonlinear Differential Problems with Smooth and Nonsmooth Constraints. DOI: 10.1016/B978-0-12-813386-6.00002-X

The range of the self-adjoint operator $I_{uu}(u,a) : X \to X^*$ is determined by

$$\begin{aligned}\operatorname{Im}(I_{uu}(u,a)) &= \overline{\operatorname{Im}(I_{uu}(u,a))} \\ &= \{z \in X^* : z(x) = 0 \text{ for all } x \in \ker I_{uu}(u,a)\}. \qquad (2.4)\end{aligned}$$

The first equality in (2.4) is valid because $I_{uu}(u,a)$ is a Fredholm operator by hypothesis (H_2), while the second equality is a classical orthogonality relation (see, e.g., [31, Corollary 2.18]). By assumption (H_2), $\ker I_{uu}(u,a)$ possesses a finite basis $\{e_i\}_{i=1}^m$. Then, from (2.3) and (2.4), it turns out that $(I_u')'(u,a)$ is surjective if and only if for every $v \in X^*$, there is an element $w \in E$ such that

$$I_{ua}(u,a)(e_i, w) = v(e_i) \text{ for all } i \in \{1, \dots, m\}.$$

The last equality is equivalent to the surjectivity of the linear operator

$$w \in E \mapsto (I_{ua}(u,a)(e_i, w))_{1 \le i \le m} \in \mathbb{R}^m,$$

which in turn is equivalent to the nonexistence of a nonzero vector $(\alpha_1, \dots, \alpha_m) \in \mathbb{R}^m$ with

$$\sum_{i=1}^m \alpha_i I_{ua}(u,a)(e_i, w) = 0 \text{ for all } w \in E \qquad (2.5)$$

(see Corollary 1.5). We have thus shown the equivalence between the surjectivity of $(I_u')'(u,a)$ and the linear independence of the linear continuous forms on E

$$\{I_{ua}(u,a)(e_i, \cdot)\}_{i=1}^m. \qquad (2.6)$$

Now, writing (2.5) as

$$I_{ua}(u,a)\Big(\sum_{i=1}^m \alpha_i e_i, w\Big) = 0 \text{ for all } w \in E,$$

the independence of the forms in (2.6) is seen to be valid if and only if property (2.2) holds true. The equivalence between (H_3) and (H_3') is proven. □

We recall a few things that are necessary in the theorem below. A residual set in a metric space means a countable intersection of open dense subsets. The Baire's category theorem ensures that in a complete metric space any residual set is dense. A critical point $u \in X$ of a C^1 function $f : X \to \mathbb{R}$ on a Banach space X is a point at which the differential vanishes, i.e., $f'(u) = 0$ (see Definition 1.41). A critical point $u \in X$ of a C^2 function $f : X \to \mathbb{R}$ is called nondegenerate if the second derivative $f''(u) : X \to X^*$ is an isomorphism. A regular value of a C^1 map $f : X \to Y$ between Banach spaces X and

Y is a point $y \in Y$ such that y does not belong to the range of f or for every $x \in X$ with $f(x) = y$ one has that the derivative $f'(x) : X \to Y$ is surjective and its kernel splits X. A C^1 map $f : X \to Y$ between Banach spaces X and Y is called a Fredholm operator if the derivative $f'(x) : X \to Y$ at each $x \in X$ is a linear Fredholm operator, that is, the kernel is finite-dimensional and the range is finite-codimensional. It was shown in [219] that if $f : X \to Y$ is a Fredholm operator, then its index exists being defined by

$$\text{ind } f = \dim \ker f'(x) - \text{codim Im } f'(x)$$

independently of $x \in X$. A fundamental result is the Sard–Smale theorem in [219] (see also [239, pp. 829–830]), which asserts that if a C^k map $f : X \to Y$ between Banach spaces X and Y is a Fredholm operator with

$$k > \max\{\text{ind } f, 0\},$$

then the set of regular values of f is residual in Y.

The next theorem provides generically essential properties of the solution set for problems in variational form (in such a case, the weak solutions coincide with the critical points of the associated Euler functionals).

Theorem 2.2. *Let U be an open subset of a separable Hilbert space X, let E be a separable Banach space, and let $I : U \times E \to \mathbb{R}$ be a function satisfying assumptions (H_1), (H_2), (H_3). Then*

$$G = \{a \in E : \; I(\cdot, a) : X \to \mathbb{R} \text{ has only nondegenerate critical points}\} \tag{2.7}$$

is a residual set in E, hence dense in E. Moreover, if in addition it holds

(H_4) *If $(u_n, a_n) \in U \times G$ satisfies (2.1) for all n and if $a_n \to a \in G$, then $\{u_n\}$ has a convergent subsequence in U,*

then the critical points of $I(\cdot, a)$ are finitely many and depend smoothly on the parameter $a \in G$ in the following sense: for every connected component G_0 of G, there exists a finite collection $\{g_i\}_{1 \le i \le n_{G_0}}$ of C^r mappings defined in the neighborhood of G_0 in E and taking values in U such that

$$\{u \in U : \; I_u(u, a) = 0\} = \{g_i(a)\}_{1 \le i \le n_{G_0}} \quad \text{for all } a \in G_0. \tag{2.8}$$

Moreover, $g_i(a) \neq g_j(a)$ for all $a \in G_0$, $i \neq j$. In particular, the critical points of $I(\cdot, a)$ are finitely many and their number is equal to n_{G_0} which is constant whenever $a \in G_0$.

Proof. Lemma 2.1 ensures that $M := (I_u)^{-1}(0)$ is a C^1 submanifold of $U \times E$. Its tangent space at a point $(u, a) \in M$ is expressed as

$$T_{(u,a)} M = I_u(u, a)^{-1}(0). \tag{2.9}$$

Denote by $P : M \to E$ the restriction to M of the projection of $U \times E$ onto E, that is,

$$P(u,a) = a \quad \text{for all } (u,a) \in M.$$

From (2.9) we can infer the equivalence

$$\begin{aligned} &a \in E \text{ is a regular value of } P : M \to E \\ &\iff 0 \in X^* \text{ is a regular value of } I_u(\cdot, a) : U \to X^*. \end{aligned} \tag{2.10}$$

Hypothesis (H_2) leads to

$$P : M \to E \text{ is a Fredholm mapping of index zero.} \tag{2.11}$$

Assertion (2.11) allows us to apply the Sard–Smale theorem to the C^1 mapping $P : M \to E$, which is possible because index $P = 0 < 1$. We infer that

$$G = \{a \in E : \ a \text{ is a regular value of } P : M \to E\} \tag{2.12}$$

is a residual set in the Banach space E. The Baire's category theorem yields the density of the set G in E. Now, on the basis of (2.10) and assumption (H_3), we can prove that the residual set G in (2.12) coincides with the set introduced in (2.7), which establishes the first part of Theorem 2.2.

For the second part of Theorem 2.2, we note from (2.12) that $\{x \in U : (x,a) \in M\}$, when $a \in G$, forms a manifold of dimension equal to index $P = 0$, so the critical points of $I(\cdot, a)$, for $a \in G$, are isolated. Fix $a \in G$ and consider the set of critical points $X_a := \{u \in X : I_u(u,a) = 0\}$ of $I(\cdot, a)$. We have just shown that X_a is a discrete space. Let us prove that the set X_a is compact. If $\{u_n\}$ is a sequence with $u_n \in X_a$, we can apply assumption (H_4) with $a_n = a$ for all n, so $\{u_n\}$ contains a strongly convergent subsequence. Therefore X_a is discrete and compact, thus it is a finite set.

For the final part of the theorem, fix a connected component G_0 of G.

Claim 1. For every $(u,a) \in U \times G_0$ satisfying (2.1), there is an open subset $V_{u,a} \subset E$ containing G_0 and a C^r mapping $g_{u,a} : V_{u,a} \to U$ satisfying

- **(i)** $g_{u,a}(a) = u$;
- **(ii)** $g_{u,a}(y)$ is a critical point of $I(\cdot, y)$ for all $y \in V_{u,a}$;
- **(iii)** for every $y \in V_{u,a}$, there are neighborhoods $V_y \subset V_{u,a}$ of y and $W_y \subset U$ of $g_{u,a}(y)$ such that $g_{u,a}(z)$ is the unique critical point of $I(\cdot, z)$ in W_y for all $z \in V_y$.

By (2.6), we know that $I_{uu}(u,a) : X \to X$ is a linear isomorphism. The implicit function theorem (see Theorem 1.36) yields the existence of a maximal open subset $V_{u,a} \subset E$ containing a and a C^r mapping $g_{u,a} : V_{u,a} \to U$ satisfying (i)–(iii). It remains to verify that $V_{u,a}$ contains G_0. In view of the connectedness of G_0, it suffices to check that $V_{u,a} \cap G_0$ is closed in G_0. If

$a_0 \in \overline{V_{u,a}} \cap G_0$, there is $\{a_n\} \subset V_{u,a} \cap G_0$ such that $a_n \to a_0$ in E, and set $u_n = g_{u,a}(a_n)$. By (H_4), up to a subsequence, we may assume that $u_n \to u_0$ for some $u_0 \in U$, thus $I_u(u_0, a_0) = 0$. Since $a_0 \in G_0$, the implicit function theorem yields neighborhoods $V_0 \subset E$ of a_0, $W_0 \subset U$ of u_0, and a C^r mapping $g_0 : V_0 \to W_0$ such that $g_0(y)$ is the unique critical point of $I(\cdot, y)$ in W_0 whenever $y \in V_0$. Up to considering V_0 smaller if necessary, we may assume that $V_{u,a} \cap V_0$ is connected. Using property (iii) of the mapping $g_{u,a}$, we see that the set $\{y \in V_{u,a} \cap V_0 : g_0(y) = g_{u,a}(y)\}$ is open and closed in $V_{u,a} \cap V_0$, and nonempty (since it contains a_n for n large enough), whence $g_0|_{V_{u,a}\cap V_0} = g_{u,a}|_{V_{u,a}\cap V_0}$. Then the maximality of $V_{u,a}$ implies $a_0 \in V_{u,a}$, so Claim 1 holds true.

Claim 2. If $(u, a), (u', a') \in U \times G_0$ satisfy (2.1), then we have either $g_{u,a} = g_{u',a'}$ on G_0 or $g_{u,a}(y) \neq g_{u',a'}(y)$ for all $y \in G_0$.

Conditions (i)–(iii) of Claim 1 imply that the set $\{y \in G_0 : g_{u,a}(y) = g_{u',a'}(y)\}$ is open and closed in G_0. Claim 2 then follows from the connectedness of the set G_0.

Set $\{g_i\}_{i\in J} := \{g_{u,a} : (u, a) \in U \times E \text{ satisfy (2.1)}\}$. By Claim 2 we know that $g_i \neq g_j$ whenever $i \neq j$, whereas by the finiteness of the solutions of (2.1) (with a fixed a) we have that J is finite.

Claim 3. For every $a \in G_0$, the map $J \to \{u \in U : I_u(u, a) = 0\}$, $i \mapsto g_i(a)$ is bijective.

The injectivity follows from Claim 2. Moreover, given $u \in U$ a critical point of $I(\cdot, a)$, we get $g_{u,a}(a) = u$ (by Claim 1). Choosing $i \in J$ such that $g_{u,a} = g_i$, we note that $u = g_i(a)$, whence the surjectivity. Claim 3 is verified.

The construction of the finite family $\{g_i\}$ of C^r mappings on the connected component G_0 of G as required in (2.8) is completed. □

We produce two applications of Theorem 2.2. The first one concerns the homoclinic solutions.

Let be given a mapping $L \in C(\mathbb{R}, \mathbb{R}^{N^2})$ such that, for $t \in \mathbb{R}$, $L(t)$ is a symmetric matrix, which is bounded and positive definite on $\mathbb{R}^N$ uniformly on $\mathbb{R}$, that is,

$$|L(t)| \leq c_0 \text{ and } L(t)u \cdot u \geq c_1|u|^2 \quad \text{for all } u \in \mathbb{R}^N \text{ and all } t \in \mathbb{R},$$

with constants $c_0, c_1 > 0$. Hereafter the symbol $|\cdot|$ stands for the Euclidean norm in any space $\mathbb{R}^M$ (for $M \geq 1$) and "$\cdot$" stands for the corresponding scalar product (also denoted by $(\cdot,\cdot)_{\mathbb{R}^M}$). We are also given a potential $V : \mathbb{R} \times \mathbb{R}^N \to \mathbb{R}$ and a function $f : \mathbb{R} \to \mathbb{R}^N$. With these data, we consider the second-order

Hamiltonian system

$$\begin{cases} -\ddot{q} + L(t)q = V_x(t,q) + f(t) & \text{for all } t \in \mathbb{R}, \\ q(t) \to 0 \ \text{ as } t \to \pm\infty, \\ \dot{q}(t) \to 0 \ \text{ as } t \to \pm\infty, \end{cases} \tag{2.13}$$

where V_x denotes the partial derivative of $V(t,x)$ with respect to the second variable $x \in \mathbb{R}^N$. The solutions of (2.13) are called homoclinics to 0.

Recall from Definition 1.9 and Theorem 1.10 that $X = H^1(\mathbb{R}, \mathbb{R}^N)$ consists of the absolutely continuous functions from $\mathbb{R}$ to $\mathbb{R}^N$ whose derivative defined almost everywhere is square integrable. The separable Hilbert space X is endowed with the scalar product $(\cdot,\cdot)_X$ introduced as

$$(q,u)_X = \int_{-\infty}^{+\infty} (\dot{q}\cdot\dot{u} + L(t)q\cdot u)\,dt \quad \text{for all } q,u \in X.$$

The associated norm on X is denoted by $\|\cdot\|_X$. Let us set $Y := L^2(\mathbb{R}, \mathbb{R}^N)$, which is a separable Hilbert space endowed with the usual norm (see Definition 1.7).

We suppose that the potential $V : \mathbb{R} \times \mathbb{R}^N \to \mathbb{R}$ in (2.13) is a function $(t,x) \mapsto V(t,x)$ for which V, V_x, V_{xx} are continuous, V_{xx} bounded,

$$L(t)x\cdot x - V_x(t,x)\cdot x \ge \alpha |x|^2, \qquad \forall t \in \mathbb{R},\ \forall x \in \mathbb{R}^N, \tag{2.14}$$

with a constant $\alpha > 0$, and

$$\begin{gathered} V(t,0) = 0, \quad V_x(t,0) = 0, \quad V_{xx}(t,0) = 0, \qquad \forall t \in \mathbb{R}, \\ \lim_{\substack{|t|\to+\infty \\ x\to 0}} V_{xx}(t,x) = 0. \end{gathered} \tag{2.15}$$

Example 2.3. A simple example of mappings $L(t)$ and $V(t,x)$ satisfying all our assumptions is $L(t)x = x$ and $V(t,x) = -\frac{1}{4}e^{-t^2}\arctan(|x|^4)$ for all $t \in \mathbb{R}$ and $x \in \mathbb{R}^N$.

Our result on problem (2.13) ensures that generically the solution set consists of finitely many homoclinic solutions. It also provides that these solutions are nondegenerate, they depend smoothly on $f \in L^2(\mathbb{R}, \mathbb{R}^N)$, and the famous hypothesis $(*)$ in the theory of homoclinics (see, e.g., [35]) is generically satisfied for problem (2.13).

Theorem 2.4. *Under the previous conditions for $L(t)$ and $V(t,x)$, there exists a dense subset A in Y such that for every $f \in A$ problem (2.13) has finitely many nondegenerate homoclinic solutions (in the sense that they are nondegenerate critical points of the Euler functional corresponding to (2.13)). Moreover, the set of solutions to (2.13) depends smoothly on $f \in A$ in the following sense: for* *every connected component A_0 of A, there is a finite collection $\{g_i\}_{1\le i\le n_{A_0}}$ of C^1 mappings defined in the neighborhood of A_0 in Y and taking values in X such that for any $f \in A_0$ the solutions of (2.13) are exactly $\{g_i(f)\}_{1\le i\le n_{A_0}}$. The number n_{A_0} of solutions to (2.13) is constant whenever $f \in A_0$.*

Proof. The idea is to apply Theorem 2.2. To this end, we define the function $I : X \times Y \to \mathbb{R}$ by

$$I(q,f) = \int_{-\infty}^{+\infty} \left[\frac{1}{2}\left(|\dot q|^2 + L(t)q\cdot q\right) - V(t,q) - f\cdot q\right] dt \quad \text{for all } (q,f) \in X \times Y, \tag{2.16}$$

that is, $I(\cdot, f)$ is the Euler functional associated to problem (2.13) whenever $f \in Y$. The function I in (2.16) is well defined because, by using the mean value theorem twice, the continuous embedding $H^1(\mathbb{R},\mathbb{R}^N) \subset L^2(\mathbb{R},\mathbb{R}^N)$, and the boundedness of V_{xx}, one finds a constant $C > 0$ such that

$$\left|\int_{-\infty}^{+\infty} V(t,q)\,dt\right| \le \sup_{(t,x)\in\mathbb{R}\times\mathbb{R}^N} |V_{xx}(t,x)|\,\|q\|_Y^2 \le C\,\|q\|_X^2 \quad \text{for all } q \in X.$$

Let us verify condition (H_1) for $U = X$ and $E = Y$. Fix $q \in X$, $f \in Y$, and $\varepsilon > 0$. Using again the mean value theorem for $V(t,\cdot)$ and $V_x(t,\cdot)$, we have for each $\varphi \in X$, with some $\lambda = \lambda(t) \in [0,1]$, the inequalities

$$\begin{aligned}
&\left|\int_{-\infty}^{+\infty} (V(t,q+\varphi) - V(t,q) - V_x(t,q)\cdot\varphi)\,dt\right| \\
&\le \int_{-\infty}^{+\infty} |V_x(t,q+\lambda\varphi) - V_x(t,q)|\,|\varphi|\,dt \\
&\le \sup_{(t,x)\in\mathbb{R}\times\mathbb{R}^N} |V_{xx}(t,x)|\,\|\varphi\|_Y^2 \le C\|\varphi\|_X^2.
\end{aligned}$$

It turns out that $I(\cdot, f)$ is Fréchet differentiable and its differential is expressed as

$$I_q(q,f)\varphi = \int_{-\infty}^{+\infty} \left(\dot q\cdot\dot\varphi + L(t)q\cdot\varphi - V_x(t,q)\cdot\varphi - f\cdot\varphi\right) dt \quad \text{for all } \varphi \in X. \tag{2.17}$$

In order to establish the continuity of the differential $I_q : X \times Y \to X^*$, let $q_n \to q$ in X and $f_n \to f$ in Y as $n \to \infty$. Then, through the mean value theorem and Cauchy–Schwarz inequality, one obtains

$$\sup_{\|\varphi\|_X\le 1} \left|\int_{-\infty}^{+\infty} (V_x(t,q_n)\cdot\varphi - V_x(t,q)\cdot\varphi)\,dt\right|$$

$$\leq \sup_{\|\varphi\|_X \leq 1} \int_{-\infty}^{+\infty} \sup_{(t,x)\in\mathbb{R}\times\mathbb{R}^N} |V_{xx}(t,x)|\,|q_n - q|\,|\varphi|\,dt \;\leq\; C\,\|q_n - q\|_X,$$

with a constant $C > 0$. Then, by (2.17), the continuity of I_q follows.

We show the existence of the second order partial Fréchet derivative I_{qq} of the functional I defined by (2.16). Let $\varepsilon > 0$ and $(q, f) \in X \times Y$ be fixed. From (2.15), one finds $r > 0$ and $\delta > 0$ to fulfill

$$|V_{xx}(t,x)| < \varepsilon \quad \text{whenever } |t| > r \text{ and } |x| < \delta. \tag{2.18}$$

Since q belongs to $X = H^1(\mathbb{R}, \mathbb{R}^N)$, there is $r > 0$ large enough so that

$$|q(t)| < \frac{\delta}{2} \quad \text{for all } |t| > r \tag{2.19}$$

(see Corollary 1.11). It is well known that the mapping

$$q \in X \mapsto \int_{-r}^{r} V_x(t,q)\,dt \in \mathbb{R}^N$$

is Fréchet differentiable. Thus, there exists $\eta > 0$ with the property

$$\left| \int_{-r}^{r} \left[V_x(t, q+\psi)\cdot\varphi - V_x(t,q)\cdot\varphi - (V_{xx}(t,q)\cdot\varphi)\cdot\psi \right] dt \right| \leq \varepsilon \|\varphi\|_X \,\|\psi\|_X \tag{2.20}$$

for all $\varphi, \psi \in X$ with $\|\psi\|_X \leq \eta$. Due to the continuous embedding $X \subset L^\infty(\mathbb{R}, \mathbb{R}^N)$ (see Corollary 1.11), for $\eta > 0$ smaller if necessary, we may assume that $\|\psi\|_{L^\infty(\mathbb{R},\mathbb{R}^N)} \leq \frac{\delta}{2}$ whenever $\|\psi\|_X \leq \eta$. Then the mean value theorem, Cauchy–Schwarz inequality, and (2.18), (2.19) yield

$$\begin{aligned}\left| \int_{|t|>r} (V_x(t,q+\psi) - V_x(t,q))\cdot\varphi\,dt \right| &\leq \sup_{\substack{|t|>r \\ |x|<\delta}} |V_{xx}(t,x)|\,\|\varphi\|_Y\,\|\psi\|_Y \\ &\leq \varepsilon C \|\varphi\|_X\,\|\psi\|_X\end{aligned} \tag{2.21}$$

for all $\varphi, \psi \in X$ with $\|\psi\|_X \leq \eta$, with a constant $C > 0$. Again by (2.18), (2.19), we obtain

$$\left| \int_{|t|>r} (V_{xx}(t,q)\cdot\varphi)\cdot\psi\,dt \right| \leq \varepsilon C \|\varphi\|_X\,\|\psi\|_X \quad \text{for all } \varphi, \psi \in X. \tag{2.22}$$

Combining (2.20)–(2.22), it is seen that

$$\begin{aligned}&\left| \int_{-\infty}^{+\infty} \left(V_x(t,q+\psi)\cdot\varphi - V_x(t,q)\varphi - (V_{xx}(t,q)\cdot\varphi)\cdot\psi \right) dt \right| \\ &\leq \varepsilon(1+2C)\|\varphi\|_X\,\|\psi\|_X\end{aligned}$$

for all $\varphi \in X$ and $\psi \in X$ with $\|\psi\|_X \leq \eta$. Therefore, there exists the second order partial derivative $I_{qq}(q, f)$, which is equal to

$$I_{qq}(q, f)(\varphi, \psi) = \int_{-\infty}^{+\infty} \left(\dot{\varphi} \cdot \dot{\psi} + L(t)\varphi \cdot \psi - (V_{xx}(t, q) \cdot \varphi) \cdot \psi\right) dt \tag{2.23}$$

for all $\varphi, \psi \in X$.

We claim that I_{qq} is continuous on $X \times Y$. Indeed, let $q_n \to q$ in X and $f_n \to f$ in Y as $n \to \infty$. Fix an $\varepsilon > 0$, and let $r > 0$, $\delta > 0$ satisfy (2.18) and (2.19). Since $q \in L^\infty(\mathbb{R}, \mathbb{R}^N)$ and $\{q_n\}$ converges uniformly to q, there is an $r > 0$ such that

$$|q_n(t)| < \delta \quad \text{whenever } |t| > r \text{ and } n \text{ is large enough.} \tag{2.24}$$

The uniform continuity of V_{xx} on $[-r, r] \times \mathcal{N}$, for a compact neighborhood $\mathcal{N}$ of $q([-r, r])$ in $\mathbb{R}^N$, and the convergence $q_n \to q$ in $L^\infty(\mathbb{R}, \mathbb{R}^N)$ imply

$$|V_{xx}(t, q_n(t)) - V_{xx}(t, q(t))| \leq \varepsilon, \quad \forall t \in [-r, r], \text{ for } n \text{ large enough.} \tag{2.25}$$

By (2.18), (2.19), (2.24), (2.25), we infer that

$$|V_{xx}(t, q_n(t)) - V_{xx}(t, q(t))| \leq 2\varepsilon \quad \text{for all } t \in \mathbb{R},$$

with any sufficiently large n. This results in

$$\sup_{\substack{\|\varphi\|_X \leq 1 \\ \|\psi\|_X \leq 1}} \left| \int_{-\infty}^{+\infty} \left(\left(V_{xx}(t, q_n(t)) - V_{xx}(t, q(t))\right) \cdot \varphi(t)\right) \cdot \psi(t)\, dt \right| \leq 2\varepsilon,$$

with any large number n. Then from (2.23), the continuity of I_{qq} is proven.

Clearly, the second order partial derivative $I_{qf}(q, f)$ exists and is given by

$$I_{qf}(q, f)(\varphi, g) = -\int_{-\infty}^{+\infty} g \cdot \varphi\, dt \quad \text{for all } f, g \in Y, \text{ all } q, \varphi \in X. \tag{2.26}$$

The continuity of I_{qf} on $X \times Y$ follows readily from (2.26).

According to what was said before, it is true that

$$I_q \in C^1(X \times Y, X^*). \tag{2.27}$$

By (2.27) we know that condition (H_1) is satisfied.

The next step in the proof is to check condition (H_2). Let $q \in X$, $f \in Y$ and $\varepsilon > 0$. Choose $r > 0$ and $\delta > 0$ such that (2.18) and (2.19) hold true. Take a bounded sequence $\{\varphi_n\}$ in X. The embedding

$$H^1([-r, r], \mathbb{R}^N) \subset L^2([-r, r], \mathbb{R}^N)$$

is compact, so we may assume that along a relabeled subsequence

$$\{\varphi_n\} \text{ converges in } L^2([-r,r],\mathbb{R}^N). \tag{2.28}$$

On the other hand, by (2.18), (2.19), and the boundedness of the sequence $\{\varphi_n\}$ in X, it follows that

$$\begin{aligned} &\sup_{\|\psi\|_X\le 1}\left|\int_{-\infty}^{+\infty}(V_{xx}(t,q)\cdot(\varphi_n-\varphi_m))\cdot\psi\,dt\right| \qquad (2.29)\\ &\le\Big(\sup_{(t,x)\in\mathbb{R}\times\mathbb{R}^N}|V_{xx}(t,x)|\Big)\sup_{\|\psi\|_X\le 1}\int_{-r}^{r}|\varphi_n-\varphi_m|\,|\psi|\,dt+C\varepsilon,\ \ \forall n,m, \end{aligned}$$

with a constant $C > 0$. By (2.28) and (2.29), we deduce that along a subsequence

$$\left\{\int_{-\infty}^{+\infty} V_{xx}(t,q)\cdot\varphi_n\,dt\right\}$$

is a Cauchy sequence in X^*, so strongly convergent in X^*. Hence the operator

$$\varphi\in X\mapsto\int_{-\infty}^{+\infty} V_{xx}(t,q)\cdot\varphi\,dt\in X^*$$

is compact. Therefore, by (2.23), $I_{qq}(q,f)$ as a linear operator from X to X^* is the sum of an isomorphism and a compact map. This implies that $I_{qq}(q,f)$ is a Fredholm operator of index zero (see, e.g., [31, p. 169]). Therefore assumption (H_2) is verified.

Now we prove that condition (H_3) holds. The kernel $\ker(I_q)'(q,f)$ splits in $X\times Y$ because X and Y are Hilbert spaces. To verify the second part of (H_3), we show that $\ker I_{qf}(q,f)=\{0\}$ for all $(q,f)\in X\times Y$. Let $(q,f)\in X\times Y$ and

$$\varphi\in\ker I_{qf}(q,f). \tag{2.30}$$

From (2.26) and (2.30) we derive that

$$\int_{-\infty}^{+\infty} g\cdot\varphi\,dt=0\quad\text{for all } g\in Y,$$

which directly yields $\varphi=0$, whence (H_3).

Consequently, we are allowed to apply Theorem 2.2 to the functional I on $X\times Y$. A dense subset A in Y is found by setting $A=G$, with G as in (2.7). Hence, for every $f\in A$, the functional $I(\cdot,f):X\to\mathbb{R}$ defined in (2.16) has only nondegenerate critical points, equivalently nondegenerate solutions to problem (2.13). Thus, for every $f\in A$, problem (2.13) possesses an at most countable discrete set of homoclinic solutions, which establishes the first part of Theorem 2.4.

Now we show that condition (H_4) is fulfilled. Let $(q_n, f_n) \in X \times A$ be such that $I_q(q_n, f_n) = 0$ for all n, and $f_n \to f \in A$. The equality $I_q(q_n, f_n) = 0$ reads as

$$\int_{-\infty}^{+\infty} \left(\dot{q}_n \cdot \dot{\varphi} + L(t)q_n \cdot \varphi - V_x(t, q_n) \cdot \varphi - f_n \cdot \varphi\right) dt = 0 \quad \text{for all } \varphi \in X. \tag{2.31}$$

Setting $\varphi = q_n$ in (2.31), by assumption (2.14) we can see that the sequence $\{q_n\}$ is bounded in X. Then along a relabeled subsequence one has that $q_n \rightharpoonup q$ in $X = H^1(\mathbb{R}, \mathbb{R}^N)$ and, by the compactness of the embedding $H^1(\mathbb{R}, \mathbb{R}^N) \subset L^2(\mathbb{R}, \mathbb{R}^N)$, that $q_n \to q$ in $L^2(\mathbb{R}, \mathbb{R}^N)$ for some $q \in H^1(\mathbb{R}, \mathbb{R}^N)$. Choosing $\varphi = q_n - q$ in (2.31) ensures

$$\lim_{n\to\infty} \|q_n\|_X^2 = \|q\|_X^2.$$

Since $q_n \rightharpoonup q$ and X is a Hilbert space, we conclude that $q_n \to q$ in X, which proves condition (H_4).

Finally, we obtain the second part of Theorem 2.4 from Theorem 2.2, which gives that for every $f \in A$ the number of solutions of system (2.13) is finite. It is constant, say equal to n_{A_0}, on a connected component A_0 of A. Moreover, Theorem 2.2 provides the description of the solution set of (2.13) with $a = f \in A_0$ in the form $\{g_i(f)\}_{1 \le i \le n_{A_0}}$, where g_i, $1 \le i \le n_{A_0}$, are C^1 mappings on a neighborhood of A_0 in A. This completes the proof. □

Our second application of Theorem 2.2 is to give verifiable criteria for obtaining a Morse function whose critical points be exactly the solutions of the semilinear elliptic boundary value problem

$$\begin{cases} -\Delta u = p(x, u) + f(x) & \text{in } \Omega, \\ u = 0 & \text{on } \partial\Omega, \end{cases} \tag{2.32}$$

for a bounded domain Ω in $\mathbb{R}^N$ with a C^2 boundary $\partial\Omega$. Here $f \in L^2(\Omega)$ is regarded as a parameter, whilst $p : \Omega \times \mathbb{R} \to \mathbb{R}$ is given. The following theorem addresses this question.

Theorem 2.5. **(a)** *Assume that the function $p : \overline{\Omega} \times \mathbb{R} \to \mathbb{R}$ in (2.32) satisfies the following conditions:*

(i) $p \in C^1(\overline{\Omega} \times \mathbb{R})$;

(ii) *there exist constants $c_1, c_2 \ge 0$ such that the partial derivative $p_t(x,t)$ satisfies*

$$|p_t(x,t)| \le c_1 + c_2|t|^{s-1} \text{ for all } (x,t) \in \Omega \times \mathbb{R},$$

with some $s \in (1, \frac{N+2}{N-2})$.

Then the set

$$G = \{f \in L^2(\Omega) : I(\cdot, f) : H_0^1(\Omega) \to \mathbb{R} \text{ has only nondegenerate critical points}\}, \tag{2.33}$$

defined with the function $I : H_0^1(\Omega) \times L^2(\Omega) \to \mathbb{R}$ *expressed as*

$$I(u, f) = \int_\Omega \left(\frac{1}{2}|\nabla u|^2 - P(x, u) - fu\right) dx \text{ for all } (u, f) \in H_0^1(\Omega) \times L^2(\Omega), \tag{2.34}$$

where ∇u *stands for the gradient of* u, *is residual, hence dense, in* $L^2(\Omega)$.

(b) *If we further suppose that the function* $p : \overline{\Omega} \times \mathbb{R} \to \mathbb{R}$ *is bounded, then the residual set* G *in (2.33) consists of functions possessing only finitely many critical points.*

(c) *If in addition to hypotheses* (i) *and* (ii) *we assume the Ambrosetti–Rabinowitz condition*

(AR) *there are constants* $\mu > 2$, $q \geq 0$, *and* $M \in \mathbb{R}$ *such that*

$$p(x, t)t - \mu P(x, t) \geq M \text{ for all } (x, t) \in \Omega \times \mathbb{R} \text{ with } |t| \geq q,$$

where $P(x, t) := \int_0^t p(x, \tau)\, d\tau$, *then the residual set* G *in (2.33) coincides with*

$$G = \{f \in L^2(\Omega) : I(\cdot, f) : H_0^1(\Omega) \to \mathbb{R} \text{ is a Morse function}\}. \tag{2.35}$$

Proof. (a) Assumptions (i) and (ii) imply that the function I introduced in (2.34) is differentiable of class C^2 (see [207, p. 94]). Consequently, condition (H_1) of Theorem 2.2 is verified. The first order partial derivative $I_u(u, f)$ at any point $(u, f) \in H_0^1(\Omega) \times L^2(\Omega)$ has the expression

$$I_u(u, f)(v) = \int_\Omega ((\nabla u, \nabla v)_{\mathbb{R}^N} - p(x, u)v - fv)\, dx \text{ for all } v \in H_0^1(\Omega). \tag{2.36}$$

Hence the critical points of $I(\cdot, f)$ coincide with the weak solutions of (2.32). Differentiating in (2.36) with respect to u, one obtains

$$I_{uu}(u, f)(v, w) = \int_\Omega ((\nabla v, \nabla w)_{\mathbb{R}^N} - p_t(x, u)vw)\, dx \text{ for all } v, w \in H_0^1(\Omega). \tag{2.37}$$

We claim that the continuous linear operator $K : H_0^1(\Omega) \to H_0^1(\Omega)$ given by

$$(Kv, w)_{H_0^1(\Omega)} = \int_\Omega p_t(x, u)vw\, dx \text{ for all } v, w \in H_0^1(\Omega)$$

is compact for each weak solution u of problem (2.32). By the assumption imposed on the exponent s in hypothesis (ii), we have

$$s < \frac{N+2}{N-2} = 1 + \frac{2N}{N-2}\Big(1 - \frac{N-2}{2N} - \frac{N-2}{2N}\Big).$$

Hence there are $q, \sigma \in [1, \frac{2N}{N-2}]$ and $\tau \in [1, \frac{2N}{N-2})$ such that

$$\frac{s-1}{q} + \frac{1}{\tau} + \frac{1}{\sigma} = 1.$$

If $\{v_n\}$ is a bounded sequence in $H_0^1(\Omega)$, then it converges in both $L^2(\Omega)$ and $L^\tau(\Omega)$ along a subsequence. Using assumption (ii), the Cauchy–Schwarz inequality, and the generalized Hölder inequality with the exponents $\frac{q}{s-1}$, τ, and σ as above, we see that for each weak solution u of (2.32) one has

$$\begin{aligned}
&\sup_{\|w\|_{H_0^1(\Omega)} \le 1} |(Kv_n - Kv_m, w)_{H_0^1(\Omega)}| \\
= &\sup_{\|w\|_{H_0^1(\Omega)} \le 1} \Big| \int_\Omega p_t(x,u)(v_n - v_m) w \, dx \Big| \\
\le\, & c_1 \sup_{\|w\|_{H_0^1(\Omega)} \le 1} \|v_n - v_m\|_{L^2(\Omega)} \|w\|_{L^2(\Omega)} \\
&+ c_2 \sup_{\|w\|_{H_0^1(\Omega)} \le 1} \|u\|_{L^q(\Omega)}^{s-1} \|v_n - v_m\|_{L^\tau(\Omega)} \|w\|_{L^\sigma(\Omega)}.
\end{aligned}$$

Theorem 1.20 (Sobolev's embedding theorem) yields the estimate

$$\sup_{\|w\|_{H_0^1(\Omega)} \le 1} |(Kv_n - Kv_m, w)_{H_0^1(\Omega)}| \le C_1 \|v_n - v_m\|_{L^2(\Omega)} + C_2 \|v_n - v_m\|_{L^\tau(\Omega)}$$

for all $n, m \ge 1$, where $C_1, C_2 > 0$ are constants. It turns out that $\{Kv_n\}$ contains a convergent subsequence in $H_0^1(\Omega)$, thus the compactness of K ensues.

Writing (2.37) in the form

$$I_{uu}(u, f) = \mathrm{id}_{H_0^1(\Omega)} - K, \tag{2.38}$$

it follows from (2.38) and the compactness of K that $I_{uu}(u, f) : H_0^1(\Omega) \to H_0^1(\Omega)$ is a Fredholm operator of index zero, whence condition (H_2) is satisfied.

Differentiating in (2.36) with respect to f (and identifying I_{uf} and I_{fu}) we find for the linear operator $I_{uf}(u, f) : H_0^1(\Omega) \to L^2(\Omega)$ that

$$I_{uf}(u, f)(v, h) = -\int_\Omega hv \, dx \quad \text{for all } (v, h) \in H_0^1(\Omega) \times L^2(\Omega).$$

Hence

$$\ker I_{uf}(u, f) = 0,$$

which shows that condition (H_3) holds. Theorem 2.2 applies in the case of the function I of (2.34) enabling us to conclude part (a) of the statement of the theorem.

(b) We need to show equality (2.35). To this end, we have to verify hypothesis (H_4) in Theorem 2.2 for the function $I : H_0^1(\Omega) \times L^2(\Omega) \to \mathbb{R}$. Let $(u_n, f_n) \in H_0^1(\Omega) \times G$, with G in (2.7), satisfy $I_u(u_n, f_n) = 0$ for all n, and $f_n \to f \in G$, that is, the convergence in $L^2(\Omega)$. Therefore we have

$$\int_\Omega ((\nabla u_n, \nabla v)_{\mathbb{R}^N} - p(x, u_n)v - f_n v)\, dx = 0 \quad \text{for all } v \in H_0^1(\Omega). \tag{2.39}$$

Setting $v = u_n$, since the function p was supposed to be bounded, the sequence $\{u_n\}$ is bounded in $H_0^1(\Omega)$. Then along a relabeled subsequence one has that $u_n \rightharpoonup u$ in $H_0^1(\Omega)$ and, by the compactness of the embedding $H_0^1(\Omega) \subset L^2(\Omega)$, that $u_n \to u$ in $L^2(\Omega)$ for some $u \in H_0^1(\Omega)$. Inserting $\varphi = u_n - u$ in (2.39) ensures

$$\lim_{n\to\infty} \|u_n\|^2_{H_0^1(\Omega)} = \|u\|^2_{H_0^1(\Omega)},$$

so $u_n \to u$ in $H_0^1(\Omega)$, which proves that (H_4) holds.

(c) From assumption (AR) we derive that the functional $I(\cdot, f) : H_0^1(\Omega) \to \mathbb{R}$ satisfies the Palais–Smale condition for every $f \in L^2(\Omega)$, that is, if $\{I(u_n, f)\}$ is bounded and $I_u(u_n, f) \to 0$, then the sequence $\{u_n\}$ is strongly convergent in $H_0^1(\Omega)$. For such a reasoning we refer to [207, pp. 10–11]. Consequently, (2.33) implies (2.35), which completes the proof. □

2.2 REGULARITY THEORY FOR NONLINEAR ELLIPTIC EQUATIONS

In this section, we consider regularity results for the weak solutions of certain nonlinear elliptic problems, which include as a particular case problems driven by the p-Laplacian. Such regularity results are established in two steps. First, we prove that the weak solution is in $L^\infty(\Omega)$ and then using this boundedness property, we show the Hölder continuous differentiability up to the boundary for the weak solutions. We do this for the Neumann problem which is more involved. Similar, and in fact simpler, proofs hold for the Dirichlet case.

Let Ω be a bounded domain in $\mathbb{R}^N$ with a C^2 boundary $\partial\Omega$. We consider the following nonlinear elliptic problem:

$$\begin{cases} -\operatorname{div} a(x, \nabla u(x)) = f(x, u(x)) & \text{in } \Omega, \\ \dfrac{\partial u}{\partial n_a} = 0 & \text{on } \partial\Omega. \end{cases} \tag{2.40}$$

Here $a : \overline{\Omega} \times \mathbb{R}^N \to \mathbb{R}^N$ is a continuous map such that, for every $x \in \overline{\Omega}$, $a(x, \cdot)$ is strictly monotone on $\mathbb{R}^N$ and continuously differentiable on $\mathbb{R}^N \setminus \{0\}$. The hypotheses allow in the map div $a(x, \nabla u(x))$ be incorporated as a special case the p-Laplacian Δ_p defined by

$$\Delta_p u = \operatorname{div}(|\nabla u(x)|^{p-2}\nabla u(x)) \text{ for all } u \in W^{1,p}(\Omega).$$

However, in contrast to the p-Laplacian, the differential operator in (2.40) need not be $(p-1)$-homogeneous. Also, $f(x, s)$ is a Carathéodory function (i.e., for all $s \in \mathbb{R}$, $x \mapsto f(x, s)$ is measurable on Ω and, for a.a. $x \in \Omega$, $s \mapsto f(x, s)$ is continuous) with subcritical growth in the second variable. Also, $\frac{\partial u}{\partial n_a}(x) = (a(x, \nabla u(x)), \nu(x))_{\mathbb{R}^N}$ for $x \in \partial\Omega$, where $\nu(x)$ is the outward unit normal on the boundary $\partial\Omega$.

The precise hypotheses on the maps $a(x, y)$ and $f(x, s)$ are the following:

H(a) $a(x, y) = h(x, |y|)y$ for all $(x, y) \in \overline{\Omega} \times \mathbb{R}^N$, where $h(x, t) > 0$ for all $(x, t) \in \overline{\Omega} \times (0, +\infty)$, and

(i) $a \in C(\overline{\Omega} \times \mathbb{R}^N, \mathbb{R}^N) \cap C^1(\overline{\Omega} \times (\mathbb{R}^N \setminus \{0\}), \mathbb{R}^N)$;

(ii) there exist $c_1 > 0$ and $1 < p < +\infty$ such that

$$\|D_y a(x, y)\| \le c_1 |y|^{p-2} \text{ for every } x \in \overline{\Omega} \text{ and } y \in \mathbb{R}^N \setminus \{0\};$$

(iii) there exists $c_0 > 0$ such that

$$(D_y a(x, y)\xi, \xi)_{\mathbb{R}^N} \ge c_0 |y|^{p-2} |\xi|^2$$
$$\text{for every } x \in \overline{\Omega} \text{ and } y \in \mathbb{R}^N \setminus \{0\}, \xi \in \mathbb{R}^N;$$

(iv) there exist $c_2 > 0$ such that

$$\|D_x a(x, y)\| \le c_2(1 + |y|^{p-1}) \text{ for every } x \in \overline{\Omega} \text{ and } y \in \mathbb{R}^N \setminus \{0\}.$$

In what follows we consider the function $G(x, y)$ defined by

$$\nabla_y G(x, y) = a(x, y) \text{ for all } (x, y) \in \overline{\Omega} \times \mathbb{R}^N, \text{ and } G(x, 0) = 0 \text{ for all } x \in \overline{\Omega}$$

(the potential which corresponds to $a(x, y)$).

Remark 2.6. Let $g(x, t) = h(x, t)t$ for all $x \in \overline{\Omega}$ and all $t \ge 0$. Hypotheses H(a) imply the one-dimensional estimate

$$c_0 t^{p-2} \le g_t'(x, t) \le c_1 t^{p-2} \text{ for all } x \in \overline{\Omega}, t > 0.$$

We set $G_0(x, t) = \int_0^t g(x, s)\, ds$ for all $(x, t) \in \overline{\Omega} \times [0, +\infty)$. Clearly, $G_0(x, \cdot)$ is strictly convex and increasing on $[0, +\infty)$. Let $G(x, y) = G_0(x, |y|)$. Then $G(x, 0) = 0$, $G(x, \cdot)$ is strictly convex for all $x \in \overline{\Omega}$, and

$$\nabla_y G(x, y) = (G_0)_t'(x, |y|)\frac{y}{|y|} = g(x, |y|)\frac{y}{|y|} = h(x, |y|)y = a(x, y)$$

for all $(x, y) \in \overline{\Omega} \times (\mathbb{R}^N \setminus \{0\})$. Therefore, the potential G is uniquely defined. Since $G(x, \cdot)$ is convex and $\nabla_y G(x, y) = a(x, y)$ for all $(x, y) \in \overline{\Omega} \times \mathbb{R}^N$, it turns out that

$$(a(x, y), y)_{\mathbb{R}^N} \geq G(x, y) \quad \text{for all } (x, y) \in \overline{\Omega} \times \mathbb{R}^N. \tag{2.41}$$

Using H(a) and (2.41), we can easily establish the next result.

Lemma 2.7. *If hypotheses* H(a) (i)–(iii) *hold, then*

(a) *for all* $x \in \overline{\Omega}$, $y \mapsto a(x, y)$ *is maximal monotone and strictly monotone;*
(b) *for all* $(x, y) \in \overline{\Omega} \times \mathbb{R}^N$, $|a(x, y)| \leq \dfrac{c_1}{p-1} |y|^{p-1}$;
(c) *for all* $(x, y) \in \overline{\Omega} \times \mathbb{R}^N$, $(a(x, y), y)_{\mathbb{R}^N} \geq \dfrac{c_0}{p-1} |y|^p$.

An immediate consequence of Lemma 2.7 is the following estimate for G.

Corollary 2.8. *If* H(a) (i)–(iii) *hold, then for all* $(x, y) \in \overline{\Omega} \times \mathbb{R}^N$ *we have*

$$\frac{c_0}{p(p-1)} |y|^p \leq G(x, y) \leq \frac{c_1}{p(p-1)} |y|^p.$$

Example 2.9. Many interesting operators fit the setting of hypotheses H(a). Given $\theta \in C^1(\overline{\Omega})$ and $\theta(x) > 0$ for all $x \in \overline{\Omega}$, the following mappings satisfy hypotheses H(a):

(e_1) $a(x, y) = \theta(x)|y|^{p-2}y$, for $p > 1$ (corresponds to the weighted p-Laplacian);
(e_2) $a(x, y) = \theta(x)[|y|^{p-2}y + \ln(1 + |y|^{p-2})y]$, $p \geq 2$;
(e_3) for $1 < \tau \leq p \leq q$ and $\tau \neq 2$,

$$a(x, y) = \begin{cases} \theta(x)\left[|y|^{p-2}y + |y|^{q-2}y\right] & \text{if } |y| \leq 1, \\ \theta(x)\left[|y|^{p-2}y + \frac{q-2}{\tau-2}|y|^{\tau-2}y - \frac{q-\tau}{\tau-2}y\right] & \text{if } |y| > 1; \end{cases}$$

(e_4) $a(x, y) = \theta(x)\left[|y|^{p-2}y + c\,\frac{|y|^{p-2}y}{1+|y|^p}\right]$ (which corresponds to the weighted sum of the p-Laplacian and a generalized mean curvature operator), with $0 < c < 4p(p-1)$ if $1 < p < 2$ and $0 < c < \frac{4p}{(p-1)^2}$ if $p \geq 2$.

The hypotheses on $f(x, s)$ are:

H(f) $f : \Omega \times \mathbb{R} \to \mathbb{R}$ is a Carathéodory function such that

$$|f(x, s)| \leq \hat{a}(x) + c|s|^{r-1} \quad \text{for a.e. } x \in \Omega \text{ and all } s \in \mathbb{R},$$

with $\hat{a} \in L^\infty(\Omega)_+$, $c > 0$, $1 \leq r < p^* = \begin{cases} \frac{Np}{N-p} & \text{if } N > p, \\ +\infty & \text{if } N \leq p. \end{cases}$

For the sake of simplicity, we denote $\|\cdot\|_q := \|\cdot\|_{L^q(\Omega)}$ for all $1 \le q \le +\infty$. First we establish the boundedness of the weak solutions for problem (2.40).

Proposition 2.10. *Assume* H(a) (i)–(iii) *and* H(f). *Let* $\bar{p}^* = p^*$ *if* $N > p$ *and* $\bar{p}^* > r$ *if* $N \le p$. *If* $u \in W^{1,p}(\Omega)$ *is a weak solution of (2.40), then* $u \in L^\infty(\Omega)$ *and* $\|u\|_\infty \le M = M(c_0, c_1, c, \hat{a}, N, p, |\Omega|, \|u\|_{\bar{p}^*})$.

Proof. Since $W^{1,p}(\Omega) \hookrightarrow W^{1,q}(\Omega)$ for all $1 \le q < p$, we only need to check the case $p < N$.

Writing $u = u^+ - u^-$ with $u^+ = \max\{u, 0\} \in W^{1,p}(\Omega)$, $u^- = \max\{-u, 0\} \in W^{1,p}(\Omega)$, it is sufficient to prove the result for u^+ because one can similarly proceed for u^-.

For every $M > 0$, we set $v_M(x) = \min\{u^+, M\}$. Let $l \ge 0$. Using hypothesis H(f), it is straightforward to get

$$\int_\Omega f(x, u) v_M^{pl+1}\, dx \le C\left(1 + \|u^+\|_{pl+r}^{pl+r}\right),$$

with a constant $C > 0$. A direct computation based on Lemma 2.7 (c) shows that

$$\int_\Omega (a(x, \nabla u), \nabla(v_M^{pl+1}))_{\mathbb{R}^N}\, dx \ge \frac{(pl+1)c_0}{p-1} \frac{1}{(l+1)^p} \int_\Omega |\nabla(v_M^{l+1})|^p\, dx.$$

Then (2.40) ensures

$$\frac{pl+1}{(l+1)^p} \|v_M^{l+1}\|_{W^{1,p}(\Omega)}^p \le C\left(1 + \frac{pl+1}{(l+1)^p}\right)\left(1 + \|u^+\|_{pl+r}^{pl+r}\right), \tag{2.42}$$

with a new constant $C > 0$. Due to the embedding $W^{1,p}(\Omega) \hookrightarrow L^{p^*}(\Omega)$ and letting $M \to +\infty$, from (2.42) it follows that

$$1 + \|u^+\|_{p^*(l+1)}^{p^*(l+1)} \le C \frac{(l+1)^{p^*}}{(pl+1)^{\frac{p^*}{p}}} \left(1 + \|u^+\|_{pl+r}^{pl+r}\right)^{\frac{p^*}{p}}, \tag{2.43}$$

with $C > 0$.

Next we construct an increasing sequence $q_i \to +\infty$ as $i \to +\infty$ as follows: $q_0 = p^*$ and $q_{i+1} = (q_i - r + p)\frac{p^*}{p}$ for $i \ge 0$. For $l = \frac{q_i - r}{p}$, (2.43) becomes

$$1 + \|u^+\|_{q_{i+1}}^{q_{i+1}} \le C \frac{(q_i - r + p)^{p^*}}{p^{p^*}(q_i - r + 1)^{\frac{p^*}{p}}} \left(1 + \|u^+\|_{q_i}^{q_i}\right)^{\frac{p^*}{p}}. \tag{2.44}$$

Set $X_i = \ln(1 + \|u^+\|_{q_i}^{q_i})$. Then (2.44) leads to

$$X_{i+1} \le p^* \ln[C(q_i - r + p)] + \frac{p^*}{p} X_i, \tag{2.45}$$

with a constant $C > 0$. Since $q_i - r + p < (\frac{p^*}{p})^i p^*$, from (2.45) we derive

$$X_{i+1} \leq C_1 + iC_2 + \frac{p^*}{p} X_i,$$

with positive constants C_1, C_2, which results in

$$X_{i+1} \leq C_1 \sum_{j=0}^{i} \left(\frac{p^*}{p}\right)^j + C_2 \sum_{j=0}^{i} (i-j) \left(\frac{p^*}{p}\right)^j + \left(\frac{p^*}{p}\right)^{i+1} X_0.$$

It turns out that

$$\|u^+\|_{q_i} \leq M, \quad \text{for some } M = M(c_0, c_1, c, \hat{a}, N, p, |\Omega|, \|u\|_{p^*}).$$

Since $q_i \to +\infty$ as $i \to +\infty$, we conclude that $\|u\|_\infty \leq M$. □

Remark 2.11. The result of Proposition 2.10 is also true for the Dirichlet problem. The proof is similar via an iterated estimate as above. In the literature this method is usually called Moser's iteration technique.

We indicate a useful regularity result up to the boundary which also establishes continuous dependence for the solution with respect to the reaction term of the nonlinear elliptic equation.

Proposition 2.12. *Let $f_n : \Omega \times \mathbb{R} \to \mathbb{R}$ be a Carathéodory function satisfying*

$$|f_n(x,t)| \leq D(1 + |t|^{r-1}) \quad \textit{for a.e. } x \in \Omega,\ t \in \mathbb{R}$$

with some positive constant D independent of n and $r \in [p, p^*)$, where $p^* = +\infty$ if $N \leq p$, $p^* = pN/(N-p)$ if $N > p$. Assume that $A_n : \Omega \times \mathbb{R}^N \to \mathbb{R}^N$ is a map satisfying H(a) *with positive constants c_1, c_0 and c_2 independent of n. If u_n is a solution for*

$$-\text{div}\, A_n(x, \nabla u) = f_n(x,u) \quad \textit{in } \Omega, \quad u = 0 \quad \textit{on } \partial\Omega,$$

and $\{u_n\}$ is bounded in $W_0^{1,p}(\Omega)$, then there exist a subsequence $\{u_{n_l}\}$ of $\{u_n\}$ and $u_0 \in C_0^1(\overline{\Omega})$ such that $u_{n_l} \to u_0$ in $C_0^1(\overline{\Omega})$ as $l \to \infty$.

Proof. Since $\{u_n\}$ is bounded in $W_0^{1,p}(\Omega)$, we may assume that $\{u_n\}$ weakly converges to some u_0 in $W_0^{1,p}(\Omega)$ by choosing a subsequence. We can show that there exists $C > 0$ depending only on $|\Omega|$, p, N, D, c_0, c_1 and $\sup_n \|u_n\|_{W_0^{1,p}(\Omega)}$ such that $\|u_n\|_\infty \leq C$ by the Moser iteration process (refer to [156, Theorem C] or Proposition 2.10 and Remark 2.11). Since D, c_1 and c_0 are independent of n, $\{\|u_n\|_\infty\}$ is bounded. Therefore, the regularity result in [129] guarantees that there exist $\gamma \in (0,1)$ and $M > 0$ of n such that $u_n \in C_0^{1,\gamma}(\overline{\Omega})$ and

$\|u_n\|_{C_0^{1,\gamma}(\overline{\Omega})} \le M$ (where we use the fact that c_2 is independent of n). Since the inclusion of $C_0^{1,\gamma}(\overline{\Omega})$ into $C_0^1(\overline{\Omega})$ is compact, $\{u_n\}$ converges to u_0 in $C_0^1(\overline{\Omega})$ (note that $u_n \rightharpoonup u_0$ in $W_0^{1,p}(\Omega)$). □

Remark 2.13. The result of Proposition 2.12 is also true for the Neumann problem.

2.3 STRONG MAXIMUM PRINCIPLE FOR NONLINEAR ELLIPTIC EQUATIONS

Consider a bounded domain $\Omega \subset \mathbb{R}^N$ with C^2 boundary $\partial\Omega$, a map $A : \overline{\Omega} \times \mathbb{R} \times \mathbb{R}^N \to \mathbb{R}^N$ for which we set

$$A(x,t,\xi) = (a_1(x,t,\xi), \dots, a_N(x,t,\xi)) \in \mathbb{R}^N$$

for all $(x,t,\xi) \in \Omega \times [0,+\infty) \times \mathbb{R}^N$, and a function $\widetilde{f} : \overline{\Omega} \times [0,+\infty) \times \mathbb{R}^N \to \mathbb{R}$.

We are concerned with the inequality problem consisting in finding $u : \Omega \to \mathbb{R}$ with $u \in C^1(\overline{\Omega})$ such that

$$-\operatorname{div} A(x,u,\nabla u) \ge \widetilde{f}(x,u,\nabla u) \quad \text{in } \Omega, \tag{2.46}$$

which is understood in the distributional sense. We will prove a general strong maximum principle for this problem (in particular, for nonlinear elliptic equations). The precise conditions that we impose on the data in (2.46) are listed below.

We formulate the required hypotheses on the mapping A in (2.46).

H(A) Given $1 < q < +\infty$, there exist positive constants C_1, C_2, C_3, C_4 and $0 < t_0 \le 1$ such that

- **(i)** $a_i \in C(\overline{\Omega} \times [0,+\infty) \times \mathbb{R}^N, \mathbb{R}) \cap C^1(\overline{\Omega} \times (0,+\infty) \times (\mathbb{R}^N \setminus \{0\}), \mathbb{R})$, $i = 1, \dots, N$;
- **(ii)** $a_i(x,t,0) = 0$ for all $(x,t) \in \overline{\Omega} \times [0,+\infty)$, and $1 \le i \le N$;
- **(iii)** $\|D_y A(x,t,y)\| \le C_1 \left(t^{q-1}(1+|\log t|) + |y|^{q-2}\right)$ for all $x \in \overline{\Omega}$, $0 \le t \le t_0$, $y \in \mathbb{R}^N$ with $0 < |y| \le t_0$;
- **(iv)** $D_y A(x,t,y)\xi \cdot \xi \ge C_2 |y|^{q-2}|\xi|^2$ for all $x \in \overline{\Omega}$, $0 \le t \le t_0$, $y \in \mathbb{R}^N$ with $0 < |y| \le t_0$ and $\xi \in \mathbb{R}^N$;
- **(v)** $|\partial_{x_i} a_i(x,t,y)| \le C_3 \left(t^{q-1}(1+|\log t|) + |y|^{q-1}(1+|\log|y||)\right)$ for all $x \in \overline{\Omega}$, $0 < t \le t_0$, $0 < |y| \le t_0$, and $1 \le i \le N$;
- **(vi)** $|\partial_t a_i(x,t,y)| \le C_4 \left(t^{q-1}(1+|\log t|) + |y|^{q-1}(1+|\log|y||)\right)$ for all $x \in \overline{\Omega}$, $0 < t \le t_0$, $0 < |y| \le t_0$, and $1 \le i \le N$.

Here the expression $t^{q-1}|\log t|$ is interpreted as equal 0 at $t = 0$.

The following consequence of hypothesis H(A) will be useful.

Lemma 2.14. *Assume condition* H(A) *and let* $d_1 > d_0 > 0$. *If* $1 < q \leq 2$, *then*

$$(A(x,t,\xi) - A(x,t,\eta)) \cdot (\xi - \eta) \geq \frac{C_2}{d_1^{2-q}} |\xi - \eta|^2$$

for all $x \in \Omega$, $t \in [0, t_0]$ *and* $\eta, \xi \in \mathbb{R}^N$ *with* $|\xi|, |\eta| < d_1$.

If $q > 2$, *then*

$$(A(x,t,\xi) - A(x,t,\eta)) \cdot (\xi - \eta) \geq \frac{C_2 d_0^{q-1}}{2^{q-1}(d_0 + d_1)} |\xi - \eta|^2$$

for all $x \in \Omega$, $t \in [0, t_0]$ *and* $\eta, \xi \in \mathbb{R}^N$ *with* $d_0 < |\xi| < d_1$ *and* $|\eta| < d_1$.

Proof. *Case* $1 < q \leq 2$. Let $\eta, \xi \in \mathbb{R}^N$ with $|\xi|, |\eta| < d_1$. If $\theta\xi + (1-\theta)\eta \neq 0$ for every $\theta \in [0, 1]$, noting $q - 2 \leq 0$ and $|\theta\xi + (1-\theta)\eta| < d_1$, by hypothesis (iv) in H(A) we obtain

$$\begin{aligned}
&(A(x,t,\xi) - A(x,t,\eta)) \cdot (\xi - \eta) \\
&= \int_0^1 D_y A(x, s, \eta + s(\xi - \eta))(\xi - \eta) \cdot (\xi - \eta)\, ds \\
&\geq C_2 \int_0^1 |\eta + s(\xi - \eta)|^{q-2} |\xi - \eta|^2\, ds \geq \frac{C_2}{d_1^{2-q}} |\xi - \eta|^2
\end{aligned} \tag{2.47}$$

whenever $x \in \Omega$ and $t \in [0, t_0]$. Similarly, if $\eta = 0$ and $\xi \neq 0$, then

$$A(x, s, \xi)\xi \geq C_2 d_1^{q-2} |\xi|^2.$$

Now we examine the case where 0 belongs to the interior of the line segment $[\xi, \eta]$ with $\xi \neq 0$, $\eta \neq 0$. Let $0 < \theta_0^n < \theta_1^n < 1$ such that $\theta\xi + (1-\theta)\eta \neq 0$ for every $\theta \in [0, \theta_0^n] \cup [\theta_1^n, 1]$ and $\theta_1^n - \theta_0^n \leq 1/n$. Set $y_i^n := \theta_i^n \xi + (1 - \theta_i^n)\eta$, $i = 0, 1$, thus

$$\xi - \eta = \frac{\xi - y_1^n}{1 - \theta_1^n} = \frac{y_0^n - \eta}{\theta_0^n} = \frac{y_1^n - y_0^n}{\theta_1^n - \theta_0^n}. \tag{2.48}$$

From (2.47) and (2.48), we have

$$\begin{aligned}
&(A(x,t,\xi) - A(x,t,y_1^n)) \cdot (\xi - \eta) \\
&= \frac{1}{1 - \theta_1^n} (A(x,t,\xi) - A(x,t,y_1^n)) \cdot (\xi - y_1^n) \\
&\geq \frac{C_2}{(1 - \theta_1^n) d_1^{2-q}} |\xi - y_1^n|^2 = \frac{C_4(1 - \theta_1^n)}{d_1^{2-q}} |\xi - \eta|^2,
\end{aligned}$$

$$
\begin{aligned}
&(A(x,t,y_0^n)-A(x,t,\eta))\cdot(\xi-\eta)\\
&=\frac{1}{\theta_0^n}(A(x,t,y_0^n)-A(x,t,\eta))\cdot(y_0^n-\eta)\\
&\geq\frac{C_2}{d_1^{2-q}\theta_0^n}|y_0^n-\eta|^2=\frac{C_2\theta_0^n}{d_1^{2-q}}|\xi-\eta|^2.
\end{aligned}
$$

By the monotonicity of A in the third variable, (2.48) and $\theta_1^n-\theta_0^n\to 0$, it turns out that

$$
\begin{aligned}
&(A(x,t,\xi)-A(x,t,\eta))\cdot(\xi-\eta)\\
&=(A(x,t,\xi)-A(x,t,y_1^n))\cdot(\xi-\eta)+(A(x,t,y_1^n)-A(x,t,y_0^n))\cdot(\xi-\eta)\\
&\quad+(A(x,t,y_0^n)-A(x,t,\eta))\cdot(\xi-\eta)\\
&\geq\frac{C_2}{d_1^{2-q}}(\theta_0^n+(1-\theta_1^n))|\xi-\eta|^2\\
&\quad+\frac{1}{\theta_1^n-\theta_0^n}(A(x,t,y_1^n)-A(x,t,y_0^n))\cdot(y_1^n-y_0^n)\\
&\geq\frac{C_2}{d_1^{2-q}}(\theta_0^n+(1-\theta_1^n))|\xi-\eta|^2\geq\frac{C_2}{d_1^{2-q}}|\xi-\eta|^2(1-o(1)).
\end{aligned}
$$

Case $q>2$. Let $\eta,\xi\in\mathbb{R}^N$ with $d_0<|\xi|<d_1$, $|\eta|<d_1$, and set

$$
\theta_0=1-\frac{d_0}{2(d_0+d_1)}\quad\text{and}\quad y_0=\theta_0\xi+(1-\theta_0)\eta.
$$

Because of $|\xi|>d_0$ and $|\eta|<d_1$, one has

$$
\begin{aligned}
|\theta\xi+(1-\theta)\eta|&\geq\theta d_0-(1-\theta)d_1=(d_0+d_1)\theta-d_1\\
&\geq\frac{d_0}{2}\quad\text{for every }\theta\in[\theta_0,1].
\end{aligned}
$$

We see that

$$
|t\xi+(1-t)y_0|=|(\theta_0+t(1-\theta_0))\xi+(1-\theta_0-t(1-\theta_0))\eta|\geq\frac{d_0}{2}
$$

for every $t\in[0,1]$. Noting that

$$
\xi-\eta=(y_0-\eta)/\theta_0=(\xi-y_0)/(1-\theta_0),
$$

by the mean value theorem with some $s\in(0,1)$ we find

$$
\begin{aligned}
&(A(x,t,\xi)-A(x,t,\eta))\cdot(\xi-\eta)\\
&=\frac{1}{\theta_0}(A(x,t,y_0)-A(x,t,\eta))\cdot(y_0-\eta)\\
&\quad+\frac{1}{1-\theta_0}D_yA(x,t,s\xi+(1-s)y_0)(\xi-y_0)\cdot(\xi-y_0)
\end{aligned}
$$

$$\geq \frac{C_2 d_0^{q-2}}{2^{q-2}(1-\theta_0)}|\xi - y_0|^2 = \frac{C_2 d_0^{q-2}}{2^{q-2}}(1-\theta_0)|\xi - \eta|^2$$
$$= \frac{C_2 d_0^{q-1}}{2^{q-1}(d_0+d_1)}|\xi - \eta|^2,$$

thanks to the monotonicity of A in the third variable and assumption H(A) (iv). □

We formulate the following hypotheses on the function $\widetilde{f}$ in problem (2.46).

H($\widetilde{f}$) $\widetilde{f} : \overline{\Omega} \times [0, +\infty) \times \mathbb{R}^N \to \mathbb{R}$ is a continuous function such that there exists a constant $\lambda > 0$ for which

$$\widetilde{f}(x, s, \xi) \geq -\lambda s^{q-1}(1 + |\log s|) - \lambda|\xi|^{q-1}(1 + |\log|\xi||) \qquad (2.49)$$

for all $x \in \overline{\Omega}$, $0 < s \leq t_0$, $0 < |\xi| \leq t_0$, where t_0 is as in H(A).

The below function will play an essential role in our approach

$$h(t) := t^{q-1}(1 + |\log t|) \quad \text{for } t \geq 0. \qquad (2.50)$$

Notice that the function h is increasing on $[0, t_1]$, where $t_1 := e^{(q-2)/(q-1)}$.

Lemma 2.15. *For any $t_2 \in (0, t_1)$ and $t_3 > t_1$, there exists $h_0 > 0$ such that*

$$h(|\xi|) - h(|\eta|) \geq -h_0|\xi - \eta|$$

for all $\xi, \eta \in \mathbb{R}^N$ with $t_1 \geq |\xi| \geq t_2$ and $|\eta| \leq t_3$.

Proof. Since h is increasing on $[0, t_1]$ (see (2.50)), if $t_1 \geq |\xi| \geq |\eta|$ holds, then $h(|\xi|) - h(|\eta|) \geq 0$. In the case of $t_2 \leq |\xi| < |\eta| \leq t_3$, the mean value theorem gives

$$h(|\xi|) - h(|\eta|) = h'(t)(|\xi| - |\eta|)$$

for some $t \in (|\xi|, |\eta|)$. Setting

$$h_0 := \max\{|h'(t)| : t_2 \leq t \leq t_3\} > 0,$$

we obtain $h(|\xi|) - h(|\eta|) \geq -h_0(|\eta| - |\xi|) \geq -h_0|\eta - \xi|$. □

We first prove a boundary point result, which represents the hard part of the proof of our strong maximum principle regarding problem (2.46).

Proposition 2.16. *Assume that Ω' is a bounded domain in $\mathbb{R}^N$ and $x_1 \in \partial\Omega'$ is such that the interior-ball condition is fulfilled at x_1. Let A and $\widetilde{f}$ satisfy the hypotheses* H(A) *and* H($\widetilde{f}$)*, respectively, with Ω' in place of Ω. If $u \in C^1(\Omega' \cup \{x_1\})$ verifies (2.46) in Ω', $u > 0$ in Ω' and $u(x_1) = 0$, then $\partial u(x_1)/\partial\nu < 0$ holds, where ν stands for the outer unit normal to $\partial\Omega$.*

Proof. The interior-ball condition of Ω' at x_1 provides $x_2 \in \Omega'$ and $T_0 > 0$ such that the open ball $B(x_2, 2T_0)$ fulfills $B(x_2, 2T_0) \subset \Omega'$ and $\partial B(x_2, 2T_0) \cap \partial\Omega' = \{x_1\}$. The assumptions $u \in C^1(\Omega' \cup \{x_1\})$ and $u(x_1) = 0$ guarantee that for x_2 and T_0 we can suppose that

$$0 < u < \min\{t_0, e^{(q-2)/(q-1)}/2\} \text{ and } |\nabla u| \le K_1 \text{ in } B(x_2, 2T_0), \tag{2.51}$$

with some constant $K_1 > 0$. Denote

$$K_0 = N(C_3 + C_4K_1 + 1) \tag{2.52}$$

and fix $T > 0$ satisfying

$$T < \min\left\{T_0, \frac{C_2}{4(\lambda + K_0)}\right\}, \tag{2.53}$$

where the constants C_2, C_3, C_4 are those in hypotheses $\mathrm{H}(A)$. Setting the convex combination

$$x_3 = \left(1 - \frac{T}{T_0}\right)x_1 + \frac{T}{T_0}x_2,$$

we have

$$B(x_3, 2T) \subset B(x_2, 2T_0) \subset \Omega' \quad \text{and} \quad \partial B(x_3, 2T) \cap \partial\Omega' = \{x_1\}.$$

Notice that $\min\{u(x) : |x - x_3| = T\} > 0$, which is true because $u > 0$ in Ω'.

Let us consider

$$Y := \{x \in \Omega' : T < |x - x_3| < 2T\} \text{ and } v_k(t) := \frac{1}{k^2}\frac{e^{kt} - 1}{e^{kT} - 1} \tag{2.54}$$

for all $t \in [0, T]$ and a constant $k > 0$, as well as the function

$$w(x) := w(r) := v_k(2T - r) = v_k(t) \ \text{ for } x \in Y,$$

with $r = |x - x_3|$ and $t = 2T - r$.

We show that for sufficiently large k the function w verifies

$$-\mathrm{div}\, A(x, u, \nabla w) + (\lambda + K_0)h(w) + \lambda h(|\nabla w|) \le K_0 h(u) \quad \text{in } Y, \tag{2.55}$$

with $u \ge w$ on ∂Y, where h and K_0 are introduced in (2.50) and (2.52), and $\lambda > 0$ is the constant in assumption $\mathrm{H}(\widetilde{f})$.

Direct computation enables us to check that

$$-\,\mathrm{div}\, A(x, u, \nabla w) = -\sum_{i=1}^{N} \frac{\partial}{\partial x_i}(a_i(x, u, \nabla w))$$

$$
\begin{aligned}
&= -\sum_{i=1}^{N} \frac{\partial a_i}{\partial x_i}(x,u,\nabla w) - \sum_{i=1}^{N} \frac{\partial a_i}{\partial t}(x,u,\nabla w)\frac{\partial u}{\partial x_i} \\
&\quad - \frac{w'}{r}\sum_{i=1}^{N} D_y A(x,u,\nabla w)e_i \cdot e_i \\
&\quad - \left(\frac{w''}{r^2} - \frac{w'}{r^3}\right) D_y A(x,u,\nabla w)(x-x_3)\cdot(x-x_3) \quad \text{in } Y,
\end{aligned}
$$

where $\{e_i\}_{1\le i\le N}$ denotes the usual orthonormal basis in $\mathbb{R}^N$.

Notice that $w'(r) = -v_k'(t) < 0$ and $w''(r) = v_k''(t) = kv_k'(t) > 0$. By taking k sufficiently large, $w \in C^\infty(\overline{Y})$ satisfies $u \ge w$ on ∂Y (note that $\min_{\partial Y} u > 0$),

$$
0 < w = v_k \le \min\{t_0, t_1/2\} \quad \text{and} \quad 0 < |w'| = v_k' \le \min\{t_0, t_1/2\} \tag{2.56}
$$

in $\overline{Y}$. Since $0 < u < t_0$ and $|\nabla u| \le K_1$ in $\overline{Y}$ (see (2.51)), through hypothesis H(A) we obtain

$$
\begin{aligned}
&-\operatorname{div} A(x,u,\nabla w) \le NC_3(h(u)+h(|\nabla w|)) + NC_4(h(u)+h(|\nabla w|))K_1 \\
&\quad - \frac{w'}{r}NC_1(h(u)+|\nabla w|^{q-2}) - \left(w'' - \frac{w'}{r}\right)C_2|\nabla w|^{q-2} \\
&= N\left(C_3 + C_4K_1 + C_1\frac{v_k'}{r}\right)h(u) + N\left(C_3 + C_4K_1\right)h(v_k') + \frac{NC_1}{r}(v_k')^{q-1} \\
&\quad - C_2(v_k')^{q-1}\left(k + \frac{1}{r}\right) \quad \text{in } Y.
\end{aligned}
$$

For a sufficiently large k, we may assume that

$$
0 < C_1\frac{v_k'}{r} \le C_1\frac{v_k'}{T} \le \frac{C_1}{T}\frac{1}{k}\left(1 + \frac{1}{e^{kT}-1}\right) \le 1 \quad \text{in } \overline{Y}
$$

(note that $T \le r \le 2T$). Hence,

$$
-\operatorname{div} A(x,u,\nabla w) \le K_0(h(u)+h(v_k')) + \frac{NC_2}{T}(v_k')^{q-1} - C_2(v_k')^{q-1}k \quad \text{in } \overline{Y}
$$

holds in view of (2.52). We get

$$
\begin{aligned}
&-\operatorname{div} A(x,u,\nabla w) + (\lambda + K_0)h(w) + \lambda h(|\nabla w|) \\
&\le K_0 h(u) + (\lambda + K_0)(h(v_k) + h(v_k')) - C_2(v_k')^{q-1}k + \frac{NC_2}{T}(v_k')^{q-1} \quad \text{in } Y.
\end{aligned}
$$

Applying Lemma 2.17 given below with a sufficiently large k shows that w satisfies (2.55).

By means of (2.55), we can prove that if k is sufficiently large then

$$u \geq w \text{ in } \overline{Y}. \tag{2.57}$$

Admit for a moment that (2.57) is valid. Note that

$$\nu(x_1) = -(x_3 - x_1)/2T.$$

Then, relying on (2.57), in conjunction with $u \in C^1(\Omega \cup \{x_1\})$ and $u(x_1) = 0$, we obtain

$$\begin{aligned} -2T \frac{\partial u}{\partial \nu}(x_1) &= \lim_{s \to 0^+} \frac{u(x_1 + s(x_3 - x_1)) - u(x_1)}{s} \\ &\geq \lim_{s \to 0^+} \frac{w(x_1 + s(x_3 - x_1)) - w(x_1)}{s} = 2T v_k'(0) > 0, \end{aligned}$$

so the desired conclusion is achieved. The proof of (2.57) is given subsequently, which will complete the proof. □

Lemma 2.17. *For sufficiently large k, the functions v_k introduced in (2.54) satisfy*

$$-C_4 k (v_k')^{q-1} + (\lambda + K_0)(h(v_k') + h(v_k)) + \frac{NC_2}{T}(v_k')^{q-1} < 0 \quad \text{in } Y,$$

with K_0 in (2.52), and

$$0 < v_k' \leq \min\{t_0, t_1/2\} \quad \text{in } Y$$

with $t_1 = e^{(q-2)/(q-1)}$ and t_0 in H(A).

Proof. On the basis of

$$v_k' \leq \frac{1}{k}\left(1 + \frac{1}{e^{kT} - 1}\right) \to 0 \quad \text{in } \overline{Y}$$

as $k \to \infty$, we may assume that

$$v_k' \leq \min\{t_0, t_1/2\}$$

in $\overline{Y}$ for sufficiently large k (see (2.54)). Since h is increasing on $[0, t_1]$, we have $h(v_k) \leq h(v_k')$ by $v_k < v_k'/k$, and so

$$h(v_k') + h(v_k) \leq 2h(v_k') = 2(v_k')^{q-1}(1 + |\log v_k'|). \tag{2.58}$$

Simple verification confirms that

$$|\log v_k'| = -\log v_k' = \log k + \log(e^{kT} - 1) - \log e^{kt} \leq kT + \log k \quad \text{in } \overline{Y}. \tag{2.59}$$

Then (2.58), (2.59) and (2.53) imply

$$
\begin{aligned}
& - C_4 k (v_k')^{q-1} + (\lambda + K_0)(h(v_k') + h(v_k)) + \frac{NC_2}{T}(v_k')^{q-1} \\
& \le (v_k')^{q-1}\Big(- \frac{C_4 k}{2} + \frac{NC_2}{T} + 2(\lambda + K_0)(1 + \log k) \\
& \quad + k\Big(- \frac{C_4}{2} + 2(\lambda + K_0)T\Big)\Big) \\
& \le k(v_k')^{q-1}\Big(- \frac{C_4}{2} + \frac{NC_2}{Tk} + \frac{2(\lambda + K_0)(1 + \log k)}{k}\Big),
\end{aligned}
$$

so our conclusion holds true for sufficiently large k. □

Proof of (2.57). To prove that $u \ge w$ in $\overline{Y}$, by way of contradiction we assume that

$$\min\{t_0, t_1/2\} \ge \max_{\overline{Y}} w \ge V := \max_{\overline{Y}} (w - u) > 0.$$

Set

$$\Sigma(l) := \{x \in Y : (V \ge)\, (w - u)(x) > l\} \quad \text{for } 0 < l < V$$

and

$$\Sigma(V) := \{x \in Y : (w - u)(x) = V\}. \tag{2.60}$$

It follows from the subsequent Lemma 2.18 that $|\Sigma(V)| = 0$, that is, the set $\Sigma(V)$ is of Lebesgue measure zero.

Fix $r > 1$ satisfying $r \le N/(N-2)$ if $N \ge 3$. Note that H($\widetilde{f}$), (2.46) and (2.50) imply

$$-\mathrm{div}\, A(x, u, \nabla u) \ge -\lambda h(u) - \lambda h(|\nabla u|) \quad \text{in } Y. \tag{2.61}$$

Since

$$0 < \min_{\overline{Y}} |\nabla w| \le \max_{\overline{Y}} |\nabla w| \le t_1/2,$$

Lemma 2.15 provides $h_0 > 0$ such that

$$h(|\nabla w|) - h(|\nabla u|) \ge -h_0 |\nabla w - \nabla u| \quad \text{in } Y.$$

Moreover, we note that

$$h(w) > h(u) \quad \text{on } \Sigma(l)$$

for any $0 < l < V$, because h is increasing in $[0, t_1]$. Using

$$\varphi := (w - u - l)^+ \in W_0^{1,\infty}(Y)$$

for $0 < l < V$ as test function (note that $u \geq w$ on ∂Y), by (2.55), (2.61), Hölder's and Sobolev's inequalities, we obtain

$$\begin{aligned}
&\int_Y (A(x,u,\nabla w) - A(x,u,\nabla u)) \cdot \nabla (w-u-l)^+ \, dx \\
&\leq -(\lambda + K_0) \int_Y (h(w) - h(u))(w-u-l)^+ \, dx \\
&\quad - \lambda \int_Y (h(|\nabla w|) - h(|\nabla u|))(w-u-l)^+ \, dx \\
&\leq \lambda h_0 \int_{\Sigma(l)} |\nabla w - \nabla u| (w-u-l) \, dx = \lambda h_0 \int_Y |\nabla \varphi| \varphi \, dx \\
&\leq \lambda h_0 \|\nabla \varphi\|_{L^2(Y)} \|\varphi\|_{L^{2r}(Y)} |\Sigma(l)|^{(r-1)/2r} \leq \lambda h_0 C_* \|\nabla \varphi\|^2_{L^2(Y)} |\Sigma(l)|^{(r-1)/2r},
\end{aligned}$$

where $C_* > 0$ is a constant determined by the embedding of $W_0^{1,2}(Y)$ into $L^{2r}(Y)$.

On the other hand, Lemma 2.14 provides a constant $C > 0$ such that

$$\begin{aligned}
&\int_Y (A(x,u,\nabla w) - A(x,u,\nabla u)) \cdot \nabla (w-u-l)^+ \, dx \\
&\geq C \int_{\{w \geq u+l\}} |\nabla w - \nabla u|^2 \, dx = C \|\nabla \varphi\|^2_{L^2(Y)}.
\end{aligned}$$

This amounts to saying that

$$C \leq \lambda h_0 C_* |\Sigma(l)|^{(r-1)/2r}$$

because $\varphi \not\equiv 0$ for any $0 < l < V$. Thus we reach a contradiction taking into account that $|\Sigma(l)| \to 0$ as $l \to V$, which concludes the proof of (2.57). □

Lemma 2.18. *There holds $|\Sigma(V)| = 0$, where $\Sigma(V)$ is the set defined by (2.60).*

Proof. Arguing by contradiction, we suppose that $|\Sigma(V)| > 0$. Let $\rho_n * \chi$ be the convolution of ρ_n and χ, where $\{\rho_n(x)\}$ is a sequence of mollifiers and $\chi(x) = 1$ if $x \in \Sigma(V)$, $\chi(x) = 0$ if $x \notin \Sigma(V)$. Note that $\operatorname{supp}(\rho_n * \chi) \subset Y$ for sufficiently large n because $u \geq w$ on ∂Y and $\Sigma(V)(\subset Y)$ is compact. Then it follows from (2.55) and (2.61) that

$$\begin{aligned}
&\int_Y A(x,u,\nabla w) \cdot \nabla (\rho_n * \chi) \, dx \\
&\leq \int_Y (K_0 h(u) - (\lambda + K_0) h(w) - \lambda h(|\nabla w|)) \rho_n * \chi \, dx
\end{aligned}$$

and

$$\int_Y A(x,u,\nabla u) \cdot \nabla (\rho_n * \chi) \, dx \geq -\lambda \int_Y (h(u) + h(|\nabla u|)) \rho_n * \chi \, dx.$$

These lead to

$$o(1) = \int_{Y \setminus \Sigma(V)} (A(x,u,\nabla w) - A(x,u,\nabla u)) \cdot \nabla(\rho_n * \chi)\, dx$$
$$\leq (\lambda + K_0) \int_Y (h(u) - h(w))\rho_n * \chi\, dx$$
$$+ \lambda \int_Y (h(|\nabla u|) - h(|\nabla w|))\rho_n * \chi\, dx,$$

taking into account in the last inequality that $\nabla u = \nabla w$ on $\Sigma(V)$. Since h is increasing in $[0, t_1]$ (note that $u + V \leq t_1$ by (2.56) and (2.51)), a contradiction arises letting $n \to \infty$ in

$$0 > (\lambda + K_0) \int_{\Sigma(V)} (h(u) - h(u+V))\, dx + o(1)$$
$$= (\lambda + K_0) \int_Y (h(u) - h(w))\rho_n * \chi\, dx + \lambda \int_Y (h(|\nabla u|) - h(|\nabla w|))\rho_n * \chi\, dx$$

because of $w = u + V$ and $\nabla u = \nabla v$ on $\Sigma(V)$. This completes the proof. □

Taking advantage of Proposition 2.16, we can establish our strong maximum principle regarding inequality (2.46).

Theorem 2.19. *Assume* H(*A*) *and* H($\widetilde{f}$). *If* $u \in C^1(\overline{\Omega})$ *satisfies* $u \geq 0$ *in* Ω, $u \not\equiv 0$ *in* Ω, *and inequality (2.46), then* $u > 0$ *in* Ω *and* $\partial u/\partial \nu < 0$ *on* $\partial\Omega$ *hold.*

Proof. We set $[u > 0] := \{x \in \Omega : u(x) > 0\}$. By hypothesis we know that $[u > 0] \neq \emptyset$. Since Ω is connected, it suffices to show that $\partial[u > 0] \cap \Omega = \emptyset$.

By way of contradiction, assume that $\partial[u > 0] \cap \Omega \neq \emptyset$. Then there exists a point $x_1 \in \partial[u > 0] \cap \Omega$ such that $[u > 0]$ fulfills the interior-ball condition at x_1. It is so because the set of $x \in \partial[u > 0]$ such that $[u > 0]$ satisfies the interior-ball condition at x is dense in $\partial[u > 0]$ (refer to [156, Appendix A]). Choose $x_2 \in [u > 0]$ and $T_0 > 0$ such that $B(x_2, T_0) \subset [u > 0]$ and $\partial B(x_2, T_0) \cap \partial[u > 0] = \{x_1\}$. Because $u > 0$ in $B(x_2, T_0)$, $u(x_1) = 0$ and u satisfies (2.46) in $B(x_2, T_0)$, applying Proposition 2.16 with $\Omega' = B(x_2, T_0)$ renders that $\nabla u(x_1) \neq 0$. However, $x_1 \in \Omega$ is a global minimizer of u, so $\nabla u(x_1) = 0$. This contradiction ensures that $\partial[u > 0] \cap \Omega = \emptyset$, thus $u > 0$ in Ω.

Since $\partial\Omega$ is of class C^2, the property $\partial u/\partial \nu < 0$ on $\partial\Omega$ results from Proposition 2.16. The proof is complete. □

2.4 NOTES

Section 2.1 takes into account the dependence of the solution sets with respect to parameters. It exploits powerful topological tools as, for instance, the transversality theory. Theorem 2.2 is taken from [162,163]. It is inspired by [214,215] regarding generic results for problems depending on parameters and applications to elliptic boundary value problems. A version of Theorem 2.2 on differentiable manifolds was given in [161]. It is worth mentioning that one can derive directly from Theorem 2.2 the density of Morse functions on a finite-dimensional manifold (see [161]). Moreover, with a suitable modification in the argument, Theorem 2.2 leads to the results of Marino and Prodi [148] of approximation by Morse functions on an infinite-dimensional Riemannian manifold (see [162]).

Theorem 2.4 provides homoclinics which are nondegenerate in the sense that they are nondegenerate critical points of the Euler functional associated to problem (2.13). Theorem 2.4 can be found in [163]. The property of a solution to be nondegenerate is an important qualitative information, guaranteeing in particular that the solution is isolated. In studying (2.13) one cannot make use of results dealing with the Cauchy problem for ordinary differential equations because the formulation in (2.13) does not involve an initial condition. Another essential feature of (2.13) is the lack of compactness, mainly caused by the fact that t runs in an unbounded set, that is reflected, for instance, in the failure of the Palais–Smale condition for Euler functional associated to (2.13). These difficulties are overcome by developing specific variational methods for the homoclinic problem (2.13), for which we refer to [35,50,59,230]. An open problem in such a context is to find verifiable conditions which guarantee that the celebrated hypothesis $(*)$ is fulfilled for (2.13) (see, e.g., [35]).

Theorem 2.4 postulates that for f belonging to a dense (actually, residual) set G in $L^2(\mathbb{R}, \mathbb{R}^N)$, problem (2.13) possesses only finitely many solutions, which in addition are nondegenerate. It also provides a partial answer to assertion $(*)$ expressing its generic validity. We drop out some requirements considered before (see [35,59]). Theorem 2.4 points out the smooth dependence of the solutions q of (2.13) with respect to $f \in L^2(\mathbb{R}, \mathbb{R}^N)$.

The abstract result presented in Theorem 2.2 is applied to deduce in Theorem 2.5 that the Morse functions related to Eq. (2.32) exist generically with respect to $f \in L^2(\Omega)$ that is regarded as a parameter. Theorem 2.5 can be found in [162]. Specifically, we show that under natural hypotheses upon the nonlinearity $p(x, u)$ the Euler functional corresponding to (2.32) (see, e.g., [207, p. 61]) is generically a Morse function, which means that it has only nondegenerate critical points and satisfies the Palais–Smale condition. It allows applying different results where the nondegeneracy of solutions is a basic assumption. For a comprehensive discussion in this direction, we refer to the related work of Mawhin [151]. Regularity information concerning the dependence of the solutions of (2.32) with respect to $f \in L^2(\Omega)$ is also available.

Section 2.2 contains regularity results for nonlinear elliptic equations, in particular applicable for problems involving the p-Laplacian. It is worth mentioning that the principal part of the considered problems is driven by nonlinear differential operators that are not required to satisfy any homogeneity condition. Proposition 2.10 represents a convenient criterion to show that the solutions are uniformly bounded. It is taken from Miyajima–Motreanu–Tanaka [156, Theorem C]. The proof is in the spirit of the method known as Moser iteration technique. The boundedness of solutions is extremely important because it is a fundamental condition to establish the regularity up to the boundary for the solutions. In the case of quasilinear elliptic problems with Dirichlet or Neumann boundary conditions, the regularity up to the boundary was obtained by Lieberman [129,130]. The interior regularity is studied in [72,126,227].

Proposition 2.12 provides a continuity property of solutions with respect to the reaction term in quasilinear elliptic problems, which enables us to pass to the limit in sequences of problems.

Section 2.3 is devoted to the study of strong maximum principle for nonlinear elliptic problems. The main result is formulated as Theorem 2.19, which is taken from Motreanu–Tanaka [183]. It establishes a strong maximum principle applicable to the differential inequality problem (2.46), in particular to the Dirichlet problem

$$\begin{cases} -\Delta_p u - \Delta_q u = f(x, u, \nabla u) & \text{in } \Omega, \\ u = 0 & \text{on } \partial\Omega \end{cases}$$

and the Neumann problem

$$\begin{cases} -\Delta_p u - \Delta_q u = f(x, u, \nabla u) & \text{in } \Omega, \\ \frac{\partial u}{\partial \nu} = 0 & \text{on } \partial\Omega \end{cases}$$

driven by (p, q)-Laplacian. Many other differential operators can be considered in the principal part of the elliptic equation. An essential feature of this result is that it allows having dependence on the solution u and its gradient ∇u on the right-hand side of the elliptic equation (often called convection term). Theorem 2.19 covers novel situations with respect to the strong maximum principles in Pucci–Serrin [206], Vázquez [229] and Felmer–Montenegro–Quaas [83]. This is demonstrated by the variety of examples mentioned in Examples 4.30 and 4.33.

Usually, the strong maximum principle is utilized jointly with the nonlinear regularity theory. The scheme starts by showing that the solution, say u, belongs to $L^\infty(\Omega)$, which can be done through the Moser iteration process as developed in Proposition 2.10. Then, by the regularity up to the boundary in [129,130], it is seen that actually $u \in C^{1,\alpha}(\overline{\Omega})$ for some $0 < \alpha < 1$. If the solution satisfies $u \geq 0$ and $u \not\equiv 0$, as well as the growth condition (2.49) along it, one can invoke the strong maximum principle in Theorem 2.19 to obtain that $u(x) > 0$ for every $x \in \Omega$ and $\partial u(x)/\partial \nu < 0$ for every $x \in \partial\Omega$, where ν denotes the outward unit normal to $\partial\Omega$.

Chapter 3

Nonlinear Elliptic Eigenvalue Problems

3.1 EIGENVALUE PROBLEM FOR p-LAPLACIAN UNDER DIFFERENT BOUNDARY CONDITIONS

The aim of this section is to give an overview on the Fučík spectrum, which is closely related to the ordinary spectrum, of the negative p-Laplacian $-\Delta_p$ with $1 < p < +\infty$ and different boundary conditions on a bounded domain Ω in $\mathbb{R}^N$. For any real-valued function u on Ω, we pose $u^+ = \max\{u, 0\}$ and $u^- = \max\{0, -u\}$.

Let V be a closed subspace of the Sobolev space $W^{1,p}(\Omega)$ such that $W_0^{1,p}(\Omega) \subset V \subset W^{1,p}(\Omega)$ and let V^* denote the dual space, with the duality pairing $\langle \cdot, \cdot \rangle$ between V and V^*. It is well known that the operator $-\Delta_p : V \to V^*$ is bounded, continuous, pseudomonotone, and has the $(S)_+$-property (i.e., from $u_n \rightharpoonup u$ in V and

$$\limsup_{n\to\infty} \langle -\Delta_p u_n, u_n - u \rangle \leq 0$$

it follows that $u_n \to u$ in V) (see Proposition 1.56). We refer to [38,169] for various nonlinear boundary problems involving $-\Delta_p$.

In order to avoid tedious arguments, as well as certain repetitive ideas, in the approach, we outline the main elements of the theory of Fučík spectrum of $-\Delta_p$ with Dirichlet, Neumann, and Steklov boundary conditions, and discuss in more details the Fučík spectrum of $-\Delta_p$ with Robin boundary condition.

Definition 3.1. The Fučík spectrum of the negative p-Laplacian $-\Delta_p$ with homogeneous Dirichlet boundary condition is defined as the set Σ_p of those $(a, b) \in \mathbb{R}^2$ such that

$$\begin{cases} -\Delta_p u = a(u^+)^{p-1} - b(u^-)^{p-1} & \text{in } \Omega, \\ u = 0 & \text{on } \partial\Omega \end{cases} \tag{3.1}$$

has a nontrivial (weak) solution u, which means that $u \in W_0^{1,p}(\Omega)$, $u \not\equiv 0$, and it satisfies the equation

$$\int_\Omega |\nabla u|^{p-2} \nabla u \cdot \nabla v \, dx = \int_\Omega (a(u^+)^{p-1} - b(u^-)^{p-1}) v \, dx, \quad \forall v \in W_0^{1,p}(\Omega).$$

Nonlinear Differential Problems with Smooth and Nonsmooth Constraints. DOI: 10.1016/B978-0-12-813386-6.00003-1

Remark 3.2. We note that if $a = b = \lambda$, problem (3.1) reduces to

$$\begin{cases} -\Delta_p u = \lambda |u|^{p-2}u & \text{in } \Omega, \\ u = 0 & \text{on } \partial\Omega, \end{cases} \tag{3.2}$$

which is called the Dirichlet eigenvalue problem with respect to $-\Delta_p$.

According to Proposition 1.57, it is known that the first eigenvalue λ_1 of (3.2) is positive, simple, and its corresponding eigenfunctions have constant sign (see also Anane [14] and Lindqvist [131]). In fact, the spectrum $\sigma(-\Delta_p)$ of the negative p-Laplacian $-\Delta_p$ associated to (3.2) includes an unbounded sequence of eigenvalues $\{\lambda_k\}$, called the variational eigenvalues, which fulfills

$$0 < \lambda_1 < \lambda_2 \le \cdots \le \lambda_k \le \cdots \to +\infty.$$

The variational eigenvalues satisfy minimax characterizations.

Proposition 3.3. *The Fučík spectrum Σ_p of the negative p-Laplacian $-\Delta_p$ with homogeneous Dirichlet boundary condition contains the two lines $\{\lambda_1\} \times \mathbb{R}$ and $\mathbb{R} \times \{\lambda_1\}$. Additionally, Σ_p contains the sequence of points (λ_k, λ_k), $k \in \mathbb{N}$.*

Proof. It can be easily seen from Definition 3.1 and Remark 3.2. □

Let us set for every $s \ge 0$,

$$J_s(u) = \int_\Omega |\nabla u|^p \, dx - s \int_\Omega (u^+)^p \, dx, \quad u \in W_0^{1,p}(\Omega).$$

The function J_s is of class C^1 on $W_0^{1,p}(\Omega)$. Consider the restriction $\tilde{J}_s = J_s|_S$, with S given by

$$S = \left\{ u \in W_0^{1,p}(\Omega) : \int_\Omega |u|^p = 1 \right\}.$$

Since S is a C^1 submanifold of $W_0^{1,p}(\Omega)$, it follows that $\tilde{J}_s$ is of class C^1 on S in the sense of manifolds.

Theorem 3.4. *The curve in $\mathbb{R}^2$ given by*

$$\mathcal{C} := \{(s + c(s), c(s)), (c(s), s + c(s)) : s \ge 0\}, \tag{3.3}$$

where $c(s)$ is described by the minimax values

$$c(s) = \inf_{\gamma \in \Gamma} \max_{u \in \gamma([-1,1])} \tilde{J}_s(u),$$

with

$$\Gamma = \{\gamma \in C([-1, 1], S) : \gamma(-1) = -\hat{u}_1 \text{ and } \gamma(1) = \hat{u}_1\}, \tag{3.4}$$

is contained in Σ_p. In (3.4), $\hat{u}_1$ denotes the eigenfunction of (3.2) corresponding to λ_1 satisfying $\hat{u}_1 > 0$ in Ω and $\|\hat{u}_1\|_p = 1$ (see Proposition 1.57). Moreover, the curve $\mathcal{C}$ in (3.3) is Lipschitz continuous and decreasing. The limit of $c(s)$ as $s \to +\infty$ is equal to the first eigenvalue λ_1 of (3.2).

Proof. First, we can prove that the curve $s \in (0, +\infty) \mapsto (s + c(s), c(s)) \in \mathbb{R}^2$ is contained in Σ_p (see [60, Theorem 2.10]). Taking into account that Σ_p is symmetric with respect to the diagonal of the plane, it turns out that the curve $\mathcal{C}$ in (3.3) is contained in Σ_p. The other properties of the first curve $\mathcal{C}$ in Σ_p that are mentioned in the statement of the proposition are proven in [60, Proposition 4.1] and [60, Proposition 4.4]. □

Remark 3.5. It is shown in [60, Theorem 3.1] that $\mathcal{C}$ given in (3.3) is the first nontrivial curve in Σ_p, which means that the first point in Σ_p belonging to the parallel to the diagonal drawn through a point of $([0, +\infty) \times \{\lambda_1\}) \times (\{\lambda_1\} \times \mathbb{R})$ must be on $\mathcal{C}$. As a consequence, we infer that the curve $\mathcal{C}$ passes through (λ_2, λ_2). In conjunction with the description of $\mathcal{C}$ in (3.3) and the minimax formula for $c(s)$, this yields that λ_2 has the variational characterization in Proposition 1.58, namely

$$\lambda_2 = \inf_{\gamma \in \Gamma} \max_{u \in \gamma([-1,1])} \int_\Omega |\nabla u|^p \, dx, \tag{3.5}$$

with Γ introduced in (3.4).

We illustrate the applicability of the Fučík spectrum Σ_p by considering the following problem with homogeneous Dirichlet boundary condition and jumping nonlinearity

$$\begin{cases} -\Delta_p u = a(u^+)^{p-1} - b(u^-)^{p-1} + g(x, u) & \text{in } \Omega, \\ u = 0 & \text{on } \partial\Omega, \end{cases} \tag{3.6}$$

where $g : \Omega \times \mathbb{R} \to \mathbb{R}$ is a Carathéodory function satisfying

$$\lim_{t \to 0} \frac{g(x, t)}{|t|^{p-1}} = 0 \quad \text{uniformly for a.a. } x \in \Omega.$$

In Carl–Perera [49] it is proven that problem (3.6) has at least three nontrivial solutions provided the point $(a, b) \in \mathbb{R}^2$ lies above the first nontrivial curve $\mathcal{C}$ in Σ_p constructed in (3.3). Moreover, a complete sign information for the three solutions is available: two solutions have opposite constant sign and the third one is sign changing (nodal solution). This information is obtained by means of the method of subsolution–supersolution whose application to problem (3.6) strongly relies on the hypothesis that the point $(a, b) \in \mathbb{R}^2$ is situated above the first nontrivial curve $\mathcal{C}$ in Σ_p.

We pass to the Fučík spectrum of the negative p-Laplacian $-\Delta_p$ with (homogeneous) Neumann boundary condition.

Definition 3.6. The Fučík spectrum of the negative p-Laplacian $-\Delta_p$ with Neumann boundary condition, denoted by Θ_p, consists of all pairs $(a, b) \in \mathbb{R}^2$ such that

$$\begin{cases} -\Delta_p u = a(u^+)^{p-1} - b(u^-)^{p-1} & \text{in } \Omega, \\ \frac{\partial u}{\partial \nu} = 0 & \text{on } \partial\Omega \end{cases} \tag{3.7}$$

is solved nontrivially, meaning that $u \in W^{1,p}(\Omega)$, $u \not\equiv 0$, and it verifies the equality

$$\int_\Omega |\nabla u|^{p-2} \nabla u \cdot \nabla v \, dx = \int_\Omega (a(u^+)^{p-1} - b(u^-)^{p-1}) v \, dx, \quad \forall v \in W^{1,p}(\Omega).$$

In (3.7), $\partial u / \partial \nu$ denotes the normal derivative, with ν being the unit outward normal to $\partial\Omega$.

Remark 3.7. In the case where $a = b = \lambda$, problem (3.7) becomes the Neumann eigenvalue problem of the negative p-Laplacian $-\Delta_p$ given by

$$\begin{cases} -\Delta_p u = \lambda |u|^{p-2} u & \text{in } \Omega, \\ \frac{\partial u}{\partial \nu} = 0 & \text{on } \partial\Omega. \end{cases} \tag{3.8}$$

The first eigenvalue $\lambda_1 = 0$ of (3.8) is simple with the corresponding eigenspace $\mathbb{R}$, so all eigenfunctions associated to λ_1 do not change sign in Ω, which does not happen for the higher order eigenvalues (see [127]).

Proposition 3.8. *The Fučík spectrum Θ_p of the negative p-Laplacian $-\Delta_p$ with Neumann boundary condition contains $(0, 0)$, (λ_2, λ_2), where λ_2 is the second eigenvalue of (3.8), and the two lines $\{0\} \times \mathbb{R}$ and $\mathbb{R} \times \{0\}$. Therefore, the nontrivial part $\tilde{\Theta}_p = \Theta_p \setminus ((\{0\} \times \mathbb{R}) \cup (\mathbb{R} \times \{0\}))$ of Θ_p is contained in $(0, +\infty) \times (0, +\infty)$.*

Proof. It is a straightforward consequence of Definition 3.6, (3.7), and Remark 3.7. □

The construction of a first nontrivial curve in $\tilde{\Theta}_p$ can be done similarly to the Dirichlet Fučík spectrum. Namely, let $J_s : W^{1,p}(\Omega) \to \mathbb{R}$ be the functional given by

$$J_s(u) = \int_\Omega |\nabla u|^p \, dx - s \int_\Omega (u^+)^p \, dx$$

and let $\tilde{J}_s$ be its restriction to

$$S = \left\{ u \in W^{1,p}(\Omega) : \int_\Omega |u|^p = 1 \right\}.$$

Notice that S is a C^1 submanifold of $W^{1,p}(\Omega)$, so $\tilde{J}_s$ is of class C^1 on S in the sense of manifolds. This enables us to consider the notions of critical points and critical values for the functional $\tilde{J}_s$.

Theorem 3.9. *The curve in* $\mathbb{R}^2$ *given by*

$$\mathcal{C} := \{(s + c(s), c(s)), (c(s), s + c(s)) : s \geq 0\},$$

where $c(s)$ *is expressed by the minimax value*

$$c(s) = \inf_{\gamma \in \Gamma} \max_{u \in \gamma([-1,1])} \tilde{J}_s(u),$$

with

$$\Gamma = \{\gamma \in C([-1,1], S) : \gamma(-1) = -\varphi_1 \text{ and } \gamma(1) = \varphi_1\}, \tag{3.9}$$

is contained in Θ_p. *In (3.9) one has* $\varphi_1 = 1/|\Omega|^{1/p}$, *with* $|\Omega|$ *denoting the Lebesgue measure of* Ω. *Moreover,*

$$\lim_{s \to +\infty} c(s) = \begin{cases} \lambda_1 = 0 & \text{if } p \leq N, \\ \hat{\lambda} & \text{if } p > N, \end{cases} \tag{3.10}$$

where

$$\hat{\lambda} = \inf\left\{ \int_\Omega |\nabla u|^p \, dx : u \in W^{1,p}(\Omega), \|u\|_{L^p(\Omega)} = 1, \right.$$
$$\left. u \text{ vanishes somewhere in } \overline{\Omega} \right\}.$$

Proof. The study for Θ_p is carried out along the same lines as in the case of the Dirichlet Fučík spectrum Σ_p in Theorem 3.4. The asymptotic properties in (3.10) are shown in [16, Theorems 2.3 and 2.6]. □

Remark 3.10. In view of Theorem 1.20 (iii), the definition of $\hat{\lambda}$ in (3.10) is meaningful because for $p > N$ the elements $u \in W^{1,p}(\Omega)$ are continuous functions on $\overline{\Omega}$.

Corollary 3.11. *The first curve* $\mathcal{C}$ *passes through* (λ_2, λ_2), *where* λ_2 *denotes the second eigenvalue of (3.8). Consequently,* λ_2 *admits the variational expression*

$$\lambda_2 = \inf_{\gamma \in \Gamma} \max_{u \in \gamma([-1,1])} \int_\Omega |\nabla u|^p \, dx,$$

with Γ *introduced in (3.9).*

Proof. This follows readily from Proposition 3.8 and Theorem 3.9. □

The next objective of this section is to briefly discuss the Fučík spectrum of $-\Delta_p$ with Steklov boundary condition.

Definition 3.12. The Steklov Fučík spectrum of $-\Delta_p$ is defined as the set $\widetilde{\Sigma}_p$ of all pairs $(a,b) \in \mathbb{R}^2$ such that

$$\begin{cases} -\Delta_p u = -|u|^{p-2}u & \text{in } \Omega, \\ \frac{\partial u}{\partial \nu} = a(u^+)^{p-1} - b(u^-)^{p-1} & \text{on } \partial\Omega \end{cases} \tag{3.11}$$

has a weak solution $u \not\equiv 0$. Explicitly, $u \in W^{1,p}(\Omega)$ is a weak solution of (3.11) if it satisfies the equality

$$\int_\Omega |\nabla u|^{p-2}\nabla u \cdot \nabla v \, dx = -\int_\Omega |u|^{p-2}uv\, dx + \int_{\partial\Omega} (a(u^+)^{p-1} - b(u^-)^{p-1})v\, d\sigma$$

for all $v \in W^{1,p}(\Omega)$. Here the notation $d\sigma$ stands for the $(N-1)$-dimensional surface measure.

Remark 3.13. If $a = b = \lambda$, (3.11) becomes the so-called Steklov eigenvalue problem, namely

$$\begin{cases} -\Delta_p u = -|u|^{p-2}u & \text{in } \Omega, \\ \frac{\partial u}{\partial \nu} = \lambda|u|^{p-2}u & \text{on } \partial\Omega. \end{cases} \tag{3.12}$$

The first eigenvalue of the Steklov eigenvalue problem (3.12) is positive, simple, and every eigenfunction corresponding to the first eigenvalue does not change sign in Ω. Actually, there is an eigenfunction associated to the first eigenvalue λ_1 belonging to $\text{int}(C^1(\overline{\Omega})_+)$, where the interior of the positive cone

$$C^1(\overline{\Omega})_+ = \{u \in C^1(\overline{\Omega}) : u(x) \geq 0, \forall x \in \Omega\}$$

in the Banach space $C^1(\overline{\Omega})$ is nonempty and is given by

$$\left\{u \in C^1(\overline{\Omega}) : u(x) > 0, \forall x \in \overline{\Omega}\right\}$$

(see [149]). It is established in [84] that there exists a sequence of eigenvalues λ_n of (3.12) such that $\lambda_n \to +\infty$ as $n \to \infty$.

In order to study variationally the Steklov Fučík spectrum defined in Definition 3.12, for each $s \geq 0$ one defines a C^1 functional $J_s : W^{1,p}(\Omega) \to \mathbb{R}$ by

$$J_s(u) = \int_\Omega |\nabla u|^p\, dx + \int_\Omega |u|^p\, dx - s\int_{\partial\Omega} (u^+)^p\, d\sigma.$$

Restricting J_s to

$$S = \left\{ u \in W^{1,p}(\Omega) : \int_{\partial\Omega} |u|^p d\sigma = 1 \right\},$$

one obtains a C^1 functional $\tilde{J}_s$ on the C^1 submanifold S of $W^{1,p}(\Omega)$.

Theorem 3.14. *The first nontrivial curve in $\tilde{\Sigma}_p$ is expressed as*

$$\mathcal{C} = \{(s + c(s), c(s)), (c(s), s + c(s)) : s \geq 0\},$$

where

$$c(s) = \inf_{\gamma \in \Gamma} \max_{u \in \gamma([-1,1])} \tilde{J}_s(u),$$
$$\Gamma = \{\gamma \in C([-1, 1], S) : \gamma(-1) = -\varphi_1 \text{ and } \gamma(1) = \varphi_1\}$$

with $\varphi_1 \in \operatorname{int}(C^1(\overline{\Omega})_+)$ being the first positive eigenfunction of (3.12) provided $\|\varphi_1\|_p = 1$. The first nontrivial curve $\mathcal{C}$ is Lipschitz continuous and decreasing. Moreover,

$$\lim_{s \to +\infty} c(s) = \begin{cases} \lambda_1 & \text{if } p \leq N, \\ \tilde{\lambda} > \lambda_1 & \text{if } p > N, \end{cases}$$

where

$$\tilde{\lambda} = \inf_{u \in L} \max_{r \in \mathbb{R}} \frac{\|r\varphi_1 + u\|^p_{W^{1,p}(\Omega)}}{\|r\varphi_1 + u\|^p_{L^p(\partial\Omega)}}$$

with

$$L = \{u \in W^{1,p}(\Omega) : u \text{ vanishes somewhere on } \partial\Omega\}.$$

The proof can be found in [150, Theorems 2.1 and 4.1].

As before, we can derive from Theorem 3.14 a variational characterization of the second eigenvalue λ_2 of the Steklov eigenvalue problem (3.12).

Corollary 3.15. *The second eigenvalue λ_2 of (3.12) admits the variational expression*

$$\lambda_2 = \inf_{\gamma \in \Gamma} \max_{u \in \gamma([-1,1])} \int_{\Omega} \left(|\nabla u|^p + |u|^p\right) dx. \tag{3.13}$$

As an application, consider the following nonlinear elliptic equation subject to Steklov-type boundary condition with perturbation:

$$\begin{cases} -\Delta_p u = f(x, u) - |u|^{p-2}u & \text{in } \Omega, \\ \frac{\partial u}{\partial \nu} = a(u^+)^{p-1} - b(u^-)^{p-1} + g(x, u) & \text{on } \partial\Omega, \end{cases} \tag{3.14}$$

for Carathéodory functions $f : \Omega \times \mathbb{R} \to \mathbb{R}$ and $g : \partial\Omega \times \mathbb{R} \to \mathbb{R}$ which are bounded on bounded sets and satisfy

$$\lim_{s\to 0} \frac{f(x,s)}{|s|^{p-1}} = 0 \quad \text{uniformly for a.a. } x \in \Omega,$$

$$\lim_{s\to 0} \frac{g(x,s)}{|s|^{p-1}} = 0 \quad \text{uniformly for a.a. } x \in \partial\Omega,$$

$$\lim_{|s|\to +\infty} \frac{f(x,s)}{|s|^{p-2}s} = -\infty \quad \text{uniformly for a.a. } x \in \Omega,$$

$$\lim_{|s|\to +\infty} \frac{g(x,s)}{|s|^{p-2}s} = -\infty \quad \text{uniformly for a.a. } x \in \partial\Omega.$$

If the point (a, b) is above the first nontrivial curve $\mathcal{C}$ in $\tilde{\Sigma}_p$, problem (3.14) possesses three nontrivial solutions: one solution with positive sign, one solution with negative sign, and the third one being sign changing (see [231]).

Finally, we discuss the Fučík spectrum of $-\Delta_p$ with a Robin boundary condition.

Definition 3.16. The Fučík spectrum of $-\Delta_p$ with Robin boundary condition is defined as the set $\widehat{\Sigma}_p$ of all pairs $(a, b) \in \mathbb{R}^2$ for which a nontrivial weak solution $u \in W^{1,p}(\Omega)$ of the problem

$$\begin{cases} -\Delta_p u = a(u^+)^{p-1} - b(u^-)^{p-1} & \text{in } \Omega, \\ \frac{\partial u}{\partial \nu} = -\beta|u|^{p-2}u & \text{on } \partial\Omega \end{cases} \tag{3.15}$$

exists, where β is a nonnegative constant. A weak solution of (3.15) means that

$$\int_\Omega |\nabla u|^{p-2}\nabla u \cdot \nabla v \, dx + \beta \int_{\partial\Omega} |u|^{p-2}uv \, d\sigma$$
$$= \int_\Omega (a(u^+)^{p-1} - b(u^-)^{p-1})v \, dx$$

for all $v \in W^{1,p}(\Omega)$.

Remark 3.17. If $\beta = 0$, $\widehat{\Sigma}_p$ reduces to the Fučík spectrum Θ_p of the negative Neumann p-Laplacian. The special case $a = b = \lambda$ leads to

$$\begin{cases} -\Delta_p u = \lambda|u|^{p-2}u & \text{in } \Omega, \\ \frac{\partial u}{\partial \nu} = -\beta|u|^{p-2}u & \text{on } \partial\Omega, \end{cases} \tag{3.16}$$

which is the Robin eigenvalue problem of the negative p-Laplacian $-\Delta_p$. The first eigenvalue in (3.16), denoted as usual by λ_1, is simple, isolated, and can be variationally characterized as

$$\lambda_1 = \inf_{u\in W^{1,p}(\Omega)} \left\{ \int_\Omega |\nabla u|^p \, dx + \beta \int_{\partial\Omega} |u|^p \, d\sigma : \int_\Omega |u|^p \, dx = 1 \right\}.$$

This infimum is attained exactly on the eigenspace associated to λ_1. It is also known that the eigenfunctions corresponding to λ_1 are of constant sign and belong to $C^{1,\alpha}(\overline{\Omega})$ for some $0 < \alpha < 1$, and in fact λ_1 is the only eigenvalue whose eigenfunctions do not change sign. For all these properties we refer to [127].

Our aim is to determine elements of the Fučik spectrum $\widehat{\Sigma}_p$ in Definition 3.16. They are found as critical points of a functional that is constructed by means of the Robin problem (3.15) following the approach in [186].

For a fixed $s \in \mathbb{R}$, $s \geq 0$, and corresponding to $\beta \geq 0$ given in problem (3.15), we introduce the functional $J_s : W^{1,p}(\Omega) \to \mathbb{R}$ by

$$J_s(u) = \int_\Omega |\nabla u|^p \, dx + \beta \int_{\partial\Omega} |u|^p \, d\sigma - s \int_\Omega (u^+)^p \, dx,$$

thus $J_s \in C^1(W^{1,p}(\Omega))$. Next we introduce the set

$$S = \left\{ u \in W^{1,p}(\Omega) : \int_\Omega |u|^p \, dx = 1 \right\}, \tag{3.17}$$

which is a C^1 submanifold of $W^{1,p}(\Omega)$, and thus $\widetilde{J}_s = J_s|_S$ is a C^1 function in the sense of manifolds. We note that $u \in S$ is a critical point of $\widetilde{J}_s$ (in the sense of manifolds) if and only if there exists $t \in \mathbb{R}$ such that

$$\begin{aligned} &\int_\Omega |\nabla u|^{p-2} \nabla u \cdot \nabla v \, dx + \beta \int_{\partial\Omega} |u|^{p-2} u v \, d\sigma - s \int_\Omega (u^+)^{p-1} v \, dx \\ &= t \int_\Omega |u|^{p-2} u v \, dx \end{aligned} \tag{3.18}$$

for all $v \in W^{1,p}(\Omega)$; in fact, inserting $v = u$ in (3.18) yields $t = J_s(u)$. Now we describe the relationship between the critical points of $\widetilde{J}_s$ and the spectrum $\widehat{\Sigma}_p$.

Lemma 3.18. *Given $r, s \in \mathbb{R}$, $s \geq 0$, one has that $(r + s, r)$ belongs to the spectrum $\widehat{\Sigma}_p$ if and only if there exists a critical point $u \in S$ of $\widetilde{J}_s$ such that $r = J_s(u)$.*

Proof. The formula for the weak solution shows that $(r + s, r) \in \widehat{\Sigma}_p$ if and only if there is $u \in S$ that solves (3.18) with $t = r$. The lemma ensues. □

Lemma 3.18 enables us to find points in $\widehat{\Sigma}_p$ through the critical points of $\widetilde{J}_s$. In order to implement this, first we look for minimizers of $\widetilde{J}_s$.

Proposition 3.19. *There hold:*

(i) *the first eigenfunction φ_1 is a global minimizer of $\widetilde{J}_s$;*
(ii) *the point $(\lambda_1, \lambda_1 - s) \in \mathbb{R}^2$ belongs to $\widehat{\Sigma}_p$.*

Proof. (i) Since $\beta, s \geq 0$, using the characterization of λ_1 we have

$$\begin{aligned}\widetilde{J}_s(u) &= \int_\Omega |\nabla u|^p\,dx + \beta\int_{\partial\Omega}|u|^p\,d\sigma - s\int_\Omega (u^+)^p\,dx \\ &\geq \lambda_1\int_\Omega |u|^p\,dx - s\int_\Omega (u^+)^p\,dx \geq \lambda_1 - s = J_s(\varphi_1), \quad \forall u \in S.\end{aligned}$$

(ii) On the basis of (i), we can apply Lemma 3.18. □

We produce a second critical point of $\widetilde{J}_s$ as a local minimizer.

Proposition 3.20. *There hold:*

- **(i)** *the negative eigenfunction $-\varphi_1$ is a strict local minimizer of $\widetilde{J}_s$;*
- **(ii)** *the point $(\lambda_1 + s, \lambda_1) \in \mathbb{R}^2$ belongs to $\widehat{\Sigma}_p$.*

Proof. (i) Arguing indirectly, let us suppose that there exists a sequence $\{u_n\} \subset S$ with $u_n \neq -\varphi_1$, $u_n \to -\varphi_1$ in $W^{1,p}(\Omega)$ and $\widetilde{J}_s(u_n) \leq \lambda_1 = \widetilde{J}_s(-\varphi_1)$. If $u_n \leq 0$ for a.a. $x \in \Omega$, we obtain

$$\widetilde{J}_s(u_n) = \int_\Omega |\nabla u_n|^p\,dx + \beta\int_{\partial\Omega}|u_n|^p\,d\sigma > \lambda_1,$$

because $u_n \neq -\varphi_1$ and $u_n \neq \varphi_1$, which contradicts the assumption $\widetilde{J}_s(u_n) \leq \lambda_1$.

Consider now the complementary situation. Hence u_n changes sign whenever n is sufficiently large, thereby we can set

$$w_n = \frac{u_n^+}{\|u_n^+\|_{L^p(\Omega)}} \quad \text{and} \quad r_n = \|\nabla w_n\|^p_{L^p(\Omega)} + \beta\|w_n\|^p_{L^p(\partial\Omega)}. \tag{3.19}$$

We claim that, along a relabeled subsequence, $r_n \to +\infty$ as $n \to \infty$. Suppose by contradiction that $\{r_n\}$ is bounded. This implies through (3.19) that $\{w_n\}$ is bounded in $W^{1,p}(\Omega)$, so there exists a subsequence denoted again by $\{w_n\}$ such that $w_n \to w$ in $L^p(\Omega)$, for some $w \in W^{1,p}(\Omega)$. Since $\|w_n\|_{L^p(\Omega)} = 1$ and $w_n \geq 0$ a.e., we get $\|w\|_{L^p(\Omega)} = 1$ and $w \geq 0$. This contradicts that $u_n \to -\varphi_1$ in $L^p(\Omega)$, thus proving the claim.

On the other hand, from (3.19) and by using the variational characterization of λ_1, we infer that

$$\begin{aligned}\widetilde{J}_s(u_n) &= (r_n - s)\int_\Omega (u_n^+)^p\,dx + \int_\Omega |\nabla u_n^-|^p\,dx + \beta\int_{\partial\Omega}(u_n^-)^p\,d\sigma \\ &\geq (r_n - s)\int_\Omega (u_n^+)^p\,dx + \lambda_1\int_\Omega (u_n^-)^p\,dx,\end{aligned}$$

whereas the choice of $\{u_n\}$ gives

$$\widetilde{J}_s(u_n) \leq \lambda_1 = \lambda_1\int_\Omega (u_n^+)^p\,dx + \lambda_1\int_\Omega (u_n^-)^p\,dx.$$

Combining the above inequalities results in

$$(\lambda_1 - r_n + s) \int_\Omega (u_n^+)^p \, dx \geq 0,$$

therefore $\lambda_1 \geq r_n - s$. This is against the unboundedness of $\{r_n\}$, which completes the proof of (i). Part (ii) follows from Lemma 3.18 because $J_s(-\varphi_1) = \lambda_1$. □

Using the two local minima obtained in Propositions 3.19 and 3.20, we seek for a third critical point of $\widetilde{J}_s$ via a version of the mountain pass theorem on C^1 manifolds.

Taking into account (3.17), we define a norm of the derivative of the restriction $\widetilde{J}_s$ of J_s at the point $u \in S$ by

$$\|\widetilde{J}_s'(u)\|_* = \min\{t \in \mathbb{R} : \|J_s'(u) - tT'(u)\|_{W^{1,p}(\Omega)^*}\},$$

where $T(u) = \int_\Omega |u|^p \, dx$. The definition of the Palais–Smale condition is here in the sense of functions on manifolds.

Definition 3.21. The functional $\widetilde{J}_s : S \to \mathbb{R}$ is said to satisfy the Palais–Smale condition on S if for any sequence $\{u_n\} \subset S$ such that $\{J_s(u_n)\}$ is bounded and $\|\widetilde{J}_s'(u_n)\|_* \to 0$ as $n \to \infty$, there exists a strongly convergent subsequence in $W^{1,p}(\Omega)$.

We can check the Palais–Smale condition for $\widetilde{J}_s$ on the manifold S.

Lemma 3.22. *The functional $\widetilde{J}_s : S \to \mathbb{R}$ satisfies the Palais–Smale condition on S.*

Proof. Let $\{u_n\} \subset S$ be a sequence such that $\{J_s(u_n)\}$ is bounded and $\|\widetilde{J}_s'(u_n)\|_* \to 0$ as $n \to \infty$, which means that there exists a sequence $\{t_n\} \subset \mathbb{R}$ such that

$$\left| \int_\Omega |\nabla u_n|^{p-2} \nabla u_n \cdot \nabla v \, dx + \beta \int_{\partial\Omega} |u_n|^{p-2} u_n v \, d\sigma - s \int_\Omega (u_n^+)^{p-1} v \, dx \right.$$
$$\left. - t_n \int_\Omega |u_n|^{p-2} u_n v \, dx \right| \leq \epsilon_n \|v\|_{W^{1,p}(\Omega)} \tag{3.20}$$

for all $v \in W^{1,p}(\Omega)$ and with $\epsilon_n \to 0^+$. Note that $J_s(u_n) \geq \|\nabla u_n\|_{L^p(\Omega)}^p - s$. Since $\{u_n\} \subset S$ and $\{J_s(u_n)\}$ is bounded, we derive that $\{u_n\}$ is bounded in $W^{1,p}(\Omega)$. Thus, along a relabeled subsequence we may suppose that $u_n \rightharpoonup u$ in $W^{1,p}(\Omega)$, $u_n \to u$ in $L^p(\Omega)$ and $u_n \to u$ in $L^p(\partial\Omega)$. Taking $v = u_n$ in (3.20) and using again the inclusion $\{u_n\} \subset S$ shows that the sequence $\{t_n\}$ is bounded. Then, if we choose $v = u_n - u$, it follows that

$$\int_\Omega |\nabla u_n|^{p-2} \nabla u_n \cdot \nabla (u_n - u) \, dx \to 0 \text{ as } n \to \infty.$$

Recalling that the negative p-Laplacian $-\Delta_p : W^{1,p}(\Omega) \to W^{1,p}(\Omega)^*$ fulfills the $(S)_+$-property (see Proposition 1.56), we conclude that $u_n \to u$ in $W^{1,p}(\Omega)$. □

The proof of the following version of the mountain pass theorem can be found in [98, Theorem 3.2].

Theorem 3.23. *Let E be a Banach space and let $g, f \in C^1(E)$. Further, let*

$$M = \{u \in E : g(u) = 0\},$$

$u_0, u_1 \in M$, and let $\epsilon > 0$ be such that $\|u_1 - u_0\|_E > \epsilon$ and

$$\inf\{f(u) : u \in M \text{ and } \|u - u_0\|_E = \epsilon\} > \max\{f(u_0), f(u_1)\}.$$

Assume that f satisfies the Palais–Smale condition on M and that

$$\Gamma = \{\gamma \in C([-1,1], M) : \gamma(-1) = u_0 \text{ and } \gamma(1) = u_1\}$$

is nonempty. Then

$$c = \inf_{\gamma\in\Gamma} \max_{u\in\gamma([-1,1])} f(u)$$

is a critical value of $f|_M$.

Now we obtain, in addition to φ_1 and $-\varphi_1$, a third critical point of $\widetilde{J}_s$ on S.

Proposition 3.24. *For each $s \geq 0$, one has:*

(i)

$$c(s) := \inf_{\gamma\in\Gamma} \max_{u\in\gamma([-1,1])} J_s(u), \tag{3.21}$$

where

$$\Gamma = \{\gamma \in C([-1,1], S) : \gamma(-1) = -\varphi_1 \text{ and } \gamma(1) = \varphi_1\},$$

is a critical value of $\widetilde{J}_s$ satisfying $c(s) > \max\{\widetilde{J}_s(-\varphi_1), \widetilde{J}_s(\varphi_1)\} = \lambda_1$. In particular, there exists a critical point of $\widetilde{J}_s$ that is different from $-\varphi_1$ and φ_1.

(ii) *The point $(s + c(s), c(s))$ belongs to $\widehat{\Sigma}_p$.*

Proof. (i) By Proposition 3.20 we know that $-\varphi_1$ is a strict local minimizer of $\widetilde{J}_s$ with $\widetilde{J}_s(-\varphi_1) = \lambda_1$, while Proposition 3.19 ensures that φ_1 is a global minimizer of $\widetilde{J}_s$ with $\widetilde{J}_s(\varphi_1) = \lambda_1 - s$. Then we can show that

$$\inf\{\widetilde{J}_s(u) : u \in S \text{ and } \|u - (-\varphi_1)\|_{W^{1,p}(\Omega)} = \epsilon\} > \max\{\widetilde{J}_s(-\varphi_1), \widetilde{J}_s(\varphi_1)\} = \lambda_1, \tag{3.22}$$

whenever $\epsilon > 0$ is sufficiently small. The proof that the inequality (3.22) is strict can be done as in [60, Lemma 2.9] (see also [150, Lemma 2.6]) on the basis of Ekeland's variational principle. In order to fulfill the mountain-pass geometry, we choose $\epsilon > 0$ even smaller if necessary to have $2\|\varphi_1\|_{W^{1,p}(\Omega)} = \|\varphi_1 - (-\varphi_1)\|_{W^{1,p}(\Omega)} > \epsilon$. Since $\widetilde{J}_s : S \to \mathbb{R}$ satisfies the Palais–Smale condition on the manifold S as shown in Lemma 3.22, we may invoke the version of mountain pass theorem on manifolds in Theorem 3.23. This guarantees that $c(s)$ introduced in (3.21) is a critical value of $\widetilde{J}_s$ with $c(s) > \lambda_1$, providing a critical point different from $-\varphi_1$ and φ_1.

(ii) Thanks to Lemma 3.18 and part (i), we infer that $(s + c(s), c(s)) \in \widehat{\Sigma}_p$. □

Proposition 3.24 provides a curve $(s + c(s), c(s)) \in \widehat{\Sigma}_p$ for $s \geq 0$. As $\widehat{\Sigma}_p$ is symmetric with respect to the diagonal, we can complete it with its symmetric part obtaining a curve in $\widehat{\Sigma}_p$, namely

$$\mathcal{C} := \{(s + c(s), c(s)), (c(s), s + c(s)) : s \geq 0\}. \tag{3.23}$$

The proposition below establishes an important sign property related to the curve $\mathcal{C}$.

Proposition 3.25. *Let $(a_0, b_0) \in \mathcal{C}$ with $\mathcal{C}$ in (3.23) and let $a, b \in L^\infty(\Omega)$ satisfy $\lambda_1 \leq a(x) \leq a_0$, $\lambda_1 \leq b(x) \leq b_0$ for a.a. $x \in \Omega$ such that $\lambda_1 < a(x)$ and $\lambda_1 < b(x)$ on subsets of positive measure. Then any nontrivial solution u of*

$$\begin{cases} -\Delta_p u = a(x)(u^+)^{p-1} - b(x)(u^-)^{p-1} & \text{in } \Omega, \\ |\nabla u|^{p-2}\frac{\partial u}{\partial \nu} = -\beta|u|^{p-2}u & \text{on } \partial\Omega \end{cases} \tag{3.24}$$

changes sign in Ω.

Proof. Let u be a nontrivial solution of Eq. (3.24). Then $-u$ is a nontrivial solution of

$$\begin{cases} -\Delta_p z = b(x)(z^+)^{p-1} - a(x)(z^-)^{p-1} & \text{in } \Omega, \\ |\nabla z|^{p-2}\frac{\partial z}{\partial \nu} = -\beta|z|^{p-2}z & \text{on } \partial\Omega. \end{cases}$$

Hence we may suppose that the point $(a_0, b_0) \in \mathcal{C}$ is such that $a_0 \geq b_0$.

We argue by contradiction and assume that u does not change sign in Ω. Without loss of generality, we may admit that $u \geq 0$ a.e. in Ω, so u is a solution of the Robin weighted eigenvalue problem with weight $a(x)$:

$$\begin{cases} -\Delta_p u = a(x)u^{p-1} & \text{in } \Omega, \\ |\nabla u|^{p-2}\frac{\partial u}{\partial \nu} = -\beta u^{p-1} & \text{on } \partial\Omega. \end{cases}$$

It means that u is an eigenfunction corresponding to the eigenvalue 1 for this problem. The first eigenvalue $\lambda_1(a)$ of the above weighted problem is expressed as

$$\lambda_1(a) = \inf_{\substack{v \in W^{1,p}(\Omega) \\ v \neq 0}} \frac{\int_\Omega |\nabla v|^p \, dx + \beta \int_{\partial\Omega} |v|^p \, d\sigma}{\int_\Omega a(x)|v|^p \, dx}.$$

The fact that $u \geq 0$ entails $\lambda_1(a) = 1$ because the only eigenfunction whose eigenfunctions do not change sign is $\lambda_1(a)$ (see [127]). Then the hypothesis that $\lambda_1 < a(x)$ on a set of positive measure leads to the contradiction

$$1 = \frac{\int_\Omega |\nabla\varphi_1|^p \, dx + \beta \int_{\partial\Omega} \varphi_1^p \, d\sigma}{\lambda_1} > \frac{\int_\Omega |\nabla\varphi_1|^p \, dx + \beta \int_{\partial\Omega} \varphi_1^p \, d\sigma}{\int_\Omega a(x)\varphi_1^p \, dx} \geq \lambda_1(a)$$
$$= 1,$$

which completes the proof. □

We use this sign information in the below proposition for establishing that the lines $\{\lambda_1\} \times \mathbb{R}$ and $\mathbb{R} \times \{\lambda_1\}$ are isolated in $\widehat{\Sigma}_p$.

Proposition 3.26. *There exists no sequence $\{(a_n, b_n)\} \subset \widehat{\Sigma}_p$ with $a_n > \lambda_1$ and $b_n > \lambda_1$ such that $(a_n, b_n) \to (a, b)$ with $a = \lambda_1$ or $b = \lambda_1$.*

Proof. Proceeding indirectly, assume that there exist sequences $\{(a_n, b_n)\} \subset \widehat{\Sigma}_p$ and $\{u_n\} \subset W^{1,p}(\Omega)$ with the properties: $a_n \to \lambda_1$, $b_n \to b$, $a_n > \lambda_1$, $b_n > \lambda_1$, $\|u_n\|_{L^p(\Omega)} = 1$, and

$$\begin{cases} -\Delta_p u_n = a_n (u_n^+)^{p-1} - b_n (u_n^-)^{p-1} & \text{in } \Omega, \\ |\nabla u_n|^{p-2} \frac{\partial u_n}{\partial \nu} = -\beta |u_n|^{p-2} u_n & \text{on } \partial\Omega. \end{cases} \tag{3.25}$$

If we act on (3.25) with the test function u_n, we get

$$\|\nabla u_n\|_{L^p(\Omega)}^p = a_n \int_\Omega (u_n^+)^p \, dx + b_n \int_\Omega (u_n^-)^p \, dx - \beta \int_{\partial\Omega} |u_n|^p \, d\sigma \leq a_n + b_n,$$

which proves the boundedness of $\{u_n\}$ in $W^{1,p}(\Omega)$. Hence, along a subsequence, $u_n \rightharpoonup u$ in $W^{1,p}(\Omega)$ and $u_n \to u$ in $L^p(\Omega)$ and $L^p(\partial\Omega)$. Now, testing (3.25) with $u_n - u$, we infer that

$$\lim_{n\to\infty} \int_\Omega |\nabla u_n|^{p-2} \nabla u_n \cdot \nabla(u_n - u) \, dx = 0.$$

The $(S)_+$-property of $-\Delta_p$ on $W^{1,p}(\Omega)$ yields that $u_n \to u$ in $W^{1,p}(\Omega)$. Thus, u is a solution of the equation

$$\begin{aligned}&\int_\Omega |\nabla u|^{p-2}\nabla u\cdot\nabla v\,dx\\&=\lambda_1\int_\Omega (u^+)^{p-1}v\,dx-b\int_\Omega (u^-)^{p-1}v\,dx-\beta\int_{\partial\Omega}|u|^{p-2}uv\,d\sigma\end{aligned}\tag{3.26}$$

for all $v\in W^{1,p}(\Omega)$. Inserting $v=u^-$ in (3.26) leads to

$$\int_\Omega |\nabla u^+|^p\,dx=\lambda_1\int_\Omega (u^+)^p\,dx-\beta\int_{\partial\Omega}(u^+)^p\,d\sigma.$$

This, in conjunction with the characterization of λ_1 and since $\|u\|_{L^p(\Omega)}=1$, ensures that either $u^+=0$ or $u^+=\varphi_1$. If $u^+=0$, then $u\le 0$ and (3.26) implies that u is an eigenfunction. Recalling from Remark 3.17 that λ_1 is the only eigenvalue whose eigenfunctions do not change sign, we deduce that $u=-\varphi_1$. This renders that $\{u_n\}$ converges either to φ_1 or to $-\varphi_1$ in $L^p(\Omega)$. Note that it forces to have

$$\text{either}\quad |\{x\in\Omega:u_n(x)<0\}|\to 0\quad\text{or}\quad |\{x\in\Omega:u_n(x)>0\}|\to 0,\tag{3.27}$$

respectively, where $|\cdot|$ denotes the Lebesgue measure. Indeed, assuming, for instance, $u_n\to\varphi_1$ in $L^p(\Omega)$, since for any compact subset $K\subset\Omega$ there holds

$$\int_{\{u_n<0\}\cap K}|u_n-\varphi_1|^p\,dx\ge\int_{\{u_n<0\}\cap K}\varphi_1^p\,dx\ge C|\{u_n<0\}\cap K|,$$

with a constant $C>0$, it is seen that the first assertion in (3.27) is fulfilled. Similarly, when $u_n\to-\varphi_1$ in $L^p(\Omega)$ we get the second assertion in (3.27).

On the other hand, since (a_n,b_n) does not belong to the trivial lines of $\widehat{\Sigma}_p$ (see Proposition 3.25), we have that u_n changes sign, hence $u_n^+\not\equiv 0$ and $u_n^-\not\equiv 0$. Using u_n^+ as test function for (3.25) in conjunction with Hölder's inequality and the continuity of the embedding $W^{1,p}(\Omega)\subset L^q(\Omega)$, with $p<q\le p^*$, we obtain the estimate

$$\begin{aligned}&\int_\Omega|\nabla u_n^+|^p\,dx+\int_\Omega(u_n^+)^p\,dx\\&=a_n\int_\Omega(u_n^+)^p\,dx-\beta\int_{\partial\Omega}(u_n^+)^p\,d\sigma+\int_\Omega(u_n^+)^p\,dx\\&\le(a_n+1)\int_\Omega(u_n^+)^p\,dx\le(a_n+1)C\,|\{x\in\Omega:u_n(x)>0\}|^{1-\frac{p}{q}}\,\|u_n^+\|^p_{W^{1,p}(\Omega)},\end{aligned}$$

with a constant $C>0$. We infer that

$$|\{x\in\Omega:u_n(x)>0\}|^{1-\frac{p}{q}}\ge\frac{1}{(a_n+1)C},$$

and in the same way

$$|\{x \in \Omega : u_n(x) < 0\}|^{1-\frac{p}{q}} \geq \frac{1}{(b_n+1)C}.$$

Hence we reach a contradiction with (3.27), which completes the proof. □

The following auxiliary fact is helpful to link with the preceding critical point results.

Lemma 3.27. *For every $r > \inf_S J_s = \lambda_1 - s$, each connected component of $\{u \in S : J_s(u) < r\}$ contains a critical point, in fact a local minimizer of $\widetilde{J}_s$.*

Proof. Let C be a connected component of $\{u \in S : J_s(u) < r\}$ and denote

$$d = \inf\{J_s(u) : u \in \overline{C}\}.$$

We claim that there exists $u_0 \in \overline{C}$ such that $\widetilde{J}_s(u_0) = d$. To this end, let $\{u_n\} \subset \overline{C}$ be a sequence such that

$$d \leq \widetilde{J}_s(u_n) \leq d + \frac{1}{n^2}.$$

Applying Ekeland's variational principle to $\widetilde{J}_s$ on $\overline{C}$ provides a sequence $\{v_n\} \subset \overline{C}$ such that

$$\widetilde{J}_s(v_n) \leq \widetilde{J}_s(u_n), \tag{3.28}$$

$$\|u_n - v_n\|_{W^{1,p}(\Omega)} \leq \frac{1}{n}, \tag{3.29}$$

$$\widetilde{J}_s(v_n) \leq \widetilde{J}_s(v) + \frac{1}{n}\|v - v_n\|_{W^{1,p}(\Omega)}, \quad \forall v \in \overline{C}. \tag{3.30}$$

If n is sufficiently large, by (3.28) we obtain

$$\widetilde{J}_s(v_n) \leq \widetilde{J}_s(u_n) \leq d + \frac{1}{n^2} < r.$$

Moreover, owing to (3.30), it can be shown that $\{v_n\}$ is a Palais–Smale sequence for $\widetilde{J}_s$. Then Lemma 3.22 and (3.29) ensure that, up to a relabeled subsequence, $u_n \to u_0$ in $W^{1,p}(\Omega)$ with $u_0 \in \overline{C}$ and $\widetilde{J}_s(u_0) = d$.

We note that $u_0 \in C$ because otherwise the maximality of C as a connected component would be contradicted, so u_0 is a local minimizer of $\widetilde{J}_s$ and we are done. □

The next result points out that $\mathcal{C}$ (see (3.23)) is the first nontrivial curve in $\widehat{\Sigma}_p$.

Theorem 3.28. *Let $s \geq 0$. Then $(s + c(s), c(s)) \in \mathcal{C}$ is the first point in the intersection between $\widehat{\Sigma}_p$ and the ray $\{(s, 0) + t(1, 1) : t > \lambda_1\}$.*

Proof. Assume, by contradiction, the existence of a point $(s + \mu, \mu) \in \widehat{\Sigma}_p$ with

$$\lambda_1 < \mu < c(s).$$

Proposition 3.26 and the fact that $\widehat{\Sigma}_p$ is closed enable us to suppose that μ is the minimum number with the required property. By virtue of Lemma 3.18, μ is a critical value of the functional $\widetilde{J}_s$ and there is no critical value of $\widetilde{J}_s$ in the interval (λ_1, μ). We complete the proof by reaching a contradiction to the definition of $c(s)$ in (3.21). To this end, it suffices to construct a path in Γ along which there holds $\widetilde{J}_s \leq \mu$.

Let $u \in S$ be a critical point of $\widetilde{J}_s$ with $\widetilde{J}_s(u) = \mu$. Then u fulfills

$$\int_\Omega |\nabla u|^{p-2}\nabla u \cdot \nabla v \, dx = (s + \mu)\int_\Omega (u^+)^{p-1} v \, dx - \mu \int_\Omega (u^-)^{p-1} v \, dx \\ - \beta \int_{\partial\Omega} |u|^{p-2} u v \, d\sigma$$

for all $v \in W^{1,p}(\Omega)$. Setting $v = u^+$ and $v = -u^-$ yields

$$\int_\Omega |\nabla u^+|^p \, dx = (s + \mu)\int_\Omega (u^+)^p \, dx - \beta \int_{\partial\Omega} (u^+)^p \, d\sigma \tag{3.31}$$

and

$$\int_\Omega |\nabla u^-|^p \, dx = \mu \int_\Omega (u^-)^p \, dx - \beta \int_{\partial\Omega} (u^-)^p \, d\sigma, \tag{3.32}$$

respectively. Since u changes sign (see Proposition 3.25), the following paths are well defined on S:

$$u_1(t) = \frac{(1-t)u + tu^+}{\|(1-t)u + tu^+\|_{L^p(\Omega)}}, \quad u_2(t) = \frac{(1-t)u^+ + tu^-}{\|(1-t)u^+ + tu^-\|_{L^p(\Omega)}},$$
$$u_3(t) = \frac{-tu^- + (1-t)u}{\|-tu^- + (1-t)u\|_{L^p(\Omega)}}$$

for all $t \in [0, 1]$. By means of direct calculations based on (3.31) and (3.32) we infer that

$$\widetilde{J}_s(u_1(t)) = \widetilde{J}_s(u_3(t)) = \mu \quad \text{for all } t \in [0, 1]$$

and

$$\widetilde{J}_s(u_2(t)) = \mu - \frac{st^p\|u^-\|_{L^p(\Omega)}}{\|(1-t)u^+ + tu^-\|_{L^p(\Omega)}} \leq \mu \quad \text{for all } t \in [0, 1].$$

Due to the minimality property of μ, the only critical points of $\widetilde{J}_s$ in the set

$$\{w \in S : \widetilde{J}_s(w) < \mu - s\}$$

are φ_1 and possibly $-\varphi_1$ provided $\mu - s > \lambda_1$. We note that, because $u^-/\|u^-\|_{L^p(\Omega)}$ does not change sign, it is not a critical point of $\widetilde{J}_s$. Therefore, there exists a C^1 path $\alpha : [-\epsilon, \epsilon] \to S$ with $\alpha(0) = u^-/\|u^-\|_{L^p(\Omega)}$ and $\frac{d}{dt}\widetilde{J}_s(\alpha(t))|_{t=0} \neq 0$. Using this path and observing from (3.32) that

$$\widetilde{J}_s(u^-/\|u^-\|_{L^p(\Omega)}) = \mu - s,$$

we can move from $u^-/\|u^-\|_{L^p(\Omega)}$ to a point v with $\widetilde{J}_s(v) < \mu - s$. Applying Lemma 3.27, we find that the connected component of $\{w \in S : \widetilde{J}_s(w) < \mu - s\}$ containing v crosses $\{\varphi_1, -\varphi_1\}$. Let us say that it passes through φ_1, otherwise the reasoning is the same employing $-\varphi_1$. Consequently, there is a path $u_4(t)$ from $u^-/\|u^-\|_{L^p(\Omega)}$ to φ_1 within the set $\{w \in S : \widetilde{J}_s(w) < \mu - s\}$. Then the path $-u_4(t)$ joins $-u^-/\|u^-\|_{L^p(\Omega)}$ and $-\varphi_1$ and, since $u_4(t) \in S$, we have

$$\widetilde{J}_s(-u_4(t)) \leq \widetilde{J}_s(u_4(t)) + s < \mu - s + s = \mu \quad \text{for all } t.$$

Connecting $u_1(t)$, $u_2(t)$ and $u_4(t)$, we construct a path joining u and φ_1, and joining $u_3(t)$ and $-u_4(t)$ we get a path which connects u and $-\varphi_1$. These yield a path $\gamma(t)$ on S joining φ_1 and $-\varphi_1$. Furthermore, in view of the above discussion, it turns out that $\widetilde{J}_s(\gamma(t)) \leq \mu$ for all t. This proves the theorem. □

Corollary 3.29. *The second eigenvalue λ_2 of (3.16) has the variational characterization*

$$\lambda_2 = \inf_{\gamma \in \Gamma} \max_{u \in \gamma([-1,1])} \left[\int_\Omega |\nabla u|^p \, dx + \beta \int_{\partial\Omega} |u|^p \, d\sigma \right], \tag{3.33}$$

with Γ and S in Proposition 3.24.

Proof. Theorem 3.28 for $s = 0$ ensures that $c(0) = \lambda_2$. Formula (3.33) follows by applying Proposition 3.24 (i) with $s = 0$. □

Proposition 3.30. *The curve $s \mapsto (s + c(s), c(s))$ is Lipschitz continuous and decreasing.*

Proof. If $s_1 < s_2$, it follows $\widetilde{J}_{s_1}(u) \geq \widetilde{J}_{s_2}(u)$ for all $u \in S$, which ensures that $c(s_1) \geq c(s_2)$. For every $\epsilon > 0$ there exists $\gamma \in \Gamma$ such that

$$\max_{u \in \gamma([-1,1])} \widetilde{J}_{s_2}(u) \leq c(s_2) + \epsilon,$$

hence

$$0 \leq c(s_1) - c(s_2) \leq \max_{u \in \gamma([-1,1])} \widetilde{J}_{s_1}(u) - \max_{u \in \gamma([-1,1])} \widetilde{J}_{s_2}(u) + \epsilon.$$

Taking $u_0 \in \gamma([-1,1])$ such that

$$\max_{u\in\gamma([-1,1])} \widetilde{J}_{s_1}(u) = \widetilde{J}_{s_1}(u_0)$$

yields

$$0 \le c(s_1) - c(s_2) \le \widetilde{J}_{s_1}(u_0) - \widetilde{J}_{s_2}(u_0) + \epsilon = s_1 - s_2 + \epsilon.$$

As $\epsilon > 0$ was arbitrary, this ensures that $s \mapsto (s + c(s), c(s))$ is Lipschitz continuous.

In order to prove that the curve is decreasing, it suffices to argue for $s > 0$. Let $0 < s_1 < s_2$. Then, since $(s_1 + c(s_1), c(s_1)), (s_2 + c(s_2), c(s_2)) \in \widehat{\Sigma}_p$, Theorem 3.28 implies that $s_1 + c(s_1) < s_2 + c(s_2)$. On the other hand, as already remarked, $c(s_1) \ge c(s_2)$, which completes the proof. □

Next we investigate the asymptotic behavior of the curve $\mathcal{C}$.

Theorem 3.31. *Let $p \le N$. Then* $\lim_{s\to+\infty} c(s) = \lambda_1$.

Proof. Let us proceed by contradiction and suppose that $c(s)$ does not converge to λ_1 as $s \to +\infty$. Then there exists $\delta > 0$ such that

$$\max_{u\in\gamma([-1,1])} \widetilde{J}_s(u) \ge \lambda_1 + \delta \quad \text{for all } \gamma \in \Gamma \text{ and all } s \ge 0.$$

Since $p \le N$, we can choose a function $\psi \in W^{1,p}(\Omega)$ which is unbounded above. Then we define $\gamma \in \Gamma$ by

$$\gamma(t) = \frac{t\varphi_1 + (1-|t|)\psi}{\|t\varphi_1 + (1-|t|)\psi\|_{L^p(\Omega)}}, \quad t \in [-1,1].$$

For every $s > 0$, let $t_s \in [-1,1]$ satisfy

$$\max_{t\in[-1,1]} \widetilde{J}_s(\gamma(t)) = \widetilde{J}_s(\gamma(t_s)).$$

Denoting $v_s = t_s\varphi_1 + (1-|t_s|)\psi$, we infer that

$$\int_\Omega |\nabla v_s|^p\,dx + \beta\int_{\partial\Omega} |v_s|^p\,d\sigma - s\int_\Omega (v_s^+)^p\,dx \ge (\lambda_1+\delta)\int_\Omega |v_s|^p\,dx. \tag{3.34}$$

Letting $s \to +\infty$, we can assume along a subsequence that $t_s \to \widetilde{t} \in [-1,1]$. The family $\{v_s : s > 0\}$ being bounded in $W^{1,p}(\Omega)$, from (3.34) one sees that

$$\int_\Omega (v_s^+)^p\,dx \to 0 \quad \text{as } s \to +\infty,$$

which forces

$$\widetilde{t}\varphi_1 + (1-|\widetilde{t}|)\psi \le 0.$$

Due to the choice of ψ, this is impossible unless $\tilde{t} = -1$. Passing to the limit in (3.34) as $s \to +\infty$ and using $\tilde{t} = -1$, we arrive at the contradiction $\delta \leq 0$, so the proof is complete. □

It remains to study the asymptotic properties of the curve $\mathcal{C}$ when $p > N$. For $\beta = 0$, problem (3.15) becomes a Neumann problem with homogeneous boundary condition that was studied in Theorem 3.9 yielding

$$\lim_{s \to +\infty} c(s) = \begin{cases} \lambda_1 = 0 & \text{if } p \leq N, \\ \hat{\lambda} & \text{if } p > N, \end{cases}$$

where

$$\hat{\lambda} = \inf \left\{ \int_\Omega |\nabla u|^p \, dx : u \in W^{1,p}(\Omega), \|u\|_{L^p(\Omega)} = 1 \text{ and } u \text{ vanishes somewhere in } \overline{\Omega} \right\}.$$

Therefore, we only have to treat the case $\beta > 0$. In this respect, the key idea is to work with an adequate equivalent norm on the space $W^{1,p}(\Omega)$. So, for $\beta > 0$ we introduce the norm

$$\|u\|_\beta = \left(\|\nabla u\|^p_{L^p(\Omega)} + \beta \|u\|^p_{L^p(\partial\Omega)} \right)^{1/p}, \tag{3.35}$$

which is an equivalent norm on $W^{1,p}(\Omega)$. Then we have the following result.

Theorem 3.32. *Let $\beta > 0$ and $p > N$. Then*

$$\lim_{s \to +\infty} c(s) = \inf_{u \in L} \max_{r \in \mathbb{R}} \frac{\|r\varphi_1 + u\|^p_\beta}{\|r\varphi_1 + u\|^p_{L^p(\Omega)}} =: \overline{\lambda},$$

where

$$L = \{u \in W^{1,p}(\Omega) : u \text{ vanishes somewhere in } \overline{\Omega}, \;\; u \not\equiv 0\}.$$

Moreover, $\overline{\lambda} > \lambda_1$.

Proof. First, we are going to prove the strict inequality $\overline{\lambda} > \lambda_1$. The characterization of λ_1 in Remark 3.17, (3.35), and the definition of $\overline{\lambda}$ yield

$$\lambda_1 = \inf_{u \in W^{1,p}(\Omega) \setminus \{0\}} \frac{\|u\|^p_\beta}{\|u\|^p_{L^p(\Omega)}} \leq \inf_{u \in L} \frac{\|u\|^p_\beta}{\|u\|^p_{L^p(\Omega)}} \leq \overline{\lambda}. \tag{3.36}$$

Let us check that the first inequality in (3.36) is strict. Assuming on the contrary, we would find a sequence $\{u_n\} \subset L$ satisfying

$$\frac{\|u_n\|_\beta^p}{\|u_n\|_{L^p(\Omega)}^p} \to \lambda_1 \text{ as } n \to \infty.$$

Set $v_n = \frac{u_n}{\|u_n\|_\beta}$, hence $\|v_n\|_\beta = 1$ and

$$\frac{1}{\|v_n\|_{L^p(\Omega)}^p} \to \lambda_1 \text{ as } n \to \infty.$$

Due to the compact embedding $W^{1,p}(\Omega) \subset C(\overline{\Omega})$, there is a subsequence of $\{v_n\}$, still denoted by $\{v_n\}$, such that $v_n \rightharpoonup v$ in $W^{1,p}(\Omega)$ and $v_n \to v$ uniformly on $\overline{\Omega}$. These convergences imply $\|v\|_\beta \leq 1$, $v \in L$, and finally

$$\frac{\|v\|_\beta^p}{\|v\|_{L^p(\Omega)}^p} \leq \lim_{n\to\infty} \frac{1}{\|v_n\|_{L^p(\Omega)}^p} = \lambda_1,$$

which ensures that v is an eigenfunction in (3.16) corresponding to the first eigenvalue λ_1 (see Remark 3.17). This is a contradiction because every eigenfunction associated to λ_1 is strictly positive or negative on $\overline{\Omega}$, whereas $v \in L$. Hence, recalling (3.36), we get $\overline{\lambda} > \lambda_1$.

Now we prove the first part in the theorem. We start by claiming that there exist $u \in L$ such that

$$\max_{r\in\mathbb{R}} \frac{\|r\varphi_1 + u\|_\beta^p}{\|r\varphi_1 + u\|_{L^p(\Omega)}^p} = \overline{\lambda}. \tag{3.37}$$

By the definition of $\overline{\lambda}$, we can find a sequence $\{u_n\} \subset L$ such that

$$\max_{r\in\mathbb{R}} \frac{\|r\varphi_1 + u_n\|_\beta^p}{\|r\varphi_1 + u_n\|_{L^p(\Omega)}^p} \to \overline{\lambda} \text{ as } n \to \infty. \tag{3.38}$$

Without loss of generality, we can assume that $\|u_n\|_{W^{1,p}(\Omega)} = 1$. Hence we may suppose that $u_n \rightharpoonup u$ in $W^{1,p}(\Omega)$, as well as $u_n \to u$ uniformly in $\overline{\Omega}$, with some $u \in L$. The weak lower semicontinuity of $\|\cdot\|_\beta$ and (3.38) yield

$$\max_{r\in\mathbb{R}} \frac{\|r\varphi_1 + u\|_\beta^p}{\|r\varphi_1 + u\|_{L^p(\Omega)}^p} \leq \lim_{n\to\infty} \max_{r\in\mathbb{R}} \frac{\|r\varphi_1 + u_n\|_\beta^p}{\|r\varphi_1 + u_n\|_{L^p(\Omega)}^p} = \overline{\lambda}.$$

Combining this with the definition of $\overline{\lambda}$, we see that (3.37) holds true.

To prove that $c(s) \to \overline{\lambda}$ as $s \to +\infty$, we argue by contradiction admitting that there exists $\delta > 0$ such that

$$\max_{t\in[-1,1]} \widetilde{J}_s(\gamma(t)) \geq \overline{\lambda} + \delta \text{ for all } \gamma \in \Gamma \text{ and all } s \geq 0.$$

Here the decreasing monotonicity of $c(s)$ has been used (see Proposition 3.30). Consider the path $\gamma \in \Gamma$ defined by

$$\gamma(t) = \frac{t\varphi_1 + (1-|t|)u}{\|t\varphi_1 + (1-|t|)u\|_{L^p(\Omega)}}, \quad t \in [-1,1],$$

with u given in (3.37). Proceeding as in the proof of Theorem 3.31, for every $s > 0$ we fix $t_s \in [-1,1]$ to satisfy

$$\max_{t\in[-1,1]} \widetilde{J}_s(\gamma(t)) = \widetilde{J}_s(\gamma(t_s))$$

and denote $v_s = t_s\varphi_1 + (1-|t_s|)u$. We have

$$\int_\Omega |\nabla v_s|^p\, dx + \beta \int_{\partial\Omega} |v_s|^p\, d\sigma - s\int_\Omega (v_s^+)^p dx \geq (\overline{\lambda} + \delta)\int_\Omega |v_s|^p\, dx. \quad (3.39)$$

From (3.39) and since $\{v_s\}$ is uniformly bounded, we obtain

$$\int_\Omega (v_s^+)^p\, dx \to 0$$

and $t_s \to \widetilde{t} \in [-1,1]$ as $s \to +\infty$, which yields $\widetilde{t}\varphi_1 \leq -(1-|\widetilde{t}|)u$. As $\varphi_1 > 0$ and u vanishes somewhere in $\overline{\Omega}$, we deduce that $\widetilde{t} \leq 0$. In addition, passing to the limit in (3.39) leads to

$$\begin{aligned}&\int_\Omega |\nabla(\widetilde{t}\varphi_1 + (1-|\widetilde{t}|)u)|^p\, dx + \beta\int_{\partial\Omega} |\widetilde{t}\varphi_1 + (1-|\widetilde{t}|)u|^p\, d\sigma \\ &\geq (\overline{\lambda}+\delta)\int_\Omega |\widetilde{t}\varphi_1 + (1-|\widetilde{t}|)u|^p\, dx.\end{aligned} \quad (3.40)$$

If $\widetilde{t} \neq -1$, (3.40) can be expressed as

$$\frac{\|\frac{\widetilde{t}}{1+\widetilde{t}}\varphi_1 + u\|_\beta^p}{\|\frac{\widetilde{t}}{1+\widetilde{t}}\varphi_1 + u\|_{L^p(\Omega)}^p} \geq \overline{\lambda} + \delta.$$

Comparing with (3.37) reveals that a contradiction is reached. If $\widetilde{t} = -1$, in view of (3.40) and $\overline{\lambda} > \lambda_1$, we also arrive at a contradiction, which establishes the result. □

3.2 EIGENVALUE PROBLEMS FOR (p,q)-LAPLACIAN WITH INDEFINITE WEIGHTS

The problem treated in this section is the following quasilinear elliptic problem involving the (p,q)-Laplacian

$$\begin{cases} -\Delta_p u - \mu\Delta_q u = \lambda(m_p(x)|u|^{p-2}u + \mu m_q(x)|u|^{q-2}u) & \text{in } \Omega, \\ u = 0 & \text{on } \partial\Omega, \end{cases} \quad (3.41)$$

on a bounded domain $\Omega \subset \mathbb{R}^N$ with C^2 boundary $\partial\Omega$, with $1 < q < p < +\infty$. Here we have two parameters $\lambda \in \mathbb{R}$ and $\mu \geq 0$, and two weights $m_p, m_q \in L^\infty(\Omega)$ such that the Lebesgue measure of $\{x \in \Omega : m_r(x) > 0\}$ is positive whenever $r \in \{p, q\}$.

A solution of problem (3.41) is understood in the weak sense, that is, any $u \in W_0^{1,p}(\Omega)$ for which it holds

$$\int_\Omega |\nabla u|^{p-2}\nabla u \cdot \nabla\varphi\, dx + \mu \int_\Omega |\nabla u|^{q-2}\nabla u \cdot \nabla\varphi\, dx$$
$$= \lambda \int_\Omega (m_p|u|^{p-2}u + \mu m_q|u|^{q-2}u)\varphi\, dx$$

whenever $\varphi \in W_0^{1,p}(\Omega)$.

In the sequel, $\|\cdot\|_r$ denotes the usual L^r-norm $(1 \leq r \leq +\infty)$. For any $u \in \mathbb{R}$, we set $u^\pm := \max\{\pm u, 0\}$. The Lebesgue measure of Ω is denoted by $|\Omega|$.

We recall from [14] (see also Proposition 1.57 in the case of nonnegative weights) that the first positive eigenvalue $\lambda_1(r, m_r)$ of $-\Delta_r$ with weight function m_r is obtained by minimizing the Rayleigh quotient:

$$\lambda_1(r, m_r) = \inf\left\{\frac{\int_\Omega |\nabla u|^r\, dx}{\int_\Omega m_r|u|^r\, dx} : u \in W_0^{1,r}(\Omega),\ \int_\Omega m_r|u|^r\, dx > 0\right\}. \quad (3.42)$$

The first positive eigenvalue is simple and isolated. Since there exist no nonnegative eigenvalues corresponding to nonpositive weights, we set

$$\lambda_1(r, m_r) = +\infty \quad \text{if} \quad m_r \leq 0. \quad (3.43)$$

We also recall that $\lambda_1(r, m_r)$ has a positive eigenfunction $\varphi_1(r, m_r) \in C_0^{1,\alpha_r}(\overline{\Omega})$ with some $\alpha_r \in (0, 1)$. It is worth mentioning that $\lambda_1(r, m_r)$ and $-\lambda_1(r, -m_r)$ are the only eigenvalues of $-\Delta_r$ with weight m_r for which there exist constant-sign eigenfunctions.

We start by pointing out that finding a solution for (3.41) is equivalent to finding a solution in the case $\mu = 1$, that is to solving the problem

$$\begin{cases} -\Delta_p u - \Delta_q u = \lambda(m_p(x)|u|^{p-2}u + m_q(x)|u|^{q-2}u) & \text{in } \Omega, \\ u = 0 & \text{on } \partial\Omega. \end{cases} \quad (3.44)$$

Indeed, if u is a solution of (3.44), then multiplying Eq. (3.44) by s^{p-1} for $s > 0$ we see that $v = su$ satisfies

$$-\Delta_p v - s^{p-q}\Delta_q v = \lambda(m_p|v|^{p-2}v + s^{p-q}m_q|v|^{q-2}v) \quad \text{in } \Omega.$$

Choosing $s^{p-q} = \mu$, we obtain that $v = su$ is a solution of (3.41). Conversely, if w is a solution of (3.41), then $v = tw$, with $t = \mu^{1/(q-p)}$, is a solution of (3.44).

We introduce the functionals Φ and Ψ on $W_0^{1,p}(\Omega)$ by

$$\Phi(u) := \frac{1}{p}\|\nabla u\|_p^p + \frac{1}{q}\|\nabla u\|_q^q, \tag{3.45}$$

$$\Psi(u) := \frac{1}{p}\int_\Omega m_p|u|^p\,dx + \frac{1}{q}\int_\Omega m_q|u|^q\,dx \tag{3.46}$$

for all $u \in W_0^{1,p}(\Omega)$.

Proposition 3.33. *Let*

$$\underline{\lambda} := \inf\left\{\frac{\Phi(u)}{\Psi(u)} : u \in W_0^{1,p}(\Omega),\ \Psi(u) > 0\right\}. \tag{3.47}$$

Then $\underline{\lambda} = \min\{\lambda_1(p, m_p), \lambda_1(q, m_q)\}$.

Proof. Let φ_1 and ψ_1 be the positive eigenfunctions corresponding to $\lambda_1(p, m_p)$ and $\lambda_1(q, m_q)$, respectively, such that $\int_\Omega m_p\varphi_1^p\,dx = 1$ and $\int_\Omega m_q\psi_1^q\,dx = 1$.

For sufficiently large $t > 0$, it turns out that

$$\Psi(t\varphi_1) = t^q\left(t^{p-q} + \int_\Omega m_q\varphi_1^q\,dx\right) > 0$$

because $p > q$. By (3.45), (3.46), (3.47), and since $q - p < 0$, we find

$$\underline{\lambda} \le \frac{\Phi(t\varphi_1)}{\Psi(t\varphi_1)} = \frac{\lambda_1(p, m_p) + pt^{q-p}\int_\Omega |\nabla\varphi_1|^q\,dx/q}{1 + pt^{q-p}\int_\Omega m_q\varphi_1^q\,dx/q} \to \lambda_1(p, m_p)$$

as $t \to +\infty$, so $\underline{\lambda} \le \lambda_1(p, m_p)$.

On the other hand, because

$$\underline{\lambda} \le \frac{\Phi(t\psi_1)}{\Psi(t\psi_1)} = \frac{\lambda_1(q, m_q) + qt^{p-q}\int_\Omega |\nabla\psi_1|^p\,dx/p}{1 + qt^{p-q}\int_\Omega m_p\psi_1^p\,dx/p} \to \lambda_1(q, m_q)$$

as $t \to 0^+$, we obtain that $\underline{\lambda} \le \lambda_1(q, m_q)$, which implies

$$\underline{\lambda} \le \min\{\lambda_1(p, m_p), \lambda_1(q, m_q)\}.$$

In order to prove the converse inequality $\underline{\lambda} \ge \min\{\lambda_1(p, m_p), \lambda_1(q, m_q)\}$, suppose that $\underline{\lambda} < \min\{\lambda_1(p, m_p), \lambda_1(q, m_q)\}$. Then, in view of (3.47), there exists an element $u \in W_0^{1,p}(\Omega)$ satisfying $\Psi(u) > 0$ and

$$\frac{\Phi(u)}{\Psi(u)} < \min\{\lambda_1(p, m_p), \lambda_1(q, m_q)\}.$$

We consider three cases:

(i) $\int_\Omega m_p|u|^p\,dx > 0$ and $\int_\Omega m_q|u|^q\,dx \le 0$;

(ii) $\int_\Omega m_p|u|^p\,dx \le 0$ and $\int_\Omega m_q|u|^q\,dx > 0$;

(iii) $\int_\Omega m_p|u|^p\,dx > 0$ and $\int_\Omega m_q|u|^q\,dx > 0$.

Case (i). There hold $p\Psi(u) \le \int_\Omega m_p|u|^p\,dx$ and $p\Phi(u) \ge \|\nabla u\|_p^p$. The definition of $\lambda_1(p, m_p)$ gives rise to the contradiction

$$\min\{\lambda_1(p,m_p),\lambda_1(q,m_q)\} > \frac{\Phi(u)}{\Psi(u)} \ge \frac{\|\nabla u\|_p^p}{\int_\Omega m_p|u|^p\,dx} \ge \lambda_1(p,m_p).$$

Case (ii). Since $q\Psi(u) \le \int_\Omega m_q|u|^q\,dx$ and $q\Phi(u) \ge \|\nabla u\|_q^q$, the contradiction arises:

$$\min\{\lambda_1(p,m_p),\lambda_1(q,m_q)\} > \frac{\Phi(u)}{\Psi(u)} \ge \frac{\|\nabla u\|_q^q}{\int_\Omega m_q|u|^q\,dx} \ge \lambda_1(q,m_q).$$

Case (iii). It follows from (3.42) that

$$\|\nabla u\|_r^r \ge \lambda_1(r,m_r)\int_\Omega m_r|u|^r\,dx$$

whenever $r \in \{q, p\}$. Hence we get

$$\begin{aligned}\Phi(u) &\ge \frac{\lambda_1(p,m_p)}{p}\int_\Omega m_p|u|^p\,dx + \frac{\lambda_1(q,m_q)}{q}\int_\Omega m_q|u|^q\,dx\\ &\ge \min\{\lambda_1(p,m_p),\lambda_1(q,m_q)\}\Psi(u)\end{aligned}$$

contradicting the assumption $\underline{\lambda} < \min\{\lambda_1(p,m_p),\lambda_1(q,m_q)\}$. □

The next result is a key step in our approach.

Proposition 3.34. *Assume that $\lambda_1(p,m_p) \ne \lambda_1(q,m_q)$ or $\varphi_1 \ne t\psi_1$ for all $t > 0$, where φ_1 and ψ_1 are positive eigenfunctions corresponding to $\lambda_1(p,m_p)$ and $\lambda_1(q,m_q)$, respectively. Then the infimum in* (3.47) *is not attained.*

Proof. Arguing by contradiction, suppose there exists $u \in W_0^{1,p}(\Omega)$ such that $\Psi(u) > 0$ and

$$\frac{\Phi(u)}{\Psi(u)} = \underline{\lambda} = \min\{\lambda_1(p,m_p),\lambda_1(q,m_q)\}. \tag{3.48}$$

Here Proposition 3.33 gives the second equality. We argue by considering the three cases in the proof of Proposition 3.33.

Case (i). By means of (3.48) and the second assumption made in Case (i) of Proposition 3.33, we may write

$$\underline{\lambda} = \frac{\Phi(u)}{\Psi(u)} \geq \frac{\|\nabla u\|_p^p + p\|\nabla u\|_q^q/q}{\int_\Omega m_p|u|^p\,dx} \geq \frac{\|\nabla u\|_p^p}{\int_\Omega m_p|u|^p\,dx} \geq \lambda_1(p, m_p) \geq \underline{\lambda}.$$

This implies

$$\|\nabla u\|_p^p = \lambda_1(p, m_p)\int_\Omega m_p|u|^p\,dx \quad \text{and} \quad \|\nabla u\|_q = 0,$$

which is against $u \neq 0$.

Case (ii). On the basis of (3.48) and the second assumption in Case (ii) of Proposition 3.33, we get

$$\|\nabla u\|_q^q = \lambda_1(q, m_q)\int_\Omega m_q|u|^q\,dx \quad \text{and} \quad \|\nabla u\|_p = 0,$$

which contradicts $u \neq 0$.

Case (iii). Reasoning as in Case (iii) of Proposition 3.33, we show that

$$\frac{\int_\Omega |\nabla u|^p\,dx}{\int_\Omega m_p|u|^p\,dx} = \lambda_1(p, m_p) = \lambda_1(q, m_q) = \frac{\int_\Omega |\nabla u|^q\,dx}{\int_\Omega m_q|u|^q\,dx}.$$

The simplicity of the eigenvalue $\lambda_1(r, m_r)$ (for $r \in \{p, q\}$) guarantees that $u = t\psi_1 = s\varphi_1$ for some $t \neq 0$ and $s \neq 0$. The hypothesis made in the statement of the proposition is contradicted. □

We are in a position to state our first main theorem, which is a nonexistence result for (3.44), hence also for (3.41) in view of the preceding comments.

Theorem 3.35. *If*

$$-\min\{\lambda_1(p, -m_p), \lambda_1(q, -m_q)\} < \lambda < \min\{\lambda_1(p, m_p), \lambda_1(q, m_q)\},$$

then (3.44) has no nontrivial solutions.

Moreover, if one of the following conditions holds:

(i) $\lambda_1(p, m_p) \neq \lambda_1(q, m_q)$ *(resp.* $\lambda_1(p, -m_p) \neq \lambda_1(q, -m_q)$*);*
(ii) $\varphi_1(p, m_p) \neq t\psi_1(q, m_q)$ *for all* $t > 0$ *(resp.* $\varphi_1(p, -m_p) \neq t\psi_1(q, -m_q)$ *(whenever* $\max\{\lambda_1(p, -m_p), \lambda_1(q, -m_q)\} < +\infty$*) for all* $t > 0$*),*

where $\varphi_1(p, \pm m_p)$ *and* $\psi_1(q, \pm m_q)$ *are positive eigenfunctions which correspond to* $\lambda_1(p, \pm m_p)$ *and* $\lambda_1(q, \pm m_q)$*, respectively, then problem (3.44) with*

$$\lambda = \min\{\lambda_1(p, m_p), \lambda_1(q, n_q)\} \quad (\textit{resp. } \lambda = -\min\{\lambda_1(p, -m_p), \lambda_1(q, -m_q)\})$$

has no nontrivial solutions.

Proof. Arguing by contradiction, let u be a nontrivial solution of (3.44). Then, for every $s > 0$, we have that $v = su$ is a nontrivial solution of the Dirichlet problem

$$-\Delta_p v - s^{p-q}\Delta_q v = \lambda\left(m_p|v|^{p-2}v + s^{p-q}m_q|v|^{q-2}v\right) \quad \text{in } \Omega,$$
$$v = 0 \quad \text{on } \partial\Omega.$$

Choosing $s^{p-q} = p/q$ and then acting with su as test function on the above equation, we obtain

$$0 < p\Phi(su) = p\lambda\Psi(su)$$

(see (3.45) and (3.46)).

First, we examine the case of $\lambda > 0$. From the above estimate and according to Proposition 3.33, we obtain the contradiction

$$\lambda = \frac{\Phi(su)}{\Psi(su)} \geq \underline{\lambda} = \min\{\lambda_1(p, m_p), \lambda_1(q, m_q)\},$$

which yields the first assertion of the theorem. The second part of the theorem follows readily by applying Proposition 3.34.

It remains to examine the case of $\lambda < 0$. We split the proof into three parts.

Case 1. $(m_p)^- \not\equiv 0$ and $(m_q)^- \not\equiv 0$. The desired conclusion follows from the above argument by replacing λ, m_p, m_q with $-\lambda$, $-m_p$, $-m_q$, respectively.

Case 2. $(m_p)^- \not\equiv 0$ and $m_q \geq 0$. On account of (3.43), we note that

$$\min\{\lambda_1(p, -m_p), \lambda_1(q, -m_q)\} = \lambda_1(p, -m_p).$$

For any $s > 0$, we obtain the estimate

$$0 < \|\nabla(su)\|_p^p < p\Phi(su) = -\lambda\left(\int_\Omega(-m_p)|su|^p\,dx + \frac{p}{q}\int_\Omega(-m_q)|su|^q\,dx\right)$$
$$\leq -\lambda\int_\Omega(-m_p)|su|^p\,dx,$$

whence $\int_\Omega(-m_p)|su|^p\,dx > 0$. Therefore, by the definition of $\lambda_1(p, -m_p)$, we have

$$-\lambda > \frac{\|\nabla(su)\|_p^p}{\int_\Omega(-m_p)|su|^p\,dx} \geq \lambda_1(p, -m_p),$$

contradicting the assumption of the theorem.

Case 3. $(m_q)^- \not\equiv 0$ and $m_p \geq 0$. As before, this situation entails

$$\min\{\lambda_1(p, -m_p), \lambda_1(q, -m_q)\} = \lambda_1(q, -m_q).$$

Since

$$\begin{aligned}0 < \frac{p}{q}\|\nabla(su)\|_q^q &< p\Phi(su)\\ &= -\lambda\left(\int_\Omega(-m_p)|su|^p\,dx + \frac{p}{q}\int_\Omega(-m_q)|su|^q\,dx\right)\\ &\le \frac{-\lambda p}{q}\int_\Omega(-m_q)|su|^q\,dx,\end{aligned}$$

we infer that

$$-\lambda > \frac{\|\nabla(su)\|_q^q}{\int_\Omega(-m_q)|su|^q\,dx} \ge \lambda_1(q,-m_q),$$

which contradicts the assumption of the theorem, completing the proof. □

Remark 3.36. If $\lambda_1(p,m_p) = \lambda_1(q,m_q)$ and $\varphi_1(p,m_p) = t\psi_1(q,m_q)$ for some $t > 0$, then $\psi_1 := \psi_1(q,m_q)$ and $\varphi_1 := \varphi_1(p,m_p)$ are both positive solutions of (3.44) with $\lambda = \lambda_1(p,m_p) = \lambda_1(q,m_q)$. Indeed, the definitions of φ_1 and ψ_1 yield

$$-\Delta_p\varphi_1 = \lambda m_p\varphi_1^{p-1} \quad\text{and}\quad -\Delta_q\psi_1 = \lambda m_q\psi_1^{q-1},$$

while the $(p-1)$-homogeneity of these equations combined with the equality $\varphi_1 = t\psi_1$ implies that

$$-\Delta_p\psi_1 = \lambda m_p\psi_1^{p-1} \quad\text{and}\quad -\Delta_q\varphi_1 = \lambda m_q\varphi_1^{q-1}.$$

For a later use, we also quote the following auxiliary result.

Lemma 3.37 ([224, Lemma 13]). *Set*

$$X(d) := \left\{u \in W_0^{1,p}(\Omega) : \|\nabla u\|_p^p \le d\|u\|_p^p\right\} \tag{3.49}$$

for $d > 0$. Then there exists $C = C(d) > 0$ such that

$$\|\nabla u\|_p \le C\|u\|_q \quad \textit{for all } u \in X(d).$$

An essential part in our approach for problem (3.44) (thus, (3.41)) is played by the underlying variational structure. We define the functional $I_\lambda = I_{(\lambda,m_p,m_q)}$ on $W_0^{1,p}(\Omega)$ by

$$\begin{aligned}I_\lambda(u) &= I_{(\lambda,m_p,m_q)} \\ &:= \frac{1}{p}\|\nabla u\|_p^p + \frac{1}{q}\|\nabla u\|_q^q - \frac{\lambda}{p}\int_\Omega m_p(u^+)^p\,dx - \frac{\lambda}{q}\int_\Omega m_q(u^+)^q\,dx\end{aligned} \tag{3.50}$$

for all $u \in W_0^{1,p}(\Omega)$.

Remark 3.38. If u is a nontrivial critical point of I_λ, then u is a positive solution of (3.44). Indeed, using $-u^-$ as a test function leads to

$$0 = \langle I'_\lambda(u), -u^- \rangle = \|\nabla u^-\|_p^p + \|\nabla u^-\|_q^q,$$

whence $u^- = 0$. This amounts to saying that u is a nonnegative solution of (3.44). The nonlinear regularity theory up to the boundary in [129, Theorem 1] and [130, p. 320] (see also Proposition 2.12) ensures that $u \in C_0^{1,\beta}(\overline{\Omega})$ with some $\beta \in (0, 1)$. Moreover, the strong maximum principle in Theorem 2.19 guarantees that $u > 0$ in Ω and $\partial u/\partial \nu < 0$ on $\partial\Omega$, because $u \not\equiv 0$ and $u \geq 0$.

Our main existence result for problem (3.44) (or (3.41)) in the nonresonant case is as follows.

Theorem 3.39. *Assume that the condition*

$$\lambda_1(p, m_p) \neq \lambda_1(q, m_q) \quad (\textit{resp. } \lambda_1(p, -m_p) \neq \lambda_1(q, -m_q))$$

holds. If

$$\begin{aligned} \min\{\lambda_1(p, m_p), \lambda_1(q, m_q)\} &< \lambda < \max\{\lambda_1(p, m_p), \lambda_1(q, m_q)\} \\ (\textit{resp.} \quad -\max\{\lambda_1(p, -m_p), \lambda_1(q, -m_q)\} &< \lambda \\ &< -\min\{\lambda_1(p, -m_p), \lambda_1(q, -m_q)\}), \end{aligned}$$

then (3.44) has at least one positive solution.

Proof. On the basis of Remark 3.38, it suffices to provide a nontrivial critical point of I_λ. According to the relevant intervals for λ, the proof will be separately developed in four cases:

(i) $\lambda_1(q, m_q) < \lambda < \lambda_1(p, m_p)$;
(ii) $\lambda_1(q, -m_q) < -\lambda < \lambda_1(p, -m_p)$;
(iii) $\lambda_1(p, m_p) < \lambda < \lambda_1(q, m_q)$;
(iv) $\lambda_1(p, -m_p) < -\lambda < \lambda_1(q, -m_q)$.

(i) I_λ is sequentially weakly lower semicontinuous since $m_p, m_q \in L^\infty(\Omega)$ and the embeddings of $W_0^{1,p}(\Omega)$ into $L^p(\Omega)$ and $L^q(\Omega)$ are compact. Fix $\varepsilon > 0$ such that

$$(1 - \varepsilon)\lambda_1(p, m_p) > \lambda, \tag{3.51}$$

which is possible due to the assumption in case (i). For every $u \in W_0^{1,p}(\Omega)$ with $\int_\Omega m_p (u^+)^p \, dx \leq 0$, through Hölder's inequality we obtain

$$\begin{aligned} I_\lambda(u) &\geq \frac{1}{p}\|\nabla u\|_p^p + \frac{1}{q}\|\nabla u\|_q^q - \frac{\lambda}{q}\|m_q\|_\infty \|u^+\|_q^q \\ &\geq \frac{1}{p}\|\nabla u\|_p^p - \frac{\lambda}{q}\|m_q\|_\infty \|u\|_p^q |\Omega|^{1-q/p} \end{aligned}$$

$$\geq \frac{1}{p}\|\nabla u\|_p^p - \frac{\lambda\|m_q\|_\infty|\Omega|^{1-q/p}}{q\lambda_1(p,1)^{q/p}}\|\nabla u\|_p^q.$$

For $u \in W_0^{1,p}(\Omega)$ with $\int_\Omega m_p(u^+)^p\,dx > 0$, by (3.42) we have

$$\|\nabla u^+\|_p^p \geq \lambda_1(p,m_p)\int_\Omega m_p(u^+)^p\,dx.$$

Then, taking into account (3.51), we derive

$$\begin{aligned}
I_\lambda(u) &\geq \frac{\varepsilon}{p}\|\nabla u\|_p^p + \frac{(1-\varepsilon)\lambda_1(p,m_p)-\lambda}{p}\int_\Omega m_p(u^+)^p\,dx - \frac{\lambda}{q}\|m_q\|_\infty\|u^+\|_q^q\\
&\geq \frac{\varepsilon}{p}\|\nabla u\|_p^p - \frac{\lambda}{q}\|m_q\|_\infty\|u\|_p^q|\Omega|^{1-q/p}\\
&\geq \frac{\varepsilon}{p}\|\nabla u\|_p^p - \frac{\lambda\|m_q\|_\infty|\Omega|^{1-q/p}}{q\lambda_1(p,1)^{q/p}}\|\nabla u\|_p^q. \qquad (3.52)
\end{aligned}$$

Since $q < p$, it follows from (3.51) and (3.52) that I_λ is coercive and bounded below. Consequently, by a standard result (see, e.g., [152, Theorem 1.1]), there exists a global minimizer u_0 of I_λ.

In order to have $u_0 \neq 0$ it suffices to show that $(I_\lambda(u_0) =) \min_{W_0^{1,p}(\Omega)} I_\lambda < 0$. Let ψ_1 be the eigenfunction corresponding to $\lambda_1(q,m_q)$ that satisfies $\int_\Omega m_q\psi_1^q\,dx = 1$. Because $\lambda > \lambda_1(q,m_q)$, for sufficiently small $t > 0$,

$$I_\lambda(t\psi_1) = t^q\left(\frac{t^{p-q}}{p}\|\nabla\psi_1\|_p^p - \frac{\lambda t^{p-q}}{p}\int_\Omega m_p\psi_1^p\,dx + \frac{\lambda_1(q,m_q)-\lambda}{q}\right) < 0,$$

which completes the proof of case (i).

(ii) Since $\lambda_1(q,-m_q) < +\infty$, by (3.43) we must have $(-m_q)^+ \not\equiv 0$. We distinguish two cases.

Case Where m_p Changes Sign. Applying the result in case (i) with $-\lambda$, $-m_p$, $-m_q$ in place of λ, m_p, m_q, respectively, permits achieving the conclusion.

Case Where $m_p \geq 0$. In this case, $\lambda_1(p,-m_p) = +\infty$. We note that $-\lambda > 0$, whence

$$-\frac{\lambda}{p}\int_\Omega m_p(u^+)^p\,dx \geq 0$$

for all $u \in W_0^{1,p}(\Omega)$. Hence we have

$$I_\lambda(u) \geq \frac{1}{p}\|\nabla u\|_p^p - \frac{\lambda\|m_q\|_\infty|\Omega|^{1-q/p}}{q\lambda_1(p,1)^{q/p}}\|\nabla u\|_p^q$$

for all $u \in W_0^{1,p}(\Omega)$, whence the functional I_λ is coercive and bounded below. As in the proof of case (i), replacing ψ_1 by the normalized positive eigenfunction $\psi_1(q,-m_q)$ corresponding to $\lambda_1(q,-m_q)$, we can show that I_λ has a global minimizer u_0 with $I_\lambda(u_0) < 0$, which completes the proof of case (ii).

(iii) We claim that there exist $\delta > 0$ and $\rho > 0$ such that

$$I_\lambda(u) \geq \delta \quad \text{whenever } \|u\|_q = \rho. \tag{3.53}$$

Choose

$$d > \max\{1, \lambda\|m_p\|_\infty, \lambda\|m_p\|_\infty/\lambda_1(p,1)\}. \tag{3.54}$$

Thanks to Lemma 3.37, there exists a constant $C > 0$ such that

$$\|\nabla u\|_p \leq C\|u\|_q \quad \text{for all } u \in X(d), \tag{3.55}$$

with $X(d)$ as in (3.49). For any $u \in X(d)$ satisfying $\int_\Omega m_q(u^+)^q\,dx \leq 0$, by (3.54) and (3.55) we have

$$\begin{aligned} I_\lambda(u) &\geq \frac{1-d}{p}\|\nabla u\|_p^p + \frac{\lambda_1(q,1)}{q}\|u\|_q^q + \frac{d}{p}\|\nabla u\|_p^p - \frac{\lambda\|m_p\|_\infty}{p}\|u\|_p^p \\ &\geq \frac{1-d}{p}\|\nabla u\|_p^p + \frac{\lambda_1(q,1)}{q}\|u\|_q^q + \frac{1}{p}\|\nabla u\|_p^p\left(d - \frac{\lambda\|m_p\|_\infty}{\lambda_1(p,1)}\right) \\ &\geq \frac{(1-d)C^p}{p}\|u\|_q^p + \frac{\lambda_1(q,1)}{q}\|u\|_q^q. \end{aligned} \tag{3.56}$$

For any $u \notin X(d)$ satisfying $\int_\Omega m_q(u^+)^q\,dx \leq 0$, owing to (3.49) and (3.54) we find

$$I_\lambda(u) \geq \frac{d - \lambda\|m_p\|_\infty}{p}\|u\|_p^p + \frac{\lambda_1(q,1)}{q}\|u\|_q^q \geq \frac{\lambda_1(q,1)}{q}\|u\|_q^q. \tag{3.57}$$

If $u \in W_0^{1,p}(\Omega)$ fulfills $\int_\Omega m_q(u^+)^q\,dx > 0$, through (3.42) we get

$$\|\nabla u\|_q^q \geq \|\nabla u^+\|_q^q \geq \lambda_1(q,m_q)\int_\Omega m_q(u^+)^q\,dx. \tag{3.58}$$

Our assumption on λ enables us to fix $0 < \varepsilon < 1$ with

$$(1-\varepsilon)\lambda_1(q,m_q) > \lambda. \tag{3.59}$$

If in addition $u \notin X(d)$, then due to (3.54), (3.58), and (3.59) we have the estimate

$$I_\lambda(u) \geq \frac{d - \lambda\|m_p\|_\infty}{p}\|u\|_p^p + \frac{\varepsilon}{q}\|\nabla u\|_q^q$$

$$
\begin{aligned}
&+\frac{1}{q}\left((1-\varepsilon)\lambda_1(q,m_q)-\lambda\right)\int_\Omega m_q(u^+)^q\,dx \\
&\geq \frac{\varepsilon\lambda_1(q,1)}{q}\|u\|_q^q. \qquad (3.60)
\end{aligned}
$$

Finally, if $u \in X(d)$ and $\int_\Omega m_q(u^+)^q\,dx > 0$, then (3.54), (3.55), (3.58), and (3.59) imply

$$
\begin{aligned}
I_\lambda(u) &\geq \frac{1-d}{p}\|\nabla u\|_p^p + \frac{\varepsilon}{q}\|\nabla u\|_q^q + \frac{1}{q}\left((1-\varepsilon)\lambda_1(q,m_q)-\lambda\right)\int_\Omega m_q(u^+)^q\,dx \\
&\quad + \frac{d}{p}\|\nabla u\|_p^p - \frac{\lambda\|m_p\|_\infty}{p}\|u\|_p^p \\
&\geq \frac{(1-d)C^p}{p}\|u\|_q^p + \frac{\varepsilon\lambda_1(q,1)}{q}\|u\|_q^q + \frac{d\lambda_1(p,1)-\lambda\|m_p\|_\infty}{p}\|u\|_p^p \\
&\geq \frac{(1-d)C^p}{p}\|u\|_q^p + \frac{\varepsilon\lambda_1(q,1)}{q}\|u\|_q^q. \qquad (3.61)
\end{aligned}
$$

Using that $q < p$, the claim in (3.53) follows from (3.56), (3.57), (3.60), and (3.61).

Now we claim that there is $R > 0$ such that

$$
\|R\varphi_1\|_q > \rho \quad \text{and} \quad I_\lambda(R\varphi_1) < 0, \qquad (3.62)
$$

where $\rho > 0$ is the constant in (3.53) and φ_1 is the positive eigenfunction corresponding to $\lambda_1(p, m_p)$ satisfying $\int_\Omega m_p\varphi_1^p\,dx = 1$. The claim in (3.62) is true because for a sufficiently large $R > 0$ we have

$$
\frac{I_\lambda(R\varphi_1)}{R^p} = \frac{\lambda_1(p,m_p)-\lambda}{p} + \frac{1}{qR^{p-q}}\left(\|\nabla\varphi_1\|_q^q - \int_\Omega m_q\varphi_1^q\,dx\right) < 0
$$

taking into account that $\lambda > \lambda_1(p, m_p)$ and $p > q$.

Since $m_p, m_q \in L^\infty(\Omega)$ and $0 \leq \lambda \neq \lambda_1(p, m_p)$, the functional I_λ satisfies the Palais–Smale condition. The proof can be done along the lines of [224, Lemma 12]. Therefore the properties stated in (3.53) and (3.62) allow us to apply the mountain pass theorem (see Theorem 1.44), which guarantees the existence of a critical value $c \geq \delta$ of I_λ, with $\delta > 0$ in (3.53), namely

$$
c = \inf_{\gamma\in\Sigma}\max_{t\in[0,1]} I_\lambda(\gamma(t)),
$$

$$
\Sigma := \left\{\gamma \in C([0,1], W_0^{1,p}(\Omega)) : \gamma(0) = 0,\ \gamma(1) = R\varphi_1\right\}.
$$

This completes the proof of case (iii).

(iv) Since $\lambda_1(p, -m_p) < +\infty$, we must have $(-m_p)^+ \not\equiv 0$. We distinguish two cases.

Case Where m_q Changes Sign. The conclusion can be achieved through the proof in case (iii) with $-\lambda$, $-m_p$, $-m_q$ in place of λ, m_p, m_q, respectively.

Case Where $m_q \geq 0$. In this case, $\lambda_1(q, -m_q) = +\infty$. Under the assumption of case (iv), we notice that $-\lambda > 0$, which results in

$$\lambda \int_\Omega m_q(u^+)^q \, dx = (-\lambda) \int_\Omega (-m_q)(u^+)^q \, dx \leq 0$$

for all $u \in W_0^{1,p}(\Omega)$. Arguing as in (3.56) and (3.57) with $-\lambda$, $-m_q$ instead of λ, m_q, respectively, enables us to find

$$I_\lambda(u) \geq \begin{cases} \dfrac{(1-d)C^p}{p}\|u\|_q^p + \dfrac{\lambda_1(q,1)}{q}\|u\|_q^q & \text{if } u \in X(d), \\ \dfrac{\lambda_1(q,1)}{q}\|u\|_q^q & \text{if } u \notin X(d), \end{cases}$$

where $d > 0$ is a constant satisfying

$$d > \max\{1, -\lambda\|m_p\|_\infty, -\lambda\|m_p\|_\infty/\lambda_1(p,1)\}$$

and $C = C(d)$ is determined by Lemma 3.37. Thus there exist $\delta > 0$ and $\rho > 0$ such that (3.53) holds true.

Moreover, proceeding as in case (iii), this time with the normalized positive eigenfunction $\widetilde{\varphi}_1 := \varphi_1(p, -m_p)$ corresponding to $\lambda_1(p, -m_p)$ in place of φ_1 provides a constant $R > 0$ such that

$$\|R\widetilde{\varphi}_1\|_q > \rho \quad \text{and} \quad I_\lambda(R\widetilde{\varphi}_1) < 0.$$

Because $I_\lambda = I_{(\lambda, m_p, m_q)} = I_{(-\lambda, -m_p, -m_q)}$ fulfills the Palais–Smale condition as can be seen on the pattern of part (iii) (note that $-\lambda > \lambda_1(p, -m_p)$), the same argument as in case (iii) ensures the existence of a nontrivial critical point of I_λ, which in view of Remark 3.38 completes the proof. □

Next we present our results for problem (3.44) in the resonant cases.

Theorem 3.40. *Assume that*

$$\lambda = \lambda_1(p, m_p) > \lambda_1(q, m_q) \quad \big(\textit{resp. } \lambda = -\lambda_1(p, -m_p) < -\lambda_1(q, -m_q)\big) \tag{3.63}$$

and

$$\begin{gathered} \int_\Omega |\nabla\varphi_1(p, m_p)|^q \, dx - \lambda \int_\Omega m_q \varphi_1(p, m_p)^q \, dx > 0 \\ \left(\textit{resp. } \int_\Omega |\nabla\varphi_1(p, -m_p)|^q \, dx - \lambda \int_\Omega m_q \varphi_1(p, -m_p)^q \, dx > 0\right). \end{gathered} \tag{3.64}$$

Then (3.44) has at least one positive solution.

Proof. We suppose that

$$\lambda = \lambda_1(p, m_p) > \lambda_1(q, m_q).$$

The situation $\lambda = -\lambda_1(p, -m_p) < -\lambda_1(q, -m_q)$ can be analogously handled taking $-\lambda$, $-m_p$, $-m_q$ instead of λ, m_p, m_q, respectively. Since the Lebesgue measure of $\{x \in \Omega : m_p(x) > 0\}$ is positive, there exists an $n_0 \in \mathbb{N}$ such that $(m_p - 1/n_0)^+ \not\equiv 0$. For $n \geq n_0$, we define the functional I_n on $W_0^{1,p}(\Omega)$ by

$$I_n(u) := I_{(\lambda, m_p, m_q)}(u) + \frac{\lambda}{pn}\|u^+\|_p^p = I_{(\lambda, m_p - 1/n, m_q)}(u)$$

for all $u \in W_0^{1,p}(\Omega)$ (see (3.50)). Using that

$$\lambda_1(p, m_p - 1/n) > \lambda_1(p, m_p) = \lambda$$

(where the first inequality follows from (3.42)), we are able to apply Theorem 3.39 obtaining a positive solution u_n of the Dirichlet problem

$$\begin{cases} -\Delta_p v - \Delta_q v = \lambda\left((m_p - 1/n)|v|^{p-2}v + m_q|v|^{q-2}v\right) & \text{in } \Omega, \\ v = 0 & \text{on } \partial\Omega. \end{cases}$$

Through the proof of Theorem 3.39 (see the case (i) therein), we may assume that u_n is a global minimizer of I_n and $I_n(u_n) < 0$. In addition, observing that $I_n \leq I_{n_0}$ provided $n \geq n_0$, we infer that

$$0 > I_{n_0}(u_{n_0}) \geq I_n(u_{n_0}) \geq \min_{W_0^{1,p}(\Omega)} I_n = I_n(u_n)$$

for all $n \geq n_0$.

At this point, we note that if $\|u_n\| := \|\nabla u_n\|_p$ is bounded, then $\{u_n\}$ is a bounded Palais–Smale sequence of I_λ because $I_n'(u_n) = 0$ and

$$\|I_\lambda'(u_n)\|_{W^{-1,p'}(\Omega)} = \|I_\lambda'(u_n) - I_n'(u_n)\|_{W_0^{-1,p'}(\Omega)} \leq \frac{\lambda\lambda_1(p,1)^{-1/p}}{n}\|(u_n)^+\|_p^{p-1}.$$

On the other hand, by a standard argument based on the $(S)_+$-property of $-\Delta_p$ (see Proposition 1.56), it can be readily shown that $\{u_n\}$ has a subsequence converging to some critical point u_0 of I_λ. We note that $u_0 \neq 0$ because

$$I_\lambda(u_0) = \lim_{n\to\infty} I_n(u_n) \leq I_{n_0}(u_{n_0}) < 0.$$

Therefore u_0 is a positive solution of (3.44) (see Remark 3.38). This ensures that the conclusion of the theorem is achieved once the sequence $\{\|u_n\|\}$ is bounded.

In order to prove the boundedness of $\{\|u_n\|\}$, suppose that $\|u_n\| \to +\infty$ along a relabeled subsequence. Setting $w_n := u_n/\|u_n\|$, we may admit that $\{w_n\}$

converges to w_0 weakly in $W_0^{1,p}(\Omega)$ and strongly in $L^p(\Omega)$ for some $w_0 \in W_0^{1,p}(\Omega)$. We observe that $w_n \geq 0$ and so by taking $(w_n - w_0)/\|u_n\|^{p-1}$ as test function, we obtain

$$\begin{aligned}
0 &= \langle I_n'(u_n), \frac{w_n - w_0}{\|u_n\|^{p-1}} \rangle \\
&= \int_\Omega |\nabla w_n|^{p-2} \nabla w_n \cdot \nabla (w_n - w_0)\, dx \\
&\quad + \frac{1}{\|u_n\|^{p-q}} \int_\Omega |\nabla w_n|^{q-2} \nabla w_n \cdot \nabla (w_n - w_0)\, dx \\
&\quad - \lambda \int_\Omega (m_p - \frac{1}{n}) w_n^{p-1} (w_n - w_0)\, dx \\
&\quad - \frac{\lambda}{\|u_n\|^{p-q}} \int_\Omega m_q w_n^{q-1} (w_n - w_0)\, dx \\
&= \int_\Omega |\nabla w_n|^{p-2} \nabla w_n \cdot \nabla (w_n - w_0)\, dx + o(1),
\end{aligned}$$

where $o(1) \to 0$ as $n \to \infty$. Due to the $(S)_+$-property of $-\Delta_p$ given in Proposition 1.56, this implies that $w_n \to w_0$ strongly in $W_0^{1,p}(\Omega)$. Then, for any $\varphi \in W_0^{1,p}(\Omega)$, by taking $\varphi/\|u_n\|^{p-1}$ as test function we have

$$\begin{aligned}
0 = &\int_\Omega |\nabla w_n|^{p-2} \nabla w_n \cdot \nabla \varphi\, dx + \frac{1}{\|u_n\|^{p-q}} \int_\Omega |\nabla w_n|^{q-2} \nabla w_n \cdot \nabla \varphi\, dx \\
&- \lambda \int_\Omega (m_p - \frac{1}{n}) w_n^{p-1} \varphi\, dx - \frac{\lambda}{\|u_n\|^{p-q}} \int_\Omega m_q w_n^{q-1} \varphi\, dx.
\end{aligned}$$

Letting $n \to \infty$, it follows that

$$-\Delta_p w_0 = \lambda m_p(x) w_0^{p-1} \quad \text{in } \Omega, \qquad w_0 = 0 \quad \text{on } \partial\Omega,$$

with $w_0 \geq 0$ and $\|\nabla w_0\|_p = 1$. According to the strong maximum principle (see Theorem 2.19), we see that $w_0 > 0$ in Ω. This entails $w_0 = \varphi_1/\|\nabla \varphi_1\|_p$ because $\lambda = \lambda_1(p, m_p)$ is a simple eigenvalue, where $\varphi_1 = \varphi_1(p, m_p)$ is a positive eigenfunction corresponding to $\lambda_1(p, m_p)$.

The facts that $I_n(u_n) < 0$ for all $n \geq n_0$ and u_n is a critical point of I_n result in

$$0 \geq \frac{I_n(u_n)}{\|u_n\|^q} = \left(\frac{1}{q} - \frac{1}{p}\right) \left(\|\nabla w_n\|_q^q - \lambda \int_\Omega m_q w_n^q\, dx \right).$$

Letting $n \to \infty$ leads to $\|\nabla \varphi_1\|_q^q - \lambda \int_\Omega m_q \varphi_1^q\, dx \leq 0$, which contradicts (3.64). The proof is thus complete. □

Theorem 3.41. *Assume that*

$$\lambda = \lambda_1(q, m_q) > \lambda_1(p, m_p) \quad \big(\textit{resp. } \lambda = -\lambda_1(q, -m_q) < -\lambda_1(p, -m_p)\big) \tag{3.65}$$

and

$$\begin{aligned} &\int_\Omega |\nabla \psi_1(q, m_q)|^p\, dx - \lambda \int_\Omega m_q \psi_1(q, m_q)^p\, dx > 0 \\ &\Big(\textit{resp.} \quad \int_\Omega |\nabla \psi_1(q, -m_q)|^p\, dx - \lambda \int_\Omega m_q \psi_1(q, -m_q)^p\, dx > 0\Big). \end{aligned} \tag{3.66}$$

Then (3.44) has at least one positive solution.

Proof. As in the proof of Theorem 3.40, we can consider according to (3.65) that $\lambda = \lambda_1(q, m_q) > \lambda_1(p, m_p)$ and choose $n_0 \in \mathbb{N}$ satisfying $(m_q - 1/n_0)^+ \not\equiv 0$.

For $n \geq n_0$, we define the functional J_n on $W_0^{1,p}(\Omega)$ by

$$J_n(u) := I_{(\lambda, m_p, m_q)}(u) + \frac{\lambda}{qn}\|u^+\|_q^q = I_{(\lambda, m_p, m_q - 1/n)}(u) \quad \text{for all } u \in W_0^{1,p}(\Omega).$$

Using that $\lambda_1(q, m_q - 1/n) > \lambda_1(q, m_q) = \lambda$ for any $n \geq n_0$, we are in a position to apply Theorem 3.39 obtaining a positive solution u_n of the Dirichlet problem

$$\begin{cases} -\Delta_p v - \Delta_q v = \lambda\left(m_p |v|^{p-2} v + (m_q - 1/n)|v|^{q-2} v\right) & \text{in } \Omega, \\ v = 0 & \text{on } \partial\Omega. \end{cases}$$

Following the pattern of the proof of Theorem 3.40, this time proceeding as in the case (iii) in the proof of Theorem 3.39, we deduce that $J_n(u_n) > 0$ for all $n \geq n_0$. Furthermore, the sequence $\{u_n\}$ is bounded in $W_0^{1,p}(\Omega)$. Namely, if $\|u_n\| := \|\nabla u_n\|_p \to +\infty$ along a subsequence, the same argument as in the proof of Theorem 3.40 yields that $w_n := u_n/\|u_n\|$ has a subsequence converging to a positive eigenfunction w_0 of $-\Delta_p$ with weight m_p, corresponding to λ. This yields that $\lambda = \lambda_1(p, m_p)$ and $w_0 = \varphi_1/\|\varphi_1\|$, where $\varphi_1 = \varphi_1(p, m_p)$ is a positive eigenfunction corresponding to $\lambda_1(p, m_p)$; indeed, any positive eigenvalue other than $\lambda_1(p, m_p)$ has no positive eigenfunctions and the eigenvalue $\lambda_1(p, m_p)$ is simple. However, this is a contradiction because we are considering the case where $\lambda = \lambda_1(q, m_q) > \lambda_1(p, m_p)$.

The bounded sequence $\{u_n\}$ is a Palais–Smale sequence for the functional I_λ as can be seen from the estimate

$$\|I'_\lambda(u_n)\|_{W^{-1,p'}(\Omega)} = \|I'_\lambda(u_n) - J'_n(u_n)\|_{W^{-1,p'}(\Omega)} \leq \frac{C\|u_n\|_q^{q-1}}{n},$$

with a positive constant C independent of n. Proceeding as in the proof of Theorem 3.40, $\{u_n\}$ possesses a subsequence converging to a critical point u_0 of I_λ. In order to complete the proof, due to Remark 3.38, it suffices to justify that $u_0 \neq 0$.

By way of contradiction we assume that up to a subsequence $u_n \to 0$ strongly in $W_0^{1,p}(\Omega)$. Set $v_n := u_n/\|u_n\|$. Then, up to a subsequence, we may suppose that $\{v_n\}$ converges to v_0 weakly in $W_0^{1,p}(\Omega)$ and strongly in $L^p(\Omega)$ for some $v_0 \in W_0^{1,p}(\Omega)$. It is clear that $v_0 \geq 0$. Using $(v_n - v_0)/\|u_n\|^{q-1}$ as test function, we obtain

$$\begin{aligned}
0 &= \langle J_n'(u_n), \frac{v_n - v_0}{\|u_n\|^{q-1}} \rangle \\
&= \|u_n\|^{p-q} \int_\Omega |\nabla v_n|^{p-2} \nabla v_n \cdot \nabla(v_n - v_0)\, dx \\
&\quad + \int_\Omega |\nabla v_n|^{q-2} \nabla v_n \cdot \nabla(v_n - v_0)\, dx \\
&\quad - \lambda \|u_n\|^{p-q} \int_\Omega m_p v_n^{p-1}(v_n - v_0)\, dx - \lambda \int_\Omega (m_q - \frac{1}{n}) v_n^{q-1}(v_n - v_0)\, dx \\
&= \int_\Omega |\nabla v_n|^{q-2} \nabla v_n \cdot \nabla(v_n - v_0)\, dx + o(1),
\end{aligned}$$

where $o(1) \to 0$ as $n \to \infty$. Invoking the $(S)_+$-property of $-\Delta_q$ on $W_0^{1,q}(\Omega)$ (see Proposition 1.56), we infer that $v_n \to v_0$ strongly in $W_0^{1,q}(\Omega)$.

We claim that $v_0 \neq 0$. By taking u_n as test function, we may write

$$\begin{aligned}
0 &= \langle J_n'(u_n), u_n \rangle \qquad (3.67) \\
&= \|\nabla u_n\|_p^p - \lambda \int_\Omega m_p u_n^p\, dx + \|\nabla u_n\|_q^q - \lambda \int_\Omega \left(m_q - \frac{1}{n}\right) u_n^q\, dx \\
&\geq \|\nabla u_n\|_p^p - \lambda \int_\Omega m_p u_n^p\, dx.
\end{aligned}$$

Here we have utilized the inequality $\|\nabla u\|_q^q - \lambda \int_\Omega m_q u^q\, dx \geq 0$ for all $u \in W_0^{1,q}(\Omega)$, which follows from the definition of $\lambda_1(q, m_q)$ and since $\lambda = \lambda_1(q, m_q)$. This renders $\|\nabla u_n\|_p^p \leq \lambda \|m_p\|_\infty \|u_n\|_p^p$, whence $u_n \in X(d)$ with $d = \lambda \|m_p\|_\infty$ (see (3.49) for the definition of $X(d)$). Therefore, Lemma 3.37 guarantees the existence of a constant $C > 0$ such that $\|\nabla u_n\|_p \leq C\|u_n\|_q \leq C\lambda_1(q,1)^{-1/q}\|\nabla u_n\|_q$ for all $n \geq n_0$. Consequently, recalling that $v_n \to v_0$ strongly in $W_0^{1,q}(\Omega)$, we derive

$$\|\nabla v_0\|_q = \lim_{n\to\infty} \|\nabla v_n\|_q = \lim_{n\to\infty} \frac{\|\nabla u_n\|_q}{\|\nabla u_n\|_p} \geq \frac{1}{C\lambda_1(q,1)^{-1/q}} > 0,$$

thus proving our claim.

For any $\varphi \in W_0^{1,p}(\Omega)$, by taking $\varphi/\|u_n\|^{q-1}$ as test function we get

$$\begin{aligned}
0 &= \langle J_n'(u_n), \frac{\varphi}{\|u_n\|^{q-1}}\rangle \\
&= \|u_n\|^{p-q} \int_\Omega |\nabla v_n|^{p-2}\nabla v_n \cdot \nabla\varphi \, dx + \int_\Omega |\nabla v_n|^{q-2}\nabla v_n \cdot \nabla\varphi \, dx \\
&\quad - \lambda\|u_n\|^{p-q} \int_\Omega m_p v_n^{p-1}\varphi \, dx - \lambda \int_\Omega \left(m_q - \frac{1}{n}\right) v_n^{q-1}\varphi \, dx \\
&= \int_\Omega |\nabla v_n|^{q-2}\nabla v_n \cdot \nabla\varphi \, dx - \lambda \int_\Omega m_q v_n^{q-1}\varphi \, dx + o(1),
\end{aligned}$$

where $o(1) \to 0$ as $n \to \infty$. Letting $n \to \infty$ shows that $-\Delta_q v_0 = \lambda m_q |v_0|^{q-2} v_0$ in Ω and $v_0 = 0$ on $\partial\Omega$. The simplicity of the eigenvalue $\lambda_1(q, m_q)$ ensures that $v_0 = \psi_1/\|\psi_1\|$, where $\psi_1 = \psi_1(q, m_q)$ stands for a positive eigenfunction corresponding to $\lambda_1(q, m_q)$.

Using that $J_n(u_n) > 0$ for all $n \geq n_0$ in conjunction with (3.67), we obtain

$$0 \leq \frac{J_n(u_n)}{\|u_n\|^p} = \left(\frac{1}{p} - \frac{1}{q}\right)\left(\|\nabla v_n\|_p^p - \lambda \int_\Omega m_p v_n^p \, dx\right).$$

Hence, on account of $1/p - 1/q < 0$, it turns out that

$$\|\nabla v_n\|_p^p - \lambda \int_\Omega m_p v_n^p \, dx \leq 0.$$

Since $\{v_n\}$ converges to $v_0 = \psi_1/\|\psi_1\|$ weakly in $W_0^{1,p}(\Omega)$ and strongly in $L^p(\Omega)$, by passing to the limit inferior we have $\|\nabla\psi_1\|_p^p - \lambda \int_\Omega m_p \psi_1^p \, dx \leq 0$. This contradicts assumption (3.66). Consequently, our claim $u_0 \neq 0$ holds true. The proof of the theorem is complete. □

We produce some examples where the hypotheses of the formulated results are fulfilled.

Example 3.42. Let us take $N = 1$, $\Omega = (0, \pi)$, $1 < q = 2 < p$, $m_q \equiv 1$ and $m_{p,n}(x) = 1 - h(x)/n$ with $0 \leq h \in L^\infty(\Omega)$ and $h \not\equiv 0$. Then it is clear that $\lambda_1(2, m_2) = 1$ and $\psi_1 = \psi_1(2, m_2) = \sin x$. By easy computation, we have

$$\int_0^\pi |\psi_1'|^p \, dx - \int_0^\pi m_{p,n}\psi_1^p \, dx = \frac{1}{n}\int_0^\pi h(x)\sin^p x \, dx > 0$$

for every $n \in \mathbb{N}$, so (3.66) holds true. On the other hand, we see that

$$\lambda_1(p, 1) < \frac{\int_0^\pi |\psi_1'|^p \, dx}{\int_0^\pi \psi_1^p \, dx} = 1$$

by the definition of $\lambda_1(p,1)$ and the facts that $\psi_1 \neq t\varphi_1(p,1)$ for any $t>0$ and $\lambda_1(p,1)$ is simple. By the continuity of $\lambda_1(p,m_p)$ with respect to m_p, it follows that for a sufficiently large n we have $\lambda_1(p,m_{p,n})<1$, so (3.65) is valid, too. Similarly, we can construct examples satisfying (3.63) and (3.64).

3.3 EIGENVALUE PROBLEMS FOR ASYMPTOTICALLY HOMOGENEOUS DIFFERENTIAL ELLIPTIC OPERATORS

The object of this section is to study the eigenvalue problem consisting in determining the existence of $\lambda \in \mathbb{R}$ called eigenvalue provided

$$\begin{cases} -\operatorname{div} A(x,\nabla u) = \lambda |u|^{p-2}u & \text{in } \Omega, \\ u=0 & \text{on } \partial\Omega \end{cases} \tag{3.68}$$

has a nontrivial solution. Here $\Omega \subset \mathbb{R}^N$ is a bounded domain with C^2 boundary $\partial\Omega$, $1<p<+\infty$, and $A:\overline{\Omega}\times\mathbb{R}^N \to \mathbb{R}^N$ is a map which is strictly monotone in the second variable and satisfies certain regularity conditions. We do not suppose that A is $(p-1)$-homogeneous in the second variable.

We assume that the map A satisfies the following conditions:

(H_A) $A(x,y)=a(x,|y|)y$, where $a(x,t)>0$ for all $(x,t)\in\overline{\Omega}\times(0,+\infty)$, and

(i) $A \in C(\overline{\Omega}\times\mathbb{R}^N,\mathbb{R}^N)\cap C^1(\overline{\Omega}\times(\mathbb{R}^N\setminus\{0\}),\mathbb{R}^N)$;

(ii) there exists $C_1>0$ such that

$$\|D_y A(x,y)\| \le C_1|y|^{p-2} \quad \text{for every } x\in\overline{\Omega}, \text{ and } y\in\mathbb{R}^N\setminus\{0\};$$

(iii) there exists $C_0>0$ such that

$$D_y A(x,y)\xi\cdot\xi \ge C_0|y|^{p-2}|\xi|^2$$
$$\text{for every } x\in\overline{\Omega},\ y\in\mathbb{R}^N\setminus\{0\} \text{ and } \xi\in\mathbb{R}^N;$$

(iv) there exists $C_2>0$ such that

$$\|D_x A(x,y)\| \le C_2(1+|y|^{p-1}) \quad \text{for every } x\in\overline{\Omega},\ y\in\mathbb{R}^N\setminus\{0\};$$

(v) there exist $C_3>0$ and $0<t_0\le 1$ such that

$$\|D_x A(x,y)\| \le C_3|y|^{p-1}\,(-\log|y|)$$

for every $x\in\overline{\Omega}$, $y\in\mathbb{R}^N$ with $0<|y|<t_0$.

Without any loss of generality we also suppose that $C_0 \le p-1 \le C_1$. Notice that the case $C_0=p-1=C_1$ characterizes the p-Laplacian.

The Banach space $W_0^{1,p}(\Omega)$ is endowed with the norm $\|u\| := \|\nabla u\|_p$, where $\|\cdot\|_q$ denotes the usual L^q-norm for $1\le q\le+\infty$ (see Remark 1.18). We also denote by $|\Omega|$ the Lebesgue measure of Ω.

Set

$$G(x, y) := \int_0^{|y|} a(x,t)t\,dt. \tag{3.69}$$

We can easily see that

$$\nabla_y G(x, y) = A(x, y) \quad \text{and} \quad G(x, 0) = 0$$

for every $x \in \overline{\Omega}$.

Remark 3.43. The following assertions hold under condition (H_A):

(i) for all $x \in \overline{\Omega}$, $A(x, y)$ is maximal monotone and strictly monotone in y;

(ii) $|A(x, y)| \leq \dfrac{C_1}{p-1}|y|^{p-1}$ for every $(x, y) \in \overline{\Omega} \times \mathbb{R}^N$;

(iii) $A(x, y) \cdot y \geq \dfrac{C_0}{p-1}|y|^p$ for every $(x, y) \in \overline{\Omega} \times \mathbb{R}^N$;

(iv) $G(x, y)$ defined in (3.69) is strictly convex in y for all x and satisfies the following inequalities:

$$A(x, y) \cdot y \geq G(x, y) \geq \frac{C_0}{p(p-1)}|y|^p \quad \text{and} \quad G(x, y) \leq \frac{C_1}{p(p-1)}|y|^p \tag{3.70}$$

for every $(x, y) \in \overline{\Omega} \times \mathbb{R}^N$,

where C_0 and C_1 are the positive constants in (H_A).

For later use we mention the following useful property.

Proposition 3.44 ([168, Proposition 1]). *Let $V : W_0^{1,p}(\Omega) \to W^{-1,p'}(\Omega)$ be the map defined by*

$$\langle V(u), v\rangle = \int_\Omega A(x, \nabla u) \cdot \nabla v\, dx$$

for $u, v \in W_0^{1,p}(\Omega)$. Then, V has the $(S)_+$-property, that is, any sequence $\{u_n\}$ weakly convergent to u is strongly convergent to u whenever

$$\limsup_{n\to\infty}\langle V(u_n), u_n - u\rangle \leq 0.$$

Let us introduce the function $J : W_0^{1,p}(\Omega) \to \mathbb{R}$ by

$$J(u) = \int_\Omega G(x, \nabla u)\, dx \quad \text{for all } u \in W_0^{1,p}(\Omega), \tag{3.71}$$

which is clearly of class C^1. We note that

$$rS := \left\{u \in W_0^{1,p}(\Omega) : \|u\|_p = r\right\} \quad \text{for } r > 0 \tag{3.72}$$

is a C^1 Finsler manifold, actually, a C^1 submanifold of $W_0^{1,p}(\Omega)$ (see [71, Sections 27.4, 27.5]) because r is a regular value of the function $u \mapsto \|u\|_p$ on $W_0^{1,p}(\Omega) \setminus \{0\}$. Hence, the norm of the derivative at $u \in rS$ of the restriction $\tilde{J}$ of J to rS is defined as

$$\begin{aligned}\|\widetilde{J}'(u)\|_* &:= \min\left\{ \|J'(u) - t\Phi'(u)\|_{W^{-1,p'}(\Omega)} : t \in \mathbb{R}\right\} \\ &= \sup\left\{ \langle J'(u), v\rangle : v \in T_u(rS),\ \|v\| = 1 \right\},\end{aligned}$$

where

$$\Phi(u) := \frac{1}{p}\|u\|_p^p$$

and $T_u(rS)$ designates the tangent space of rS at u, that is,

$$T_u(rS) = \{v \in W_0^{1,p}(\Omega) : \int_\Omega |u|^{p-2}uv\,dx = 0\}.$$

The restriction $\tilde{J} = J|_{rS}$ is a C^1 function on rS in the sense of manifolds (see (3.71), (3.72)).

Recall that λ_1 is the first eigenvalue of $-\Delta_p$ on $W_0^{1,p}(\Omega)$ (see Proposition 1.57).

Proposition 3.45. *For $r > 0$, the infimum*

$$\mu_1(A, r) = \inf_{u \in rS} \int_\Omega G(x, \nabla u)\,dx \tag{3.73}$$

is attained at points $\pm\hat{u}_r \in rS$ with $\hat{u}_r \in C^{1,\alpha}(\overline{\Omega})$ and $\hat{u}_r > 0$ in Ω. Moreover, $\pm\hat{u}_r$ are solutions of (3.68) with $\lambda = \lambda_1(A, \hat{u}_r)/r^p$, where

$$\lambda_1(A, \hat{u}_r) = \int_\Omega A(x, \nabla\hat{u}_r) \cdot \nabla\hat{u}_r\,dx \geq \frac{c_0}{p-1}\lambda_1 r^p. \tag{3.74}$$

Proof. Let $\{u_n\} \subset rS$ be a minimizing sequence for (3.73). Using (3.70), it follows that $\{u_n\}$ is bounded in $W_0^{1,p}(\Omega)$, so along a relabeled subsequence there hold $u_n \rightharpoonup \hat{u}_r$ in $W_0^{1,p}(\Omega)$ and $u_n \to \hat{u}_r$ in $L^p(\Omega)$ for some $\hat{u}_r \in W_0^{1,p}(\Omega)$, thus $\hat{u}_r \in rS$. Since $G(x, \cdot)$ is convex and continuous for all $x \in \Omega$, J is weakly lower semicontinuous on $W_0^{1,p}(\Omega)$ (see, e.g., [152, Theorem 1.2]). Therefore, we derive that

$$\mu_1(A, r) \leq \int_\Omega G(x, \nabla\hat{u}_r)\,dx \leq \liminf_{n\to\infty} \int_\Omega G(x, \nabla u_n)\,dx,$$

which yields

$$\mu_1(A,r) = \int_\Omega G(x, \nabla \hat{u}_r)\, dx.$$

The fact that the functional J is even implies that $|\hat{u}_r|$ is also a global minimizer of $\tilde{J}$. Consequently, we may assume that $\hat{u}_r \geq 0$. On the other hand, the Lagrange multiplier rule in Theorem 1.37 leads to the existence of $t \in \mathbb{R}$ such that

$$\int_\Omega A(x, \nabla \hat{u}_r) \cdot \nabla v\, dx = t \int_\Omega \hat{u}_r^{p-1} v\, dx \quad \text{for all } v \in W_0^{1,p}(\Omega). \tag{3.75}$$

It follows that $\hat{u}_r$ is a solution of (3.68) with $\lambda = t$. By the nonlinear regularity theory, we obtain that $\hat{u}_r \in C^{1,\alpha}(\overline{\Omega})$ $(0 < \alpha < 1)$ and $\hat{u}_r > 0$ in Ω. Moreover, inserting $v = \hat{u}_r$ in (3.75) entails

$$tr^p = \int_\Omega A(x, \nabla \hat{u}_r) \cdot \nabla \hat{u}_r\, dx \geq \frac{C_0}{p-1} \|\nabla \hat{u}_r\|_p^p \geq \frac{C_0 \lambda_1}{p-1} \|\hat{u}_r\|_p^p = \frac{C_0 \lambda_1}{p-1} r^p. \tag{3.76}$$

Therefore, we have

$$t = \frac{\lambda_1(A, \hat{u}_r)}{r^p} \geq \frac{C_0 \lambda_1}{p-1}.$$

Since J is even, we derive that

$$J(-\hat{u}_r) = J(\hat{u}_r) = \mu_1(A,r)$$

and $-\hat{u}_r$ is a negative solution of (3.68) with $\lambda = \lambda_1(A, \hat{u}_r)/r^p$. From (3.76), we see that (3.74) ensues. □

Set

$$K_1(A,r) := \left\{ \hat{u}_r \in rS : J(u) = \mu_1(A,r) \right\},$$

which is nonempty for each $r > 0$ as shown in Proposition 3.45. We also introduce

$$\underline{\lambda}_1(A,r) := \inf_{u \in K_1(A,r)} \lambda_1(A,u) \quad \text{and} \quad \overline{\lambda}_1(A,r) := \sup_{u \in K_1(A,r)} \lambda_1(A,u),$$

where $\lambda_1(A,u) = \int_\Omega A(x, \nabla u) \cdot \nabla u\, dx$.

Lemma 3.46. *For every $r > 0$, $\underline{\lambda}_1(A,r)$ and $\overline{\lambda}_1(A,r)$ are attained.*

Proof. We only prove for $\underline{\lambda}_1(A,r)$ because the corresponding formula for $\overline{\lambda}_1(A,r)$ can be shown similarly. Fix any $r>0$. Let $u_n \in K_1(A,r)$ satisfy $\lambda_1(A,u_n) \to \underline{\lambda}_1(A,r)$ as $n\to\infty$. We see that $\|\nabla u_n\|_p$ is bounded because

$$\frac{C_0}{p(p-1)}\|\nabla u_n\|_p^p \le \int_\Omega G(x,\nabla u_n)\,dx = \mu_1(A,r) \le \int_\Omega G(x,\nabla w)\,dx$$

for all $w\in rS$, where we use (3.70).

Recall that each u_n is a solution of (3.68) with $\lambda = \lambda_1(A,u_n)/r^p$. Moreover, we have

$$\frac{C_0}{p-1}\lambda_1 r^p \le \lambda_1(A,u_n) \le \frac{C_1}{p-1}\|\nabla u_n\|_p^p$$

by Remark 3.43 (ii) (see (3.74) for the first inequality), so $\lambda_1(A,u_n)/r^p$ is bounded. As a result (see Proposition 2.12), we may assume that there exists $u_0 \in W_0^{1,p}(\Omega)$ such that $u_n \to u_0$ in $C_0^1(\overline{\Omega})$ by choosing a subsequence if necessary. Since J in (3.71) and $\lambda_1(A,\cdot)$ are continuous on $W_0^{1,p}(\Omega)$, it is seen that

$$J(u_0) = \lim_{n\to\infty} J(u_n) = \mu_1(A,r), \quad u_0 \in K_1(A,r)$$

and

$$\lambda_1(A,u_0) = \lim_{n\to\infty}\lambda_1(A,u_n) = \underline{\lambda}_1(A,r).$$

Thus the conclusion follows. □

Set

$$\lambda_1(A) := \inf_{u\in W_0^{1,p}(\Omega)\setminus\{0\}} \int_\Omega \frac{A(x,\nabla u)\cdot\nabla u}{\|u\|_p^p}\,dx \quad \text{and}$$

$$\mu_1(A) := \inf_{u\in W_0^{1,p}(\Omega)\setminus\{0\}} \int_\Omega \frac{G(x,\nabla u)}{\|u\|_p^p}\,dx.$$

Recall that λ_1 stands for the first eigenvalue of $-\Delta_p$ on $W_0^{1,p}(\Omega)$.

Lemma 3.47.

$$\frac{C_0}{p-1}\lambda_1 \le \lambda_1(A) \le \min\left\{\inf_{r>0}\frac{\underline{\lambda}_1(A,r)}{r^p}, \frac{C_1}{p-1}\lambda_1\right\} \quad \textit{and} \tag{3.77}$$

$$\mu_1(A) = \inf_{r>0}\frac{\mu_1(A,r)}{r^p}.$$

Proof. First, we establish the result for $\lambda_1(A)$. For every $u\in W_0^{1,p}(\Omega)$, $u\ne 0$, by (ii) and (iii) in Remark 3.43 we have

$$\frac{C_0}{p-1}\frac{\|\nabla u\|_p^p}{\|u\|_p^p} \le \int_\Omega \frac{A(x,\nabla u)\cdot\nabla u}{\|u\|_p^p}\,dx \le \frac{C_1}{p-1}\frac{\|\nabla u\|_p^p}{\|u\|_p^p},$$

thus,

$$\frac{C_0}{p-1}\lambda_1 \le \lambda_1(A) \le \frac{C_1}{p-1}\lambda_1$$

by passing to the infimum with respect to u.

Now fix any $\varepsilon > 0$. Then there exists $r_\varepsilon > 0$ such that

$$\underline{\lambda}_1(A, r_\varepsilon)/r_\varepsilon^p \le \inf_{r>0} \frac{\underline{\lambda}_1(A,r)}{r^p} + \varepsilon.$$

By Lemma 3.46, we can choose $u_\varepsilon \in r_\varepsilon S$ such that

$$\int_\Omega A(x, \nabla u_\varepsilon) \cdot \nabla u_\varepsilon \, dx = \lambda_1(A, u_\varepsilon) = \underline{\lambda}_1(A, r_\varepsilon).$$

By the definition of $\lambda_1(A)$, we obtain

$$\lambda_1(A) \le \int_\Omega \frac{A(x, \nabla u_\varepsilon) \cdot \nabla u_\varepsilon}{\|u_\varepsilon\|_p^p} dx = \frac{\underline{\lambda}_1(A, r_\varepsilon)}{r_\varepsilon^p} \le \inf_{r>0} \frac{\underline{\lambda}_1(A,r)}{r^p} + \varepsilon.$$

Since $\varepsilon > 0$ is arbitrary, we infer that

$$\lambda_1(A) \le \inf_{r>0} \frac{\underline{\lambda}_1(A,r)}{r^p}.$$

Next, we prove the formula for $\mu_1(A)$ in (3.77). Fix $\varepsilon > 0$. There exists an $r_\varepsilon > 0$ such that

$$\frac{\mu_1(A, r_\varepsilon)}{r_\varepsilon^p} \le \inf_{r>0} \frac{\mu_1(A,r)}{r^p} + \varepsilon.$$

On the other hand, because $\mu_1(A, r_\varepsilon)$ is attained at some $u_\varepsilon \in r_\varepsilon S$ (by Proposition 3.45),

$$\begin{aligned}\inf_{u \in W_0^{1,p}(\Omega)\setminus\{0\}} \int_\Omega \frac{G(x, \nabla u)}{\|u\|_p^p} dx &\le \int_\Omega \frac{G(x, \nabla u_\varepsilon)}{\|u_\varepsilon\|_p^p} dx = \frac{\mu_1(A, r_\varepsilon)}{r_\varepsilon^p} \\ &\le \inf_{r>0} \frac{\mu_1(A,r)}{r^p} + \varepsilon.\end{aligned}$$

Since $\varepsilon > 0$ is arbitrary, this yields $\mu_1(A) \le \inf_{r>0} \frac{\mu_1(A,r)}{r^p}$.

For any $\varepsilon > 0$, we take $v_\varepsilon \ne 0$ such that

$$\int_\Omega \frac{G(x, \nabla v_\varepsilon)}{\|v_\varepsilon\|_p^p} dx \le \mu_1(A) + \varepsilon.$$

Then we obtain $r_\varepsilon := \|v_\varepsilon\|_p > 0$ and

$$\frac{\mu_1(A, r_\varepsilon)}{r_\varepsilon^p} \le \int_\Omega \frac{G(x, \nabla v_\varepsilon)}{\|v_\varepsilon\|_p^p}\, dx \le \mu_1(A) + \varepsilon.$$

This leads to $\mu_1(A) \ge \inf\limits_{r>0} \frac{\mu_1(A,r)}{r^p}$. □

The following nonexistence result is valid.

Proposition 3.48. *If $\lambda < \lambda_1(A)$, then problem (3.68) has no nontrivial solutions.*

Proof. If u is a nontrivial solution of (3.68), by the definition of $\lambda_1(A)$ we have $\lambda_1(A) \le \int_\Omega \frac{A(x,\nabla u)\cdot\nabla u}{\|u\|_p^p}\, dx = \lambda$. □

Set

$$A_p := \frac{C_1}{p-1}\left(\frac{C_1}{C_0}\right)^{p-1} \ge 1, \tag{3.78}$$

which is equal to 1 exactly in the case of $A(x, y) = |y|^{p-2}y$ (i.e., the special case of the p-Laplacian).

Lemma 3.49 ([223, Lemma 16]). *Given $\varepsilon > 0$, for every $u, \varphi \in W^{1,p}(\Omega) \cap C^1(\Omega) \cap L^\infty(\Omega)$ with $u \ge 0$ and $\varphi \ge 0$ in Ω we have*

$$\int_\Omega A(x, \nabla u) \cdot \nabla \left(\frac{\varphi^p}{(u+\varepsilon)^{p-1}}\right) dx \le A_p \|\nabla\varphi\|_p^p.$$

We deduce the following qualitative property. Hereafter, $\lambda_1 < \lambda_2$ are the first two eigenvalues of $-\Delta_p$ on $W_0^{1,p}(\Omega)$ and $\hat{u}_1$ stands for the positive eigenfunction of $-\Delta_p$ corresponding to λ_1 such that $\|\hat{u}_1\|_p = 1$ (whose existence is ensured by Proposition 1.57).

Proposition 3.50. *If u is a nontrivial solution of (3.68) with $\lambda > A_p\lambda_1$, then u changes sign.*

Proof. On the contrary, we may assume that $u \ge 0$ because A is odd. Due to the strong maximum principle in Theorem 2.19, we infer that $u \in C_0^1(\overline{\Omega})$ and $u > 0$ in Ω. According to Lemma 3.49, we obtain

$$\begin{aligned} A_p\lambda_1 = A_p\|\nabla\hat{u}_1\|_p^p &\ge \int_\Omega A(x, \nabla u) \cdot \nabla \left(\frac{\hat{u}_1^p}{(u+\varepsilon)^{p-1}}\right) dx \\ &= \lambda \int_\Omega \left(\frac{u}{u+\varepsilon}\right)^{p-1} \hat{u}_1^p\, dx \end{aligned}$$

whenever $\varepsilon > 0$. Letting $\varepsilon \to 0^+$, we have $\lambda \le A_p\lambda_1$. This is a contradiction. □

Proposition 3.51. *Assume that* $A_p\lambda_1 < \frac{C_0}{p-1}\lambda_2$. *If* $A_p\lambda_1 < \lambda < \frac{C_0}{p-1}\lambda_2$, *then problem (3.68) has no nontrivial solutions.*

Proof. By way of contradiction, assume that problem (3.68) has a nontrivial solution u. Then, it follows from Proposition 3.50 that u changes sign. Moreover, by taking u^+ and u^- as test function in (3.68), we may write

$$\frac{C_0}{p-1}\|\nabla u^\pm\|_p^p \le \int_\Omega A(x, \nabla u) \cdot (\pm\nabla u^\pm)\, dx = \lambda\|u^\pm\|_p^p,$$

whence

$$\|\nabla u^\pm\|_p^p < \lambda_2\|u^\pm\|_p^p. \tag{3.79}$$

The inequality (3.79) guarantees the existence of a continuous path γ_0 on $S := \{u \in W_0^{1,p}(\Omega) : \|u\|_p = 1\}$ such that $\gamma_0(0) = \hat{u}_1$, $\gamma_0(1) = -\hat{u}_1$ and

$$\max_{t\in[0,1]} \|\nabla\gamma_0(t)\|_p^p < \lambda_2$$

as known from the proof of [60, Lemma 5.3]. This contradicts Proposition 1.58, which proves the result. □

We consider the case where A is asymptotically $(p-1)$-homogeneous near zero in the following sense:

H$(A)_0$ there exist a positive function $a_0 \in C^1(\overline{\Omega})$ and a function $\tilde{a}_0 \in C(\overline{\Omega} \times [0, +\infty))$ such that

$$A(x, y) = a_0(x)|y|^{p-2}y + \tilde{a}_0(x, |y|)y \quad \text{for every } x \in \Omega,\ y \in \mathbb{R}^N,$$
$$\lim_{t\to 0^+} \frac{\tilde{a}_0(x, t)}{t^{p-2}} = 0 \quad \text{uniformly in } x \in \overline{\Omega}.$$

Related to the weight function a_0, we define

$$\lambda_{1,a_0} := \inf\left\{\int_\Omega a_0(x)|\nabla u|^p\, dx : \|u\|_p = 1\right\}.$$

Because

$$0 < \min_{x\in\overline{\Omega}} a_0(x) \le \max_{x\in\overline{\Omega}} a_0(x) < +\infty,$$

in analogy with Proposition 1.57 we can prove that λ_{1,a_0} is the first eigenvalue of

$$-\operatorname{div}\left(a_0(x)|\nabla u|^{p-2}\nabla u\right) = \lambda|u|^{p-2}u \quad \text{in } \Omega, \quad u = 0 \quad \text{on } \partial\Omega. \tag{3.80}$$

Moreover, λ_{1,a_0} has a positive eigenfunction $\varphi_{a_0} \in C^1(\overline{\Omega})$, it is simple, and (3.80) has no constant-sign solutions other than 0 provided $\lambda \neq \lambda_{1,a_0}$.

Theorem 3.52. *Assume* H$(A)_0$. *For every* $\varepsilon > 0$ *there exists* $r_0 > 0$ *such that problem (3.68) has no nontrivial solutions in*

$$B_p(r_0) := \{v \in W_0^{1,p}(\Omega) : \|v\|_p < r_0\}$$

provided $\lambda < \lambda_{1,a_0} - \varepsilon$.

Proof. We argue by contradiction, thus we assume that there exist $\varepsilon_0 > 0$, $\{\lambda_n\}$ and $\{u_n\}$ such that $\lambda_n < \lambda_{1,a_0} - \varepsilon_0$, $u_n \in B_p(1/n)$ and u_n is a nontrivial solution of problem (3.68) with $\lambda = \lambda_n$. By taking u_n as test function, we have

$$\frac{C_0}{p-1}\|\nabla u_n\|_p^p \leq \int_\Omega A(x,\nabla u_n)\cdot \nabla u_n\,dx = \lambda_n\|u_n\|_p^p < \frac{\lambda_{1,a_0} - \varepsilon_0}{n^p} \to 0 \tag{3.81}$$

as $n \to \infty$. Therefore, $u_n \to 0$ in $W_0^{1,p}(\Omega)$. In addition, since $0 \leq \lambda_n < \lambda_{1,a_0} - \varepsilon_0$ and u_n is a nontrivial solution of (3.68) with $\lambda = \lambda_n$, Proposition 2.12 yields that $\{u_n\}$ converges to 0 in $C^1(\overline{\Omega})$.

Set $v_n := u_n/\|u_n\|_p$. It follows from (3.81) and the boundedness of $\{\lambda_n\}$ that $\{v_n\}$ is bounded in $W_0^{1,p}(\Omega)$. Hence, we may assume by choosing a subsequence that $\{v_n\}$ converges to some v_0 weakly in $W_0^{1,p}(\Omega)$ and strongly in $L^p(\Omega)$. Again by taking $u_n/\|u_n\|_p^p$ as test function in (3.68) with $\lambda = \lambda_n$, we obtain

$$\begin{aligned}\lambda_{1,a_0} - \varepsilon_0 > \lambda_n &= \int_\Omega \frac{a_0(x)|\nabla u_n|^p}{\|u_n\|_p^p}dx + \int_\Omega \frac{\tilde{a}_0(x,|\nabla u_n|)|\nabla u_n|^2}{\|u_n\|_p^p}dx \\ &= \int_\Omega a_0(x)|\nabla v_n|^p\,dx + \int_\Omega \frac{\tilde{a}_0(x,|\nabla u_n|)|\nabla u_n|^2}{\|u_n\|_p^p} \\ &\geq \lambda_{1,a_0} + \int_\Omega \frac{\tilde{a}_0(x,|\nabla u_n|)|\nabla u_n|^2}{\|u_n\|_p^p} =: \lambda_{1,a_0} + I\end{aligned}$$

because of the characterization of λ_{1,a_0}. Hypothesis H$(A)_0$ guarantees that for every $\delta > 0$ there exists $\rho_0 > 0$ such that $|\tilde{a}_0(x,t)| \leq \delta|t|^{p-2}$ if $|t| \leq \rho_0$. Since $\|u_n\|_{C^1(\overline{\Omega})} \to 0$ and using (3.81), we get

$$|I| \leq \delta\int_\Omega |\nabla v_n|^p\,dx \leq \frac{\delta(p-1)}{C_0}\lambda_n \leq \frac{\delta(p-1)}{C_0}(\lambda_{1,a_0} - \varepsilon_0)$$

for sufficiently large n. As a result, by taking a sufficiently small $\delta > 0$, we have a contradiction for sufficiently large n. □

Theorem 3.53. *Assume* $H(A)_0$. *Then for every* $\varepsilon > 0$ *there exists* $r_1 > 0$ *such that problem (3.68) has no constant-sign solutions in* $B_p(r_1) \setminus \{0\}$ *provided* $\lambda > \lambda_{1,a_0} + \varepsilon$.

Proof. Arguing by contradiction, we assume that there exist $\varepsilon_0 > 0$, $\{\mu_n\}$ and $\{u_n\}$ such that $\mu_n > \lambda_{1,a_0} + \varepsilon_0$, $u_n \in B_p(1/n)$ and u_n is a nontrivial, constant-sign solution of (3.68) with $\lambda = \mu_n$. Because A is odd, we may suppose that $u_n \geq 0$ by considering $-u_n$ if necessary. Thus, by nonlinear regularity theory and strong maximum principle, $u_n \in C^1(\overline{\Omega})$ and $u_n > 0$ in Ω. Here we note that $\mu_n \leq A_p\lambda_1$ holds by Proposition 3.50, where λ_1 denotes the first eigenvalue of $-\Delta_p$ (see (3.78) for the definition of A_p), and so the sequence $\{\mu_n\}$ is bounded. Therefore, we may assume that $\{\mu_n\}$ converges to some μ_0 along a relabeled subsequence. In addition, by the same argument as in Theorem 3.52, we can show that $u_n \to 0$ in $C^1(\overline{\Omega})$.

Set $A_n(x, y) := A(x, \|u_n\|_p y)/\|u_n\|_p^{p-1}$ and $f_n(x,t) := \mu_n|t|^{p-2}t$. Then, A_n satisfies conditions (H_A) (i)–(iv) with the same constants C_0, C_1 and C_2. Moreover,

$$|f_n(x,t)| \leq \mu_n|t|^{p-1} \leq A_p\lambda_1|t|^{p-1}$$

for every $t \in \mathbb{R}$, a.e. $x \in \Omega$. Note also that we have the boundedness of $\{\|u_n\|\}$ due to the inequality

$$C_0\|\nabla u_n\|_p^p/(p-1) \leq \int_\Omega A(x, \nabla u_n) \cdot \nabla u_n \, dx = \mu_n \|u_n\|_p^p.$$

Since $v_n := u_n/\|u_n\|_p$ is a positive solution of

$$-\operatorname{div} A_n(x, \nabla u) = f_n(x, u) \quad \text{in } \Omega, \quad u = 0 \quad \text{on } \partial\Omega,$$

Proposition 2.12 guarantees that $\{v_n\}$ has a convergent subsequence in $C^1(\overline{\Omega})$. By choosing a subsequence, we may suppose that $v_n \to v_0 \neq 0$ in $C^1(\overline{\Omega})$ (note that $\|v_0\|_p = 1$). Observe that for every $w \in W_0^{1,p}(\Omega)$, we obtain

$$\int_\Omega \frac{\tilde{a}_0(x, |\nabla u_n|)\nabla u_n}{\|u_n\|_p^{p-1}} \cdot \nabla w \, dx = \int_\Omega \frac{\tilde{a}_0(x, |\nabla u_n|)\nabla u_n}{|\nabla u_n|^{p-1}} \cdot \nabla w |\nabla v_n|^{p-1} \, dx \to 0$$

as $n \to \infty$ by $H(A)_0$, $u_n \to 0$ and $v_n \to v_0$ in $C^1(\overline{\Omega})$. As a result, letting $n \to \infty$ in the equality

$$\int_\Omega a_0(x)|\nabla v_n|^{p-2}\nabla v_n \cdot \nabla w \, dx + \int_\Omega \frac{\tilde{a}_0(x, |\nabla u_n|)\nabla u_n}{\|u_n\|_p^{p-1}} \cdot \nabla w \, dx$$
$$= \mu_n \int_\Omega |v_n|^{p-2} v_n w \, dx$$

for each $w \in W_0^{1,p}(\Omega)$, we see that v_0 is a positive solution of (3.80) with $\lambda = \mu_0$. This confirms that $\mu_0 = \lambda_{1,a_0}$ because otherwise (3.80) has no positive solutions. Therefore, we have a contradiction because $\mu_0 = \lim_{n\to\infty} \mu_n \geq \lambda_{1,a_0} + \varepsilon_0$. □

Proposition 3.54. *Suppose* H$(A)_0$. *Then for every* $\varepsilon > 0$ *there exists* $r_0 > 0$ *such that*

$$\frac{\underline{\lambda}_1(A,r)}{r^p} \geq \lambda_{1,a_0} - \varepsilon \quad \textit{for every } 0 < r < r_0.$$

Proof. Assume that there exist $\varepsilon > 0$ and $r_n > 0$ such that $r_n \to 0$ as $n \to \infty$ and

$$\underline{\lambda}_1(A,r_n)/r_n^p < \lambda_{1,a_0} - \varepsilon$$

for every $n \in \mathbb{N}$. By Proposition 3.45 and Lemma 3.46 (note that A is odd in the second variable), we can choose a positive function $u_n \in (r_n S) \cap C^1(\overline{\Omega})$ satisfying

$$\int_\Omega A(x, \nabla u_n) \cdot \nabla u_n \, dx = \underline{\lambda}_1(A, r_n),$$
$$\min_{v \in r_n S} \int_\Omega G(x, \nabla v)\, dx = \int_\Omega G(x, \nabla u_n)\, dx.$$

Observe that

$$\frac{C_0}{p-1}\|\nabla u_n\|_p^p \leq \int A(x, \nabla u_n)\cdot \nabla u_n \, dx = \underline{\lambda}_1(A, r_n) < (\lambda_{1,a_0} - \varepsilon) r_n^p \to 0. \tag{3.82}$$

By (3.82), we obtain $u_n \to 0$ in $W_0^{1,p}(\Omega)$. Because u_n is a solution of (3.68) with $\lambda = \underline{\lambda}_1(A,r_n)/r_n^p$ (see Proposition 3.45), by combining the inequality

$$\lambda_{1,a_0} - \varepsilon > \frac{\underline{\lambda}_1(A, r_n)}{r_n^p} = \int_\Omega a_0(x)|\nabla v_n|^p\, dx + \int_\Omega \frac{\tilde{a}_0(x, |\nabla u_n|)|\nabla u_n|^2}{\|u_n\|_p^p}\, dx,$$

and an argument as in Theorem 3.52 with $\lambda_n = \underline{\lambda}_1(A,r_n)/r_n^p$, we arrive at a contradiction. □

Proposition 3.55. *Suppose* H$(A)_0$. *Then for every* $\varepsilon > 0$ *there exists* $r_1 > 0$ *such that*

$$\frac{\overline{\lambda}_1(A,r)}{r^p} \leq \lambda_{1,a_0} + \varepsilon \quad \textit{for every } 0 < r < r_1.$$

Proof. Assume that there exist $\varepsilon_0 > 0$ and $r_n > 0$ such that $r_n \to 0$ as $n \to \infty$ and $\overline{\lambda}_1(A, r_n)/r_n^p > \lambda_{1, a_0} + \varepsilon_0$ for every $n \in \mathbb{N}$. According to Lemma 3.46 and Proposition 3.45, we can find a positive function $u_n \in (r_n S) \cap C^1(\overline{\Omega})$ satisfying

$$\int_\Omega A(x, \nabla u_n) \cdot \nabla u_n \, dx = \overline{\lambda}_1(A, r_n),$$

$$\min_{v \in r_n S} \int_\Omega G(x, \nabla v) \, dx = \int_\Omega G(x, \nabla u_n) \, dx.$$

Recalling that φ_{a_0} denotes the L^p-normalized positive eigenfunction corresponding to λ_{1, a_0}, it turns out that

$$\begin{aligned} \frac{C_0}{p(p-1)} \|\nabla u_n\|_p^p \le \int_\Omega G(x, \nabla u_n) \, dx \le \int_\Omega G(x, r_n \nabla \varphi_{a_0}) \, dx \\ \le \frac{C_1 r_n^p}{p(p-1)} \|\nabla \varphi_{a_0}\|_p^p. \end{aligned}$$

We see that $u_n \to 0$ in $C^1(\overline{\Omega})$ due to Proposition 2.12 because u_n is a positive solution of (3.68) with $\lambda = \overline{\lambda}_1(A, r_n)/r_n^p$ and

$$\lambda_{1, a_0} + \varepsilon_0 < \frac{\overline{\lambda}_1(A, r_n)}{r_n^p} \le A_p \lambda_1$$

by Proposition 3.50, where λ_1 denotes the first eigenvalue of $-\Delta_p$ (see (3.78) for the definition of A_p). Therefore, through the same argument as in Theorem 3.53 with $\mu_n = \overline{\lambda}_1(A, r_n)/r_n^p$, we get at a contradiction. □

Noting that $\underline{\lambda}_1(A, r) \le \overline{\lambda}_1(A, r)$ for every $r > 0$, from Propositions 3.54 and 3.55 the following result is readily obtained.

Corollary 3.56. *Under* H$(A)_0$,

$$\lim_{r \to 0^+} \frac{\overline{\lambda}_1(A, r)}{r^p} = \lim_{r \to 0^+} \frac{\underline{\lambda}_1(A, r)}{r^p} = \lambda_{1, a_0}.$$

Next we get the following.

Proposition 3.57. *Under* H$(A)_0$,

$$\lim_{r \to 0^+} \frac{\mu_1(A, r)}{r^p} = \frac{\lambda_{1, a_0}}{p}.$$

Proof. Due to Proposition 3.45, for every $r > 0$ there exists a positive solution $u_r \in (rS) \cap C^1(\overline{\Omega})$ of (3.68) with $\lambda = \lambda_1(A, u_r)/r^p$ and $\mu_1(A, r) = J(u_r)$. Then we can prove that $u_r \to 0$ in $C^1(\overline{\Omega})$ as $r \to 0^+$ and $u_r/\|u_r\|_p$ is bounded in $W_0^{1,p}(\Omega)$ as $r \to 0^+$ by a similar reasoning to the one in Proposition 3.55 on the basis of Corollary 3.56.

Set

$$\tilde{G}_0(x, y) := \int_0^{|y|} \tilde{a}_0(x, t)t\, dt$$

for $y \in \mathbb{R}^N$. We point out that

$$\underline{\lambda}_1(A, r) \le \lambda_1(A, u_r) \le \overline{\lambda}_1(A, r)$$

and

$$\begin{aligned}\mu_1(A, r) &= \int_\Omega G(x, \nabla u_r)\, dx = \frac{1}{p}\int_\Omega a_0(x)|\nabla u_r|^p\, dx + \int_\Omega \tilde{G}_0(x, \nabla u_r)\, dx \\ &= \frac{\lambda_1(A, u_r)}{p} - \frac{1}{p}\int_\Omega \tilde{a}_0(x, |\nabla u_r|)|\nabla u_r|^2\, dx + \int_\Omega \tilde{G}_0(x, \nabla u_r)\, dx.\end{aligned}$$

Thus, by Corollary 3.56 and $r = \|u_r\|_p$, it suffices to prove that

$$\lim_{r\to 0^+} \int_\Omega \frac{\tilde{a}_0(x, |\nabla u_r|)|\nabla u_r|^2}{\|u_r\|_p^p}\, dx = 0 \quad \text{and} \quad \lim_{r\to 0^+} \int_\Omega \frac{\tilde{G}_0(x, \nabla u_r)}{\|u_r\|_p^p}\, dx = 0.$$

Now we fix any $\varepsilon > 0$. Then, by $\mathrm{H}(A)_0$, there exists $\delta > 0$ such that

$$|\tilde{a}_0(x, t)| \le \varepsilon t^{p-2} \quad \text{and} \quad |\tilde{G}_0(x, y)| \le \varepsilon \frac{|y|^p}{p} \quad \text{for every } 0 < t \le \delta,\ |y| \le \delta.$$

Because $u_r \to 0$ in $C^1(\overline{\Omega})$ as $r \to 0^+$, we may assume that $\|u_r\|_{C^1(\overline{\Omega})} \le \delta$ for sufficiently small $r > 0$. Therefore, we obtain

$$\left|\int_\Omega \frac{\tilde{a}_0(x, |\nabla u_r|)|\nabla u_r|^2}{\|u_r\|_p^p}\, dx\right| \le \varepsilon \frac{\|\nabla u_r\|_p^p}{\|u_r\|_p^p},$$

$$\left|\int_\Omega \frac{\tilde{G}_0(x, \nabla u_r)}{\|u_r\|_p^p}\, dx\right| \le \varepsilon \frac{\|\nabla u_r\|_p^p}{p\|u_r\|_p^p}.$$

Since $\|\nabla u_r\|_p / \|u_r\|_p$ is bounded as $r \to 0^+$ and $\varepsilon > 0$ is arbitrary, the desired conclusion follows. □

Our next objective is to examine the asymptotic homogeneity near $+\infty$.

We consider the case where A is asymptotically $(p-1)$-homogeneous near $+\infty$ in the following sense:

$\mathrm{H}(A)_\infty$ there exist a positive function $a_\infty \in C^1(\overline{\Omega})$ and a function $\tilde{a} \in C(\overline{\Omega} \times \mathbb{R})$ such that

$$A(x, y) = a_\infty(x)|y|^{p-2}y + \tilde{a}(x, |y|)y \quad \text{for every } x \in \Omega,\ y \in \mathbb{R}^N,$$

$$\lim_{t\to+\infty} \frac{\tilde{a}(x, t)}{t^{p-2}} = 0 \quad \text{uniformly in } x \in \overline{\Omega}.$$

For the weight function a_∞, we define

$$\lambda_{1,a_\infty} := \inf\left\{\int_\Omega a_\infty(x)|\nabla u|^p\,dx \;:\; \|u\|_p = 1\right\}.$$

Because

$$0 < \min_{x\in\overline{\Omega}} a_\infty(x) \le \max_{x\in\overline{\Omega}} a_\infty(x) < +\infty,$$

as for the first eigenvalue of $-\Delta_p$ we can prove:

(i) λ_{1,a_∞} is the first eigenvalue of

$$-\text{div}\left(a_\infty(x)|\nabla u|^{p-2}\nabla u\right) = \lambda|u|^{p-2}u \quad \text{in } \Omega, \quad u = 0 \quad \text{on } \partial\Omega; \tag{3.83}$$

(ii) λ_{1,a_∞} has a positive eigenfunction $\varphi_{a_\infty} \in C^1(\overline{\Omega})$ with $\|\varphi_{a_\infty}\|_p = 1$ and it is simple;

(iii) problem (3.83) has no nontrivial constant-sign solutions provided $\lambda \neq \lambda_{1,a_\infty}$.

Example 3.58. An example covering the desired asymptotical $(p-1)$-homogeneity near 0 and near $+\infty$ can be considered as follows:

$$A(x,y) = (a_0(x)|y|^{p-2} + a_\infty(x)|y|^{q-2})(1+|y|^q)^{\frac{p-q}{q}}\,y$$

for $1 < p \le q < +\infty$, $a_0, a_\infty \in C^1(\overline{\Omega})$ with $\min_{\overline{\Omega}} a_0 > 0$ and $\min_{\overline{\Omega}} a_\infty > 0$. This map A satisfies

$$\begin{aligned} A(x,y) - a_0(x)|y|^{p-2}y &= o(|y|^{p-1}) \quad \text{as } |y| \to 0, \\ A(x,y) - a_\infty(x)|y|^{p-2}y &= o(|y|^{p-1}) \quad \text{as } |y| \to +\infty. \end{aligned}$$

An auxiliary result is needed.

Lemma 3.59. *Assume* $\mathrm{H}(A)_\infty$ *and let* $\{u_n\} \subset W_0^{1,p}(\Omega)$ *be a sequence satisfying* $\|u_n\|_p \to +\infty$ *as* $n \to \infty$. *If* $v_n := u_n/\|u_n\|_p$ *is bounded in* $W_0^{1,p}(\Omega)$, *then the following assertions hold:*

(i) $\displaystyle\lim_{n\to\infty}\int_\Omega \frac{\tilde{a}(x,|\nabla u_n|)|\nabla u_n|^2}{\|u_n\|_p^p}\,dx = 0;$

(ii) $\displaystyle\lim_{n\to\infty}\int_\Omega \frac{\tilde{a}(x,|\nabla u_n|)\nabla u_n\cdot\nabla w}{\|u_n\|_p^{p-1}}\,dx = 0$ *for every* $w \in W_0^{1,p}(\Omega)$;

(iii) $\displaystyle\lim_{n\to\infty}\int_\Omega \frac{\tilde{G}(x,\nabla u_n)}{\|u_n\|_p^p}\,dx = 0$, *where* $\tilde{G}(x,y) := \int_0^{|y|}\tilde{a}(x,t)t\,dt$ *for* $y \in \mathbb{R}^N$.

Proof. (i) Fix $\varepsilon > 0$. By the property of $\tilde{a}$, there exist constants $R > 0$ and $C > 0$ such that

$$|\tilde{a}(x,t)| \le \varepsilon |t|^{p-2} \text{ if } t \ge R \quad \text{and} \quad |\tilde{a}(x,t)| \le C \text{ if } 0 \le t \le R. \tag{3.84}$$

Therefore, by (3.84) we deduce

$$\begin{aligned}\left| \int_\Omega \frac{\tilde{a}(x,|\nabla u_n|)|\nabla u_n|^2}{\|u_n\|_p^p} dx \right| &\le \int_{\{|\nabla u_n| > R\}} \varepsilon |\nabla v_n|^p \, dx + \int_{\{|\nabla u_n| \le R\}} \frac{C|\nabla u_n|^2}{\|u_n\|_p^p} dx \\ &\le \varepsilon \|\nabla v_n\|_p^p + \frac{CR^2|\Omega|}{\|u_n\|_p^p} \le \varepsilon D^p + \frac{CR^2|\Omega|}{\|u_n\|_p^p},\end{aligned}$$

where $D := \sup_n \|\nabla v_n\|_p$. Letting $n \to \infty$, we have

$$\limsup_{n\to\infty} \left| \int_\Omega \frac{\tilde{a}(x,|\nabla u_n|)|\nabla u_n|^2}{\|u_n\|_p^p} dx \right| \le \varepsilon D^p$$

because $\|u_n\|_p \to +\infty$ as $n \to \infty$. Since $\varepsilon > 0$ is arbitrary, the conclusion is achieved.

(ii) For any $\varepsilon > 0$ and $w \in W_0^{1,p}(\Omega)$, by Hölder's inequality and (3.84) we have

$$\begin{aligned}&\left| \int_\Omega \frac{\tilde{a}(x,|\nabla u_n|)\nabla u_n \cdot \nabla w}{\|u_n\|_p^{p-1}} dx \right| \\ &\le \int_{\{|\nabla u_n| > R\}} \varepsilon |\nabla v_n|^{p-1} |\nabla w| \, dx + \int_{\{|\nabla u_n| \le R\}} \frac{C|\nabla u_n||\nabla w|}{\|u_n\|_p^{p-1}} dx \\ &\le \varepsilon \|\nabla v_n\|_p^{p-1} \|\nabla w\|_p + \frac{CR|\nabla w\|_p |\Omega|^{(p-1)/p}}{\|u_n\|_p^{p-1}}.\end{aligned}$$

Combining this inequality and the argument in (i) leads to the conclusion.

(iii) It is straightforward to show that for every $\varepsilon > 0$ there exists a constant $C > 0$ such that

$$|\tilde{G}(x,y)| \le \varepsilon |y|^p + C \quad \text{for every } y \in \mathbb{R}^N.$$

Consequently,

$$\left| \int_\Omega \tilde{G}(x,\nabla u_n) \, dx \right| \le \varepsilon \|\nabla u_n\|_p^p + C|\Omega|$$

holds. This estimate implies the conclusion. □

Now we are able to set forth our existence and nonexistence results under condition $H(A)_\infty$. Recall that $B_p(R) = \{u \in W_0^{1,p}(\Omega) : \|u\|_p < R\}$.

Theorem 3.60. *Assume* $H(A)_\infty$. *Then for every* $\varepsilon > 0$ *there exists* $R_0 > 0$ *such that problem (3.68) has no solutions in* $W_0^{1,p}(\Omega) \setminus B_p(R_0)$ *provided* $\lambda < \lambda_{1,a_\infty} - \varepsilon$.

Proof. By way of contradiction, we assume that there exist $\varepsilon_0 > 0$, $\{\lambda_n\}$ and $\{u_n\}$ such that $\lambda_n < \lambda_{1,a_\infty} - \varepsilon_0$, $\lim_{n\to\infty} \|u_n\|_p = +\infty$ and u_n is a solution of (3.68) with $\lambda = \lambda_n$. Inserting u_n as a test function gives

$$\frac{C_0}{p-1}\|\nabla u_n\|_p^p \le \int_\Omega A(x,\nabla u_n)\cdot\nabla u_n\,dx = \lambda_n\|u_n\|_p^p \le (\lambda_{1,a_\infty} - \varepsilon_0)\|u_n\|_p^p$$

(refer to (iii) in Remark 3.43). Therefore, $v_n := u_n/\|u_n\|_p$ is bounded in $W_0^{1,p}(\Omega)$.

Again by taking $u_n/\|u_n\|_p^p$ as test function in (3.68) with $\lambda = \lambda_n$, we obtain

$$\begin{aligned}\lambda_{1,a_\infty} - \varepsilon_0 > \lambda_n &= \int_\Omega \frac{a_\infty(x)|\nabla u_n|^p}{\|u_n\|_p^p}\,dx + \int_\Omega \frac{\tilde{a}(x,|\nabla u_n|)|\nabla u_n|^2}{\|u_n\|_p^p}\,dx \\ &= \int a_\infty(x)|\nabla v_n|^p\,dx + \int_\Omega \frac{\tilde{a}(x,|\nabla u_n|)|\nabla u_n|^2}{\|u_n\|_p^p} \\ &\ge \lambda_{1,a_\infty} + o(1)\end{aligned}$$

due to the definition of λ_{1,a_∞} and Lemma 3.59 (i), which yields a contradiction. $\square$

Theorem 3.61. *Assume condition* H$(A)_\infty$. *Then for every* $\varepsilon > 0$ *there exists* $R_1 > 0$ *such that problem (3.68) has no constant-sign solutions in* $W_0^{1,p}(\Omega) \setminus B_p(R_1)$ *provided* $\lambda > \lambda_{1,a_\infty} + \varepsilon$.

Proof. Arguing indirectly, assume that there exist $\varepsilon_0 > 0$, $\{\mu_n\}$ and $\{u_n\}$ such that $\mu_n > \lambda_{1,a_\infty} + \varepsilon_0$, $\lim_{n\to\infty} \|u_n\|_p = +\infty$ and u_n is a constant-sign solution of (3.68) with $\lambda = \mu_n$. Because A is odd, we may suppose that $u_n \ge 0$ by considering $-u_n$ if necessary. Thus, by Proposition 2.12 and Theorem 2.19, we know that $u_n \in C^1(\overline{\Omega})$ and $u_n > 0$ in Ω. Here, we note that $\mu_n \le A_p\lambda_1$ holds by Proposition 3.50, where λ_1 denotes the first eigenvalue of $-\Delta_p$ (see (3.78) for the definition of A_p), and so $\{\mu_n\}$ is bounded. Hence, we may assume that $\{\mu_n\}$ converges to some $\mu_0 \in [\lambda_{1,a_\infty} + \varepsilon_0, A_p\lambda_1]$ passing to a relabeled subsequence. In addition, taking u_n as test function in (3.68) with $\lambda = \mu_n$, we obtain

$$\frac{C_0}{p-1}\|\nabla u_n\|_p^p \le \int_\Omega A(x,\nabla u_n)\cdot\nabla u_n\,dx = \mu_n\|u_n\|_p^p,$$

hence, letting $v_n := u_n/\|u_n\|_p$, the sequence $\{v_n\}$ is bounded in $W_0^{1,p}(\Omega)$. Thus, along a relabeled subsequence, we may suppose that $\{v_n\}$ converges to some v weakly in $W_0^{1,p}(\Omega)$ and strongly in $L^p(\Omega)$.

We claim that v is a positive solution of

$$-\mathrm{div}\left(a_\infty(x)|\nabla v|^{p-2}\nabla v\right) = \mu_0|v|^{p-2}v \quad \text{in } \Omega, \qquad v = 0 \quad \text{on } \partial\Omega. \tag{3.85}$$

If the claim holds, then necessarily $\mu_0 = \lambda_{1,a_\infty}$ is true because (3.83) has no positive solutions in the case of $\lambda \neq \lambda_{1,a_\infty}$. Hence, this contradicts

$$\lambda_{1,a_\infty} + \varepsilon_0 \le \lim_{n\to\infty} \mu_n = \mu_0.$$

Now we prove the claim. First, we show that $\{v_n\}$ converges to v strongly in $W_0^{1,p}(\Omega)$. Indeed, taking $(v_n - v)/\|u_n\|_p^{p-1}$ as test function in (3.68) with $\lambda = \mu_n$ results in

$$\begin{aligned}\mu_n \int_\Omega v_n^{p-1}(v_n - v)\,dx &= \int_\Omega a_\infty(x)|\nabla v_n|^{p-2}\nabla v_n \cdot \nabla(v_n - v)\,dx \\ &\quad + \int_\Omega \frac{\tilde{a}(x, |\nabla u_n|)\nabla u_n}{\|u_n\|_p^{p-1}} \cdot \nabla(v_n - v)\,dx \\ &= \int_\Omega a_\infty(x)|\nabla v_n|^{p-2}\nabla v_n \cdot \nabla(v_n - v)\,dx + o(1)\end{aligned}$$

as $n \to \infty$ due to (i) and (ii) in Lemma 3.59. Since $v_n \to v$ in $L^p(\Omega)$, this implies that

$$\int_\Omega a_\infty(x)|\nabla v_n|^{p-2}\nabla v_n \cdot \nabla(v_n - v)\,dx \to 0$$

as $n \to \infty$. Noting that

$$\begin{aligned}o(1) &= \int_\Omega a_\infty(x)\left(|\nabla v_n|^{p-2}\nabla v_n - |\nabla v|^{p-2}\nabla v\right) \cdot \nabla(v_n - v)\,dx \\ &\ge \left(\min_{\overline{\Omega}} a_\infty\right)\int_\Omega \left(|\nabla v_n|^{p-2}\nabla v_n - |\nabla v|^{p-2}\nabla v\right) \cdot \nabla(v_n - v)\,dx \\ &\ge \left(\min_{\overline{\Omega}} a_\infty\right)(\|\nabla v_n\|_p^{p-1} - \|\nabla v\|_p^{p-1})(\|\nabla v_n\|_p - \|\nabla v\|_p) \ge 0\end{aligned}$$

and

$$0 < \min_{\overline{\Omega}} a_\infty \le \max_{\overline{\Omega}} a_\infty < +\infty,$$

we have $v_n \to v$ in $W_0^{1,p}(\Omega)$. As a result, v is a solution of (3.85) by letting $n \to \infty$ in the equality

$$\begin{aligned}&\int_\Omega a_\infty(x)|\nabla v_n|^{p-2}\nabla v_n \cdot \nabla w\,dx + \int_\Omega \frac{\tilde{a}(x, |\nabla u_n|)\nabla u_n \cdot \nabla w}{\|u_n\|_p^{p-1}}\,dx \\ &= \mu_n \int_\Omega v_n^{p-1} w\,dx\end{aligned}$$

for every $w \in W_0^{1,p}(\Omega)$ (note that the second term converges to 0 by Lemma 3.59 (ii)). Since $v_n = u_n/\|u_n\|_p > 0$ in Ω, v is nonnegative, and so $\|v\|_p = 1$

and v is positive by Theorem 2.19. The claim is proven, which completes the proof. □

Proposition 3.62. *Suppose* H$(A)_\infty$. *Then for every* $\varepsilon > 0$ *there exists* $R_0 > 0$ *satisfying*

$$\frac{\underline{\lambda}_1(A,r)}{r^p} \geq \lambda_{1,a_\infty} - \varepsilon \quad \textit{for every } r > R_0.$$

Proof. Assume that there exist $\varepsilon_0 > 0$ and $r_n > 0$ such that $r_n \to +\infty$ as $n \to \infty$ and

$$\frac{\underline{\lambda}_1(A,r_n)}{r_n^p} < \lambda_{1,a_\infty} - \varepsilon_0$$

for every n. Using Proposition 3.45 and Lemma 3.46, we can choose a positive function $u_n \in (r_n S) \cap C^1(\overline{\Omega})$ satisfying

$$\int_\Omega A(x, \nabla u_n) \cdot \nabla u_n \, dx = \underline{\lambda}_1(A, r_n),$$

$$\min_{v \in r_n S} \int_\Omega G(x, \nabla v)\, dx = \int_\Omega G(x, \nabla u_n)\, dx.$$

Notice that

$$\frac{C_0}{p-1} \|\nabla u_n\|_p^p \leq \int_\Omega A(x, \nabla u_n) \cdot \nabla u_n \, dx = \underline{\lambda}_1(A, r_n) < (\lambda_{1,a_\infty} - \varepsilon_0) r_n^p,$$

so $\{u_n/r_n\}$ is bounded in $W_0^{1,p}(\Omega)$. Since u_n is a solution of (3.68) with $\lambda = \underline{\lambda}_1(A,r_n)/r_n^p$ (see Proposition 3.45), the same argument as in Theorem 3.60 with $\mu_n = \underline{\lambda}_1(A,r_n)/r_n^p$ yields a contradiction. □

Proposition 3.63. *Suppose* H$(A)_\infty$. *Then for every* $\varepsilon > 0$ *there exists* $R_1 > 0$ *such that*

$$\frac{\overline{\lambda}_1(A,r)}{r^p} \leq \lambda_{1,a_\infty} + \varepsilon \quad \textit{for every } r > R_1.$$

Proof. Assume that there exist $\varepsilon_0 > 0$ and $r_n > 0$ satisfying $r_n \to +\infty$ as $n \to \infty$ and $\overline{\lambda}_1(A,r_n)/r_n^p > \lambda_{1,a_\infty} + \varepsilon_0$ for every n. According to Proposition 3.45 and Lemma 3.46, there is a positive function $u_n \in (r_n S) \cap C^1(\overline{\Omega})$ with the property

$$\int_\Omega A(x, \nabla u_n) \cdot \nabla u_n \, dx = \overline{\lambda}_1(A, r_n),$$

$$\min_{v \in r_n S} \int_\Omega G(x, \nabla v)\, dx = \int_\Omega G(x, \nabla u_n)\, dx.$$

Using φ_{a_∞} introduced below (3.83), we note that

$$\frac{C_0}{p(p-1)}\|\nabla u_n\|_p^p \le \int_\Omega G(x,\nabla u_n)\,dx \le \int_\Omega G(x, r_n\nabla\varphi_{a_\infty})\,dx$$
$$\le \frac{C_1 r_n^p}{p(p-1)}\|\nabla\varphi_{a_\infty}\|_p^p.$$

Consequently, we obtain that $\{u_n/r_n\}$ is bounded in $W_0^{1,p}(\Omega)$. Since u_n is a positive solution of (3.68) with $\lambda = \overline{\lambda}_1(A, r_n)/r_n^p$, the same argument as in Theorem 3.61 with $\mu_n = \overline{\lambda}_1(A, r_n)/r_n^p$ leads to a contradiction. □

Propositions 3.62 and 3.63 imply the following result.

Corollary 3.64. *Under* H$(A)_\infty$,

$$\lim_{r\to+\infty}\frac{\overline{\lambda}_1(A,r)}{r^p} = \lim_{r\to+\infty}\frac{\underline{\lambda}_1(A,r)}{r^p} = \lambda_{1,a_\infty}.$$

Proposition 3.65. *Under* H$(A)_\infty$, *we have* $\displaystyle\lim_{r\to+\infty}\frac{\mu_1(A,r)}{r^p} = \frac{\lambda_{1,a_\infty}}{p}$.

Proof. Due to Proposition 3.45, for every $r > 0$ there exists a positive solution $u_r \in (rS)\cap C^1(\overline{\Omega})$ of (3.68) with $\lambda = \lambda_1(A,u_r)/r^p$ and $\mu_1(A,r) = \int_\Omega G(x,\nabla u_r)\,dx$. Then, $\{u_r/r\}$ is bounded in $W_0^{1,p}(\Omega)$ as seen from

$$\frac{C_0}{p(p-1)}\|\nabla u_r\|_p^p \le \int_\Omega G(x,\nabla u_r)\,dx \le \int_\Omega G(x, r\nabla w)\,dx$$
$$\le \frac{r^p C_1}{p(p-1)}\|\nabla w\|_p^p$$

for any $w \in W_0^{1,p}(\Omega)$ with $\|w\|_p = 1$.

Set

$$\tilde{G}(x,y) := \int_0^{|y|}\tilde{a}(x,t)t\,dx$$

for all $y \in \mathbb{R}^N$. Notice that

$$\underline{\lambda}_1(A,r) \le \lambda_1(A,u_r) \le \overline{\lambda}_1(A,r)$$

and
$$\mu_1(A,r) = \int_\Omega G(x,\nabla u_r)\,dx$$
$$= \frac{1}{p}\int_\Omega a_\infty(x)|\nabla u_r|^p\,dx + \int_\Omega \tilde{G}(x,\nabla u_r)\,dx$$
$$= \frac{\lambda_1(A,u_r)}{p} - \frac{1}{p}\int_\Omega \tilde{a}(x,|\nabla u_r|)|\nabla u_r|^2\,dx + \int_\Omega \tilde{G}(x,\nabla u_r)\,dx.$$

According to Corollary 3.64 and Lemma 3.59 (i) and (iii) (note that $\|u_r\|_p = r \to +\infty$), our conclusion is achieved. □

The rest of this section is devoted to existence results of positive solutions to the equation

$$\begin{cases} -\operatorname{div} A(x, \nabla u) = f(x, u) & \text{in } \Omega, \\ u = 0 & \text{on } \partial\Omega. \end{cases} \tag{3.86}$$

The nonlinear term f is supposed to satisfy:

H(f) f is a Carathéodory function on $\Omega \times \mathbb{R}$ with $f(x, 0) = 0$ for a.e. $x \in \Omega$, f is bounded on bounded sets and

$$\lim_{s \to 0^+} \frac{f(x, s)}{s^{p-1}} = \alpha_0 \quad \text{uniformly for a.e. } x \in \Omega, \tag{3.87}$$

$$\lim_{s \to +\infty} \frac{f(x, s)}{s^{p-1}} = \alpha \quad \text{uniformly for a.e. } x \in \Omega, \tag{3.88}$$

for some constants α_0 and α.

Theorem 3.66. *Assume* H$(A)_0$, H$(A)_\infty$, *and* H(f). *If one of the following conditions:*

(i) $\alpha_0 > \lambda_{1, a_0}$ *and* $\alpha < \lambda_{1, a_\infty}$*;*
(ii) $\alpha_0 < \lambda_{1, a_0}$ *and* $\alpha > \lambda_{1, a_\infty}$*,*

where λ_{1, a_0} *and* λ_{1, a_∞} *are the first eigenvalues obtained by* (3.80) *and* (3.83), *respectively, is true, then problem (3.86) has at least one positive solution.*

In particular, this ensures the existence of a whole interval of eigenvalues for our eigenvalue problem (3.68) because we can apply Theorem 3.66 to the function $f(x, s) = \lambda |s|^{p-2} s$.

Corollary 3.67. *Assume* H$(A)_0$, H$(A)_\infty$, *and* $\lambda_{1, a_0} \neq \lambda_{1, a_\infty}$. *Then for every* λ *between* λ_{1, a_0} *and* λ_{1, a_∞}, *(3.68) has a nontrivial (positive) solution, therefore,* λ *is an eigenvalue.*

In order to prove Theorem 3.66, we develop a variational approach relying on the C^1 functional I on $W_0^{1,p}(\Omega)$ defined by

$$I(u) := \int_\Omega G(x, \nabla u)\, dx - \int_\Omega F_+(x, u)\, dx \quad \text{for } u \in W_0^{1,p}(\Omega),$$

where

$$F_+(x, t) := \int_0^t f_+(x, s)\, ds \quad \text{and}$$
$$f_+(x, t) = f(x, t) \text{ if } t \geq 0, \quad f_+(x, t) = 0 \text{ if } t \leq 0.$$

Remark 3.68. If $u \in W_0^{1,p}(\Omega)$ is a nontrivial critical point of I, then u is a positive solution of problem (3.86). Indeed, by acting with $-u^-$ as a test function, we obtain

$$0 = \langle I'(u), -u^- \rangle = \int_\Omega A(x, \nabla u) \cdot (-\nabla u^-)\, dx - \int_\Omega f_+(x, u)(-u^-)\, dx$$
$$= \int_\Omega A(x, \nabla u) \cdot (-\nabla u^-)\, dx \geq \frac{C_0}{p-1} \|\nabla u^-\|_p^p.$$

Thus, $u \geq 0$ holds. By Theorem 2.19, since $u \not\equiv 0$ and $f_+(x, u) = f(x, u)$, we see that u is a positive solution of (3.86).

From now on in this section, we assume that hypothesis H(f) is satisfied.

Lemma 3.69. *If $\alpha \neq \lambda_{1, a_\infty}$, then the functional I satisfies the Palais–Smale condition.*

Proof. Let $\{u_n\}$ be a Palais–Smale sequence of I, that is,

$$I(u_n) \to c \quad \text{and} \quad \|I'(u_n)\|_{W^{-1,p'}(\Omega)} \to 0 \quad \text{as } n \to \infty$$

for some $c \in \mathbb{R}$. It is sufficient to prove the boundedness of $\{u_n\}$ in $W_0^{1,p}(\Omega)$ because the operator $V(u) = -\operatorname{div} A(x, \nabla u)$ on $W_0^{1,p}(\Omega)$ has the $(S)_+$-property (see Proposition 3.44) and the embedding of $W_0^{1,p}(\Omega)$ into $L^p(\Omega)$ is compact. Then, thanks to the inequality

$$\frac{C_0}{p(p-1)} \|\nabla u_n\|_p^p \leq \int_\Omega G(x, \nabla u_n)\, dx = I(u_n) + \int_\Omega F_+(x, u_n)\, dx$$
$$\leq I(u_n) + C\|u_n\|_p^p, \tag{3.89}$$

it is sufficient to prove the boundedness of $\{u_n\}$ in $L^p(\Omega)$. Seeking a contradiction, assume that $\|u_n\|_p \to +\infty$ as $n \to \infty$ by choosing a subsequence if necessary. Set $v_n := u_n/\|u_n\|_p$. Inequality (3.89) provides that $\{v_n\}$ is bounded in $W_0^{1,p}(\Omega)$. Hence, we may suppose, by choosing a subsequence, that $v_n \rightharpoonup v_0$ in $W_0^{1,p}(\Omega)$ and $v_n \to v_0$ in $L^p(\Omega)$ for some v_0.

We see that $v_0 \geq 0$ a.e. in Ω. Indeed, acting with $-(u_n)^-$ as a test function shows that

$$o(1)\|\nabla(u_n)^-\|_p = \langle I'(u_n), -(u_n)^- \rangle = \int_\Omega A(x, \nabla u_n) \cdot (-\nabla(u_n)^-)\, dx$$
$$\geq \frac{C_0}{p-1} \|\nabla(u_n)^-\|_p^p.$$

Since $p > 1$, we get $\|\nabla(u_n)^-\|_p \to 0$ as $n \to \infty$. Thus, $(v_n)^- \to 0$ in $W_0^{1,p}(\Omega)$, and hence $(v_0)^- = 0$ a.e. $x \in \Omega$.

Let us prove that

$$\lim_{n\to\infty} \frac{\|f_+(\cdot, u_n) - \alpha((u_n)^-)^{p-1}\|_{p'}}{\|u_n\|_p^{p-1}} = 0, \tag{3.90}$$

where $p' = p/(p-1)$. Let an arbitrary $\varepsilon > 0$. It follows from (3.88) that there exists a constant $C_\varepsilon > 0$ such that

$$|f(x,s) - \alpha s^{p-1}| \le \varepsilon |s|^{p-1} + C_\varepsilon \quad \text{for every } s \ge 0, \text{ a.e. } x \in \Omega.$$

Then we obtain

$$\int_\Omega |f_+(x,u_n) - \alpha((u_n)^+)^{p-1}|^{p'}\,dx \le 2^{p'-1}(\varepsilon^{p'-1}\|(u_n)^+\|_p^p + C_\varepsilon^{p'-1}|\Omega|).$$

Since $\|u_n\|_p \to +\infty$ as $n \to \infty$, this shows that (3.90) holds true because $\varepsilon > 0$ is arbitrary.

We know from Lemma 3.59 that

$$\begin{aligned} &\lim_{n\to\infty} \int_\Omega \frac{\tilde{a}(x, |\nabla u_n|)\nabla u_n}{\|u_n\|_p^{p-1}} \cdot \nabla(v_n - v_0)\,dx \\ &= \lim_{n\to\infty} \int_\Omega \frac{\tilde{a}(x, |\nabla u_n|)\nabla u_n}{\|u_n\|_p^{p-1}} \cdot \nabla\varphi\,dx = 0 \end{aligned} \tag{3.91}$$

for every $\varphi \in W_0^{1,p}(\Omega)$. Thus, by considering

$$o(1) = \frac{\langle I'(u_n), v_n - v_0\rangle}{\|u_n\|_p^{p-1}} = \int_\Omega a_\infty(x)|\nabla v_n|^{p-2}\nabla v_n \cdot \nabla(v_n - v_0)\,dx + o(1),$$

we see that $\{v_n\}$ strongly converges to v_0 in $W_0^{1,p}(\Omega)$ (refer to the $(S)_+$-property of $u \mapsto -\mathrm{div}\,(a_\infty(x)|\nabla u|^{p-2}\nabla u)$). Therefore, passing to the limit in $o(1) = \langle I'(u_n), \varphi\rangle/\|u_n\|_p^{p-1}$ for any $\varphi \in W_0^{1,p}(\Omega)$ and taking into account (3.90) and (3.91), we infer that v_0 is a nontrivial solution (note that $\|v_0\|_p = 1$) of

$$-\mathrm{div}\,(a_\infty(x)|\nabla u|^{p-2}\nabla u) = \alpha|u|^{p-2}u \quad \text{in } \Omega, \quad u = 0 \quad \text{on } \partial\Omega.$$

Since $v_0 \ge 0$ a.e. $x \in \Omega$, v_0 is a positive solution of (3.83) with $\lambda = \alpha$. This implies that $\alpha = \lambda_{1,a_\infty}$ because (3.83) has no positive solutions if $\lambda \ne \lambda_{1,a_\infty}$, which contradicts that $\alpha \ne \lambda_{1,a_\infty}$. Hence, $\{u_n\}$ is bounded in $L^p(\Omega)$, which completes the proof. □

Lemma 3.70. *Assume* $\mathrm{H}(A)_\infty$ *and that* $\alpha < \lambda_{1,a_\infty}$. *Then the functional* I *is coercive, bounded below, and weakly lower semicontinuous on* $W_0^{1,p}(\Omega)$.

Proof. Because of $\alpha < \lambda_{1,a_\infty}$, we can find sufficiently small constants $\varepsilon > 0$ and $0 < \delta < 1$ satisfying

$$(1-\delta)(\lambda_{1,a_\infty} - \varepsilon) > \alpha + \varepsilon. \tag{3.92}$$

By (3.88), there exists $C > 0$ such that $|F_+(x,s)| \le (\alpha+\varepsilon)s^p/p + C$ for every $s \ge 0$, a.e. $x \in \Omega$. Due to Proposition 3.65 and the definition of $\mu_1(A,r)$, there exists $R > 0$ such that for every $u \in W_0^{1,p}(\Omega)$ with $\|u\|_p \ge R$ it holds

$$\int_\Omega G(x,\nabla u)\,dx \ge \mu_1(A,\|u\|_p) \ge \frac{\lambda_{1,a_\infty} - \varepsilon}{p}\|u\|_p^p. \tag{3.93}$$

Hence, for every $u \in W_0^{1,p}(\Omega)$ with $|u\|_p \ge R$, by (3.70), (3.92), and (3.93) we obtain

$$\begin{aligned} I(u) &\ge \frac{(1-\delta)(\lambda_{1,a_\infty}-\varepsilon)}{p}\|u\|_p^p + \frac{\delta C_0}{p(p-1)}\|\nabla u\|_p^p - \frac{\alpha+\varepsilon}{p}\|u^+\|_p^p - C|\Omega| \\ &\ge \frac{\delta C_0}{p(p-1)}\|\nabla u\|_p^p - C|\Omega|. \end{aligned}$$

This shows that I is coercive. Moreover, because I is bounded below on $B_p(R)$, we see that I is bounded below on $W_0^{1,p}(\Omega)$. Since J is weakly lower semicontinuous (see the proof of Proposition 3.45) and $W_0^{1,p}(\Omega) \hookrightarrow L^p(\Omega)$ is compact, I is weakly lower semicontinuous on $W_0^{1,p}(\Omega)$. □

Lemma 3.71. *Assume* H$(A)_0$ *and* $\alpha_0 < \lambda_{1,a_0}$ *with* α_0 *in (3.87). Let* $p < q \le p^*$, *where* $p^* = Np/(N-p)$ *if* $N > p$, $p^* = +\infty$ *if* $N \le p$. *Then there exists* $\rho_0 > 0$ *such that*

$$\inf\{I(u) : \|u\|_q = \rho\} > 0 \quad \textit{for every } 0 < \rho < \rho_0.$$

Proof. In view of the hypothesis $\alpha_0 < \lambda_{1,a_0}$, we can take sufficiently small constants $\varepsilon > 0$ and $0 < \delta < 1$ satisfying

$$(1-\delta)(\lambda_{1,a_0} - \varepsilon) > \alpha_0 + \varepsilon. \tag{3.94}$$

According to Proposition 3.57, there exists $r_0 > 0$ such that

$$\frac{\mu_1(A,r)}{r^p} \ge \frac{\lambda_{1,a_0} - \varepsilon}{p} \quad \text{for every } 0 < r < r_0. \tag{3.95}$$

In addition, hypothesis H(f) guarantees the existence of $D_q > 0$ satisfying

$$F_+(x,s) \le \frac{\alpha_0+\varepsilon}{p}s^p + D_q s^q \quad \text{for every } s \ge 0, \text{ a.e. } x \in \Omega. \tag{3.96}$$

Since the embedding $W_0^{1,p}(\Omega) \hookrightarrow L^q(\Omega)$ is continuous, there is a constant $C_q > 0$ such that $\|u\|_q \le C_q \|\nabla u\|_p$ for every $u \in W_0^{1,p}(\Omega)$. We choose a constant $\rho > 0$ satisfying

$$\rho < \min\left\{ r_0|\Omega|^{1/q-1/p}, \left(\frac{\delta C_0}{2p(p-1)D_q C_q^p} \right)^{1/(q-p)} \right\} =: \rho_0. \tag{3.97}$$

Note that $\|u\|_p < r_0$ if $\|u\|_q = \rho$ by Hölder's inequality and (3.97). Therefore, whenever $\|u\|_q = \rho$, we have

$$\begin{aligned}
I(u) &= (1-\delta)\int_\Omega G(x,\nabla u)\,dx + \delta \int_\Omega G(x,\nabla u)\,dx - \int_\Omega F_+(x,u)\,dx \\
&\ge (1-\delta)\frac{\mu_1(A,\|u\|_p)}{\|u\|_p^p}\|u\|_p^p + \frac{\delta C_0}{p(p-1)}\|\nabla u\|_p^p \\
&\quad - \frac{\alpha_0+\varepsilon}{p}\|u^+\|_p^p - D_q\|u^+\|_q^q \\
&\ge \frac{1}{p}\left((1-\delta)(\lambda_{1,a_0}-\varepsilon)-\alpha_0-\varepsilon\right)\|u\|_p^p \\
&\quad + \left(\frac{\delta C_0}{p(p-1)C_q^p} - D_q\|u\|_q^{q-p}\right)\|u\|_q^p \\
&\ge \frac{\delta C_0}{2p(p-1)C_q^p}\rho^p
\end{aligned}$$

by the definition of $\mu_1(A,r)$, (3.70), (3.96), (3.95), (3.94), and (3.97). The conclusion is ensued. □

Proof of Theorem 3.66 (i). Lemma 3.70 guarantees the existence of a global minimizer of the functional I. Thus it suffices to prove that

$$\min_{W_0^{1,p}(\Omega)} I < 0$$

in order to show the existence of a nontrivial critical point of I. Choose a positive constant $\varepsilon > 0$ such that $\alpha_0 > \lambda_{1,a_0} + 2\varepsilon$. Let $\varphi_{a_0} \in C^1(\overline{\Omega})$ be a positive eigenfunction corresponding to λ_{1,a_0} with $\|\varphi_{a_0}\|_p = 1$ (refer to the text below (3.80)). It is seen that

$$\int_\Omega \tilde{G}_0(x, r\nabla\varphi_{a_0})\,dx/r^p \to 0$$

as $r \to 0^+$ (refer to the proof of Proposition 3.57). Hence there exists $r_0 > 0$ such that

$$\int_\Omega G(x, r\nabla\varphi_{a_0})\,dx = \frac{r^p}{p}\int_\Omega a_0(x)|\nabla\varphi_{a_0}|^p\,dx + r^p\int_\Omega \frac{\tilde{G}_0(x, r\nabla\varphi_{a_0})}{r^p}\,dx$$

$$\leq \frac{\lambda_{1,a_0}+\varepsilon}{p} r^p = \frac{\lambda_{1,a_0}+\varepsilon}{p} \|r\varphi_{a_0}\|_p^p \tag{3.98}$$

for every $0 < r < r_0$. On the other hand, it follows from (3.87) that there exists $\delta > 0$ such that

$$F_+(x,s) \geq \frac{\alpha_0 - \varepsilon}{p} s^p \quad \text{for every } s \in [0,\delta], \text{ a.e. } x \in \Omega. \tag{3.99}$$

Therefore, by (3.98) and (3.99), for every $0 < r < \min\{r_0, \delta/\|\varphi_{a_0}\|_\infty\}$ we have

$$I(ru_0) \leq \frac{r^p}{p}(\lambda_{1,a_0} + 2\varepsilon - \alpha_0)\|\varphi_{a_0}\|_p^p < 0$$

(recall that $\lambda_{1,a_0} + 2\varepsilon - \alpha_0 < 0$), whence $\min_{W_0^{1,p}(\Omega)} I < 0$. □

Proof of Theorem 3.66 (ii). Let $p < q \leq p^*$. By Lemma 3.71, we find $\rho > 0$ for which one has

$$\delta_0 := \inf\{I(u) : \|u\|_q = \rho\} > 0.$$

We claim the existence of $w \in W_0^{1,p}(\Omega)$ such that

$$\|w\|_q > \rho \quad \text{and} \quad I(w) < \delta_0. \tag{3.100}$$

Admitting the validity of this claim, we introduce

$$\Gamma := \left\{\gamma \in C([0,1], W_0^{1,p}(\Omega)) : \gamma(0) = 0,\ \gamma(1) = w\right\} \quad \text{and}$$
$$c := \inf_{\gamma\in\Gamma} \max_{t\in[0,1]} I(\gamma(t)).$$

Obviously, $\Gamma \neq \emptyset$ and $\gamma([0,1]) \cap \{u \in W_0^{1,p}(\Omega) : \|u\|_q = \rho\} \neq \emptyset$ for every $\gamma \in \Gamma$ since $W_0^{1,p}(\Omega) \hookrightarrow L^q(\Omega)$ is continuous. Thus, the mountain pass theorem (see Theorem 1.44) guarantees that $c(\geq \delta_0)$ is a nontrivial critical value of I because I satisfies the Palais–Smale condition by Lemma 3.69.

Finally, we prove the existence of w satisfying (3.100). Since $\alpha > \lambda_{1,a_\infty}$, we can choose $\varepsilon_0 > 0$ such that

$$\alpha > \lambda_{1,a_\infty} + 2\varepsilon_0. \tag{3.101}$$

Let $\varphi_{a_\infty} \in C^1(\overline{\Omega})$ be a positive eigenfunction corresponding to λ_{1,a_∞} with $\|\varphi_{a_\infty}\|_p = 1$ (refer to the text below (3.83)). It follows from Lemma 3.59 (iii) that

$$\int_\Omega \tilde{G}(x, r\nabla\varphi_{a_\infty})\,dx/r^p \to 0$$

as $r \to +\infty$ (note that $\|r\varphi_{a_\infty}\|_p = r$). Hence there exists $R_0 > 0$ such that

$$\int_\Omega G(x, r\nabla\varphi_{a_\infty})\,dx = \frac{r^p}{p}\int_\Omega a_\infty(x)|\nabla\varphi_{a_\infty}|^p\,dx + r^p\int_\Omega \frac{\tilde{G}_0(x, r\nabla\varphi_{a_\infty})}{r^p}\,dx$$
$$\le \frac{\lambda_{1,a_\infty} + \varepsilon_0}{p} r^p = \frac{\lambda_{1,a_\infty} + \varepsilon_0}{p}\|r\varphi_{a_\infty}\|_p^p \tag{3.102}$$

for every $r \ge R_0$. In addition, it follows from (3.88) that there exists $D > 0$ such that

$$F_+(x, u) \ge \frac{\alpha - \varepsilon_0}{p} u^p - D \quad \text{for every } u \ge 0, \text{ a.e. } x \in \Omega. \tag{3.103}$$

Consequently, from (3.101), (3.102), and (3.103), we obtain

$$I(r\varphi_{a_0}) \le \frac{r^p}{p}(\lambda_{1,a_\infty} + 2\varepsilon_0 - \alpha)\|\varphi_{a_0}\|_p^p + D|\Omega| \to -\infty$$

as $r \to +\infty$. This implies the existence of w satisfying (3.100). □

In order to consider resonant cases, we introduce the following hypotheses on

$$\tilde{G}(x, y) = \int_0^{|y|} \tilde{a}(x, t)t\,dt$$

and

$$\tilde{G}_0(x, y) = \int_0^{|y|} \tilde{a}_0(x, t)t\,dt,$$

where $\tilde{a}$ and $\tilde{a}_0$ are the functions in hypotheses $\mathrm{H}(A)_\infty$ and $\mathrm{H}(A)_0$, respectively.

$H(\mathcal{R})^+$ There exist constants $1 \le q < p$ and $H_0 > 0$ such that

$$\lim_{|y|\to+\infty} \frac{p\tilde{G}(x, y) - \tilde{a}(x, |y|)|y|^2}{|y|^q} = +\infty \quad \text{for a.e. } x \in \Omega,$$
$$p\tilde{G}(x, y) - \tilde{a}(x, |y|)|y|^2 \ge -H_0(1 + |y|^q) \quad \text{for a.e. } x \in \Omega, \text{ every } y \in \mathbb{R}^N,$$
$$\text{and} \quad f(x, t)t - pF(x, t) \ge -H_0(1 + t^q) \quad \text{for a.e. } x \in \Omega, \text{ every } t \ge 0.$$

$H(\mathcal{R})^-$ There exist constants $1 \le q < p$ and $H_0 > 0$ such that

$$\lim_{|y|\to+\infty} \frac{p\tilde{G}(x, y) - \tilde{a}(x, |y|)|y|^2}{|y|^q} = -\infty \quad \text{for a.e. } x \in \Omega,$$
$$p\tilde{G}(x, y) - \tilde{a}(x, |y|)|y|^2 \le H_0(1 + |y|^q) \quad \text{for a.e. } x \in \Omega, \text{ every } y \in \mathbb{R}^N,$$
$$\text{and} \quad f(x, t)t - pF(x, t) \le H_0(t^q + 1) \quad \text{for a.e. } x \in \Omega, \text{ every } t \ge 0.$$

$\widetilde{H}(\mathcal{R})^+$ There exist constants $1 \le q < p$ and $H_0 > 0$ such that

$$p\tilde{G}(x, y) - \tilde{a}(x, |y|)|y|^2 \ge -H_0(1 + |y|^q) \quad \text{for a.e. } x \in \Omega, \text{ every } y \in \mathbb{R}^N,$$
$$f(x,t)t - pF(x,t) \ge -H_0(1 + t^q) \quad \text{for a.e. } x \in \Omega, \text{ every } t \ge 0,$$
$$\text{and} \quad \lim_{t \to +\infty} \frac{f(x,t)t - pF(x,t)}{t^q} = +\infty \quad \text{for a.e. } x \in \Omega.$$

$\widetilde{H}(\mathcal{R})^-$ There exist constants $1 \le q < p$ and $H_0 > 0$ such that

$$p\tilde{G}(x, y) - \tilde{a}(x, |y|)|y|^2 \le H_0(1 + |y|^q) \quad \text{for a.e. } x \in \Omega, \text{ every } y \in \mathbb{R}^N,$$
$$f(x,t)t - pF(x,t) \le H_0(1 + t^q) \quad \text{for a.e. } x \in \Omega, \text{ every } t \ge 0,$$
$$\text{and} \quad \lim_{t \to +\infty} \frac{f(x,t)t - pF(x,t)}{t^q} = -\infty \quad \text{for a.e. } x \in \Omega.$$

$H(\mathcal{R})_0^+$ There exist constants $p \le r < p^*$ and $H_0 > 0$ such that

$$\lim_{|y| \to 0} \frac{p\tilde{G}_0(x, y) - \tilde{a}_0(x, |y|)|y|^2}{|y|^r} = +\infty \quad \text{for a.e. } x \in \Omega,$$
$$p\tilde{G}_0(x, y) - \tilde{a}_0(x, |y|)|y|^2 \ge -H_0|y|^r \quad \text{for a.e. } x \in \Omega, \text{ every } |y| \le 1,$$
$$\text{and} \quad f(x,t)t - pF(x,t) \ge -H_0 t^r \quad \text{for a.e. } x \in \Omega, \text{ every } t \in [0, 1].$$

$H(\mathcal{R})_0^-$ There exist constants $p \le r < p^*$ and $H_0 > 0$ such that

$$\lim_{|y| \to 0} \frac{p\tilde{G}_0(x, y) - \tilde{a}_0(x, |y|)|y|^2}{|y|^r} = -\infty \quad \text{for a.e. } x \in \Omega,$$
$$p\tilde{G}_0(x, y) - \tilde{a}_0(x, |y|)|y|^2 \le H_0|y|^r \quad \text{for a.e. } x \in \Omega, \text{ every } |y| \le 1,$$
$$\text{and} \quad f(x,t)t - pF(x,t) \le H_0 t^r \quad \text{for a.e. } x \in \Omega, \text{ every } t \in [0, 1].$$

$\widetilde{H}(\mathcal{R})_0^+$ There exist constants $p \le r < p^*$ and $H_0 > 0$ such that

$$p\tilde{G}_0(x, y) - \tilde{a}_0(x, |y|)|y|^2 \ge -H_0|y|^r \quad \text{for a.e. } x \in \Omega, \text{ every } |y| \le 1,$$
$$f(x,t)t - pF(x,t) \ge -H_0 t^r \quad \text{for a.e. } x \in \Omega, \text{ every } t \in [0, 1],$$
$$\text{and} \quad \lim_{t \to 0^+} \frac{f(x,t)t - pF(x,t)}{t^r} = +\infty \quad \text{for a.e. } x \in \Omega.$$

$\widetilde{H}(\mathcal{R})_0^-$ There exist constants $p \le r < p^*$ and $H_0 > 0$ such that

$$p\tilde{G}_0(x, y) - \tilde{a}_0(x, |y|)|y|^2 \le H_0|y|^r \quad \text{for a.e. } x \in \Omega, \text{ every } |y| \le 1,$$
$$f(x,t)t - pF(x,t) \le H_0 t^r \quad \text{for a.e. } x \in \Omega, \text{ every } t \in [0, 1],$$
$$\text{and} \quad \lim_{t \to 0^+} \frac{f(x,t)t - pF(x,t)}{t^r} = -\infty \quad \text{for a.e. } x \in \Omega.$$

We discuss preliminary results needed in the sequel. Set

$$f_{\pm n}(x,t) := f(x,t) \pm \frac{p}{n}|t|^{p-2}t,$$

and define approximate functionals on $W_0^{1,p}(\Omega)$ by

$$I_{\pm n}(u) := \int_\Omega G(x,\nabla u)\,dx - \int_\Omega (F_{\pm n})_+(x,u)\,dx = I(u) \mp \frac{1}{n}\|u^+\|_p^p.$$

From now on, we assume that f satisfies H(f) (in particular (3.87)–(3.88) hold).

Lemma 3.72. *If either $H(\mathcal{R})^+$ or $\widetilde{H}(\mathcal{R})^+$* (resp. either $H(\mathcal{R})^-$ or $\widetilde{H}(\mathcal{R})^-$) *holds and $\{u_n\}$ satisfies*

$$\sup_{n\in\mathbb{N}} I_{\pm n}(u_n) < +\infty \quad \textit{and} \quad \lim_{n\to\infty} \|I'_{\pm n}(u_n)\|_{W^{-1,p'}(\Omega)} = 0,$$

$$\left(\textit{resp.} \ \inf_{n\in\mathbb{N}} I_{\pm n}(u_n) > -\infty \quad \textit{and} \quad \lim_{n\to\infty} \|I'_{\pm n}(u_n)\|_{W^{-1,p'}(\Omega)} = 0 \right),$$

then $\{u_n\}$ is bounded in $W_0^{1,p}(\Omega)$.

Proof. We note that the boundedness of $\{\|u_n\|_p\}$ guarantees that $\{\|u_n\|\}$ is bounded because

$$o(1)\|u_n\| = \langle I'_{\pm n}(u_n), u_n\rangle \geq \frac{C_0}{p-1}\|u_n\|^p - C(1+\|u_n\|_p^p) \mp \frac{1}{n}\|(u_n)^+\|_p^p$$

for some $C>0$ independent of n. Reasoning by contradiction, assume that $\|u_n\|_p \to +\infty$ as $n\to\infty$. Let $v_n := u_n/\|u_n\|_p$. Then by the same argument as in Lemma 3.69, we see that $\{v_n\}$ has a subsequence strongly convergent to a positive solution v_0 of

$$-\mathrm{div}\,(a_\infty(x)|\nabla u|^{p-2}\nabla u) = \alpha|u|^{p-2}u \quad \text{in } \Omega, \quad u=0 \quad \text{on } \partial\Omega. \tag{3.104}$$

Problem (3.104) has positive solutions only if $\alpha = \lambda_{1,a_\infty}$. So, we have $\alpha = \lambda_{1,a_\infty}$ and $v_0 = \varphi_{a_\infty}$ (note that $\|v_0\|_p = 1$). To simplify the notation, we denote the above subsequence again by $\{v_n\}$. Thus, $u_n(x) \to +\infty$ as $n\to\infty$ for a.e. $x\in\Omega$ (note that $v_0 = \varphi_{a_\infty} > 0$ in Ω).

Assume $\widetilde{H}(\mathcal{R})^+$ or $\widetilde{H}(\mathcal{R})^-$. Then it turns out that

$$\mathcal{K}_1 := \int_\Omega \frac{f_+(x,u_n)u_n - pF_+(x,u_n)}{\|u_n\|_p^q}\,dx \to \pm\infty \qquad \text{if } \widetilde{H}(\mathcal{R})^\pm \text{ holds, respectively,} \tag{3.105}$$

where q is the constant in $\widetilde{H}(\mathcal{R})^\pm$. Indeed, it follows from $\widetilde{H}(\mathcal{R})^+$ that $(f_+(x,t)t - pF_+(x,t))/t^q$ is bounded below on $\Omega\times[1,+\infty)$. Since $u_n(x)\to$

$+\infty$ a.e. $x \in \Omega$, we have (3.105) if $\widetilde{H}(\mathcal{R})^+$ holds, by applying Fatou's lemma to the inequality

$$\mathcal{K}_1 \geq \int_{\{u_n(x)\geq 1\}} \frac{f_+(x,u_n)u_n - pF_+(x,u_n)}{u_n^q} v_n^q \, dx - \frac{2H_0}{\|u_n\|_p^p}|\Omega|,$$

where $H_0 > 0$ is the constant in $\widetilde{H}(\mathcal{R})^+$. In the case of $\widetilde{H}(\mathcal{R})^-$, by considering $-f$ instead of f, we can check the claim in (3.105), too.

Furthermore, in the case of $\widetilde{H}(\mathcal{R})^-$, through Hölder's inequality we get

$$\begin{aligned}\mathcal{K}_2 &:= \int_\Omega \frac{p\tilde{G}(x,\nabla u_n) - \tilde{a}(x,|\nabla u_n|)|\nabla u_n|^2}{\|u_n\|_p^q} dx \\ &\leq H_0 \int_\Omega (|\nabla v_n|^q + \frac{1}{\|u_n\|_p^q}) \, dx \leq H_0 \|\nabla v_n\|_p^q |\Omega|^{(p-q)/p} + o(1) \\ &\leq H_0 \|\nabla v_0\|_p^q |\Omega|^{(p-q)/p} + o(1) \qquad (3.106)\end{aligned}$$

because $v_n \to v_0$ in $W^{1,p}(\Omega)$, where $q \in [1,p)$ and $H_0 > 0$ are the constants in $\widetilde{H}(\mathcal{R})^-$. Similarly, in the case of $\widetilde{H}(\mathcal{R})^+$, we obtain

$$\mathcal{K}_2 \geq -H_0 \|\nabla v_0\|_p^q |\Omega|^{(p-q)/p} + o(1). \qquad (3.107)$$

Hence, due to (3.105), (3.106) or (3.107), we reach a contradiction passing to the limit inferior or superior in the equality

$$\frac{pI_{\pm n}(u_n) - \langle I'_{\pm n}(u_n), u_n\rangle}{\|u_n\|_p^q} = \mathcal{K}_2 + \mathcal{K}_1.$$

Now assume $H(\mathcal{R})^+$ or $H(\mathcal{R})^-$. Since v_0 is a positive solution of (3.104), we have $|\nabla u_n(x)| \to +\infty$ as $n \to \infty$ for a.e. $x \in \Omega_0 := \{x' \in \Omega : |\nabla v_0(x')| \neq 0\}$. Because $|\Omega_0| > 0$, we can show that

$$\int_\Omega \frac{p\tilde{G}(x,\nabla u_n) - \tilde{a}(x,|\nabla u_n|)|\nabla u_n|^2}{\|u_n\|_p^q} dx \to \pm\infty \quad \text{if } H(\mathcal{R})^\pm \text{ holds, respectively,}$$

by a similar argument to the one for f as above. In addition, we can easily obtain the inequality

$$\pm \int_\Omega \frac{f_+(x,u_n)u_n - pF_+(x,u_n)}{\|u_n\|_p^q} dx \geq -H_0 \|v_n\|_q^q + o(1) = -H_0 \|v_0\|_q^q + o(1)$$

in the case of $H(\mathcal{R})^\pm$, respectively. Thereby, a contradiction arises by considering the limit of $(pI_{\pm n}(u_n) - \langle I'_{\pm n}(u_n), u_n\rangle)/\|u_n\|_p^q$. □

The following preliminary result is related to the case of $\alpha_0 = \lambda_{1,a_0}$.

Lemma 3.73. *Assume $H(\mathcal{R})_0^-$ or $\widetilde{H}(\mathcal{R})_0^-$ (resp. $H(\mathcal{R})_0^+$ or $\widetilde{H}(\mathcal{R})_0^+$). Let $u_n \in W_0^{1,p}(\Omega) \setminus \{0\}$ satisfy $I'_{\pm n}(u_n) = 0$ for every n and $\inf_n I_{\pm n}(u_n) \geq 0$ (resp. $\sup_n I_{\pm n}(u_n) \leq 0$). Then $\liminf_{n\to\infty} \|u_n\|_p > 0$.*

Proof. For otherwise, assume that $\lim_{n\to\infty} \|u_n\|_p = 0$ along a relabeled subsequence. Notice that the boundedness of $\{\|u_n\|_p\}$ yields that $\{\|u_n\|\}$ and $\{\|u_n\|/\|u_n\|_p\}$ are bounded owing to

$$o(1)\|u_n\| = \langle I'_{\pm n}(u_n), u_n\rangle \geq \frac{C_0}{p-1}\|u_n\|^p - C(1 + \|(u_n)^+\|_p^p) \mp \frac{p}{n}\|(u_n)^+\|_p^p \tag{3.108}$$

for some $C > 0$ independent of n. Then, since u_n is a positive solution of

$$-\mathrm{div}\, A(x, \nabla u) = f_{\pm n}(x, u_n)$$

in Ω (refer to Remark 3.68), it follows from Proposition 2.12 that $u_n \to 0$ in $C^1(\overline{\Omega})$ (note that $|(f_{\pm n})_+(x,t)| \leq C(t^+)^{p-1}$, and use H$(f)$ and the fact that $u_n \to 0$ in $L^p(\Omega)$). Therefore, we may assume that $\|u_n\|_{C^1(\overline{\Omega})} \leq 1$ by considering a sufficiently large n.

Since

$$\left|\frac{f_{\pm n}(x, \|u_n\|_p t)}{\|u_n\|_p^{p-1}}\right| \leq Ct^p$$

for a.e. $x \in \Omega$, every $t \geq 0$ ($C > 0$ is independent of n and use H(f) and (3.108)), by a similar argument to Theorem 3.53, letting $v_n := u_n/\|u_n\|_p$, we see that $\{v_n\}$ has a subsequence convergent to a positive solution v_0 in $C^1(\overline{\Omega})$ of the problem

$$-\mathrm{div}\,(a_0(x)|\nabla u|^{p-2}\nabla u) = \alpha_0|u|^{p-2}u \quad \text{in } \Omega, \quad u = 0 \quad \text{on } \partial\Omega. \tag{3.109}$$

If $\alpha_0 \neq \lambda_{1,a_0}$, then we arrive at a contradiction because (3.80) does not have a positive solution unless $\lambda = \lambda_{1,a_0}$. So, we may assume that $\alpha_0 = \lambda_{1,a_0}$ and $v_0 = \varphi_{a_0}$ (recalling that $\|v_0\|_p = 1$). To simplify the notation, we denote the above subsequence by $\{v_n\}$.

Assume $H(\mathcal{R})_0^+$ or $H(\mathcal{R})_0^-$. Then we can prove that

$$\mathcal{L}_1 := \int_\Omega \frac{p\tilde{G}_0(x, \nabla u_n) - \tilde{a}_0(x, |\nabla u_n|)|\nabla u_n|^2}{\|u_n\|_p^r}\, dx \to \pm\infty \tag{3.110}$$

if $H(\mathcal{R})_0^\pm$ holds, respectively,

where $r \in [p, p^*)$ is the constant in $H(\mathcal{R})_0^+$ or $H(\mathcal{R})_0^-$. Indeed, since $\|\nabla v_0\|_p > 0$, we can choose a constant $\varepsilon_0 > 0$ such that $|\{x \in \Omega : |\nabla v_0| >$

$2\varepsilon_0\}| > 0$. Under assumption $H(\mathcal{R})_0^+$, we may write

$$\begin{aligned}
\mathcal{L}_1 &\geq \int_{\{|\nabla v_n|>\varepsilon_0\}} \frac{p\tilde{G}_0(x,\nabla u_n) - \tilde{a}_0(x,|\nabla u_n|)|\nabla u_n|^2}{|\nabla u_n|^r} |\nabla v_n|^r \, dx \\
&\quad - \int_{\{|\nabla v_n|\leq\varepsilon_0\}} H_0 |\nabla v_n|^r \, dx \\
&\geq \int_{\{|\nabla v_n|>\varepsilon_0\}} \frac{p\tilde{G}_0(x,\nabla u_n) - \tilde{a}_0(x,|\nabla u_n|)|\nabla u_n|^2}{|\nabla u_n|^r} |\nabla v_n|^r \, dx - \varepsilon_0^r H_0 |\Omega|,
\end{aligned} \tag{3.111}$$

where H_0 is the positive constant in $H(\mathcal{R})_0^+$. Applying Fatou's lemma, the claim is proven because the Lebesgue measure of $\{x \in \Omega : |\nabla v_0| > 2\varepsilon_0\}$ is positive. Similarly, by considering $\tilde{a}_0(x,|\nabla u_n|)|\nabla u_n|^2 - p\tilde{G}_0(x,\nabla u_n)$, we can prove (3.110) under $H(\mathcal{R})_0^-$.

On the other hand, letting

$$\mathcal{L}_2 := \int_\Omega \frac{f_+(x,u_n)u_n - pF_+(x,u_n)}{\|u_n\|_p^r},$$

by using $H(\mathcal{R})_0^+$ or $H(\mathcal{R})_0^-$, we are able to find that

$$\pm\mathcal{L}_2 \geq -H_0 \int_\Omega ((v_n)^+)^r \, dx \geq -H_0 \|v_n\|_r^r = -H_0\|v_0\|_r^r + o(1) \tag{3.112}$$

(note that $\|u_n\|_{C^1(\overline{\Omega})} \leq 1$ and $v_n \to v_0$ in $C^1(\overline{\Omega})$). If $\sup_n(\pm I_{\pm n}(u_n)) \leq 0$, observing that

$$\pm(\mathcal{L}_1 + \mathcal{L}_2) = \pm\frac{pI_{\pm n}(u_n) - \langle I'_{\pm n}(u_n), u_n\rangle}{\|u_n\|_p^r} = \pm\frac{pI_{\pm n}(u_n)}{\|u_n\|_p^r} \leq 0, \tag{3.113}$$

respectively, we get a contradiction by taking the limit superior or inferior in (3.113) because of (3.111) and (3.112).

Assume $\widetilde{H}(\mathcal{R})_0^+$ or $\widetilde{H}(\mathcal{R})_0^-$. By an argument similar to the one used for $\mathcal{L}_1$ under $H(\mathcal{R})_0^\pm$, we can show that

$$\begin{aligned}
&\int_\Omega \frac{f_+(x,u_n)u_n - pF_+(x,u_n)}{\|u_n\|_p^r} \, dx \\
&\quad = \int_{\{v_n>0\}} \frac{f_+(x,u_n)u_n - pF_+(x,u_n)}{((u_n)^+)^r} ((v_n)^+)^r \, dx \to \pm\infty
\end{aligned}$$

if $\widetilde{H}(\mathcal{R})_0^\pm$ holds,

respectively. Moreover, under $\widetilde{H}(\mathcal{R})_0^{\pm}$, it is easily seen that

$$\pm \int_\Omega \frac{p\tilde{G}_0(x, \nabla u_n) - \tilde{a}_0(x, |\nabla u_n|)|\nabla u_n|^2}{\|u_n\|_p^r} \, dx \geq \mp H_0 \|\nabla v_n\|_r^r$$

$$= \mp H_0 \|\nabla v_0\|_r^r + o(1),$$

respectively (note that $\|u_n\|_{C^1(\overline{\Omega})} \leq 1$ and $v_n \to v_0$ in $C^1(\overline{\Omega})$). The conclusion follows proceeding as above. □

Next we prepare the investigation of resonant cases with other preliminary results.

Choose smooth nonnegative functions φ and ψ on $[0, +\infty)$ satisfying $\varphi(t) = 1$ if $0 \leq t \leq 2$, $\varphi(t) = 0$ if $t \geq 4$, $\psi(t) = 0$ if $t \leq 5$ and $\psi(t) = 1$ if $t \geq 10$. We define approximate functionals on $W_0^{1,p}(\Omega)$ by

$$\tilde{I}_{\pm n}(u) := I(u) \mp \frac{1}{n} \psi(\|u\|_p^p) \|u^+\|_p^p \pm \frac{1}{n} \varphi(\|u\|_p^p) \|u^+\|_p^p.$$

Lemma 3.74. *If $\alpha \pm p/n \neq \lambda_{1, a_\infty}$, then $\tilde{I}_{\pm n}$ satisfies the Palais–Smale condition, respectively.*

Proof. Let $\{u_m\}$ be a Palais–Smale sequence of $\tilde{I}_{+n}$ or $\tilde{I}_{-n}$. If $\|u_m\|_p \to +\infty$, then the equality $\tilde{I}_{\pm n}(u_m) = I_{\pm n}(u_m)$ holds for sufficiently large m. So, by applying Lemma 3.69 to $f_{\pm n}$ (due to the fact that $\alpha \pm p/n \neq \lambda_{1, a_\infty}$), we get a contradiction if $\|u_m\|_p \to +\infty$. Consequently, we see that $\{\|u_m\|_p\}$ is bounded. Then, the same proof as in Lemma 3.69 provides that $\{u_m\}$ has a convergent subsequence in $W_0^{1,p}(\Omega)$. □

Lemma 3.75. *Assume conditions $H(\mathcal{R})_0^-$ or $\widetilde{H}(\mathcal{R})_0^-$ (respectively, $H(\mathcal{R})_0^+$ or $\widetilde{H}(\mathcal{R})_0^+$). Let $u_n \in W_0^{1,p}(\Omega) \setminus \{0\}$, satisfy $\tilde{I}'_{\pm n}(u_n) = 0$ for every n and $\inf_n \tilde{I}_{\pm n}(u_n) \geq 0$ (respectively, $\sup_n \tilde{I}_{\pm n}(u_n) \leq 0$). Then $\liminf_{n \to \infty} \|u_n\|_p > 0$.*

Proof. Notice that $\tilde{I}_{\pm n}(u) = I_{\mp n}(u)$ provided $\|u\|_p \leq 2$. Then the result can be established by the same argument as in Lemma 3.73. □

Lemma 3.76. *If either $H(\mathcal{R})^+$ or $\widetilde{H}(\mathcal{R})^+$* (resp. either $H(\mathcal{R})^-$ or $\widetilde{H}(\mathcal{R})^-$) *holds and $\{u_n\}$ satisfies*

$$\sup_{n \in \mathbb{N}} \tilde{I}_{\pm n}(u_n) < +\infty \quad \text{and} \quad \lim_{n \to \infty} \|\tilde{I}'_{\pm n}(u_n)\|_{W^{-1,p'}(\Omega)} = 0,$$

$$\left(\text{resp. } \inf_{n \in \mathbb{N}} \tilde{I}_{\pm n}(u_n) > -\infty \quad \text{and} \quad \lim_{n \to \infty} \|\tilde{I}'_{\pm n}(u_n)\|_{W^{-1,p'}(\Omega)} = 0 \right),$$

then $\{u_n\}$ is bounded in $W_0^{1,p}(\Omega)$.

Proof. Because $\tilde{I}_{\pm n}(u) = I_{\pm n}(u)$ provided $\|u\|_p \geq 10$, the result can be proven along the same argument as in Lemma 3.72. □

Our results regarding the resonant cases are as follows.

Theorem 3.77. *Assume* H(f), H$(A)_0$, *and* H$(A)_\infty$. *If one of the following conditions holds:*

(i) $\alpha_0 > \lambda_{1,a_0}$, $\alpha = \lambda_{1,a_\infty}$ *and* $\widetilde{H}(\mathcal{R})^+$ *or* $H(\mathcal{R})^+$;
(ii) $\alpha_0 < \lambda_{1,a_0}$, $\alpha = \lambda_{1,a_\infty}$ *and* $\widetilde{H}(\mathcal{R})^-$ *or* $H(\mathcal{R})^-$;
(iii) $\alpha_0 = \lambda_{1,a_0}$, $\alpha < \lambda_{1,a_\infty}$ *and* $\widetilde{H}(\mathcal{R})_0^+$ *or* $H(\mathcal{R})_0^+$;
(iv) $\alpha_0 = \lambda_{1,a_0}$, $\alpha > \lambda_{1,a_\infty}$ *and* $\widetilde{H}(\mathcal{R})_0^-$ *or* $H(\mathcal{R})_0^-$;
(v) $\alpha_0 = \lambda_{1,a_0}$, $\alpha = \lambda_{1,a_\infty}$, $(\widetilde{H}(\mathcal{R})_0^+$ *or* $H(\mathcal{R})_0^+)$ *and* $(\widetilde{H}(\mathcal{R})^+$ *or* $H(\mathcal{R})^+)$;
(vi) $\alpha_0 = \lambda_{1,a_0}$, $\alpha = \lambda_{1,a_\infty}$, $(\widetilde{H}(\mathcal{R})_0^-$ *or* $H(\mathcal{R})_0^-)$ *and* $(\widetilde{H}(\mathcal{R})^-$ *or* $H(\mathcal{R})^-)$,

then problem (3.86) has at least one positive solution.

Proof. (i) Since $\alpha_0 > \lambda_{1,a_0}$, there exists $n_0 \in \mathbb{N}$ such that $\alpha_0 - p/n_0 > \lambda_{1,a_0}$. Note that

$$\frac{f_{-n}(x,t)}{t^{p-1}} \to \alpha_0 - \frac{p}{n} > \lambda_{1,a_0} \quad \text{as } t \to 0^+$$

whenever $n \geq n_0$, and

$$\frac{f_{-n}(x,t)}{t^{p-1}} \to \alpha - \frac{p}{n} = \lambda_{1,a_\infty} - \frac{p}{n} < \lambda_{1,a_\infty} \quad \text{as } t \to +\infty.$$

Hence, by using the proof of Theorem 3.66 (i) for f_{-n}, we can find a global minimizer u_n of I_{-n} with $I_{-n}(u_n) < 0$ for each $n \geq n_0$. Here, we remark that $\sup_{n \geq n_0} I_{-n}(u_n) < 0$. In fact, for every $n \geq n_0$, we have

$$\begin{aligned} I_{-n}(u_n) &\leq I_{-n}(u_{n_0}) = I(u_{n_0}) + \frac{1}{n}\|u_{n_0}\|_p^p \leq I(u_{n_0}) + \frac{1}{n_0}\|u_{n_0}\|_p^p \\ &= I_{-n_0}(u_{n_0}) < 0, \end{aligned}$$

where in the first inequality we use the fact that u_n is a global minimizer of I_{-n}. Now, due to Lemma 3.72, we see that $\{u_n\}$ is bounded in $W_0^{1,p}(\Omega)$. Therefore,

$$\|I'(u_n)\|_{W^{-1,p'}(\Omega)} = \|I'(u_n) - I'_{-n}(u_n)\|_{W^{-1,p'}(\Omega)} \leq \frac{p}{n\lambda_1^p}\|u_n\|^{p-1} \to 0$$

as $n \to \infty$, where λ_1 denotes the first eigenvalue of $-\Delta_p$ on $W_0^{1,p}(\Omega)$. Since I is bounded on bounded sets, we may assume that $\{u_n\}$ is a bounded Palais–Smale sequence of I. Then, by Proposition 3.44, it follows that $\{u_n\}$ has a convergent subsequence to some v_0. It is clear that

$$I(v_0) \leq \sup_{n \geq n_0} I_{-n}(u_n) = I_{-n_0}(u_{n_0}) < 0,$$

and so v_0 is a nontrivial critical point of I.

(ii) By Lemma 3.71 and $\alpha_0 < \lambda_{1,a_0}$, we can choose $q_0 \in (p, p^*]$ and $\rho > 0$ such that

$$\inf\{I(u) : \|u\|_{q_0} = \rho\} > 0.$$

Since

$$I_{+n}(u) \geq I(u) - \frac{\|u\|_{q_0}^p |\Omega|^{1-p/q_0}}{n} \quad \text{for every } u \in W_0^{1,p}(\Omega),$$

we can take $n_0 \in \mathbb{N}$ such that

$$\alpha_0 + \frac{p}{n_0} < \lambda_{1,a_0} \quad \text{and} \quad \delta_0 := \inf\{I_{+n_0}(u) : \|u\|_{q_0} = \rho\} > 0.$$

Hence

$$\inf\{I_{+n}(u) : \|u\|_{q_0} = \rho\} \geq \delta_0$$

for every $n \geq n_0$ because $I_{+n}(u) \geq I_{+n_0}(u)$ for all $n \geq n_0$ and $u \in W_0^{1,p}(\Omega)$. By means of

$$\frac{f_{+n}(x,t)}{t^{p-1}} \to \alpha + \frac{p}{n} > \alpha = \lambda_{1,a_\infty} \quad \text{as } t \to +\infty$$

and applying Lemma 3.69 to f_{+n} in place of f, I_{+n} satisfies the Palais–Smale condition. Therefore, the proof of Theorem 3.66 (ii) implies that for every $n \geq n_0$ there exists a critical point $u_n \in W_0^{1,p}(\Omega)$ of I_{+n} such that $I_{+n}(u_n) \geq \delta_0$. According to Lemma 3.72, the sequence $\{u_n\}$ is bounded in $W_0^{1,p}(\Omega)$. Thus, because we have a bounded Palais–Smale sequence of I, a similar proof as in the case of (i) yields a nontrivial critical point of I taking into account that

$$\inf_{n \geq n_0} I(u_n) \geq \inf_{n \geq n_0} I_{+n}(u_n) \geq \delta_0 > 0.$$

(iii) Fix $n_0 \in \mathbb{N}$ so that $\alpha + p/n_0 < \lambda_{1,a_\infty}$. The proof of Theorem 3.66 (i) guarantees that, for every $n \geq n_0$, I_{+n} has a global minimizer u_n satisfying $I_{+n}(u_n) < 0$ because

$$\frac{f_{+n}(x,t)}{t^{p-1}} \to \alpha_0 + \frac{p}{n} > \alpha_0 = \lambda_{1,a_0} \quad \text{as } t \to 0^+$$

and

$$\frac{f_{+n}(x,t)}{t^{p-1}} \to \alpha + \frac{p}{n} < \lambda_{1,a_\infty} \quad \text{as } t \to +\infty \text{ whenever } n \geq n_0.$$

Noting that $I_{+n}(u) \geq I_{+n_0}(u)$ for every $u \in W_0^{1,p}(\Omega)$ and $n \geq n_0$, $\{u_n\}$ is bounded in $W_0^{1,p}(\Omega)$ since I_{+n_0} is coercive on $W_0^{1,p}(\Omega)$ by Lemma 3.70. Thus

$\{u_n\}$ is a bounded Palais–Smale sequence of I by the same argument as in case (i). Therefore, as before, $\{u_n\}$ has a convergent subsequence to some u_0 in $W_0^{1,p}(\Omega)$. On the other hand, Lemma 3.73 and the fact that $\sup_{n\geq n_0} I_{+n}(u_n) \leq 0$ guarantee that $u_0 \neq 0$. Consequently, u_0 is a nontrivial critical point of I.

(iv) Let $n_0 \in \mathbb{N}$ be so that $\alpha - p/n_0 > \lambda_{1,a_\infty}$. By applying Lemma 3.71 to f_{-n} for $n \geq n_0$ in view of $\alpha_0 - p/n < \lambda_{1,a_0}$, we can choose $q_0 \in (p, p^*]$ and $\rho_n > 0$ such that

$$\delta_n := \inf\{I_{-n}(u) : \|u\|_{q_0} = \rho_n\} > 0.$$

By noting that

$$\frac{f_{-n}(x,t)}{t^{p-1}} \to \alpha - \frac{p}{n} > \lambda_{1,a_\infty} \quad \text{as } t \to +\infty$$

and applying, for every $n \geq n_0$, Lemma 3.69 to f_{-n} in place of f, I_{-n} satisfies the Palais–Smale condition. Therefore, the proof of Theorem 3.66 (ii) implies that for every $n \geq n_0$ there exists a critical point $u_n \in W_0^{1,p}(\Omega)$ of I_{-n} such that $I_{-n}(u_n) \geq \delta_n > 0$. According to Lemma 3.72, the sequence $\{u_n\}$ is bounded in $W_0^{1,p}(\Omega)$. Thus, as in case (i), $\{u_n\}$ has a convergent subsequence to some u_0 in $W_0^{1,p}(\Omega)$. On the other hand, Lemma 3.73 ensures that $u_0 \neq 0$ (note that $\inf_{n\geq n_0} I_{-n}(u_n) \geq 0$). Thereby, u_0 is a nontrivial critical point of I.

(v) Note that $\tilde{I}_{-n}(u) = I_{-n}(u)$ provided $\|u\|_p \geq 10$ and $\tilde{I}_{-n}(u) = I_{+n}(u)$ if $\|u\|_p \leq 2$. So, by an argument similar to (i), $\tilde{I}_{-n}$ has a global minimizer u_n. Moreover, as in (iii), observing that

$$\frac{f_{+n}(x,t)}{t^{p-1}} \to \alpha_0 + \frac{p}{n} > \lambda_{1,a_0} \text{ as } t \to 0^+ \quad \text{and}$$
$$\frac{f_{-n}(x,t)}{t^{p-1}} \to \alpha - \frac{p}{n} < \lambda_{1,a_\infty} \text{ as } t \to +\infty,$$

we have $\tilde{I}_{-n}(u_n) < 0$, whence $u_n \neq 0$. Lemma 3.76 implies the boundedness of $\{\|u_n\|\}$, through the same justification as in (i). We conclude that $\{u_n\}$ is a bounded Palais–Smale sequence of I. Therefore, we may assume that $\{u_n\}$ converges to some u_0 in $W_0^{1,p}(\Omega)$ by choosing a relabeled subsequence. On the other hand, Lemma 3.73 yields $\liminf_{n\to\infty} \|u_n\|_p > 0$. Hence $u_0 \neq 0$, which means that u_0 is a nontrivial critical point of I.

(vi) We point out that $\tilde{I}_{+n}(u) = I_{+n}(u)$ provided $\|u\|_p \geq 10$ and $\tilde{I}_{+n}(u) = I_{-n}(u)$ if $\|u\|_p \leq 2$. So, because

$$\frac{f_{-n}(x,t)}{t^{p-1}} \to \alpha_0 - \frac{p}{n} < \lambda_{1,a_0} \text{ as } t \to 0^+ \quad \text{and}$$
$$\frac{f_{+n}(x,t)}{t^{p-1}} \to \alpha + \frac{p}{n} > \lambda_{1,a_\infty} \text{ as } t \to +\infty,$$

carrying out a similar argument to (ii) and (iv) for each n, this time on the basis of Lemmas 3.74, 3.75, and 3.76, we obtain a nontrivial critical point u_n of $\tilde{I}_{+n}$ with $\tilde{I}_{+n}(u_n) > 0$. Then, as in (v), we can find a nontrivial critical point of I. $\square$

3.4 NOTES

Section 3.1 is based on the references [186,187]. The general idea of the Fučík spectrum originates in the following context. Given a bounded domain $\Omega \subset \mathbb{R}^N$, let T be a self-adjoint linear operator on $L^2(\Omega)$ with compact resolvent whose eigenvalues satisfy

$$0 < \lambda_0 < \lambda_1 < \cdots < \lambda_k < \cdots .$$

The Fučík spectrum Σ of T is defined as the set of all pairs $(a, b) \in \mathbb{R}^2$ such that the equation

$$Tu = au^+ - bu^- \tag{3.114}$$

has a nontrivial solution. In (3.114) we denoted $u^+ = \max\{u, 0\}$ (the positive part of u) and $u^- = \min\{-u, 0\}$ (the negative part of u). Fučík [88] and Dancer [65] recognized that the set Σ plays an important part in the study of semilinear equations of type

$$Tu = g(x, u),$$

where $g : \Omega \times \mathbb{R} \to \mathbb{R}$ is a Carathéodory function with jumping nonlinearities

$$\frac{g(x,s)}{s} \to a \quad \text{as } s \to +\infty, \qquad \frac{g(x,s)}{s} \to b \quad \text{as } s \to -\infty.$$

A systematic study of this spectrum was developed by Fučík [89] in the case of the negative Laplacian in one-dimension, i.e., for $N = 1$, with periodic boundary condition proving that this spectrum is composed of two families of curves in $\mathbb{R}^2$ emanating from the points (λ_k, λ_k) determined by the eigenvalues λ_k of the negative periodic Laplacian in one-dimension. Afterwards, many authors studied the Fučík spectrum Σ_2 for the negative Laplacian $-\Delta$ with Dirichlet boundary condition on a bounded domain $\Omega \subset \mathbb{R}^N$ (see, e.g., [15,34,66,69,147, 216]). In this respect, Dancer [66] proved that the lines $\mathbb{R} \times \{\lambda_1\}$ and $\{\lambda_1\} \times \mathbb{R}$ are isolated in Σ_2, whereas de Figueiredo–Gossez [69] constructed a first nontrivial curve in Σ_2 passing through (λ_2, λ_2). Here λ_1 and λ_2 denote the first and second eigenvalue of $-\Delta$ with Dirichlet boundary condition, respectively. The next step in this direction was to investigate the Fučík spectrum Σ_p of the negative p-Laplacian

$$-\Delta_p u = -\operatorname{div}(|\nabla u|^{p-2}\nabla u), \quad 1 < p < +\infty,$$

which is a nonlinear operator if $p \neq 2$, aiming to extend the results known for $-\Delta$. Drábek [74] has shown that, for $p \neq 2$, in the case of a one-dimensional domain, Σ_p has similar properties as in the linear case, i.e., for $p = 2$.

In Section 3.1 we discuss the Fučík spectrum of $-\Delta_p$ under different boundary conditions. It is shown that there exists a close relationship between these Fučík spectra and the ordinary spectra of $-\Delta_p$ subject to different boundary conditions. A fundamental fact is that every such Fučík spectrum contains a first nontrivial curve $\mathcal{C}$ which is Lipschitz continuous and decreasing. However, the asymptotic behavior of these curves is different relatively to the imposed boundary condition. In the applications of these spectra to certain nonlinear elliptic problems with jumping nonlinearities, subtle phenomena can occur due to the interaction of the involved nonlinearities with the spectra, in particular resonance to spectral elements. Problems of this type can be considered beyond the setting of quasilinear elliptic equations.

As already mentioned, the Fučík spectrum of $-\Delta_p$ depends strongly on the choice of the boundary condition related to Ω. Specifically, we provide essential results related to the Fučík spectrum of $-\Delta_p$ with Dirichlet, Neumann, Steklov, and Robin boundary conditions.

The work of Cuesta–de Figueiredo–Gossez [60] was the first that gave a complete study of the beginning of the Fučík spectrum of $-\Delta_p$ in the general case of $1 < p < +\infty$ with homogeneous Dirichlet boundary condition. The variational approach through the mountain pass theorem built in [60] was the starting point for investigating the Fučík spectrum under other boundary conditions (Neumann, Steklov, Robin). An important consequence is the variational characterization of the second eigenvalue λ_2 of $-\Delta_p$ on $W_0^{1,p}(\Omega)$ as stated in (3.5). Generally, the complete description of the Fučík spectrum is an open problem.

Multiple solutions results concerning problems of type (3.6) and using the representation of the first nontrivial curve $\mathcal{C}$, in particular the characterization of the second eigenvalue λ_2 of $-\Delta_p$ on $W_0^{1,p}(\Omega)$ in (3.5), can be found in numerous publications; see, for example, [40,44,178]. We also refer to versions of such results in the case of nonsmooth potential associated to (3.6) (see, e.g., [42,45,46]).

The basic paper dealing with the Fučík spectrum of the negative Neumann p-Laplacian is due to Arias–Campos–Gossez [16]. A striking difference between the Dirichlet Fučík spectrum Σ_p and the Neumann Fučík spectrum Θ_p consists in the asymptotic behavior of the first nontrivial curve $\mathcal{C}$. In the Neumann case, to describe the asymptotic properties of the curve $\mathcal{C}$, it is required to consider the situations $p \leq N$ and $p > N$ separately. An extension to the Fučík spectrum of the negative Neumann p-Laplacian with weights has been achieved by Arias–Campos–Cuesta–Gossez [17]. For the weights given by the measurable functions $m(x)$ and $n(x)$ on Ω, the authors consider the set Σ of all pairs $(a, b) \in \mathbb{R}^2$ such that

$$\begin{cases} -\Delta_p u = am(x)(u^+)^{p-1} - bn(x)(u^-)^{p-1} & \text{in } \Omega, \\ \frac{\partial u}{\partial \nu} = 0 & \text{on } \partial\Omega \end{cases}$$

has a nontrivial solution. Under suitable assumptions on the data it is shown that Σ contains a first nontrivial curve.

In Motreanu–Tanaka [179] the Fučík spectrum of the negative Neumann p-Laplacian is used to study quasilinear elliptic equations of the form

$$\begin{cases} -\operatorname{div} A(x, \nabla u) = f(x, u) & \text{in } \Omega, \\ \frac{\partial u}{\partial \nu} = 0 & \text{on } \partial\Omega, \end{cases}$$

where in the principal part of the equation one has an operator $A \in C(\overline{\Omega} \times \mathbb{R}^N, \mathbb{R}^N) \cap C^1(\overline{\Omega} \times (\mathbb{R}^N \setminus \{0\}), \mathbb{R}^N)$ of the form $A(x, y) = a(x, |y|)y$, with $a(x, t) > 0$ for all $(x, t) \in \overline{\Omega} \times (0, +\infty)$, which is strictly monotone with respect to the second variable and fulfills some further regularity assumptions, while $f : \Omega \times \mathbb{R} \to \mathbb{R}$ is a Carathéodory function having a representation similar to (3.6). The obtained results apply in particular to the case of the Neumann p-Laplacian, i.e., when $\operatorname{div} A(x, \nabla u) = \Delta_p u$.

The Steklov eigenvalue problem (3.12) was first studied by Martínez–Rossi [149] (see also Lê [127]). The Steklov Fučík spectrum defined in (3.11) has been studied by Martínez–Rossi [150]. The fundamental difference of it with respect to the Dirichlet and Neumann Fučík spectra is that in the Steklov case a boundary integral is involved, a fact that substantially modifies the analysis regarding the relevant values a and b. Similarly to the Neumann Fučík spectrum, in order to derive the asymptotic properties of the first curve in the Fučík spectrum of $-\Delta_p$ with Steklov boundary condition, it is needed to take into account two cases, $p \leq N$ and $p > N$. In particular, one obtains the variational characterization of the second eigenvalue λ_2 in (3.13). Applications to quasilinear elliptic equations are given in Winkert [231,232].

The Fučík spectrum $-\Delta_p$ on $W^{1,p}(\Omega)$ with Robin boundary condition as introduced in (3.15) was investigated in [187]. The corresponding eigenvalue problem (3.16) was studied before by Lê [127] proving that similar results hold as for the eigenvalue problems for the negative p-Laplacian with Dirichlet, Neumann, and Steklov boundary conditions. In comparison to the corresponding functionals related to the Fučík spectra for the Dirichlet and Steklov problems, the functional associated to the Fučík spectrum of $-\Delta_p$ with Robin boundary condition exhibits an essential difference because its expression does not incorporate the norm of the space $W^{1,p}(\Omega)$, and it is also different from the functional used to treat the Neumann problem because it has the additional boundary term involving β.

It is shown that the Robin Fučík spectrum of $-\Delta_p$ on $W^{1,p}(\Omega)$ contains a first nontrivial curve expressed as

$$\mathcal{C} = \{(s + c(s), c(s)), (c(s), s + c(s)) : s \geq 0\},$$

where $c(s)$ has an explicit minimax description in terms of the positive normalized eigenfunction of (3.16) associated to the first eigenvalue λ_1. The asymptotic behavior of $\mathcal{C}$ requires, as in the Neumann and Steklov cases, some more involved considerations distinguishing the cases $p \leq N$ and $p > N$. One can suppose that $\beta > 0$ since the case $\beta = 0$ is included in (3.10). The key idea is to argue with an adequate equivalent norm on the space $W^{1,p}(\Omega)$, namely

$$\|u\|_\beta = \|\nabla u\|_{L^p(\Omega)} + \beta\|u\|_{L^p(\partial\Omega)}$$

(see also Deng [70, Theorem 2.1]). The required asymptotic properties are achieved in Theorems 3.31 and 3.32.

Section 3.2 is cast according to Motreanu–Tanaka [182] focusing on the eigenvalue problem (3.41) which involves the (p, q)-Laplacian operator.

Letting $\mu \to 0^+$, problem (3.41) becomes the following $(p-1)$-homogeneous equation known as the usual weighted eigenvalue problem for the p-Laplacian:

$$\begin{cases} -\Delta_p u = \lambda m_p(x)|u|^{p-2}u & \text{in } \Omega, \\ u = 0 & \text{on } \partial\Omega. \end{cases} \tag{3.115}$$

Moreover, after multiplying our equation by $1/\mu$ and then letting $\mu \to +\infty$, (3.41) becomes the equation

$$\begin{cases} -\Delta_q u = \lambda m_q(x)|u|^{q-2}u & \text{in } \Omega, \\ u = 0 & \text{on } \partial\Omega. \end{cases} \tag{3.116}$$

Thus, we can understand that there is a close relationship between (3.41) and the usual weighted eigenvalue problems for the p-Laplacian and q-Laplacian operators, i.e., problems (3.115) and (3.116).

We briefly discuss the motivation for the given formulation of the eigenvalue problem (3.41). Let us observe that setting $A_r(t) := |t|^{r-2}$, where $r \in \{p, q\}$, the basic eigenvalue problems (3.115), (3.116) can be written as

$$-\text{div}\,(A_r(|\nabla u|)\nabla u) = \lambda m_r(x)A_r(u)u \quad \text{in } \Omega, \quad u = 0 \quad \text{on } \partial\Omega.$$

Now in line with this, if $m := m_p \equiv m_q$ and $\mu = 1$, then setting $A_{p,q}(t) := A_p(t) + A_q(t)$ there corresponds the problem

$$-\text{div}\,(A_{p,q}(|\nabla u|)\nabla u) = \lambda m(x)A_{p,q}(u)u \quad \text{in } \Omega, \quad u = 0 \quad \text{on } \partial\Omega.$$

We have thus arrived at the statement of problem (3.41) encompassing the natural formulation for the eigenvalue problem for the (p, q)-Laplacian operator.

Nonlinear eigenvalue problems for elliptic equations have been thoroughly studied (see [210] for a comprehensive survey of different developments). Among the papers that studied (p, q)-Laplacian equations we cite [81,145,

218,233–235]. There are few results on the eigenvalue problems for the (p,q)-Laplacian, that is, the operator $\Delta_p + \Delta_q$. Benouhiba–Belyacine [25] showed the existence of the principal eigenvalue and of a continuous family of eigenvalues for the equation

$$-\Delta_p u - \Delta_q u = \lambda g(x)|u|^{p-2}u \quad \text{in } \mathbb{R}^N.$$

In [54, Theorem 4.2], Cingolani and Degiovanni proved the existence of a nontrivial solution for the problem

$$-\Delta_p u - \mu\Delta u = \lambda|u|^{p-2}u + g(u) \quad \text{in } \Omega, \qquad u=0 \quad \text{on } \partial\Omega$$

in the case where $p > 2 = q$, $g \in C^1(\mathbb{R})$ and $\lambda \notin \sigma(-\Delta_p)$ (the spectrum of $-\Delta_p$ on $W_0^{1,p}(\Omega)$). Under the Neumann boundary condition, Mihăilescu [157] determined the set of eigenvalues for the equation $-\Delta_p u - \Delta u = \lambda u$ in Ω, where $p > 2 = q$. Tanaka [224] has completely described the eigenvalues λ for which the problem

$$\begin{cases} -\Delta_r u - \mu\Delta_\rho u = \lambda m_r(x)|u|^{r-2}u & \text{in } \Omega, \\ u=0 & \text{on } \partial\Omega \end{cases}$$

has a positive solution, where $1 < r \neq \rho < +\infty$ and $\mu > 0$. The above eigenvalue problems are incorporated in (3.41) by posing $m_p \equiv 0$ or $m_q \equiv 0$.

We emphasize that problem (3.41) (or, equivalently, (3.44) in the case of $\mu = 1$) exhibits striking differences with respect to the classical eigenvalue problem (3.115). This is mainly caused by the fact that the elliptic problem (3.44), as well as (3.41), is driven by a nonhomogeneous operator $-\Delta_p - \Delta_q$, whereas (3.115) is driven by a homogeneous, precisely $(p-1)$-homogeneous, operator. As an example of such a difference, we mention that for the homogeneous problem (3.115) the infimum in (3.42) is attained by a positive eigenfunction, whereas in our nonhomogeneous problem (3.41) (or, equivalently, (3.44)) there exist no such functions except in the special case where $\lambda_1(p,m_p) = \lambda_1(q,m_q)$ and the first eigenvalues $\lambda_1(p,m_p)$ and $\lambda_1(q,m_q)$ have the same eigenspace (see Proposition 3.34 and Theorem 3.35). We refer to [75, 180,211,225] for related results on generalized eigenvalue problems involving nonhomogeneous operators. Existence of positive solutions in such cases can be found in [90] and [203].

An essential point in our approach is that problem (3.41) is equivalent to another eigenvalue problem (3.44) where we have only one parameter λ and just $\mu = 1$, which evidently simplifies the setting. This equivalence allows us to make an efficient use of the Rayleigh quotient. In this respect, we recall that a standard argument relying on the Rayleigh quotient permits to show that if $-\lambda_1(p,-m_p) < \lambda < \lambda_1(p,m_p)$, then (3.115) has no nontrivial solutions. In line with this classical result, we obtain through the Rayleigh quotient for our problem a nonexistence result formulated as Theorem 3.35, which is inspired from [224, Theorem 1]. Our main result that guarantees the existence of positive solutions for problem (3.41) is stated as Theorem 3.39. It covers the range $-\min\{\lambda_1(p,-m_p),\lambda_1(q,-m_q)\}<\lambda<\min\{\lambda_1(p,m_p),\lambda_1(q,m_q)\}$ for the parameter λ in problem (3.44). Since we have both $\lambda_1(p,m_p)$ and $\lambda_1(q,m_q)$ well-defined (although $\lambda_1(r,0)=+\infty$, see (3.43)), here the resonant cases read as $\lambda=\lambda_1(p,m_p)>\lambda_1(q,m_q)$ and $\lambda=\lambda_1(q,m_q)>\lambda_1(p,m_p)$. Our existence results in the resonant cases $\lambda=\lambda_1(p,m_p)>\lambda_1(q,m_q)$ and $\lambda=\lambda_1(q,m_q)>\lambda_1(p,m_p)$ related to problem (3.44) are presented in Theorems 3.40 and 3.41, respectively.

Section 3.3 presents the results established by Motreanu–Tanaka [180]. Hypotheses closely related to assumption (H_A) are already considered in the study of quasilinear elliptic problems (see [176, Example 2.2], [64,156,168,223]). We also refer to [92,120,121,211] for generalized p-Laplacian operators. In particular, for $A(x,y)=|y|^{p-2}y$, that is, $\operatorname{div} A(x,\nabla u)$ stands for the usual p-Laplacian $\Delta_p u$, we can take $C_0=C_1=p-1$ in (H_A). Conversely, in the case where $C_0=C_1=p-1$ holds in (H_A), by the inequalities in Remark 3.43 (ii) and (iii), we see that $a(x,t)=|t|^{p-2}$ whence $A(x,y)=|y|^{p-2}y$. Hence problem (3.68) contains the corresponding p-Laplacian eigenvalue problem as a special case.

In the p-Laplacian case, the first eigenvalue λ_1 is obtained by the Rayleigh quotient, namely

$$\lambda_1=\inf_{u\in W_0^{1,p}(\Omega)\setminus\{0\}}\frac{\int_\Omega|\nabla u|^p\,dx}{\|u\|_p^p}$$

as stated in Proposition 1.57. In the case of our operator $A(x,\cdot)$, possibly nonhomogeneous, in general

$$\inf\{\lambda\in\mathbb{R}:\lambda \text{ is eigenvalue of A}\}$$

is not obtained by such Rayleigh quotient corresponding to A. However, the Rayleigh quotient continues to play a significant role. We examine its behavior as $\|u\|_p\to 0$ or $\|u\|_p\to+\infty$ under an additional condition describing certain asymptotic $(p-1)$-homogeneity. Precisely, under conditions $H(A)_0$ and $H(A)_\infty$ we prove that

$$\min\left\{\int_\Omega\int_0^{|\nabla u(x)|}\frac{a(x,t)t}{r^p}\,dt\,dx\,:\,\|u\|_p=r\right\}\to\frac{\lambda_{1,a_0}}{p}\quad\text{or}\quad\frac{\lambda_{1,a_\infty}}{p}$$

as $r\to 0^+$ or $r\to+\infty$ respectively, where

$$\lambda_{1,a_0}=\min\left\{\int_\Omega a_0(x)|\nabla u|^p\,dx\,:\,\|u\|_p=1\right\},$$

$$\lambda_{1,a_\infty}=\min\left\{\int_\Omega a_\infty(x)|\nabla u|^p\,dx\,:\,\|u\|_p=1\right\}.$$

Concerning eigenvalue problems with a nonhomogeneous operator under the Neumann boundary condition, we can refer to [211,225].

A new insight of eigenvalue problem (3.68) is earned by incorporating it in the problem of seeking a positive solution for the quasilinear elliptic problem

$$\begin{cases} -\operatorname{div} A(x, \nabla u) = f(x, u) & \text{in } \Omega, \\ u = 0 & \text{on } \partial\Omega, \end{cases} \tag{3.117}$$

where f satisfies the assumption H(f), which demands that $f(x, \cdot)$ be asymptotically $(p-1)$-linear near 0^+ and $+\infty$ meaning

$$\lim_{s \to 0^+} \frac{f(x,s)}{s^{p-1}} = \alpha_0 \quad \text{uniformly for a.e. } x \in \Omega,$$
$$\lim_{s \to +\infty} \frac{f(x,s)}{s^{p-1}} = \alpha \quad \text{uniformly for a.e. } x \in \Omega,$$

with some constants α_0 and α.

Early results regarding the existence of a positive solution for problem (3.117) with nonhomogeneous operators are available, e.g., in [90] and [203], as well as [92] for the existence of a positive radial solution. However, we cannot apply these results to our nonlinear term which is only asymptotically $(p-1)$-linear near 0^+ and $+\infty$, and furthermore with possibly different weights.

The starting point of our analysis is that for the p-Laplacian equation, if $(\alpha - \lambda_1)(\alpha_0 - \lambda_1) < 0$ (where λ_1 denotes the first eigenvalue of $-\Delta_p$ under Dirichlet boundary condition), then

$$-\Delta_p u = f(x, u) \quad \text{in } \Omega, \quad u = 0 \quad \text{on } \partial\Omega,$$

has a positive solution (see [67]). In Theorem 3.66 we extend this existence result from the p-Laplacian equation to the corresponding problem involving our nonhomogeneous operator A. We also mention that in the special case of $A(x, y) = A(y)$, the result in [125] provides the existence of a positive solution if there hold $\alpha < \lambda_1 C_0/(p-1)$ and $\lambda_1 C_1/(p-1) < \alpha_0$ (note that we can apply this result only to the case where $\alpha < \alpha_0$). We emphasize that for our general operator the case $\lambda_{1,a_0} \neq \lambda_{1,a_\infty}$ can occur. Note that in such a situation, contrary to the case of the p-Laplacian, we can still apply our theorem when $\alpha_0 = \alpha$ provided this number is between λ_{1,a_0} and λ_{1,a_∞}. The known result for the case of the p-Laplacian is obtained from our theorem simply by setting $a_0 \equiv 1$ and $a_\infty \equiv 1$.

Theorem 3.66 implies that if $\lambda_{1,a_0} \neq \lambda_{1,a_\infty}$ is true, then every λ within the interval determined by the numbers λ_{1,a_0} and λ_{1,a_∞} is an eigenvalue of A. This is pointed out in Corollary 3.67. Besides there is a positive eigenfunction. This shows that generally the spectrum of the operator $-\operatorname{div} A(x, \nabla \cdot)$ on $W_0^{1,p}(\Omega)$ is not discrete. We cannot extend this conclusion to the case of the p-Laplacian because there we have the equality $\lambda_{1, a_0} = \lambda_{1, a_\infty}$.

The final part of Section 3.3 deals with the study of problem (3.117) at resonance. In this respect we treat the one-sided resonant and doubly resonant cases under additional conditions on the term $f(x, s)$. For the p-Laplacian equation under resonance and doubly resonance, we refer to [222]. Theorem 3.77 provides the existence of a positive solution in all cases of resonance for problem (3.117) with a nonhomogeneous operator in the principal part.

Chapter 4

Nonlinear Elliptic Equations with General Dependence on the Gradient of the Solution

4.1 GRADIENT DEPENDENT PROBLEMS WITH p-LAPLACIAN

The problem considered in this section is stated as follows:

$$\begin{cases} -\Delta_p u = f(x, u, \nabla u) & \text{in } \Omega, \\ u = 0 & \text{on } \partial\Omega. \end{cases} \tag{4.1}$$

In the problem statement Ω is a bounded domain in $\mathbb{R}^N$ of class C^2, $1 < p < N$, Δ_p is the p-Laplacian operator and $f : \Omega \times \mathbb{R} \times \mathbb{R}^N \to \mathbb{R}$ is a Carathéodory function, that is, $f(\cdot, s, \xi)$ is measurable for every $s \in \mathbb{R}$ and $\xi \in \mathbb{R}^N$, $f(x, \cdot, \cdot)$ is continuous for almost every $x \in \Omega$. We note that the right-hand side of the equation depends on the point x in the domain Ω, on the solution $u(x)$ and on the gradient $\nabla u(x)$ of the solution. Here our main concern is the dependence on the gradient $\nabla u(x)$, which prevents the application of variational methods. Our main goal is to obtain positive solutions to problem (4.1).

As before, we denote by $|\Omega|$ the Lebesgue measure of Ω, whereas p' stands for the conjugate of p, i.e., $\frac{1}{p} + \frac{1}{p'} = 1$. We endow the Sobolev space $W_0^{1,p}(\Omega)$ with the standard norm

$$\|u\| = \left(\int_\Omega |\nabla u|^p \, dx \right)^{\frac{1}{p}}$$

(see Remark 1.18). As usual, we set $u^+ = \max\{u, 0\}$ and $u^- = \max\{0, -u\}$. It is well known that if $u \in W_0^{1,p}(\Omega)$, then $u^+, u^- \in W_0^{1,p}(\Omega)$ (see Corollary 1.16).

In what follows, λ_1 denotes the first eigenvalue of the negative p-Laplacian operator on $W_0^{1,p}(\Omega)$ with the associated eigenfunction $\hat{u}_1 \in \operatorname{int}(C_0^1(\overline{\Omega})_+)$ (see (1.39)) satisfying $\|\hat{u}_1\|_p = 1$ as provided in Proposition 1.57. Our first assumptions on the reaction term $f(x, u, \nabla u)$ are:

H$(f)_1$ for every $M > 0$, there exist constants $k_M > 0$ and $0 < \theta_M < \lambda_1$ such that

$$|f(x, s, \xi)| \le k_M + \theta_M |s|^{p-1}$$

for a.e. $x \in \Omega$, $s \in \mathbb{R}$ and $\xi \in \mathbb{R}^N$ with $|\xi| \le M$;

Nonlinear Differential Problems with Smooth and Nonsmooth Constraints. DOI: 10.1016/B978-0-12-813386-6.00004-3

$H(f)_2^+$ for every $M > 0$ there exists a constant $\eta_M > \lambda_1$ such that

$$\liminf_{s\to 0^+} \frac{f(x,s,\xi)}{s^{p-1}} \geq \eta_M > \lambda_1$$

uniformly for a.e. $x \in \Omega$ and all $\xi \in \mathbb{R}^N$ with $|\xi| \leq M$;

$H(f)_3^+$ for every $M > 0$ there exists a constant $\zeta_M > 0$ such that

$$\limsup_{s\to 0^+} \frac{f(x,s,\xi)}{s^{p-1}} \leq \zeta_M$$

uniformly for a.e. $x \in \Omega$ and all $\xi \in \mathbb{R}^N$ with $|\xi| \leq M$.

Remark 4.1. Assumptions $H(f)_2^+$ and $H(f)_3^+$ imply that $f(x,0,\xi) = 0$ for a.e. $x \in \Omega$ and every $\xi \in \mathbb{R}^N$. So, in particular, $u = 0$ is a solution of (4.1). Since we seek positive solutions for problem (4.1), without loss of generality we may assume that $f(x,s,\xi) = 0$ for a.e. $x \in \Omega$, $s \leq 0$, $\xi \in \mathbb{R}^N$.

An indication that our growth conditions are justified in the context of looking for positive solutions is offered by the below remark.

Remark 4.2. Notice that if u is a positive solution of (4.1), then

$$f(x,u(x),\nabla u(x)) \leq \lambda_1 u^{p-1}(x)$$

on a set of positive measure in Ω. Indeed, otherwise u would solve the problem

$$\begin{cases} -\Delta_p u = m(x)u^{p-1} & \text{in } \Omega, \\ u = 0 & \text{on } \partial\Omega, \end{cases}$$

for a.e. $x \in \Omega$, where

$$m(x) = \frac{f(x,u(x),\nabla u(x))}{u^{p-1}(x)} > \lambda_1.$$

Then u has to change sign (see Proposition 1.57), which contradicts that u is positive.

We are going to consider an auxiliary problem by freezing the gradient in (4.1). Namely, for every $w \in C_0^1(\overline{\Omega})$, let us formulate the Dirichlet problem

$$\begin{cases} -\Delta_p u = f(x,u,\nabla w) & \text{in } \Omega, \\ u = 0 & \text{on } \partial\Omega. \end{cases} \tag{4.2}$$

By a solution of problem (4.2) we mean a weak solution, i.e., any $u \in W_0^{1,p}(\Omega)$ such that

$$\int_\Omega |\nabla u|^{p-2}\nabla u \cdot \nabla v\, dx = \int_\Omega f(x,u,\nabla w)v\, dx$$

for all $v \in W_0^{1,p}(\Omega)$. More generally, a function $\overline{u}_w \in W^{1,p}(\Omega)$, with $\overline{u}_w \geq 0$ on $\partial\Omega$ (in the sense of trace), is a supersolution for problem (4.2) if

$$\int_\Omega |\nabla \overline{u}_w|^{p-2} \nabla \overline{u}_w \cdot \nabla v \, dx \geq \int_\Omega f(x, \overline{u}_w, \nabla w) v \, dx$$

for all $v \in W_0^{1,p}(\Omega)$, $v \geq 0$ a.e. in Ω. Similarly, $\underline{u}_w \in W^{1,p}(\Omega)$, with $\underline{u}_w \leq 0$ on $\partial\Omega$ (in the sense of trace), is a subsolution for problem (4.2) if

$$\int_\Omega |\nabla \underline{u}_w|^{p-2} \nabla \underline{u}_w \nabla v \, dx \leq \int_\Omega f(x, \underline{u}_w, \nabla w) v \, dx$$

for all $v \in W_0^{1,p}(\Omega)$, $v \geq 0$ a.e. in Ω.

Lemma 4.3. *Assuming* $\mathrm{H}(f)_1$, *for every* $w \in C_0^1(\overline{\Omega})$ *there exists* $\overline{u}_w \in \operatorname{int}(C_0^1(\overline{\Omega})_+)$, *which is a supersolution of (4.2).*

Proof. Fix $w \in C_0^1(\overline{\Omega})$ and set $M = \|w\|_{C^1(\overline{\Omega})}$. From assumption $\mathrm{H}(f)_1$, we have that

$$|f(x, s, \nabla w(x))| \leq k_M + \theta_M |s|^{p-1} \tag{4.3}$$

for a.e. $x \in \Omega$ and all $s \in \mathbb{R}$, where $k_M > 0$ and $0 < \theta_M < \lambda_1$.

Consider the following Dirichlet problem:

$$\begin{cases} -\Delta_p u = k_M + \theta_M |u|^{p-1} & \text{in } \Omega, \\ u = 0 & \text{on } \partial\Omega. \end{cases}$$

Since the embedding of $W_0^{1,p}(\Omega)$ into $L^p(\Omega)$ is compact (see Theorem 1.20), the superposition operator $K_M : W_0^{1,p}(\Omega) \to W^{-1,p'}(\Omega)$ defined by $K_M(u(\cdot)) = k_M + \theta_M |u(\cdot)|^{p-1}$ is completely continuous. Taking into account from Proposition 1.56 that the p-Laplacian operator is strictly monotone, continuous, and bounded, we have that $-\Delta_p - K_M$ is pseudomonotone and bounded. Thanks to the estimate

$$\begin{aligned} &\frac{1}{p} \int_\Omega |\nabla u|^p \, dx - \int_\Omega \left(k_M |u| + \frac{1}{p} \theta_M |u|^p \right) dx \\ &\geq \frac{1}{p} \left(1 - \frac{\theta_M}{\lambda_1} \right) \|u\|^p - k_M |\Omega|^{\frac{1}{p'}} \lambda_1^{-\frac{1}{p}} \|u\|, \end{aligned}$$

and using $\theta_M < \lambda_1$ (see $\mathrm{H}(f)_1$), it follows that $-\Delta_p - K_M$ is coercive, hence surjective by Theorem 1.55. Therefore, there exists a function $\overline{u}_w \in W_0^{1,p}(\Omega)$ such that

$$\begin{cases} -\Delta_p \overline{u}_w = k_M + \theta_M |\overline{u}_w|^{p-1} & \text{in } \Omega, \\ \overline{u}_w = 0 & \text{on } \partial\Omega. \end{cases}$$

Next we prove that $\overline{u}_w \geq 0$. Using as test function $-\overline{u}_w^-$, we get

$$\begin{aligned}\int_\Omega |\nabla \overline{u}_w^-|^p \, dx &= -\int_\Omega |\nabla \overline{u}_w|^{p-2} \nabla \overline{u}_w \cdot \nabla \overline{u}_w^- \, dx \\ &= -\int_\Omega (k_M + \theta_M |\overline{u}_w|^{p-1}) \overline{u}_w^- \, dx \leq 0,\end{aligned}$$

which implies that $\overline{u}_w^- = 0$, so $\overline{u}_w \geq 0$. Notice also that $\overline{u}_w \neq 0$. By nonlinear regularity theory [129] and strong maximum principle [229], we have that $\overline{u}_w \in \operatorname{int}(C_0^1(\overline{\Omega})_+)$. Then from (4.3) we infer that $\overline{u}_w$ is a supersolution of (4.2), which completes the proof. □

Lemma 4.4. *Assume that condition* $\mathrm{H}(f)_2^+$ *holds. Then, for every* $w \in C_0^1(\overline{\Omega})$, *there exists* $\delta = \delta(w) > 0$ *such that if* $0 < \varepsilon < \delta$, *then* $\varepsilon \hat{u}_1$ *is a subsolution of (4.2).*

Proof. Let us fix $w \in C_0^1(\overline{\Omega})$ and set $M = \|w\|_{C^1(\overline{\Omega})}$. From assumption $\mathrm{H}(f)_2^+$, for $\sigma > 0$ so small that $\eta_M - \sigma > \lambda_1$, there exists $\gamma = \gamma(w) > 0$ such that

$$f(x, s, \nabla w(x)) \geq (\eta_M - \sigma) s^{p-1} > \lambda_1 s^{p-1} \tag{4.4}$$

for a.e. $x \in \Omega$ and all $0 < s < \gamma$. Set $\delta = \gamma \|\hat{u}_1\|_\infty^{-1}$. Then, for every $0 < \varepsilon < \delta$ and all $\varphi \in W_0^{1,p}(\Omega)$, $\varphi \geq 0$, from (4.4) we find that

$$\begin{aligned}\int_\Omega |\nabla(\varepsilon \hat{u}_1)|^{p-2} \nabla(\varepsilon \hat{u}_1) \cdot \nabla \varphi \, dx &= \lambda_1 \int_\Omega (\varepsilon \hat{u}_1)^{p-1} \varphi \, dx \\ &\leq \int_\Omega f(x, \varepsilon \hat{u}_1, \nabla w) \varphi \, dx,\end{aligned}$$

that is, $\varepsilon \hat{u}_1$ is a subsolution of (4.2). □

Now we are ready to prove the main result on the auxiliary problem (4.2).

Theorem 4.5. *Assume* $\mathrm{H}(f)_1$, $\mathrm{H}(f)_2^+$, *and* $\mathrm{H}(f)_3^+$. *Then for every* $w \in C_0^1(\overline{\Omega})$ *there exists* $u_w \in \operatorname{int}(C_0^1(\overline{\Omega})_+)$ *which is the smallest positive solution of (4.2).*

Proof. Let us fix $w \in C_0^1(\overline{\Omega})$ and set $M = \|w\|_{C^1(\overline{\Omega})}$. From Lemmas 4.3 and 4.4, we get that there exist a supersolution $\overline{u}_w$ and, for $0 < \varepsilon < \delta(w)$, a subsolution $\varepsilon \hat{u}_1$ of (4.2). Since $\overline{u}_w$ and $\varepsilon \hat{u}_1$ belong to $\operatorname{int}(C_0^1(\overline{\Omega})_+)$, it is possible to choose ε small enough in order that $\overline{u}_w - \varepsilon \hat{u}_1 \in \operatorname{int}(C_0^1(\overline{\Omega})_+)$.

Consider now the following truncation of f:

$$f_+(x, s) = \begin{cases} f(x, \varepsilon \hat{u}_1(x), \nabla w(x)) & \text{if } s < \varepsilon \hat{u}_1(x), \\ f(x, s, \nabla w(x)) & \text{if } \varepsilon \hat{u}_1(x) \leq s \leq \overline{u}_w(x), \\ f(x, \overline{u}_w(x), \nabla w(x)) & \text{if } s > \overline{u}_w(x). \end{cases}$$

Denote by $\Phi_+ : W_0^{1,p}(\Omega) \to \mathbb{R}$ the associated energy functional, that is,

$$\Phi_+(u) = \frac{1}{p}\|u\|^p - \int_\Omega \int_0^{u(x)} f_+(x,t)\,dt\,dx \quad \text{for all } u \in W_0^{1,p}(\Omega).$$

Clearly, Φ_+ is sequentially weakly lower semicontinuous, coercive, and continuously differentiable. Hence, it has a global minimizer $\tilde{u}_w^\varepsilon$ which is a critical point of Φ_+, thus a weak solution of the Dirichlet problem

$$\begin{cases} -\Delta_p u = f_+(x,u) & \text{in } \Omega, \\ u = 0 & \text{on } \partial\Omega. \end{cases}$$

Moreover, a standard comparison argument yields

$$\varepsilon \hat{u}_1 \le \tilde{u}_w^\varepsilon \le \overline{u}_w$$

a.e. in Ω, so that $\tilde{u}_w^\varepsilon$ is a solution of problem (4.2). Also Theorem 2.19 (see also the strong maximum principle in [229]) entails $\tilde{u}_w^\varepsilon \in \operatorname{int}(C_0^1(\overline{\Omega})_+)$.

Denote by S_ε the set of $C_0^1(\overline{\Omega})$-solutions of (4.2) which lie in the ordered interval $[\varepsilon\hat{u}_1, \overline{u}_w]$. As seen before, $S_\varepsilon \neq \emptyset$. If we consider on S_ε the pointwise order, through Zorn's lemma, it has a minimal element u_w^ε (see [165] for more details). Let us prove that u_w^ε is the smallest solution of (4.2) in S_ε. To this end, take $v \in S_\varepsilon$. The function $\min\{v, u_w^\varepsilon\}$ is a supersolution of (4.2) and clearly, $\min\{v, u_w^\varepsilon\} \ge \varepsilon\hat{u}_1$. Then, there exists a solution z of (4.2) such that $\varepsilon\hat{u}_1 \le z \le \min\{v, u_w^\varepsilon\}$. In particular, it turns out that $z \in S_\varepsilon$, $z \le u_w^\varepsilon$, and from the minimality of u_w^ε in S_ε, we conclude that $z = u_w^\varepsilon$. So $u_w^\varepsilon \le v$, as desired.

Fix now a decreasing sequence $\{\varepsilon_n\}$ of positive numbers such that $\varepsilon_n \to 0$. For every n there exists $u_w^{\varepsilon_n}$ which is the smallest solution of (4.2) in the ordered interval $[\varepsilon_n\hat{u}_1, \overline{u}_w]$. The sequence $\{u_w^{\varepsilon_n}\}$ is bounded in $W_0^{1,p}(\Omega)$, so there exists $u_w \in W_0^{1,p}(\Omega)$ such that, along a subsequence, $u_w^{\varepsilon_n} \rightharpoonup u_w$ in $W_0^{1,p}(\Omega)$. In particular, $u_w^{\varepsilon_n} \to u_w$ in $L^p(\Omega)$ and $u_w^{\varepsilon_n}(x) \to u_w(x)$ for a.e. $x \in \Omega$. Noticing that $\{u_w^{\varepsilon_n}\}$ is decreasing by construction and invoking Proposition 2.12, we get that the convergence $u_w^{\varepsilon_n} \to u_w$ is uniform.

We want to prove that $u_w \neq 0$. Assume by contradiction that $u_w = 0$ and set

$$z_n = \frac{u_w^{\varepsilon_n}}{\|u_w^{\varepsilon_n}\|}.$$

Hence $z_n \in W_0^{1,p}(\Omega)$ and $\|z_n\| = 1$, thereby up to a subsequence $\{z_n\}$ converges weakly in $W_0^{1,p}(\Omega)$ to some $z \in W_0^{1,p}(\Omega)$. On account of $z_n \ge 0$ a.e. in Ω, we have $z \ge 0$ a.e. in Ω. Denote

$$h_n(x) = \frac{f(x, u_w^{\varepsilon_n}(x), \nabla w(x))}{(u_w^{\varepsilon_n}(x))^{p-1}}.$$

The fact that $u_w^{\varepsilon_n}$ is a solution of (4.2) reads as

$$-\Delta_p u_w^{\varepsilon_n} = f(x, u_w^{\varepsilon_n}(x), \nabla w(x)),$$

so we get

$$-\Delta_p z_n = h_n(x) z_n^{p-1}. \tag{4.5}$$

Notice also that by invoking assumptions $\mathrm{H}(f)_2^+$ and $\mathrm{H}(f)_3^+$ and using the uniform convergence of $\{u_w^{\varepsilon_n}\}$ to zero, we deduce

$$\lambda_1 < \eta_M - \sigma \le h_n(x) \le \zeta_M$$

for a.e. $x \in \Omega$, whenever n is sufficiently large. From this inequality we obtain that the sequence $\{h_n\}$ is bounded in $L^\infty(\Omega)$. Consequently, there exists a function $h \in L^{p'}(\Omega)$ such that $h_n \rightharpoonup h$ in $L^{p'}(\Omega)$. By Mazur's lemma (see, e.g., [73, Chapter II]) we derive that

$$\lambda_1 < \eta_M - \sigma \le h(x) \le \zeta_M \tag{4.6}$$

for a.e. $x \in \Omega$. On the other hand, (4.5) implies that $\{z_n\}$ strongly converges to some z in $W_0^{1,p}(\Omega)$. Here we use the boundedness of $\{h_n\}$ in $L^\infty(\Omega)$ and the $(S)_+$-property of the p-Laplacian operator (see Proposition 1.56). The strong convergence ensures that $\|z\| = 1$, so $z \neq 0$. Passing to the limit in (4.5) shows that z verifies

$$\begin{cases} -\Delta_p z = h(x) z^{p-1} & \text{in } \Omega, \\ z = 0 & \text{on } \partial\Omega. \end{cases}$$

Therefore 1 is an eigenvalue of this weighted eigenvalue problem. From (4.6) and the monotonicity of the first eigenvalue $\hat{\lambda}_1(\cdot)$ of $-\Delta_p$ with respect to the weight, we note that $1 = \hat{\lambda}_1(\lambda_1) > \hat{\lambda}_1(h)$ which, from Proposition 1.57, implies that z changes sign, a contradiction. So, we have established that $u_w \neq 0$.

Nonlinear regularity theory up to the boundary (see Proposition 2.12 and Theorem 2.19) enables us to derive that $u_w \in \mathrm{int}(C_0^1(\overline{\Omega})_+)$. In order to conclude the proof, it remains to show that u_w is the smallest positive solution of (4.2). It is enough to prove that it is the smallest positive solution in the ordered interval $[0, \overline{u}_w]$. To this end, fix a solution v of (4.2) in $[0, \overline{u}_w]$. In particular, v is a supersolution of (4.2), and we can choose $\varepsilon_n > 0$ such that $v - \varepsilon_n \hat{u}_1 \in \mathrm{int}(C_0^1(\overline{\Omega})_+)$. Then we infer that $\varepsilon_n \hat{u}_1 \le u_w^{\varepsilon_n} \le v \le \overline{u}_w$ for $n \in \mathbb{N}$ large enough. Letting $n \to \infty$, we are led to $u_w \le v$, as desired. □

In view of Theorem 4.5, it is well defined the map $T : C_0^1(\overline{\Omega}) \to C_0^1(\overline{\Omega})$ given by

$$T(w) = u_w, \tag{4.7}$$

where u_w is the smallest positive solution of (4.2) as guaranteed by Theorem 4.5. It is clear that any fixed point for the map T will provide a positive solution to the original problem (4.1). In order to prove the continuity and the compactness of the map T, we will need some preliminary results. For any $w \in C_0^1(\overline{\Omega})$, denote by S_w the set of all functions $u \in \operatorname{int}(C_0^1(\overline{\Omega})_+)$ (see (1.39)) that are solutions of problem (4.2).

Lemma 4.6. *If $\{w_n\}$ is a bounded sequence in $C_0^1(\overline{\Omega})$ and $\{u_n\}$ is a sequence in $C_0^1(\overline{\Omega})$ with $u_n \in S_{w_n}$ for all n, then the sequence $\{u_n\}$ is relatively compact in $C_0^1(\overline{\Omega})$.*

Proof. Let $M > 0$ satisfy $\|w_n\|_{C_0^1(\overline{\Omega})} \le M$ for all n. From assumption $\mathrm{H}(f)_1$ it follows that

$$\begin{aligned}\|u_n\|^p = \int_\Omega f(x, u_n, \nabla w_n) u_n \, dx &\le k_M |\Omega|^{\frac{1}{p'}} \|u_n\|_p + \theta_M \lambda_1^{-1} \|u_n\|^p \\ &\le k_M |\Omega|^{\frac{1}{p'}} \lambda_1^{-\frac{1}{p}} \|u_n\| + \theta_M \lambda_1^{-1} \|u_n\|^p,\end{aligned}$$

which implies that $\{u_n\}$ is bounded in $W_0^{1,p}(\Omega)$ because $\theta_M < \lambda_1$. On the basis of regularity up to the boundary in [129], we infer that $\{u_n\}$ is bounded in $C^{1,\alpha}(\overline{\Omega})$ for some $\alpha \in (0,1)$ independent of n (due to the assumption that $\{w_n\}$ is bounded in $C_0^1(\overline{\Omega})$). Since $C^{1,\alpha}(\overline{\Omega})$ is compactly embedded into $C^1(\overline{\Omega})$, we achieve the conclusion. □

Lemma 4.7. *If $\{w_n\}$ is a sequence in $C_0^1(\overline{\Omega})$ such that $w_n \to w$ in $C_0^1(\overline{\Omega})$ and if $\{u_n\}$ is a sequence in $C_0^1(\overline{\Omega})$ with $u_n \in S_{w_n}$ for all n, then there exist a subsequence $\{u_{n_k}\}$ and an element $u \in S_w$ such that $u_{n_k} \to u$ in $C_0^1(\overline{\Omega})$.*

Proof. From Lemma 4.6, there exist a subsequence $\{u_{n_k}\}$ and some $u \in C_0^1(\overline{\Omega})$ such that $u_{n_k} \to u$ in $C_0^1(\overline{\Omega})$. We note that

$$\begin{cases} -\Delta_p u_{n_k} = f(x, u_{n_k}, \nabla w_{n_k}) & \text{in } \Omega, \\ u_{n_k} = 0 & \text{on } \partial\Omega \end{cases}$$

for all k. So, for every $\varphi \in W_0^{1,p}(\Omega)$ and $k \in \mathbb{N}$ we obtain

$$\int_\Omega |\nabla u_{n_k}|^{p-2} \nabla u_{n_k} \cdot \nabla \varphi \, dx = \int_\Omega f(x, u_{n_k}, \nabla w_{n_k}) \varphi \, dx.$$

Passing to the limit as $k \to \infty$ yields

$$\int_\Omega |\nabla u|^{p-2} \nabla u \cdot \nabla \varphi \, dx = \int_\Omega f(x, u, \nabla w) \varphi \, dx,$$

i.e., u is a solution of (4.2). In order to prove that $u \neq 0$ it is enough to make use of the same argument as in the corresponding part of the proof of Theorem 4.5. Then the regularity up to the boundary and strong maximum principle imply $u \in \operatorname{int}(C_0^1(\overline{\Omega})_+)$, whence the conclusion. □

The following lemma is the key step in our construction.

Lemma 4.8. *If $\{w_n\}$ is a sequence in $C_0^1(\overline{\Omega})$ such that $w_n \to w$ in $C_0^1(\overline{\Omega})$, then for each $v \in S_w$ there exists $v_n \in S_{w_n}$ for all $n \in \mathbb{N}$ such that*

$$v_n \to v \quad \text{in } C_0^1(\overline{\Omega}).$$

Proof. Fix $n \in \mathbb{N}$, and let z_n^0 be the unique solution of the problem

$$\begin{cases} -\Delta_p u = f(x, v, \nabla w_n) & \text{in } \Omega, \\ u = 0 & \text{on } \partial\Omega. \end{cases}$$

Exploiting assumption $\mathrm{H}(f)_2^+$ and bearing in mind that $v \in \operatorname{int}(C_0^1(\overline{\Omega})_+)$, observe that $f(\cdot, v(\cdot), \nabla w_n(\cdot)) \not\equiv 0$, which leads to $z_n^0 \neq 0$.

Let us prove that z_n^0 lies in $\operatorname{int}(C_0^1(\overline{\Omega})_+)$. First, it is straightforward to check that $\{z_n^0\}$ is bounded in $W_0^{1,p}(\Omega)$, thus by Proposition 2.12 there exists a subsequence $\{z_{n_q}^0\}$ convergent in $C_0^1(\overline{\Omega})$ to a solution of the problem

$$\begin{cases} -\Delta_p u = f(x, v, \nabla w) & \text{in } \Omega, \\ u = 0 & \text{on } \partial\Omega. \end{cases}$$

Taking into account that v is the unique solution of the above problem, we get

$$\lim_{q \to \infty} z_{n_q}^0 = v \quad \text{in } C_0^1(\overline{\Omega}).$$

Actually, as readily seen, the strong convergence is true for the whole sequence

$$\lim_{n \to \infty} z_n^0 = v \quad \text{in } C_0^1(\overline{\Omega}),$$

which implies the above assertion.

Next let us consider the unique (positive) solution z_n^1 of the following problem:

$$\begin{cases} -\Delta_p u = f(x, z_n^0, \nabla w_n) & \text{in } \Omega, \\ u = 0 & \text{on } \partial\Omega. \end{cases}$$

As before it turns out that

$$\lim_{n \to \infty} z_n^1 = v \quad \text{in } C_0^1(\overline{\Omega}).$$

Inductively, we can define, for each $n \in \mathbb{N}$, z_n^k in $C_0^1(\overline{\Omega})$ as the unique solution of the problem

$$\begin{cases} -\Delta_p u = f(x, z_n^{k-1}, \nabla w_n) & \text{in } \Omega, \\ u = 0 & \text{on } \partial\Omega, \end{cases}$$

and for each $k \in \mathbb{N}$ it follows that

$$\lim_{n\to\infty} z_n^k = v \quad \text{in } C_0^1(\overline{\Omega}).$$

Let us now prove that there exists a constant $c > 0$ such that $\|z_n^k\| \le c$ for all $n, k \in \mathbb{N}$. Indeed, setting $M = \max\{\sup_n \|w_n\|_{C^1(\overline{\Omega})}, \|w\|_{C^1(\overline{\Omega})}\}$, by hypothesis $\mathrm{H}(f)_1$ we have

$$\begin{aligned} \|z_n^k\|^p &= \int_\Omega f(x, z_n^{k-1}, \nabla w_n) z_n^k \, dx \\ &\le \left(k_M |\Omega|^{\frac{1}{p'}} + \theta_M \left(\int_\Omega (z_n^{k-1})^p \, dx\right)^{\frac{p-1}{p}}\right) \lambda_1^{-\frac{1}{p}} \|z_n^k\|, \end{aligned}$$

so

$$\begin{aligned} \|z_n^k\|^{p-1} &\le \lambda_1^{-\frac{1}{p}} \left(k_M |\Omega|^{\frac{1}{p'}} + \theta_M \|z_n^{k-1}\|_{L^p(\Omega)}^{p-1}\right) \\ &\le \lambda_1^{-\frac{1}{p}} k_M |\Omega|^{\frac{1}{p'}} + \lambda_1^{-1} \theta_M \|z_n^{k-1}\|^{p-1}. \end{aligned}$$

If we set $c_1 = \lambda_1^{-\frac{1}{p}} k_M |\Omega|^{\frac{1}{p'}}$ and $c_2 = \lambda_1^{-1}\theta_M$, it is easy to see by induction that

$$\|z_n^k\|^{p-1} \le c_1 \sum_{j=0}^{k-1} c_2^j + c_2^k \|z_n^0\|^{p-1}$$

for every $k \ge 1$. Since $c_2 < 1$ (according to assumption $\mathrm{H}(f)_1$) and $\{z_n^0\}$ is bounded in $W_0^{1,p}(\Omega)$, we deduce that, for every $n, k \in \mathbb{N}$,

$$\|z_n^k\|^{p-1} \le c_1 \sum_{j=0}^{\infty} c_2^j + c_3,$$

with a constant $c_3 > 0$. This entails our claim.

By standard arguments, we have that the net $\{z_n^k\}_{n,k}$ is relatively compact in $C_0^1(\overline{\Omega})$. Then up to a subnet, we can assume that there exists $u \in C_0^1(\overline{\Omega})$ such that

$$\lim_{n,k\to\infty} z_n^k = u,$$

that is, there exists $N \in \mathbb{N}$ such that

$$\|z_n^k - u\|_{C^1(\overline{\Omega})} < \varepsilon \tag{4.8}$$

for all $n, k > N$. At this point, for every $k > N$, we obtain that

$$\limsup_{n\to\infty} \|z_n^k - u\|_{C^1(\overline{\Omega})} \le \varepsilon.$$

We are thus able to select a subsequence $\{z_{n_q}^k\}$ with the property

$$\lim_{q\to\infty} \|z_{n_q}^k - u\|_{C^1(\overline{\Omega})} = \|\lim_{q\to\infty} z_{n_q}^k - u\|_{C^1(\overline{\Omega})} = \|v - u\|_{C^1(\overline{\Omega})} \le \varepsilon,$$

thereby $u = v$. Consequently, according to (4.8), for all $n, k > N$ we have

$$\|z_n^k - v\|_{C^1(\overline{\Omega})} < \varepsilon. \tag{4.9}$$

Now, for each $n \in \mathbb{N}$, there exist a subsequence $\{k_s = k_s(n)\}_s \subset \mathbb{N}$ and $v_n \in C_0^1(\overline{\Omega})$ satisfying

$$\lim_{s\to\infty} z_n^{k_s} = v_n \text{ in } C_0^1(\overline{\Omega}).$$

Clearly, v_n is a solution of (P_{w_n}). Let us prove that

$$\lim_{n\to\infty} v_n = v \text{ in } C_0^1(\overline{\Omega}).$$

If not, we can construct a subsequence $\{n_q\}$ with $n_q > N$ such that $\|v_{n_q} - v\|_{C^1(\overline{\Omega})} > \varepsilon$ for all $q \in \mathbb{N}$ and some $\varepsilon > 0$. This amounts to saying that

$$\varepsilon < \|v_{n_q} - v\|_{C^1(\overline{\Omega})} = \|\lim_{s\to\infty} z_{n_q}^{k_s} - v\|_{C^1(\overline{\Omega})} = \lim_{s\to\infty} \|z_{n_q}^{k_s} - v\|_{C^1(\overline{\Omega})}.$$

In particular, for any $q \in \mathbb{N}$ there exists $s_q = s_q(n_q) \in \mathbb{N}$ such that $k_{s_q} > N$ and

$$\|z_{n_q}^{k_{s_q}} - v\|_{C^1(\overline{\Omega})} > \varepsilon,$$

against (4.9).

Recall that $v \in \operatorname{int}(C_0^1(\overline{\Omega})_+)$. Since the convergence of $\{v_n\}$ to v is uniform on the compact subsets of Ω, it follows that $v_n > 0$ in Ω whenever n is sufficiently large. Also, since $\{\frac{\partial v_n}{\partial \nu}\}$ converges uniformly to $\frac{\partial v}{\partial \nu}$ on $\partial\Omega$, we obtain that $v_n \in \operatorname{int}(C_0^1(\overline{\Omega})_+)$ for n large enough. This completes the proof. □

In the rest of this section we rely on the following well known result (see, e.g., [141, Corollary 4.4.12], [169, Theorem 4.27]).

Theorem 4.9 (Schaefer's fixed point theorem). *Let X be a Banach space and let $T : X \to X$ be a continuous and compact map. Assume that the set*

$$\{u \in X : u = \lambda T(u) \text{ for some } \lambda \in [0, 1]\}$$

is bounded. Then T has a fixed point.

In order to apply Theorem 4.9 to the map introduced in (4.7), we need to strengthen the growth condition $\mathrm{H}(f)_1$ by requiring

$\mathrm{H}(f)'_1$ there exist positive constants k_0, θ_0, θ_1 with $\theta_0 + \theta_1\lambda_1^{1/p'} < \lambda_1$ such that

$$|f(x, s, \xi)| \le k_0 + \theta_0|s|^{p-1} + \theta_1|\xi|^{p-1}$$

for a.a. $x \in \Omega$, all $s \in \mathbb{R}$, and $\xi \in \mathbb{R}^N$.

Our main result on problem (4.1) reads as follows:

Theorem 4.10. *Assume that conditions $\mathrm{H}(f)'_1$, $\mathrm{H}(f)_2^+$, and $\mathrm{H}(f)_3^+$ hold. Then problem (4.1) has at least one (positive) solution $u \in \operatorname{int}(C_0^1(\overline{\Omega})_+)$.*

Proof. First, let us note that the map T in (4.7) is compact, that is, for each sequence $\{w_n\}$ bounded in $C_0^1(\overline{\Omega})$, $\{T(w_n)\}$ is relatively compact in $C_0^1(\overline{\Omega})$. This assertion follows readily from Lemma 4.6 applied to the sequence $u_n = T(w_n) \in S_{w_n}$.

Next let us prove that the map T in (4.7) is continuous. Towards this let $\{w_n\}$ be a sequence in $C_0^1(\overline{\Omega})$ such that

$$\lim_{n\to\infty} w_n = w \quad \text{in } C_0^1(\overline{\Omega}).$$

For every $n \in \mathbb{N}$, set $u_n = T(w_n)$. By Lemma 4.7 there exist a subsequence $\{u_{n_k}\}$ and an element $u \in S_w$ that fulfill

$$\lim_{k\to\infty} u_{n_k} = u \quad \text{in } C_0^1(\overline{\Omega}).$$

We claim that u is the smallest positive solution of (P_w). In this respect, fix a positive solution v of (4.2). According to Lemma 4.8, there exists a (positive) solution $v_n \in \operatorname{int}(C_0^1(\overline{\Omega})_+)$ of (P_{w_n}) such that

$$\lim_{n\to\infty} v_n = v \quad \text{in } C_0^1(\overline{\Omega}).$$

Since u_{n_k} is the smallest positive solution of $(P_{w_{n_k}})$, we have

$$u_{n_k} \le v_{n_k} \quad \text{for all } k \in \mathbb{N}.$$

Passing to the limit yields $u \le v$, which means that

$$\lim_{k\to\infty} T(w_{n_k}) = T(w) \quad \text{in } C_0^1(\overline{\Omega}).$$

In fact, the whole sequence $\{T(w_n)\}$ converges to $T(w)$. If not, there would be a subsequence $\{w_{n_q}\}$ such that $\|T(w_{n_q}) - T(w)\|_{C^1(\overline{\Omega})} \ge \varepsilon$ for every $q \in \mathbb{N}$ and for some $\varepsilon > 0$. Arguing as above, we find a subsequence $u_{n_{q_r}} = T(w_{n_{q_r}})$ converging to $T(w)$, which gives rise to a contradiction. It is thus proven that the map T in (4.7) is continuous.

Now we check that the set

$$\{w \in C_0^1(\overline{\Omega}) : \ w = \lambda T(w) \ \text{ for some } \lambda \in [0,1]\} \tag{4.10}$$

is bounded in $C_0^1(\overline{\Omega})$. To this end, let $w \in C_0^1(\overline{\Omega})$ and $\lambda \in [0,1]$ satisfy $w = \lambda T(w)$. We may assume that $\lambda > 0$ (otherwise, $w = 0$). Then w is a solution of the Dirichlet problem

$$\begin{cases} -\Delta_p w = \tilde{f}_\lambda(x, w, \nabla w) & \text{in } \Omega, \\ w = 0 & \text{on } \partial\Omega, \end{cases}$$

where $\tilde{f}_\lambda(x,s,\xi) = \lambda^{p-1} f(x, \lambda^{-1}s, \xi)$. By assumption $\mathrm{H}(f)_1'$, the Carathéodory function $\tilde{f}_\lambda$ satisfies the growth condition

$$|\tilde{f}_\lambda(x,s,\xi)| \le k_0 + \theta_0 |s|^{p-1} + \theta_1 |\xi|^{p-1} \tag{4.11}$$

for a.a. $x \in \Omega$, all $s \in \mathbb{R}$, and $\xi \in \mathbb{R}^N$. Notice that the coefficients in (4.11) are independent of $\lambda \in (0,1]$.

We claim that there is $M > 0$ independent of w and λ such that

$$\|w\| \le M. \tag{4.12}$$

Indeed, acting with the test function w and using (4.11), we obtain the estimate

$$\begin{aligned} \|w\|^p &= \int_\Omega \tilde{f}_\lambda(x, w, \nabla w) w \, dx \le \int_\Omega \left(k_0|w| + \theta_0 |w|^p + \theta_1 |\nabla w|^{p-1}|w|\right) dx \\ &\le k_0 |\Omega|^{\frac{1}{p'}} \lambda_1^{-\frac{1}{p}} \|w\| + \theta_0 \lambda_1^{-1} \|w\|^p + \theta_1 \lambda_1^{-\frac{1}{p}} \|w\|^p. \end{aligned}$$

Since $1 > \theta_0 \lambda_1^{-1} + \theta_1 \lambda_1^{-\frac{1}{p}}$ (see $\mathrm{H}(f)_1'$) and $p > 1$, we arrive at (4.12).

In view of (4.11) and (4.12), the nonlinear regularity theory up to the boundary in [129] (see also Proposition 2.12) provides a constant $M' > 0$ independent of w and λ such that $\|w\|_{C^1(\overline{\Omega})} \le M'$, which justifies the boundedness of the set in (4.10) as claimed.

Observe that any fixed point of the map T in (4.7) is a solution of problem (4.1) belonging to $\mathrm{int}\,(C_0^1(\overline{\Omega})_+)$. Therefore, through Theorem 4.9, the proof is complete. □

We set forth the counterpart of Theorem 4.10 on the negative half-line. We formulate in a symmetric way the corresponding hypotheses:

$H(f)_2^-$ for every $M > 0$ there exists a constant $\eta_M > \lambda_1$ such that

$$\liminf_{s \to 0^-} \frac{f(x,s,\xi)}{|s|^{p-2}s} \geq \eta_M > \lambda_1$$

uniformly for a.e. $x \in \Omega$ and all $\xi \in \mathbb{R}^N$ with $|\xi| \leq M$;

$H(f)_3^-$ for every $M > 0$ there exists a constant $\zeta_M > 0$ such that

$$\limsup_{s \to 0^-} \frac{f(x,s,\xi)}{|s|^{p-2}s} \leq \zeta_M$$

uniformly for a.e. $x \in \Omega$ and all $\xi \in \mathbb{R}^N$ with $|\xi| \leq M$.

On the pattern of Theorem 4.10 we can prove the existence of a negative solution to problem (4.1). More precisely, we can show that the map assigning to every $w \in C_0^1(\overline{\Omega})$ the biggest negative solution of (4.2) (its existence can be established similarly to that of the smallest positive solution) possesses a fixed point resulting in the following statement.

Theorem 4.11. *Assume that conditions* $H(f)_1'$, $H(f)_2^-$, *and* $H(f)_3^-$ *hold. Then problem (4.1) has at least one (negative) solution* $v \in -\mathrm{int}\,(C_0^1(\overline{\Omega})_+)$.

By combining Theorems 4.10 and 4.11, we obtain the following multiplicity result.

Theorem 4.12. *Assume* $H(f)_1'$, $H(f)_2^{\pm}$, *and* $H(f)_3^{\pm}$. *Then problem (4.1) has at least two solutions u and v of opposite constant sign with* $u \in \mathrm{int}\,(C_0^1(\overline{\Omega})_+)$ *and* $v \in -\mathrm{int}\,(C_0^1(\overline{\Omega})_+)$.

A consequence of Theorem 4.12 is the existence of extremal constant-sign solutions for problem (4.1).

Corollary 4.13. *Under the assumptions of Theorem 4.12, problem (4.1) has the smallest positive solution and the biggest negative solution.*

Proof. From Theorem 4.10 we know that there exists a solution $u \in \mathrm{int}(C_0^1(\overline{\Omega})_+)$ which can be regarded as a supersolution of problem (4.1).

By virtue of condition $H(f)_2^+$ we can find $\varepsilon_0 > 0$ such that

$$-\Delta_p(\varepsilon\hat{u}_1) = \lambda_1(\varepsilon\hat{u}_1)^{p-1} \leq f(x, \varepsilon\hat{u}_1, \varepsilon\nabla\hat{u}_1)$$

provided $0 < \varepsilon \leq \varepsilon_0$. This expresses that $\varepsilon\hat{u}_1$ is a subsolution of problem (4.1) for all $0 < \varepsilon \leq \varepsilon_0$. In addition, since $u \in \mathrm{int}\,(C_0^1(\overline{\Omega})_+)$, we may choose ε_0 so small that $u - \varepsilon\hat{u}_1 \in \mathrm{int}\,(C_0^1(\overline{\Omega})_+)$ for every $0 < \varepsilon \leq \varepsilon_0$.

We are thus in a position to apply [38, Theorem 3.22], which ensures the existence of the smallest positive solution u_ε of problem (4.1) in the ordered interval $[\varepsilon\hat{u}_1, u]$ for all $0 < \varepsilon \leq \varepsilon_0$. Set $u_n := u_{\frac{1}{n}}$. Thanks to the choice of u_n, we note that the sequence $\{u_n\}$ is decreasing. So, there exists $u_0 \in C_0^1(\overline{\Omega})_+$ such that $u_n \to u_0$ in $C_0^1(\overline{\Omega})$ and u_0 is a solution of (4.1). On the basis of hypotheses $\mathrm{H}(f)_2^+$ and $\mathrm{H}(f)_3^+$ we also infer that $u_0 \neq 0$ (see the proof of Theorem 4.5). This enables us to apply the strong maximum principle in [229] (or Theorem 2.19) to obtain that $u_0 \in \mathrm{int}\,(C_0^1(\overline{\Omega})_+)$.

It remains to show that u_0 is the smallest positive solution of problem (4.1). Let v be a positive solution of problem (4.1). The nonlinear regularity theory up to the boundary and strong maximum principle ensure that $v \in \mathrm{int}(C_0^1(\overline{\Omega})_+)$. It follows that

$$\frac{1}{n}\hat{u}_1 \leq \min\{u, v\} \leq u$$

whenever n is sufficiently large. By applying [38, Theorem 3.22], we see that there exists a solution v_n of (4.1) with $v_n \in [\frac{1}{n}\hat{u}_1, \min\{u, v\}]$ because $\min\{u, v\}$ is a supersolution. Then the minimality property of u_n entails that $u_n \leq v_n \leq v$. Letting $n \to \infty$ leads to $u_0 \leq v$, whence the desired conclusion.

In an analogous way we can proceed for the existence of the biggest negative solution of problem (4.1). □

Finally, we illustrate the applicability of our results with a simple example of nonlinearity $f(x, s, \xi)$ that fulfills all our hypotheses.

Example 4.14. Let $g : \overline{\Omega} \times \mathbb{R}^N \to \mathbb{R}$ be a continuous, positive function. Then, for any continuous function $f : \overline{\Omega} \times \mathbb{R} \times \mathbb{R}^N \to \mathbb{R}$ satisfying the growth condition $\mathrm{H}(f)_1'$ and

$$f(x, s, \xi) = |s|^{p-2}s(\lambda_1 + g(x, \xi)) \quad \text{for } |s| \text{ small},$$

it is readily seen that hypotheses $\mathrm{H}(f)_2^\pm$ and $\mathrm{H}(f)_3^\pm$ hold true, so Theorem 4.12 and Corollary 4.13 apply.

4.2 GRADIENT DEPENDENT PROBLEMS WITH (p, q)-LAPLACIAN

This section is devoted to the following quasilinear elliptic problem with gradient dependence

$$\begin{cases} -\Delta_p u - \mu\Delta_q u = f(x, u, \nabla u) & \text{in } \Omega, \\ u = 0 & \text{on } \partial\Omega \end{cases} \tag{4.13}$$

on a bounded domain Ω in $\mathbb{R}^N$ $(N \geq 2)$ with C^2 boundary $\partial\Omega$, where Δ_p and Δ_q stand for the p-Laplacian and q-Laplacian on $W_0^{1,p}(\Omega)$ and $W_0^{1,q}(\Omega)$, respectively, with $1 < q < p < +\infty$, and a constant $\mu \geq 0$. Problem (4.13) has a strong unifying effect. If $\mu = 0$, the equation in (4.13) is driven by the p-Laplacian, which is a $(p-1)$-homogeneous operator, whereas for $\mu = 1$ the leading operator is the (p, q)-Laplacian $\Delta_p + \Delta_q$, which is a nonhomogeneous operator. The C^2 smoothness for the boundary $\partial\Omega$ is necessary later on to manage appropriate regularity results. In order to get more insight on the main ideas, we suppose that $N > p$. The case $N \leq p$ is actually simpler and consequently is omitted.

The right-hand side in the equation of (4.13) is determined by a Carathéodory function $f : \Omega \times \mathbb{R} \times \mathbb{R}^N \to \mathbb{R}$, that is, $f(\cdot, s, \xi)$ is measurable for every $(s, \xi) \in \mathbb{R} \times \mathbb{R}^N$, and $f(x, \cdot, \cdot)$ is continuous for a.e. $x \in \Omega$, satisfying the growth condition

H$(f)_1$ there exist constants $a_1, a_2 \geq 0$, $0 \leq \alpha < p^* - 1$, $0 \leq \beta < \frac{(p^*-1)p}{p^*}$, $1 \leq \gamma < p^*$, and a function $\sigma \in L^{\gamma'}(\Omega)$, $\sigma \geq 0$, such that

$$|f(x, s, \xi)| \leq \sigma(x) + a_1|s|^\alpha + a_2|\xi|^\beta$$
$$\text{for a.e. } x \in \Omega, \text{ all } s \in \mathbb{R},\ \xi \in \mathbb{R}^N.$$

Here, as usual, $p^* := \frac{Np}{N-p}$ is the Sobolev critical exponent. In the present section for every $r \geq 1$ we set $r' := \frac{r}{r-1}$, that is, the Hölder conjugate of r.

For a later use, we write an additional, this time unilateral growth condition:

H$(f)_2$ there exist constants $b_1, b_2 \geq 0$ such that $b_1\lambda_{1,p}^{-1} + b_2 < 1$, and a function $\tau \in L^1(\Omega)$ such that

$$f(x, s, \xi)s \leq \tau(x) + b_1|s|^p + b_2|\xi|^p$$
$$\text{for a.e. } x \in \Omega, \text{ all } s \in \mathbb{R},\ \xi \in \mathbb{R}^N.$$

The notation $\lambda_{1,p}$ stands for the first eigenvalue of $-\Delta_p$ on $W_0^{1,p}(\Omega)$, which is expressed by

$$\lambda_{1,p} = \inf_{u \in W_0^{1,p}(\Omega) \setminus \{0\}} \frac{\|\nabla u\|_p^p}{\|u\|_p^p}, \tag{4.14}$$

where $\|\cdot\|_p$ denotes the usual L^p-norm (see Proposition 1.57). We emphasize that condition H$(f)_2$ is independent of condition H$(f)_1$. For instance, H$(f)_2$ is fulfilled if the sign requirement $f(x, s, \xi)s \leq 0$ holds, without any growth restriction for $f(x, s, \xi)$.

Later on we shall need the L^p-normalized, positive eigenfunction $\phi_{1,p}$ of $-\Delta_p$ on $W_0^{1,p}(\Omega)$ corresponding to the eigenvalue $\lambda_{1,p}$ (see (4.14)), so that

$$-\Delta_p \phi_{1,p} = \lambda_{1,p}\phi_{1,p}^{p-1} \text{ in } \Omega,\ \phi_{1,p} = 0 \text{ on } \partial\Omega,\ \int_\Omega \phi_{1,p}^p\, dx = 1,$$

and the L^q-normalized positive eigenfunction $\phi_{1,q}$ of $-\Delta_q$ on $W_0^{1,q}(\Omega)$ corresponding to $\lambda_{1,q}$, that is,

$$-\Delta_q \phi_{1,q} = \lambda_{1,q}\phi_{1,q}^{q-1} \text{ in } \Omega,\ \phi_{1,q} = 0 \text{ on } \partial\Omega,\ \text{with} \int_\Omega \phi_{1,q}^q\, dx = 1.$$

By the nonlinear regularity theory and the strong maximum principle, we know that $\phi_{1,p}, \phi_{1,q} \in \operatorname{int}(C_0^1(\overline{\Omega})_+)$ and, for positive constants c_1 and c_2,

$$c_1\phi_{1,p}(x) \le \phi_{1,q}(x) \le c_2\phi_{1,p}(x) \quad \text{for all } x \in \Omega.$$

A solution $u \in W_0^{1,p}(\Omega)$ of problem (4.13) is understood in the weak sense meaning that

$$\int_\Omega (|\nabla u|^{p-2} + \mu|\nabla u|^{q-2})\nabla u \cdot \nabla h\, dx = \int_\Omega f(x,u,\nabla u)h\, dx \tag{4.15}$$

for all $h \in W_0^{1,p}(\Omega)$. The definition makes sense in view of condition H$(f)_1$.
Later, we will need to strengthen the conditions H$(f)_1$ and H$(f)_2$ as follows:

H(f) $f : \Omega \times \mathbb{R} \times \mathbb{R}^N \to \mathbb{R}$ is a Carathéodory function for which there exist constants $c_1 \ge 0$, $c_2 \ge 0$, $r > \max\{\frac{p^*}{p-1}, N\}$, and a function $\omega \in L^\infty(\Omega)$, $\omega \ge 0$, such that

$$|f(x,s,\xi)| \le \omega(x) + c_1|s|^{\frac{p^*}{r}} + c_2|\xi|^{\frac{p}{r}}$$
$$\text{for a.e. } x \in \Omega,\ \text{all } s \in \mathbb{R},\ \xi \in \mathbb{R}^N.$$

We point out that condition H(f) is stronger than both H$(f)_1$ and H$(f)_2$.

Lemma 4.15. *Hypothesis* H(f) *implies* H$(f)_1$ *and* H$(f)_2$.

Proof. By the choice of r in H(f), we have that $\frac{p^*}{r} < p^* - 1$ and $\frac{p}{r} < \frac{(p^*-1)p}{p^*}$. Consequently, H$(f)$ yields H$(f)_1$ by taking $\sigma(x) = \omega(x)$, $\alpha = \frac{p^*}{r}$ and $\beta = \frac{p}{r}$.
From H(f) and Young's inequality we obtain that

$$\begin{aligned}|f(x,s,\xi)s| &\le \omega(x)|s| + c_1|s|^{\frac{p^*}{r}+1} + c_2|\xi|^{\frac{p}{r}}|s| \\ &\le \omega(x)|s| + c_1|s|^{\frac{p^*}{r}+1} + c(\varepsilon)|s|^{\frac{r}{r-1}} + \varepsilon|\xi|^p\end{aligned}$$

for a.e. $x \in \Omega$ and all $s \in \mathbb{R}$, $\xi \in \mathbb{R}^N$, with an arbitrary $\varepsilon > 0$ and a corresponding constant $c(\varepsilon)$. We note that the choice of r in H(f) results in $\frac{p^*}{r} + 1 < p$ and $\frac{r}{r-1} < p$. A further application of Young's inequality yields

$$|f(x,s,\xi)s| \le c(\delta)\big(\omega(x)^{p'} + 1\big) + \delta|s|^p + \varepsilon|\xi|^p$$

for a.e. $x \in \Omega$ and all $s \in \mathbb{R}$, $\xi \in \mathbb{R}^N$, with $\delta > 0$ arbitrary and some constant $c(\delta)$. Therefore, choosing δ, ε sufficiently small, we can infer that $\mathrm{H}(f)_2$ holds true, which completes the proof. □

As usual, the space $W_0^{1,p}(\Omega)$ is endowed with the norm

$$\|u\| = \left(\int_\Omega |\nabla u|^p \, dx \right)^{\frac{1}{p}}.$$

The continuity of the embedding of $W_0^{1,p}(\Omega)$ in $L^r(\Omega)$ for $1 \leq r \leq p^*$ (see Theorem 1.20) guarantees the existence of a constant $c_r > 0$ such that

$$\|u\|_r \leq c_r \|u\| \quad \text{for all } u \in W_0^{1,p}(\Omega). \tag{4.16}$$

In view of the growth assumption $\mathrm{H}(f)_1$, we can consider the Nemytskii operator $N_f : W_0^{1,p}(\Omega) \to W^{-1,p'}(\Omega)$ (where $W^{-1,p'}(\Omega) = W_0^{1,p}(\Omega)^*$) defined by

$$N_f(u) = f(\cdot, u, \nabla u) \quad \text{for all } u \in W_0^{1,p}(\Omega). \tag{4.17}$$

Then problem (4.13) can be equivalently expressed

$$-\Delta_p u - \mu \Delta_q u = N_f(u) \quad \text{in } W^{-1,p'}(\Omega).$$

We are going to introduce the notions of supersolution and of subsolution for problem (4.13). A function $u \in W^{1,p}(\Omega)$ is called a supersolution of problem (4.13) if it satisfies $u \geq 0$ on $\partial\Omega$ and

$$\int_\Omega (|\nabla u|^{p-2} + \mu |\nabla u|^{q-2}) \nabla u \cdot \nabla h \, dx \geq \int_\Omega f(x, u, \nabla u) h \, dx$$

for all $h \in W_0^{1,p}(\Omega)$, $h \geq 0$ a.e. in Ω. A function $u \in W^{1,p}(\Omega)$ is called a subsolution of problem (4.13) if it satisfies $u \leq 0$ on $\partial\Omega$ and

$$\int_\Omega (|\nabla u|^{p-2} + \mu |\nabla u|^{q-2}) \nabla u \cdot \nabla h \, dx \leq \int_\Omega f(x, u, \nabla u) h \, dx$$

for all $h \in W_0^{1,p}(\Omega)$, $h \geq 0$ a.e. in Ω. Owing to the growth condition $\mathrm{H}(f)_1$, the notions of supersolution and subsolution of problem (4.13) are correctly defined.

The following useful property of subsolution–supersolutions is shown in [38, Theorem 3.20].

Lemma 4.16. **(a)** *If v_1, v_2 are subsolutions of problem (4.13), then so is* $\max\{v_1, v_2\}$.

(b) *If u_1, u_2 are supersolutions of problem (4.13), then so is* $\min\{u_1, u_2\}$.

We also quote an existence result whose proof can be found in [19, Theorem 1].

Lemma 4.17. *Assume that conditions* $\mathrm{H}(f)_1$ *and* $\mathrm{H}(f)_2$ *hold. Then problem (4.13) has at least one (weak) solution* $u \in W_0^{1,p}(\Omega)$.

The regularity of solutions and a priori estimates for problem (4.13) are now addressed.

Lemma 4.18. *Assume that condition* $\mathrm{H}(f)$ *holds. Then there are constants* $R > 0$ *and* $\gamma \in (0,1)$ *such that every solution* $u \in W_0^{1,p}(\Omega)$ *of problem (4.13) belongs to* $C^{1,\gamma}(\overline{\Omega})$ *and satisfies the estimate*

$$\|u\|_{C^{1,\gamma}(\overline{\Omega})} \leq R.$$

Proof. The fact that the solution set of problem (4.13) is nonempty is ensured by Lemmas 4.15 and 4.17. Let $u \in W_0^{1,p}(\Omega)$ be any solution of (4.13). Inserting $h = u$ in (4.15) and using $\mathrm{H}(f)_2$ and (4.14), we have

$$\begin{aligned}\int_\Omega (|\nabla u|^p + \mu|\nabla u|^q)\,dx &= \int_\Omega f(x,u,\nabla u)u\,dx \\ &\leq \int_\Omega (\tau(x) + b_1|u|^p + b_2|\nabla u|^p)\,dx \\ &\leq \int_\Omega \tau(x)\,dx + (b_1\lambda_{1,p}^{-1} + b_2)\int_\Omega |\nabla u|^p\,dx.\end{aligned}$$

Since $\mu \geq 0$ and $b_1\lambda_{1,p}^{-1} + b_2 < 1$, there exists a constant $c > 0$ independent of the solution u such that

$$\|u\| \leq c. \tag{4.18}$$

Then by (4.16) and (4.18) we infer that $\|u\|_{p^*} \leq c_{p^*}c$ and $\|\nabla u\|_p \leq c$ for every solution u of problem (4.13). Using these estimates through the growth condition $\mathrm{H}(f)$, it follows that

$$\|f(\cdot,u,\nabla u)\|_r \leq c_0, \tag{4.19}$$

with a constant $c_0 > 0$ independent of the solution u. At this point we make use of the assumption that $r > N$ in $\mathrm{H}(f)$, which enables us to refer to the gradient bound in [53, Theorem 3.1; Remark 3.3]. Consequently, from (4.19) we infer that

$$\|\nabla u\|_\infty \leq C\|f(\cdot,u,\nabla u)\|_r^{\frac{1}{p-1}} \leq C\,c_0^{\frac{1}{p-1}} \tag{4.20}$$

whenever u is a solution of (4.13), where the constant C depends only on p and Ω. In particular, the solution set of (4.13) is uniformly bounded, that is, there exists a constant $C_0 > 0$ independent of any solution u such that

$$\|u\|_\infty \le C_0. \tag{4.21}$$

Due to (4.20), (4.21), the nonlinear regularity up to the boundary in [129,130] (see also Proposition 2.12) applies, which leads to the desired conclusion. ☐

Remark 4.19. A careful reading of the above proof shows that the constant c in (4.18) is explicitly given by

$$c = \frac{\|\tau\|_1}{1 - (b_1\lambda_{1,p}^{-1} + b_2)}.$$

Moreover, the a priori estimate (4.18) holds under hypotheses $\mathrm{H}(f)_1$ and $\mathrm{H}(f)_2$ without need of the stronger hypothesis $\mathrm{H}(f)$.

We now focus on the location of nodal (that is, sign-changing) solutions for problem (4.13), which means solutions that are positive and negative on (disjoint) parts of nonzero measure.

Theorem 4.20. *Assume that conditions* $\mathrm{H}(f)_1$ *and* $\mathrm{H}(f)_2$ *are satisfied.*

(i) *If $f(x,0,0) \ge 0$ for a.e. $x \in \Omega$, then for every nodal solution u_0 of problem (4.13) there exists a nontrivial solution u_+ of (4.13) such that $u_0 \le u_+$ and $u_+ \ge 0$ on Ω.*

(ii) *If $f(x,0,0) \le 0$ for a.e. $x \in \Omega$, then for every nodal solution u_0 of problem (4.13) there exists a nontrivial solution u_- of (4.13) such that $u_0 \ge u_-$ and $u_- \le 0$ on Ω.*

(iii) *If $f(x,0,0) = 0$ for a.e. $x \in \Omega$, then for every nodal solution u_0 of problem (4.13) there exist two other nontrivial solutions u_+ and u_- of (4.13) such that $u_- \le u_0 \le u_+$, $u_+ \ge 0$ and $u_- \le 0$ on Ω.*

Proof. (i) Let u_0 be a nodal solution of problem (4.13). The assumption $f(x,0,0) \ge 0$ for a.e. $x \in \Omega$ ensures that 0 is a subsolution of problem (4.13). By Lemma 4.16 (a), we infer that

$$\underline{u} := \max\{0, u_0\} \text{ is a subsolution of problem (4.13).} \tag{4.22}$$

Let $T : W_0^{1,p}(\Omega) \to W_0^{1,p}(\Omega)$ be the truncation operator defined by

$$(Tu)(x) = \begin{cases} u(x) & \text{if } u(x) \ge \underline{u}(x), \\ \underline{u}(x) & \text{if } u(x) < \underline{u}(x) \end{cases} \tag{4.23}$$

for a.e. $x \in \Omega$. It is clear that the operator $T : W_0^{1,p}(\Omega) \to W_0^{1,p}(\Omega)$ is bounded and continuous. We also consider the following cut-off function by setting for a.e. $x \in \Omega$, every $s \in \mathbb{R}$:

$$b(x,s) = \begin{cases} 0 & \text{if } s \geq \underline{u}(x), \\ -(\underline{u}(x) - s)^{p-1} & \text{if } s < \underline{u}(x). \end{cases}$$

Let $B : W_0^{1,p}(\Omega) \to W^{-1,p'}(\Omega)$ denote the corresponding Nemytskii operator, that is, $Bu(x) = b(x, u(x))$, which is completely continuous in view of the compact embedding of $W_0^{1,p}(\Omega)$ into $L^p(\Omega)$ as stated in Theorem 1.20.

Now we formulate an auxiliary problem

$$\begin{cases} -\Delta_p u - \mu \Delta_q u + Bu = f(x, Tu, \nabla(Tu)) & \text{in } \Omega, \\ u = 0 & \text{on } \partial\Omega. \end{cases} \tag{4.24}$$

Using the Nemytskii operator $N_f : W_0^{1,p}(\Omega) \to W^{-1,p'}(\Omega)$ given by (4.17), problem (4.24) can be equivalently expressed as

$$-\Delta_p u - \mu \Delta_q u + Bu = N_f \circ T(u) \text{ in } W^{-1,p'}(\Omega). \tag{4.25}$$

By assumption H$(f)_1$, the operator $-\Delta_p - \mu \Delta_q + B - N_f \circ T : W_0^{1,p}(\Omega) \to W^{-1,p'}(\Omega)$ is bounded.

We claim that $-\Delta_p - \mu \Delta_q + B - N_f \circ T$ is a pseudomonotone operator on $W_0^{1,p}(\Omega)$. In order to show this, let a sequence $\{u_n\} \subset W_0^{1,p}(\Omega)$ be such that $u_n \rightharpoonup u$ in $W_0^{1,p}(\Omega)$ and

$$\limsup_{n \to \infty} \langle -\Delta_p u_n - \mu \Delta_q u_n + Bu_n - N_f(Tu_n), u_n - u \rangle \leq 0. \tag{4.26}$$

By assumption H$(f)_1$, we have that $\frac{p^*}{p^*-\alpha} < p^*$. Invoking the boundedness of the operator T and Rellich–Kondrachov compactness embedding theorem (see Theorem 1.20), it follows that the sequence $\{|Tu_n|^\alpha\}$ is bounded in $L^{(\frac{p^*}{p^*-\alpha})'}(\Omega) = L^{\frac{p^*}{\alpha}}(\Omega)$ and that $u_n \to u$ in $L^{\frac{p^*}{p^*-\alpha}}(\Omega)$. Therefore Hölder's inequality implies

$$\int_\Omega |Tu_n|^\alpha |u_n - u| \, dx \leq \|Tu_n\|_{p^*}^\alpha \|u_n - u\|_{\frac{p^*}{p^*-\alpha}} \to 0. \tag{4.27}$$

Also, by H$(f)_1$ we have that $\frac{p}{p-\beta} < p^*$. Using again the boundedness of the operator T and Rellich–Kondrachov compactness embedding theorem (see Theorem 1.20), we infer that the sequence $\{|\nabla(Tu_n)|^\beta\}$ is bounded in $L^{(\frac{p}{p-\beta})'}(\Omega) = L^{\frac{p}{\beta}}(\Omega)$ and that $u_n \to u$ in $L^{\frac{p}{p-\beta}}(\Omega)$. Consequently, Hölder's inequality yields

$$\int_\Omega |\nabla(Tu_n)|^\beta |u_n - u| \, dx \leq \|\nabla(Tu_n)\|_p^\beta \|u_n - u\|_{\frac{p}{p-\beta}} \to 0. \tag{4.28}$$

Therefore, thanks to assumption $\mathrm{H}(f)_1$ and (4.27), (4.28), we derive

$$\lim_{n\to\infty}\langle N_f(Tu_n), u_n - u\rangle = \lim_{n\to\infty}\int_\Omega f(x, Tu_n, \nabla(Tu_n))(u_n - u)\,dx = 0. \tag{4.29}$$

Combining (4.26), (4.29), and the complete continuity of B, it turns out that

$$\limsup_{n\to\infty}\langle -\Delta_p u_n - \mu\Delta_q u_n, u_n - u\rangle \le 0.$$

Using the $(S)_+$-property of $-\Delta_p - \mu\Delta_q$ (see [169, Proposition 2.70]), we conclude that $u_n \to u$ in $W_0^{1,p}(\Omega)$. Thus

$$\begin{aligned}&\lim_{n\to\infty}\langle -\Delta_p u_n - \mu\Delta_q u_n + Bu_n - N_f(Tu_n), u_n - v\rangle\\ &= \langle -\Delta_p u - \mu\Delta_q u + Bu - N_f(Tu), u - v\rangle\end{aligned}$$

for all $v \in W_0^{1,p}(\Omega)$, which proves that the operator $-\Delta_p - \mu\Delta_q + B - N_f \circ T$ is pseudomonotone.

Now we verify that the operator $-\Delta_p - \mu\Delta_q + B - N_f \circ T$ is coercive. Let $\{u_n\}$ be a sequence in $W_0^{1,p}(\Omega)$ such that $\|u_n\| \to +\infty$ as $n \to \infty$. By $\mathrm{H}(f)_2$, Hölder's inequality and the definition of T, we get the estimate

$$\begin{aligned}&\langle(-\Delta_p - \mu\Delta_q + B - N_f \circ T)(u_n), u_n\rangle\\ &= \|u_n\|^p + \mu\|\nabla u_n\|_q^q - \int_{\{u_n<\underline{u}\}}(\underline{u} - u_n)^{p-1}u_n\,dx\\ &\quad - \int_{\{u_n<\underline{u}\}} f(x, \underline{u}, \nabla\underline{u})u_n\,dx - \int_{\{\underline{u}\le u_n\}} f(x, u_n, \nabla u_n)u_n\,dx\\ &\ge \|u_n\|^p - \ell_1\|u_n - \underline{u}\|_p^{p-1} - \ell_2\|u_n\|_p - \int_\Omega(\tau(x) + b_1|u_n|^p + b_2|\nabla u_n|^p)\,dx\\ &\ge (1 - b_1\lambda_{1,p}^{-1} - b_2)\|u_n\|^p - \ell_1 c_p^{p-1}\|u_n - \underline{u}\|^{p-1} - \ell_2 c_p\|u_n\| - \|\tau\|_1,\end{aligned}$$

with positive constants ℓ_1, ℓ_2. In view of the condition $b_1\lambda_{1,p}^{-1} + b_2 < 1$ supposed in assumption $\mathrm{H}(f)_2$, we obtain that the operator $-\Delta_p - \mu\Delta_q + B - N_f \circ T$ is coercive. According to the above properties, we are in a position to apply Theorem 1.55 (the main theorem for pseudomonotone operators) to the operator $-\Delta_p - \mu\Delta_q + B - N_f \circ T$. It entails the existence of $u_+ \in W_0^{1,p}(\Omega)$ solving (4.25). Therefore u_+ is a solution of the auxiliary problem (4.24).

Let us show the inequality

$$u_+(x) \ge \underline{u}(x) \quad \text{for a.e. } x \in \Omega. \tag{4.30}$$

Acting as test function on (4.25) (or (4.24)) with $(u_+ - \underline{u})^- = \max\{0, -(u_+ - \underline{u})\} \in W_0^{1,p}(\Omega)$ and using (4.23) and (4.22) lead to

$$\begin{aligned}
&\langle -\Delta_p u_+ - \mu\Delta_q u_+, (u_+ - \underline{u})^- \rangle \\
&= \int_\Omega (\underline{u} - u_+)^{p-1}(u_+ - \underline{u})^- \, dx \\
&\quad + \int_\Omega f(x, Tu_+, \nabla(Tu_+))(u_+ - \underline{u})^- \, dx \\
&= \int_{\{u_+ < \underline{u}\}} (\underline{u} - u_+)^p \, dx + \int_\Omega f(x, \underline{u}, \nabla\underline{u})(u_+ - \underline{u})^- \, dx \\
&\geq \int_{\{u_+ < \underline{u}\}} (\underline{u} - u_+)^p \, dx + \langle -\Delta_p \underline{u} - \mu\Delta_q \underline{u}, (u_+ - \underline{u})^- \rangle,
\end{aligned}$$

whence

$$\begin{aligned}
&\int_{\{u_+ < \underline{u}\}} (|\nabla u_+|^{p-2}\nabla u_+ - |\nabla\underline{u}|^{p-2}\nabla\underline{u}, \nabla u_+ - \nabla\underline{u})_{\mathbb{R}^N} \, dx \\
&\quad + \mu \int_{\{u_+ < \underline{u}\}} (|\nabla u_+|^{q-2}\nabla u_+ - |\nabla\underline{u}|^{q-2}\nabla\underline{u}, \nabla u_+ - \nabla\underline{u})_{\mathbb{R}^N} \, dx \\
&\quad + \int_{\{u_+ < \underline{u}\}} (\underline{u} - u_+)^p \, dx \leq 0.
\end{aligned}$$

By means of the (strict) monotonicity of the map $\xi \mapsto (|\xi|^{p-2} + \mu|\xi|^{q-2})\xi$ on $\mathbb{R}^N$, we infer that the set $\{u_+ < \underline{u}\} = \{x \in \Omega : u_+(x) < \underline{u}(x)\}$ has zero Lebesgue measure, so (4.30) holds true.

Due to (4.25), we have $Tu_+ = u_+$. This, in conjunction with the fact that u_+ is a solution of problem (4.24), guarantees that u_+ is a solution of problem (4.13). Since the solution u_0 of (4.13) is nodal, its positive part $u_0^+ = \max\{0, u_0\}$ is strictly positive on a subset of Ω of positive measure, so from (4.30) and (4.22) we deduce that $u_+ \geq 0$ and $u_+ \not\equiv 0$.

(ii) Let u_0 be a nodal solution of problem (4.13). The inequality $f(x,0,0) \leq 0$ for a.e. $x \in \Omega$ expresses that 0 is a supersolution of problem (4.13). Then Lemma 4.16 (*b*) provides that

$$\overline{u} := \min\{0, u_0\} \text{ is a supersolution of problem (4.13).} \tag{4.31}$$

We introduce the truncation operator $T : W_0^{1,p}(\Omega) \to W_0^{1,p}(\Omega)$ by

$$(Tu)(x) = \begin{cases} u(x) & \text{if } u(x) \leq \overline{u}(x), \\ \overline{u}(x) & \text{if } u(x) > \overline{u}(x) \end{cases} \tag{4.32}$$

for a.e. $x \in \Omega$. We also consider the cut-off function

$$b(x,s) = \begin{cases} 0 & \text{if } s \le \overline{u}(x), \\ (s - \overline{u}(x))^{p-1} & \text{if } s > \overline{u}(x) \end{cases}$$

for a.e. $x \in \Omega$, every $s \in \mathbb{R}$.

As before denote by $B : W_0^{1,p}(\Omega) \to W^{-1,p'}(\Omega)$ the associated Nemytskii operator, namely $Bu(x) = b(x, u(x))$. Corresponding to these new operators T and B, we state the auxiliary problem (4.24).

As in part (i) we prove that the operator $-\Delta_p - \mu\Delta_q + B - N_f \circ T : W_0^{1,p}(\Omega) \to W^{-1,p'}(\Omega)$ is bounded, pseudomonotone, and coercive. Through the main theorem for pseudomonotone operators (i.e., Theorem 1.55), this results in the existence of $u_- \in W_0^{1,p}(\Omega)$ satisfying

$$-\Delta_p u_- - \mu\Delta_q u_- + Bu_- = N_f \circ T(u_-). \tag{4.33}$$

We claim that

$$u_-(x) \le \overline{u}(x) \quad \text{for a.e. } x \in \Omega. \tag{4.34}$$

In order to prove the claim, we develop a comparison argument acting on (4.33) with the test function $(u_- - \overline{u})^+ = \max\{0, u_- - \overline{u}\} \in W_0^{1,p}(\Omega)$ and make use of (4.31) and (4.32). We are led to

$$\begin{aligned}
&\langle -\Delta_p u_- - \mu\Delta_q u_-, (u_- - \overline{u})^+\rangle + \int_\Omega (u_- - \overline{u})^{p-1}(u_- - \overline{u})^+ \, dx \\
&= \int_\Omega f(x, Tu_-, \nabla(Tu_-))(u_- - \overline{u})^+ \, dx = \int_\Omega f(x, \overline{u}, \nabla\overline{u})(u_- - \overline{u})^+ \, dx \\
&\le \langle -\Delta_p \overline{u} - \mu\Delta_q \overline{u}, (u_- - \overline{u})^+\rangle.
\end{aligned}$$

Following the reasoning as in (i), we can show that the claim in (4.34) is valid.

From (4.33), (4.34), and (4.32) we infer that $u_- \in W_0^{1,p}(\Omega)$ is a solution of problem (4.13). Recalling that the solution u_0 of (4.13) is nodal, we have that $\overline{u} = \min\{0, u_0\}$ is strictly negative on a subset of Ω of positive measure. Then (4.34) and (4.31) enable us to conclude that $u_- \le 0$ and $u_- \not\equiv 0$.

(iii) If $f(x, 0, 0) = 0$ for a.e. $x \in \Omega$, then we can apply simultaneously assertions (i) and (ii), which gives rise to two nontrivial opposite constant-sign solutions u_+ and u_- of problem (4.13) with the properties required in the statement. □

Our next goal is to provide a criterion for having in Theorem 4.20 the enclosure of the nodal solution u_0 within the ordered interval determined by a positive solution u_+ and a negative solution u_-.

Theorem 4.21. *Assume that condition* H(f) *is satisfied.*

(i) *If there exist constants* $\delta \in (0,1)$, $k_0 > 0$, *and* $k_1 > 0$ *such that*

$$f(x,s,\xi) \geq -k_0 s^{p-1} - k_1(|\xi|^{p-1} + \mu|\xi|^{q-1}) \tag{4.35}$$

for a.e. $x \in \Omega$, *all* $0 \leq s \leq \delta$, $|\xi| \leq \delta$, *then for every nodal solution* u_0 *of problem (4.13) there exists a (positive) solution* $u_+ \in \operatorname{int}(C_0^1(\overline{\Omega})_+)$ *of (4.13) such that* $u_0 \leq u_+$ *on* Ω.

(ii) *If there exist constants* $\delta \in (0,1)$, $k_0 > 0$, *and* $k_1 > 0$ *such that*

$$f(x,s,\xi) \leq k_0|s|^{p-1} + k_1(|\xi|^{p-1} + \mu|\xi|^{q-1}) \tag{4.36}$$

for a.e. $x \in \Omega$, *all* $-\delta \leq s \leq 0$, $|\xi| \leq \delta$, *then for every nodal solution* u_0 *of problem (4.13) there exists a (negative) solution* $u_- \in -\operatorname{int}(C_0^1(\overline{\Omega})_+)$ *of problem (4.13) such that* $u_0 \geq u_-$ *on* Ω.

(iii) *If there exist constants* $\delta \in (0,1)$, $k_0 > 0$, *and* $k_1 > 0$ *such that*

$$|f(x,s,\xi)| \leq k_0|s|^{p-1} + k_1(|\xi|^{p-1} + \mu|\xi|^{q-1}) \tag{4.37}$$

for a.e. $x \in \Omega$, *all* $-\delta \leq s \leq \delta$, $|\xi| \leq \delta$, *then for every nodal solution* u_0 *of problem (4.13) there exist a (positive) solution* $u_+ \in \operatorname{int}(C_0^1(\overline{\Omega})_+)$ *and a (negative) solution* $u_- \in -\operatorname{int}(C_0^1(\overline{\Omega})_+)$ *of problem (4.13) such that* $u_- \leq u_0 \leq u_+$ *on* Ω.

Proof. (i) Owing to Lemma 4.15, we know that the present hypotheses are stronger than those in Theorem 4.20. Since assumption (4.35) entails that $f(x,0,0) \geq 0$ for a.e. $x \in \Omega$, by Theorem 4.20 (i) there exists a nontrivial nonnegative solution u_+ of (4.13) with $u_0 \leq u_+$ on Ω. It suffices to prove that the solution u_+ fulfills the properties required in the statement.

Due to assumption H(f), we are able to invoke the regularity result in Lemma 4.18, which ensures that $u_+ \in C_0^1(\overline{\Omega})$, so in particular $0 \leq u_+(x) \leq M$ and $|\nabla u_+(x)| \leq M$ for all $x \in \Omega$, with a constant $M > 1$.

Now we rely on the strong maximum principle in [205, Theorem 5.4.1] and on Hopf's boundary point lemma in [205, Theorem 5.5.1] (see also [169, Section 8.2]) applied to the function $u_+ \in C_0^1(\overline{\Omega})$ that satisfies

$$-\Delta_p u_+ - \mu\Delta_q u_+ = f(x,u_+,\nabla u_+) \text{ in } \Omega.$$

In order to meet the needed requirements, let us denote $A(s) := s^{p-2} + \mu s^{q-2}$ for $s > 0$. We find that

$$c := \lim_{t\to 0^+} \frac{tA'(t)}{A(t)} > -1,$$

since $c = q-2$ if $\mu > 0$ and $c = p-2$ if $\mu = 0$.

From assumption H(f) and the inequality

$$\frac{p^*}{r} < p - 1$$

we derive that there are constants $k_2 = k_2(\delta, M) > 0$ and $k_3 = k_3(\delta, M) > 0$ with

$$f(x, s, \xi) \geq -k_2 s^{p-1} \text{ for a.e. } x \in \Omega, \text{ all } s \geq \delta, \ |\xi| \leq M,$$
$$f(x, s, \xi) \geq -k_3(|\xi|^{p-1} + \mu|\xi|^{q-1})$$
$$\text{for a.e. } x \in \Omega, \text{ all } 0 \leq s \leq \delta, \ \delta \leq |\xi| \leq M.$$

Combining with (4.35) we arrive at

$$f(x, s, \xi) \geq -C_1 \Phi(|\xi|) - f(s) \text{ for a.e. } x \in \Omega, \text{ all } s \geq 0, \ |\xi| \leq M,$$

where $\Phi(s) := sA(s) = s^{p-1} + \mu s^{q-1}$ and $f(s) = C_2 s^{p-1}$, with positive constants $C_1 = C_1(\delta, M)$ and $C_2 = C_2(\delta, M)$. Taking into account that $M > 1$, conditions (B1) and (F2) in [205] are fulfilled.

We note that

$$H(s) := s\Phi(s) - \int_0^s \Phi(t)\, dt = \frac{p-1}{p} s^p + \mu \frac{q-1}{q} s^q$$

and

$$F(s) := \int_0^s f(t)\, dt = \frac{C_2}{p} s^p$$

for all $s \geq 0$. The function H is increasing on $[0, +\infty)$ and we have

$$F(s) \leq H\big(C_2^{\frac{1}{p}} (p-1)^{-\frac{1}{p}} s\big)$$

hence

$$H^{-1}(F(s)) \leq C_2^{\frac{1}{p}} (p-1)^{-\frac{1}{p}} s \text{ for all } s > 0.$$

Thus the condition

$$\int_{0+} \frac{ds}{H^{-1}(F(s))} = +\infty$$

required in (1.1.5) of [205] holds true. All the hypotheses of [205, Theorem 5.4.1] and [205, Theorem 5.5.1] are satisfied (see also [205, Remark 3, p. 117]), thereby $u_+ \in \operatorname{int}(C_0^1(\overline{\Omega})_+)$ as seen from formula (1.39).

(ii) Observe that hypothesis (4.36) implies that $f(x, 0, 0) \leq 0$ for a.e. $x \in \Omega$. Hence, by means of Lemma 4.15, we can refer to Theorem 4.20 (ii) for obtaining a nontrivial nonpositive solution u_- of (4.13) such that $u_0 \geq u_-$ on Ω. Consequently, it is sufficient to show that the solution u_- has the properties required in the statement.

First, due to H(f) and Lemma 4.18, we know that $u_- \in C_0^1(\overline{\Omega})$. Then we utilize the strong maximum principle in [205, Theorem 5.4.1] and Hopf's boundary point lemma in [205, Theorem 5.5.1] applied to the nonnegative function $-u_- \in C_0^1(\overline{\Omega})$ verifying

$$-\Delta_p(-u_-) - \mu\Delta_q(-u_-) = \tilde{f}(x, -u_-, \nabla(-u_-)) \quad \text{in } \Omega,$$

where $\tilde{f}(x, s, \xi) := -f(x, -s, -\xi)$. From now on the proof proceeds in a manner similar to (i) with $-u_-$ in place of u_+.

(iii) We note that assumption (4.37) implies both (4.35) and (4.36). Then part (iii) can be readily obtained from assertions (i) and (ii), which completes the proof. □

Remark 4.22. Part (iii) of Theorem 4.21 holds more generally whenever both unilateral conditions (4.35) and (4.36) are satisfied, which is weaker than assuming the bilateral condition (4.37). Note that (4.35) and (4.36) are fulfilled by a function f that satisfies the local sign condition

$$f(x, s, \xi)s \geq 0 \text{ for a.e. } x \in \Omega, \text{ all } -\delta \leq s \leq \delta, |\xi| \leq \delta,$$

for some $\delta > 0$.

Remark 4.23. Denoting by $\lambda_{2,p}$ the second eigenvalue of $-\Delta_p$ on $W_0^{1,p}(\Omega)$, one shows in [165, Theorem 2 (b)] the existence of a nodal solution $u_0 \in C_0^1(\overline{\Omega})$ between a negative solution $u_- \in -\text{int}\,(C_0^1(\overline{\Omega})_+)$ and a positive solution $u_+ \in \text{int}\,(C_0^1(\overline{\Omega})_+)$ of problem (4.13) provided f does not depend on $\xi \in \mathbb{R}^N$ and there exist $\eta_2, \theta \in L^\infty(\Omega)$, with $\eta_2(x) \geq \lambda_{2,p}$ for a.e. $x \in \Omega$, $\eta_2 \neq \lambda_{2,p}$ on a set of positive measure, such that

$$\eta_2(x) \leq \liminf_{s\to 0} \frac{f(x,s)}{|s|^{p-2}s} \leq \limsup_{s\to 0} \frac{f(x,s)}{|s|^{p-2}s} \leq \theta(x) \text{ uniformly for a.e. } x \in \Omega.$$

Theorem 4.21 complements this result offering a verifiable criterion for the converse property, namely that every nodal solution of problem (4.13) is located within the ordered interval determined by a negative solution and a positive solution of problem (4.13).

Remark 4.24. The result in Theorem 4.21 is sharp. For example, every nontrivial solution of the problem $-\Delta_p u = \lambda|u|^{p-2}u$ in Ω, $u = 0$ on $\partial\Omega$, with $\lambda > \lambda_{1,p}$, is nodal. Notice that in this case hypothesis H(f)$_2$ is not fulfilled.

Our next aim is to prove the existence of extremal solutions, i.e., the biggest and smallest solutions, for problem (4.13). This will permit bounding the whole set of nodal solutions by a positive solution and a negative solution, thus complementing the location principle in Theorem 4.21 where the bounds with constant-sign solutions depend on the fixed nodal solution.

We start with the problem obtained for $\mu = 0$, that is,

$$\begin{cases} -\Delta_p u = f(x, u, \nabla u) & \text{in } \Omega, \\ u = 0 & \text{on } \partial\Omega. \end{cases} \tag{4.38}$$

Recall that $\Omega \subset \mathbb{R}^N$ is a bounded domain with C^2 boundary $\partial\Omega$. Then for any $\delta > 0$ sufficiently small, say $\delta \leq \delta_0$, the δ-neighborhood Ω_δ of $\overline{\Omega}$, i.e.,

$$\Omega_\delta = \{x \in \mathbb{R}^N : d(x, \overline{\Omega}) < \delta\},$$

has a C^2 boundary $\partial\Omega_\delta$. Let $\lambda_{1,p,\delta}$ be the first eigenvalue of $-\Delta_p$ on $W_0^{1,p}(\Omega_\delta)$ with the associated L^p-normalized positive eigenfunction $\phi_{1,p,\delta}$, that is,

$$-\Delta_p \phi_{1,p,\delta} = \lambda_{1,p,\delta}\phi_{1,p,\delta}^{p-1} \text{ in } \Omega_\delta, \ \phi_{1,p,\delta} = 0 \text{ on } \partial\Omega_\delta, \ \int_{\Omega_\delta} \phi_{1,p,\delta}^p \, dx = 1. \tag{4.39}$$

The below result describes a continuity property of the first eigenvalue $\lambda_{1,p}$ with respect to the domain.

Lemma 4.25. *There holds* $\lim_{\delta \to 0^+} \lambda_{1,p,\delta} = \lambda_{1,p}$.

Proof. The variational characterizations of $\lambda_{1,p,\delta}$ and $\lambda_{1,p}$ (see (4.14)) give

$$\lambda_{1,p,\delta} = \inf_{u \in W_0^{1,p}(\Omega_\delta)\setminus\{0\}} \frac{\|u\|^p_{W_0^{1,p}(\Omega_\delta)}}{\|u\|^p_{L^p(\Omega_\delta)}} \leq \inf_{u \in W_0^{1,p}(\Omega)\setminus\{0\}} \frac{\|u\|^p}{\|u\|_p^p} = \lambda_{1,p},$$

whence

$$\limsup_{\delta \to 0^+} \lambda_{1,p,\delta} \leq \lambda_{1,p}. \tag{4.40}$$

Take a sequence $\delta_n \to 0^+$ as $n \to \infty$. From (4.39) and (4.40) we derive

$$\int_{\Omega_{\delta_0}} |\nabla \phi_{1,p,\delta_n}|^p \, dx = \int_{\Omega_{\delta_n}} |\nabla \phi_{1,p,\delta_n}|^p \, dx = \lambda_{1,p,\delta_n} \leq M$$

with a constant $M > 0$, so up to a subsequence, $\phi_{1,p,\delta_n} \rightharpoonup u$ in $W_0^{1,p}(\Omega_{\delta_0})$ and $\phi_{1,p,\delta_n} \to u$ in $L^p(\Omega_{\delta_0})$ as $n \to \infty$ with some $u \in W_0^{1,p}(\Omega_{\delta_0})$.

We have that $u = 0$ a.e. in $\Omega_{\delta_0} \setminus \overline{\Omega}$ because, for any $v \in C_c^\infty(\Omega_{\delta_0} \setminus \overline{\Omega})$, the inclusion $\operatorname{supp} v \subset \Omega_{\delta_0} \setminus \overline{\Omega_{\delta_n}}$ holds provided n is sufficiently large, thus

$$\int_{\Omega_{\delta_0}\setminus\overline{\Omega}} uv \, dx = \int_{\Omega_{\delta_0}} uv \, dx = \lim_{n \to \infty} \int_{\Omega_{\delta_0}} \phi_{1,p,\delta_n} v \, dx = 0.$$

Consequently, since $u \in W_0^{1,p}(\Omega)$ and $\int_\Omega u^p\,dx = 1$, we find that

$$\lambda_{1,p} \le \int_{\Omega_{\delta_0}} |\nabla u|^p\,dx \le \liminf_{n\to\infty} \int_{\Omega_{\delta_0}} |\nabla\phi_{1,p,\delta_n}|^p\,dx = \liminf_{n\to\infty} \lambda_{1,p,\delta_n}.$$

In view of (4.40), the proof is complete. □

The next result provides subsolutions and supersolutions for problem (4.38).

Proposition 4.26. *Assume* H$(f)_1$ *with* $\sigma \in L^\infty(\Omega)$, $\beta < p-1$, *and* ($\alpha = p-1$ *and* $a_1 < \lambda_{1,p}$) *or* $\alpha < p-1$. *Then there exists* $\hat\delta > 0$ *with the property that, for every* $\delta \in (0,\hat\delta]$, *there is* $t_0 = t_0(\delta) > 0$ *such that* $-t\phi_{1,p,\delta}$ *and* $t\phi_{1,p,\delta}$ *are subsolution and supersolution, respectively, of problem (4.38) for all* $t \ge t_0$.

Proof. If $a_1 < \lambda_{1,p}$, Lemma 4.25 yields $\hat\delta > 0$ such that $a_1 < \lambda_{1,p,\delta}$ for all $\delta \in (0,\hat\delta]$. Because $\phi_{1,p,\delta} \in \operatorname{int}(C_0^1(\overline{\Omega_\delta})_+)$, there exist constants $\rho = \rho(\delta) > 0$ and $t_0 = t_0(\delta) > 0$ with $\phi_{1,p,\delta}(x) \ge \rho$ and

$$\frac{\sigma(x)}{(\rho t_0)^{p-1}} + \frac{a_1}{(\rho t_0)^{p-\alpha-1}} + \frac{a_2}{\rho^{p-1} t_0^{p-\beta-1}} |\nabla\phi_{1,p,\delta}(x)|^\beta \le \lambda_{1,p,\delta}$$

for all $x \in \overline{\Omega}$ (this holds in both situations ($\alpha = p-1$ and $a_1 < \lambda_{1,p}$) and ($\alpha < p-1$) of the statement). Combining with our hypothesis leads to

$$\begin{aligned}
&f(x, t\phi_{1,p,\delta}(x), \nabla(t\phi_{1,p,\delta})(x)) \\
&\le \sigma(x) + a_1(t\phi_{1,p,\delta}(x))^\alpha + a_2(t|\nabla\phi_{1,p,\delta}(x)|)^\beta \\
&\le \Big(\frac{\sigma(x)}{(\rho t_0)^{p-1}} + \frac{a_1}{(\rho t_0)^{p-\alpha-1}} + \frac{a_2}{\rho^{p-1} t_0^{p-\beta-1}} |\nabla\phi_{1,p,\delta}(x)|^\beta\Big)(t\phi_{1,p,\delta}(x))^{p-1} \\
&\le \lambda_{1,p,\delta}(t\phi_{1,p,\delta}(x))^{p-1} = -\Delta_p(t\phi_{1,p,\delta})(x)
\end{aligned}$$

in Ω, for all $t \ge t_0$. Therefore $t\phi_{1,p,\delta}$ is a supersolution of problem (4.38). The proof for the subsolution is similar. □

Our existence result of extremal solutions for (4.38) is as follows.

Theorem 4.27. *Assume* H(f). *Then:*

(i) *Problem (4.38) has the biggest solution* $\hat u \in C_0^1(\overline{\Omega})$ *and the smallest solution* $\hat v \in C_0^1(\overline{\Omega})$. *In particular, every solution of (4.38) is contained in the ordered interval* $[\hat v, \hat u]$.

(ii) *If (4.35)–(4.36) hold and the set of nodal solutions of problem (4.38) is nonempty, then we have* $\hat u \in \operatorname{int}(C_0^1(\overline{\Omega})_+)$ *and* $\hat v \in -\operatorname{int}(C_0^1(\overline{\Omega})_+)$.

Proof. Denote by $\mathcal{S}$ the set of solutions of problem (4.38) endowed with the pointwise partial order $\le$. Lemmas 4.15 and 4.17 guarantee that $\mathcal{S}$ is nonempty.

Let us show that the ordered set $\mathcal{S}$ is upward directed, that is, if $u_1, u_2 \in \mathcal{S}$, then there is $u \in \mathcal{S}$ such that $\max\{u_1, u_2\} \le u$. Indeed, if $u_1, u_2 \in \mathcal{S}$, by Lemma 4.16 (a) we know that $\underline{u} = \max\{u_1, u_2\}$ is a subsolution of problem (4.38), whereas Lemma 4.18 entails that $u_1, u_2 \in C_0^1(\overline{\Omega})$. Note that H($f$) allows us to apply Proposition 4.26, which ensures that there is $\delta > 0$ such that $t\phi_{1,p,\delta}$ is a supersolution of problem (4.38) for $t > 0$ large enough. Moreover, for a possibly larger t, we can suppose that $t\phi_{1,p,\delta} \ge \underline{u}$ in Ω because $u_1, u_2 \in C_0^1(\overline{\Omega})$ and $\phi_{1,p,\delta} \in \operatorname{int}(C_0^1(\overline{\Omega_\delta})_+)$. Now invoking the general theory of subsolutions–supersolutions (see [38, Theorem 3.17] or [169, Proposition 11.8]), the existence of $u \in \mathcal{S}$ such that $\underline{u}(x) \le u(x) \le t\phi_{1,p,\delta}(x)$ for all $x \in \Omega$ follows, which proves our claim.

Next we show that each chain $\mathcal{C}$ has an upper bound in $\mathcal{S}$. On the basis of [76, p. 336], there is a sequence $\{u_n\} \subset \mathcal{C}$ such that

$$\sup \mathcal{C} = \sup_{n \ge 1} u_n = \lim_{n \to \infty} u_n. \tag{4.41}$$

According to Lemma 4.18, the sequence $\{u_n\}$ is bounded in $C^{1,\gamma}(\overline{\Omega})$ for some $\gamma \in (0, 1)$. Since $C^{1,\gamma}(\overline{\Omega})$ is compactly embedded in $C^1(\overline{\Omega})$, along a relabeled subsequence we have that $u_n \to u$ in $C_0^1(\overline{\Omega})$ as $n \to \infty$. In view of (4.41) it turns out that $u = \sup \mathcal{C}$. In addition, we can pass to the limit in the equation $-\Delta_p u_n = f(x, u_n, \nabla u_n)$ in $W^{-1,p'}(\Omega)$, obtaining that $u \in \mathcal{S}$.

The preceding discussion enables us to apply Zorn's lemma, which provides a maximal element $\hat{u} \in \mathcal{S}$. Actually, $\hat{u}$ is the biggest element in $\mathcal{S}$. Indeed, if $u \in \mathcal{S}$, as shown before we can find $v \in \mathcal{S}$ such that $v \ge \max\{\hat{u}, u\}$. Then the maximality of $\hat{u}$ renders $\hat{u} = v$, so $\hat{u} \ge u$ in Ω.

Under the additional assumption (4.35), we may invoke Theorem 4.21 (i) ensuring that if problem (4.38) possesses nodal solutions, then $\hat{u} \in \operatorname{int}(C_0^1(\overline{\Omega})_+)$.

Arguing in the same way, we can establish the existence of the smallest solution $\hat{v}$ of problem (4.38). Furthermore, through Theorem 4.21 (ii) we get that $\hat{v} \in -\operatorname{int}(C_0^1(\overline{\Omega})_+)$ provided that (4.36) is valid and there are nodal solutions. □

Now we turn to extremal solutions for problem (4.13) with $\mu > 0$, which basically is the case driven by the (p, q)-Laplacian. Here we cannot proceed as for problem (4.38), i.e., when $\mu = 0$, because if $\mu > 0$, the nonhomogeneity of the operator $-\Delta_p - \mu\Delta_q$ prevents constructing subsolutions and supersolutions by means of the first eigenfunctions of $-\Delta_p$ or $-\Delta_q$. We obtain subsolutions and supersolutions for problem (4.13) with $\mu > 0$ relying on the following assumption:

$$\liminf_{s \to +\infty} f(x, s, 0) \le \gamma_1 < 0 < \gamma_2 \le \limsup_{s \to -\infty} f(x, s, 0) \tag{4.42}$$

uniformly for a.e. $x \in \Omega$, with constants γ_1 and γ_2.

We have the following existence and enclosure result.

Theorem 4.28. *Assume that conditions* H(f) *and* (4.42) *are verified, and that* $\mu > 0$ *in problem (4.13). Then:*

(i) *Problem (4.13) has the biggest solution* $\hat{u} \in C_0^1(\overline{\Omega})$ *and the smallest solution* $\hat{v} \in C_0^1(\overline{\Omega})$. *Thus every solution of (4.13) is contained in the ordered interval* $[\hat{v}, \hat{u}]$.

(ii) *If (4.35)–(4.36) hold and the set of nodal solutions of problem (4.13) is nonempty, then* $\hat{u} \in \operatorname{int}(C_0^1(\overline{\Omega})_+)$ *and* $\hat{v} \in -\operatorname{int}(C_0^1(\overline{\Omega})_+)$.

Proof. Denote by $\mathcal{S}$ the set of solutions of problem (4.13). As proven in Lemma 4.17, the set $\mathcal{S}$ is nonempty. Due to assumption H(f), it is allowed to apply Lemma 4.18 for obtaining the uniform bound

$$M := \sup_{u \in \mathcal{S}} \|u\|_\infty < +\infty.$$

By virtue of (4.42), there exists $s_0 > M$ satisfying

$$f(x, s_0, 0) < 0 \text{ for a.e. } x \in \Omega.$$

It follows that s_0 is a (constant) supersolution of problem (4.13) such that $s_0 \geq u$ for all $u \in \mathcal{S}$.

We claim that the set $\mathcal{S}$ endowed with the pointwise order $\leq$ is upward directed. To this end let $u_1, u_2 \in \mathcal{S}$. From Lemma 4.16 (a) we know that $\underline{u} := \max\{u_1, u_2\}$ is a subsolution for problem (4.13). Since $s_0 \geq \underline{u}$, the general theory of subsolutions–supersolutions (see [38, Theorem 3.17] or [169, Proposition 11.8]) provides a solution $u \in \mathcal{S}$ within the ordered interval $[\underline{u}, s_0]$, which proves the claim.

As in the proof of Theorem 4.27, we can check that Zorn's lemma can be applied on the set $\mathcal{S}$, thus getting a maximal element $\hat{u} \in \mathcal{S}$. The fact that $\hat{u}$ is the biggest element of $\mathcal{S}$ can be seen following the pattern of the corresponding part in the proof of Theorem 4.27. Similar arguments lead to the existence of the smallest solution of problem (4.13). Part (i) of the statement is established.

Finally, combining with Theorem 4.21, we deduce part (ii) of the statement of the theorem. The proof is thus complete. □

Remark 4.29. Theorem 4.28 is also valid if, in place of (4.42), we assume the existence of constants $\underline{c}, \overline{c}$ such that

$$\underline{c} \leq -\sup_{u \in \mathcal{S}} \|u\|_\infty, \quad \overline{c} \geq \sup_{u \in \mathcal{S}} \|u\|_\infty,$$
$$f(x, \overline{c}, 0) \leq 0 \leq f(x, \underline{c}, 0) \text{ for a.e. } x \in \Omega.$$

4.3 GRADIENT DEPENDENT PROBLEMS DRIVEN BY NONHOMOGENEOUS DIFFERENTIAL OPERATORS

Here we study the nonlinear elliptic Dirichlet problem

$$\begin{cases} -\sum_{i=1}^{N} \frac{\partial}{\partial x_i} a_i(x, u, \nabla u) = f(x, u, \nabla u) & \text{in } \Omega, \\ u = 0 & \text{on } \partial\Omega \end{cases} \tag{4.43}$$

on a bounded domain $\Omega \subset \mathbb{R}^N$ with C^2 boundary $\partial\Omega$. This is an extension of problem (4.1), (4.13) in the preceding two sections because the left-hand side of the equation is governed by a more general operator.

Set

$$A(x, t, \xi) := (a_1(x, t, \xi), \ldots, a_N(x, t, \xi)) \in \mathbb{R}^N$$

for all $(x, t, \xi) \in \Omega \times [0, +\infty) \times \mathbb{R}^N$.

With this notation, problem (4.43) becomes

$$\begin{cases} -\operatorname{div} A(x, u, \nabla u) = f(x, u, \nabla u) & \text{in } \Omega, \\ u = 0 & \text{on } \partial\Omega. \end{cases}$$

Let $1 < q \le p < +\infty$. We state the following hypotheses:

H(A) There exist positive constants C_i $(1 \le i \le 6)$, $0 < \alpha \le 1$, $0 < t_0 \le 1$ and a continuous positive function g_0 on $[0, +\infty)$ such that

(i) $a_i \in C(\overline{\Omega} \times [0, +\infty) \times \mathbb{R}^N, \mathbb{R}) \cap C^1(\overline{\Omega} \times (0, +\infty) \times (\mathbb{R}^N \setminus \{0\}), \mathbb{R})$ for all $1 \le i \le N$;

(ii) $a_i(x, t, 0) = 0$ for all $(x, t) \in \overline{\Omega} \times [0, +\infty)$, and $1 \le i \le N$;

(iii) $\|D_y A(x, t, y)\| \le g_0(t)(C_1|y|^{p-2} + C_2|y|^{q-2})$ for all $x \in \overline{\Omega}$, $t \ge 0$ and $y \in \mathbb{R}^N \setminus \{0\}$;

(iv) $D_y A(x, t, y)\xi \cdot \xi \ge (C_3|y|^{p-2} + C_4|y|^{q-2})|\xi|^2$ for all $x \in \overline{\Omega}$, $t \ge 0$, $y \in \mathbb{R}^N \setminus \{0\}$, and $\xi \in \mathbb{R}^N$;

(v) for each $M > 0$ there exists $C_M > 0$ such that

$$\big|A(x, t, y) - A(x', t', y)\big| \le C_M(1 + |y|^{p-1})\left(|x - x'|^\alpha + |t - t'|^\alpha\right)$$

whenever $x, x' \in \Omega$, $y \in \mathbb{R}^N$, and $t, t' \in [0, M]$;

(vi) $|\partial_{x_i} a_i(x, t, y)| \le C_5\left(t^{q-1}(1 + |\log t|) + |y|^{q-1}(1 + |\log|y||)\right)$ for all $x \in \overline{\Omega}$, all $0 < t < t_0$, $0 < |y| < t_0$, and $1 \le i \le N$;

(vii) $|\partial_t a_i(x, t, y)| \le C_6\left(t^{q-1}(1 + |\log t|) + |y|^{q-1}(1 + |\log|y||)\right)$ for all $x \in \overline{\Omega}$, all $0 < t < t_0$, $0 < |y| < t_0$, and $1 \le i \le N$.

The prototype of the differential operator constructed through the map A is the (p,q)-Laplacian $\Delta_p + \Delta_q$. Many other situations are incorporated. We provide a characteristic example.

Example 4.30. Assume $1 < q < r < p$ and that

(i) $a \in C^1(\overline{\Omega})$ such that $0 < \inf_{\Omega} a \leq \sup_{\Omega} a < +\infty$ and $\sup_{\Omega} |a'| < +\infty$;

(ii) $\varphi_j \in C([0,+\infty)) \cap C^1((0,+\infty))$ with $\sup_{t>0} |\varphi_j'(t)| < +\infty$ for $j = p, q, r$;

(iii) $\inf_{t\geq 0} \varphi_p(t) > 0$ and $\inf_{t\geq 0} \varphi_q(t) > 0$;

(iv) $\inf_{t\geq 0} \big(\varphi_p(t) + \varphi_r(t)\big) > 0$ and $\inf_{t\geq 0} \big(\varphi_q(t) + \varphi_r(t)\big) > 0$.

Then the map A defined by

$$A(x,t,y) = a(x)\left(\varphi_q(t)|y|^{q-2} + \varphi_r(t)|y|^{r-2} + \varphi_p(t)|y|^{p-2}\right) y$$

satisfies assumption H(A).

Remark 4.31. If H(A) holds, then the map A satisfies H(A) in Section 2.3.

We will also consider the class of operators A which are asymptotically $(q-1)$-homogeneous near zero. This relevant property is described in the following condition:

H$(A)_0$ There exist $a_0 \in C^1(\overline{\Omega}, (0,+\infty))$, $A_0 \in C(\overline{\Omega} \times [0,+\infty) \times \mathbb{R}^N, \mathbb{R}^N)$, and $s_0 > 0$ such that

$$A(x,t,y) = a_0(x)|y|^{q-2}y + A_0(x,t,y)$$
$$\text{for all } x \in \Omega,\ t \in [0,+\infty),\ y \in \mathbb{R}^N$$
$$\text{and} \quad \lim_{y\to 0} \frac{|A_0(x,t,y)|}{|y|^{q-1}} = 0 \text{ uniformly in } x \in \overline{\Omega},\ t \in (0,s_0].$$

For later use, in connection with condition H$(A)_0$, we denote by $\lambda_{1,a_0}(m)$ the first positive eigenvalue of $-\mathrm{div}(a_0(x)|\nabla u|^{q-2}\nabla u)$ with weight function m, namely

$$\lambda_{1,a_0}(m) := \inf\left\{\int_\Omega a_0(x)|\nabla u|^q\,dx : u \in W_0^{1,q}(\Omega) \text{ and } \int_\Omega m|u|^q\,dx = 1\right\}. \tag{4.44}$$

Now we present some useful consequences of assumption H(A).

Proposition 4.32. *Assume that condition* H(A) *holds. Then we have:*

(i) *for every $x \in \overline{\Omega}$ and $t \in [0,+\infty)$, $A(x,t,y)$ is maximal monotone and strictly monotone in y;*

(ii) $|A(x,t,y)| \le g_0(t)\left(\frac{C_1}{p-1}|y|^{p-1} + \frac{C_2}{q-1}|y|^{q-1}\right)$ *for all* $(x,t,y) \in \overline{\Omega} \times [0,+\infty) \times \mathbb{R}^N$;

(iii) $A(x,t,y) \cdot y \ge \frac{C_3}{p-1}|y|^p + \frac{C_4}{q-1}|y|^q$ *for all* $(x,t,y) \in \overline{\Omega} \times [0,+\infty) \times \mathbb{R}^N$;

(iv) *there exist constants* $C_p > 0$ *and* $C_q > 0$ *such that*

$$(A(x,t,\xi) - A(x,t,\eta)) \cdot (\xi - \eta) \ge \begin{cases} C_3|\xi-\eta|^2(|\xi|+|\eta|)^{p-2} + C_4|\xi-\eta|^2(|\xi|+|\eta|)^{q-2} \\ \quad \text{if } 1 < q \le p \le 2,\ |\xi|+|\eta| > 0, \\ C_3C_p|\xi-\eta|^p + C_4|\xi-\eta|^2(|\xi|+|\eta|)^{q-2} \\ \quad \text{if } 1 < q \le 2 \le p,\ |\xi|+|\eta| > 0, \\ C_3C_p|\xi-\eta|^p + C_4C_q|\xi-\eta|^q \\ \quad \text{if } 2 \le q \le p \end{cases}$$

for all $(x,t) \in \overline{\Omega} \times [0,+\infty)$, *where* C_i $(1 \le i \le 4)$ *are the positive constants in assumption* H(A).

Proof. Assertions (i), (ii), and (iii) are direct consequences of hypotheses H(A) (i)–(iv). We focus on the proof of assertion (iv). Since the other cases can be handled similarly, we suppose that $s\xi + (1-s)\eta \ne 0$ for every $s \in [0,1]$. From hypothesis H(A) (iv), we obtain

$$\begin{aligned}&(A(x,t,\xi) - A(x,t,\eta)) \cdot (\xi - \eta) \\ &\ge C_3|\xi-\eta|^2 \int_0^1 |\eta + s(\xi-\eta)|^{p-2}\,ds + C_4|\xi-\eta|^2 \int_0^1 |\eta + s(\xi-\eta)|^{q-2}\,ds.\end{aligned}$$

If $1 < r \le 2$, it is clear that

$$\int_0^1 |\eta + s(\xi-\eta)|^{r-2}\,ds \ge \int_0^1 (|\xi|+|\eta|)^{r-2}\,ds = (|\xi|+|\eta|)^{r-2}.$$

If $r > 2$, [64, Lemma 2.1] gives

$$\int_0^1 |\eta + s(\xi-\eta)|^{r-2}\,ds \ge C_r(|\xi|+|\eta|)^{r-2},$$

with a constant $C_r > 0$ provided $|\xi|+|\eta| > 0$. If $r \ge 2$, we also note that

$$|\xi-\eta|^r \le |\xi-\eta|^2(|\xi|+|\eta|)^{r-2},$$

so property (iv) follows. ☐

We formulate the assumptions on the right-hand side of the equation in (4.43).

H(f) $f : \overline{\Omega} \times [0, +\infty) \times \mathbb{R}^N \to \mathbb{R}$ is a continuous function satisfying

$$f(x, 0, \xi) \geq 0 \text{ for every } (x, \xi) \in \Omega \times \mathbb{R}^N,$$

as well as

(i) there exist constants $b_1 > 0$ and $r_1 > 1$ such that

$$|f(x, t, \xi)| \leq b_1(1 + t^{r_1 - 1} + |\xi|^{p+1}) \tag{4.45}$$

for all $(x, t, \xi) \in \Omega \times [0, +\infty) \times \mathbb{R}^N$;

(ii) there exist constants $b_2 > 0$ and $1 < r_2 < p$ such that

$$f(x, t, \xi)t \leq b_2(1 + t^{r_2} + |\xi|^p) \tag{4.46}$$

for all $(x, t, \xi) \in \Omega \times [0, +\infty) \times \mathbb{R}^N$.

A relevant example of such function is as follows.

Example 4.33. The function f defined below verifies H(f):

$$f(x, t, \xi) = t^{r_3 - 1}(2 - |\xi|^{p_1}) + \frac{|\xi|^{p_2}}{1 + t^{r_4}} - t^{r_5 - 1} + |\xi|^{p_3}$$

with $0 \leq p_1 < p + 1$, $1 \leq r_3 < p$, $0 \leq p_2 \leq p$, $r_4 \geq 1$, $r_5 > 1$ and $0 \leq p_3 < p - 1$.

Our approach is based on the construction of approximate solutions, which are solutions of the auxiliary problem

$$\begin{cases} -\operatorname{div} A(x, u, \nabla u) = f(x, u, \nabla u) + \varepsilon & \text{in } \Omega, \\ u = 0 & \text{on } \partial\Omega \end{cases} \tag{4.47}$$

corresponding to each $0 < \varepsilon \leq 1$.

Fix a number β with

$$\frac{1}{p} < \beta < 1 \tag{4.48}$$

and introduce for any $M > 0$ the truncations $t^M := \max\{0, \min\{M, t\}\}$,

$$A_M(x, t, \xi) := A(x, t^M, \xi),$$

$$f_M(x, t, \xi) := \begin{cases} f\left(x, t^M, M^\beta \frac{\xi}{|\xi|}\right) & \text{if } |\xi| \geq M^\beta, \\ f(x, t^M, \xi) & \text{if } |\xi| < M^\beta \end{cases}$$

for all $x \in \Omega$, $t \in \mathbb{R}$, and $\xi \in \mathbb{R}^N$.

Proposition 4.34. *Let the map* $\mathcal{A}_M : W_0^{1,p}(\Omega) \to W^{-1,p'}(\Omega)$ *be defined by*

$$\langle \mathcal{A}_M(u), v\rangle = \int_\Omega A_M(x,u,\nabla u)\cdot \nabla v\, dx \text{ for all } u, v \in W_0^{1,p}(\Omega), \tag{4.49}$$

with some constant $M > 0$. *Then* $\mathcal{A}_M$ *has the* $(S)_+$*-property.*

Proof. Let a sequence $\{u_n\}$ weakly convergent to u in $W_0^{1,p}(\Omega)$ with

$$\limsup_{n\to\infty} \langle \mathcal{A}_M(u_n), u_n - u\rangle \le 0.$$

By assumption H(A) (v), Hölder's inequality, and Lebesgue's dominated convergence theorem, we infer that

$$\begin{aligned}
&\left|\int_\Omega (A(x,u_n^M,\nabla u) - A(x,u^M,\nabla u))\cdot(\nabla u_n - \nabla u)\, dx\right| \\
&\le C_M \int_\Omega (1+|\nabla u|^{p-1})|u_n^M - u^M|^\alpha\, |\nabla u_n - \nabla u|\, dx \\
&\le C_M \|\nabla u_n - \nabla u\|_p \left(\int_\Omega (1+|\nabla u|^{p-1})^{\frac{p}{p-1}} |u_n^M - u^M|^{\frac{p\alpha}{p-1}}\, dx\right)^{\frac{p-1}{p}} \to 0
\end{aligned}$$

as $n\to\infty$. This leads to

$$\limsup_{n\to\infty} \int_\Omega (A(x,u_n^M,\nabla u_n) - A(x,u_n^M,\nabla u))\cdot(\nabla u_n - \nabla u)\, dx \le 0. \tag{4.50}$$

If $p \ge 2$, from (4.50) and Proposition 4.32 (iv) we find

$$\limsup_{n\to\infty} \|\nabla u_n - \nabla u\|_p^p \le 0,$$

whence $u_n \to u$ in $W_0^{1,p}(\Omega)$. If $1 < p < 2$, by Hölder's inequality we get for every $v \in W_0^{1,p}(\Omega)$ that

$$\begin{aligned}
&\|\nabla u - \nabla v\|_p^p \\
&= \int_{\{|\nabla u| + |\nabla v| > 0\}} |\nabla u - \nabla v|^p (|\nabla u| + |\nabla v|)^{\frac{p(p-2)}{2}} (|\nabla u| + |\nabla v|)^{\frac{p(2-p)}{2}}\, dx \\
&\le \left(\int_{\{|\nabla u| + |\nabla v| > 0\}} (|\nabla u| + |\nabla v|)^{p-2} |\nabla u - \nabla v|^2\, dx\right)^{\frac{p}{2}} \\
&\quad \times (\|\nabla u\|_p + \|\nabla v\|_p)^{p(1-\frac{p}{2})}.
\end{aligned}$$

Setting $v = u_n$, from (4.50) and Proposition 4.32 (iv), we infer that $\{u_n\}$ strongly converges to u in $W_0^{1,p}(\Omega)$. ☐

Proposition 4.35. *Let* $B_{M,\varepsilon} : W_0^{1,p}(\Omega) \to W^{-1,p'}(\Omega)$ *be the map defined by*

$$\langle B_{M,\varepsilon}(u), v\rangle = \int_\Omega f_M(x, u, \nabla u) v \, dx + \varepsilon \int_\Omega v \, dx$$

for all u, $v \in W_0^{1,p}(\Omega)$, *with* $M > 0$ *and* $\varepsilon \in (0, 1]$. *Then, for* $\mathcal{A}_M$ *in (4.49),* $\mathcal{A}_M - B_{M,\varepsilon}$ *is a pseudomonotone operator.*

Proof. Let $\{u_n\}$ be a sequence such that $u_n \rightharpoonup u$ in $W_0^{1,p}(\Omega)$ and

$$\limsup_{n\to\infty} \langle \mathcal{A}_M(u_n) - B_{M,\varepsilon}(u_n), u_n - u\rangle \le 0.$$

Since the truncated function f_M is bounded, we infer that

$$\left|\langle B_{M,\varepsilon}(u_n), u_n - u\rangle\right| \le (\tilde{C}_M + 1)\|u_n - u\|_1 \to 0,$$

where

$$\tilde{C}_M := \max\{|f_M(x, t, \xi)| : (x, t, \xi) \in \Omega \times \mathbb{R} \times \mathbb{R}^N\},$$

which implies

$$\limsup_{n\to\infty} \langle \mathcal{A}_M(u_n), u_n - u\rangle \le 0.$$

Then the $(S)_+$-property of Proposition 4.34 implies that $\{u_n\}$ strongly converges to u in $W_0^{1,p}(\Omega)$, hence

$$\lim_{n\to\infty} \langle B_{M,\varepsilon}(u_n), u_n - v\rangle = \langle B_{M,\varepsilon}(u), u - v\rangle$$

for every $v \in W_0^{1,p}(\Omega)$. On account of

$$\left|A(x, u_n^M, \nabla u_n)\right| \le G(M) \left(\frac{C_1}{p-1}|\nabla u_n|^{p-1} + \frac{C_2}{q-1}|\nabla u_n|^{q-1}\right),$$

with $G(M) := \max\{g_0(t) : t \in [0, M]\}$ (see Proposition 4.32 (ii)), Lebesgue's dominated convergence theorem yields, as $n \to \infty$, that

$$\begin{aligned}\langle \mathcal{A}_M(u_n), u_n - v\rangle &= \int_\Omega A(x, u_n^M, \nabla u_n) \cdot (\nabla u_n - \nabla v) \, dx \\ &\to \int_\Omega A(x, u^M, \nabla u) \cdot (\nabla u - \nabla v) \, dx = \langle \mathcal{A}_M(u), u - v\rangle\end{aligned}$$

for all $v \in W_0^{1,p}(\Omega)$, thus $\mathcal{A}_M - B_{M,\varepsilon}$ is pseudomonotone. □

Proposition 4.36. *For every $M > 0$ and $\varepsilon \in (0, 1]$, the problem*

$$\begin{cases} -\operatorname{div} A_M(x, u, \nabla u) = f_M(x, u, \nabla u) + \varepsilon & \text{in } \Omega, \\ u = 0 & \text{on } \partial\Omega \end{cases} \tag{4.51}$$

has a solution $u_{M,\varepsilon} \in W_0^{1,p}(\Omega)$.

Proof. Fix $M > 0$ and $\varepsilon \in (0, 1]$. The operator $\mathcal{A}_M - B_{M,\varepsilon}$ is bounded because $\mathcal{A}_M$ is bounded by Proposition 4.32 (ii), the function f_M is bounded and $0 \le u^M \le M$. The boundedness of f_M and Proposition 4.32 (iii) entail that $\mathcal{A}_M - B_{M,\varepsilon}$ is coercive, which reads as

$$\lim_{\|u\| \to +\infty} \frac{\langle \mathcal{A}_M(u) - B_{M,\varepsilon}(u), u \rangle}{\|u\|} = +\infty,$$

where we recall that $\|\cdot\|$ stands for the usual norm on $W_0^{1,p}(\Omega)$. Recalling that $\mathcal{A}_M - B_{M,\varepsilon}$ is pseudomonotone (by Proposition 4.35), the main surjectivity result on pseudomonotone operators (Theorem 1.55) ensures that there exists $u_{M,\varepsilon} \in W_0^{1,p}(\Omega)$ such that $\mathcal{A}_M(u_{M,\varepsilon}) - B_{M,\varepsilon}(u_{M,\varepsilon}) = 0$ in $W^{-1,p'}(\Omega)$, thereby $u_{M,\varepsilon}$ is a solution of (4.51). □

We further discuss certain properties of the approximate solution $u_{M,\varepsilon}$.

Lemma 4.37. *If $u_{M,\varepsilon} \in W_0^{1,p}(\Omega)$ solves problem (4.51), then $u_{M,\varepsilon}(x) \ge 0$ for a.e. $x \in \Omega$.*

Proof. Acting with $-(u_{M,\varepsilon})^-$ as a test function in (4.51) shows that

$$\begin{aligned} &\frac{C_3}{p-1} \|\nabla(u_{M,\varepsilon})^-\|_p^p + \frac{C_4}{q-1} \|\nabla(u_{M,\varepsilon})^-\|_q^q \\ &\le \int_{\{u_{M,\varepsilon} \le 0\}} A(x, 0, \nabla u_{M,\varepsilon}) \cdot \nabla u_{M,\varepsilon}\, dx \\ &= \int_\Omega A_M(x, u_{M,\varepsilon}, \nabla u_{M,\varepsilon}) \cdot (-\nabla(u_{M,\varepsilon})^-)\, dx \\ &= \int_{\{u_{M,\varepsilon} \le 0\}} f_M(x, u_{M,\varepsilon}, \nabla u_{M,\varepsilon})\, u_{M,\varepsilon}\, dx - \varepsilon \int_\Omega (u_{M,\varepsilon})^-\, dx \le 0 \end{aligned}$$

due to the fact that $f_M(x, t, \xi) \ge 0$ if $t \le 0$ and Proposition 4.32 (iii). □

Lemma 4.38. *If $u_{M,\varepsilon} \in W_0^{1,p}(\Omega)$ is a solution of problem (4.51), then $u_{M,\varepsilon} \in L^\infty(\Omega)$.*

Proof. For sake of simplicity, we pose $u := u_{M,\varepsilon}$. By Lemma 4.37 we know that $u^M = \min\{u, M\}$. Take $\overline{p}^* = Np/(N - p)$ if $N > p$, and any $\overline{p}^* > p$

otherwise. Letting $u^L := \min\{u, L\}$ and using $(u^L)^\alpha u$ as a test function, where $L > M$ and $\alpha > 0$, we obtain from Proposition 4.32 (iii) that

$$\begin{aligned}
&\int_\Omega (f_M(x,u,\nabla u)+\varepsilon)\,(u^L)^\alpha u\,dx\\
&= \int_{\{u\ge L\}} L^\alpha A(x,u^M,\nabla u)\cdot\nabla u\,dx + \int_{\{u<L\}} (1+\alpha)u^\alpha A(x,u^M,\nabla u)\cdot\nabla u\,dx\\
&\ge (1+\alpha)\int_{\{u<L\}} u^\alpha\Big(\frac{C_3}{p-1}|\nabla u|^p + \frac{C_4}{q-1}|\nabla u|^q\Big)\,dx\\
&\ge \frac{(1+\alpha)C_3}{p-1}\int_\Omega |\nabla u^L|^p (u^L)^\alpha\,dx = \frac{(1+\alpha)C_3 p^p}{(p-1)(p+\alpha)^p}\|(u^L)^{1+\alpha/p}\|^p\\
&\ge C_*\frac{(1+\alpha)C_3 p^p}{(p-1)(p+\alpha)^p}\|(u^L)^{1+\alpha/p}\|^p_{\overline{p}^*} = C_*\frac{(1+\alpha)C_3 p^p}{(p-1)(p+\alpha)^p}\|u^L\|^{\alpha+p}_{\overline{p}^*(1+\alpha/p)}.
\end{aligned}$$

The constant C_* comes from the continuous embedding of $W_0^{1,p}(\Omega)$ into $L^{\overline{p}^*}(\Omega)$. If $u \in L^{1+\alpha}(\Omega)$, the boundedness of f_M implies

$$\int_\Omega (f_M(x,u,\nabla u)+\varepsilon)\,(u^L)^\alpha u\,dx \le (\overline{C}_M + 1)\|u\|^{1+\alpha}_{1+\alpha},$$

where

$$\overline{C}_M := \max\left\{|f(x,t,\xi)| \,:\, x\in\Omega,\ t\in[0,M],\ |\xi|\le M^\beta\right\}.$$

Letting $L \to +\infty$, through Fatou's lemma we find

$$\|u\|^{\alpha+p}_{\overline{p}^*(1+\alpha/p)} \le C'\frac{(p+\alpha)^p}{1+\alpha}\|u\|^{1+\alpha}_{1+\alpha}, \tag{4.52}$$

with a constant $C' > 0$ independent of α.

Consider the sequence $\{\alpha_m\}$ inductively defined by $\alpha_0 := \overline{p}^* - 1$ and

$$\alpha_{m+1} := \overline{p}^*\Big(1+\frac{\alpha_m}{p}\Big) - 1.$$

It follows from (4.52) that

$$u \in L^{\overline{p}^*(p+\alpha_m)/p}(\Omega) = L^{1+\alpha_{m+1}}(\Omega)$$

provided $u \in L^{1+\alpha_m}(\Omega)$. Since

$$\alpha_{m+1} = \overline{p}^*\frac{\alpha_m}{p} + \overline{p}^* - 1 \ge \left(\frac{\overline{p}^*}{p}\right)^{m+1}\alpha_0 \to +\infty$$

as $m \to \infty$, we may conclude that $u \in L^\infty(\Omega)$. □

The next step is to establish a priori estimates for problem (4.51).

Lemma 4.39. *Denote $\beta_0 := \max\{r_2, p\beta\}$. Then there exist constants $\alpha^* > 0$ and $D_0 > 0$ independent of M and ε such that $\|u_{M,\varepsilon}\|_{\beta_0+\alpha^*} \le D_0$ for every solution $u_{M,\varepsilon}$ of problem (4.51).*

Proof. Set, for simplicity, $u := u_{M,\varepsilon}$ and use $u^{1+\alpha} \in W_0^{1,p}(\Omega)$, with $\alpha > 0$, as a test function in (4.51), which is possible because by Lemma 4.38 we know that $u \in L^\infty(\Omega)$. From assumption H(f) (ii) (see (4.46)) and the fact that $u^M = \min\{u, M\}$, we see that

$$
\begin{aligned}
&\int_\Omega (f_M(x,u,\nabla u)+\varepsilon)\,u^{1+\alpha}\,dx\\
&\le b_2\int_{\{|\nabla u|<M^\beta,\,u<M\}} (1+(u^M)^{r_2}+|\nabla u|^p)u^\alpha\,dx\\
&\quad + b_2\int_{\{|\nabla u|<M^\beta,\,u\ge M\}} (1+M^{r_2}+|\nabla u|^p)\frac{u^{1+\alpha}}{M}\,dx\\
&\quad + b_2\int_{\{|\nabla u|\ge M^\beta,\,u<M\}} (1+(u^M)^{r_2}+M^{p\beta})u^\alpha\,dx\\
&\quad + b_2\int_{\{|\nabla u|\ge M^\beta,\,u\ge M\}} (1+M^{r_2}+M^{p\beta})\frac{u^{1+\alpha}}{M}\,dx + \int_\Omega u^{1+\alpha}\,dx.
\end{aligned}
$$

Then, through (4.48) and Hölder's inequality, we derive that

$$
\begin{aligned}
&\int_\Omega (f_M(x,u,\nabla u)+\varepsilon)\,u^{1+\alpha}\,dx\\
&\le b_2\int_\Omega \left(u^\alpha+u^{1+\alpha}+u^{\alpha+r_2}+u^{\alpha+p\beta}\right)dx + \int_\Omega u^{1+\alpha}\,dx + b_2\int_\Omega u^\alpha|\nabla u|^p\,dx\\
&\le (b_2+1)(1+|\Omega|)\left(1+\|u\|_{\beta_0+\alpha}+\|u\|_{\beta_0+\alpha}^{r_2}+\|u\|_{\beta_0+\alpha}^{p\beta}\right)\|u\|_{\beta_0+\alpha}^{\alpha}\\
&\quad + b_2\int_\Omega u^\alpha|\nabla u|^p\,dx. \qquad (4.53)
\end{aligned}
$$

By Proposition 4.32 (iii), we have

$$
\int_\Omega A_M(x,u,\nabla u)\cdot\nabla(u^{1+\alpha})\,dx \ge \frac{(1+\alpha)C_3}{p-1}\int_\Omega |\nabla u|^p u^\alpha\,dx. \qquad (4.54)
$$

Fix a number α^* so that

$$
\frac{(1+\alpha^*)C_3}{p-1} > 2b_2.
$$

Combining (4.53) and (4.54) yields

$$\frac{(1+\alpha^*)C_3}{2(p-1)}\int_\Omega |\nabla u|^p u^{\alpha^*}\,dx \tag{4.55}$$
$$\leq (b_2+1)(1+|\Omega|)\big(1+\|u\|_{\beta_0+\alpha^*}+\|u\|^{r_2}_{\beta_0+\alpha^*}+\|u\|^{p\beta}_{\beta_0+\alpha^*}\big)\|u\|^{\alpha^*}_{\beta_0+\alpha^*}.$$

On the other hand, using the constant $C_* > 0$ ensured by the continuous embedding of $W_0^{1,p}(\Omega)$ into $L^{\overline{p}^*}(\Omega)$, Holder's inequality and

$$\overline{p}^*\Big(1+\frac{\alpha^*}{p}\Big) > p+\alpha^* > \beta_0+\alpha^*$$

results in

$$\int_\Omega |\nabla u|^p u^{\alpha^*}\,dx$$
$$\geq \frac{C_* p^p}{(p+\alpha^*)^p}\|u^{1+\alpha^*/p}\|^p_{\overline{p}^*} = \frac{C_* p^p}{(p+\alpha^*)^p}\|u\|^{\alpha^*+p}_{\overline{p}^*(1+\alpha^*/p)} \tag{4.56}$$
$$\geq \frac{C_* p^p}{(1+|\Omega|)(p+\alpha^*)^p}\|u\|^{\alpha^*+p}_{\beta_0+\alpha^*}. \tag{4.57}$$

From (4.55) and (4.57) we get

$$\frac{C_* p^p(1+\alpha^*)C_3}{2(p-1)(1+|\Omega|)(p+\alpha^*)^p}\|u\|^p_{\beta_0+\alpha^*}$$
$$\leq 4(b_2+1)(1+|\Omega|)\max\{1, \|u\|^{\beta_0}_{\beta_0+\alpha^*}\}.$$

Since $\beta_0 < p$, this reads as

$$\|u\|_{\beta_0+\alpha^*} \leq \max\{C^{1/(p-\beta_0)}, C^{1/p}\}$$

with $C = \frac{8(b_2+1)(1+|\Omega|)^2(p+\alpha^*)^p(p-1)}{C_* p^p(1+\alpha^*)C_3}$. □

Now we are in a position to obtain the uniform bounds.

Proposition 4.40. *There exists a constant $D_1 > 0$ independent of M and ε such that for every solution u of problem (4.51) one has $\|u\|_\infty \leq D_1$.*

Proof. Let $u \in W_0^{1,p}(\Omega)$ be a solution of (4.51) with $M > 0$ and $\varepsilon \in (0, 1]$. For each $\alpha > 0$, we act with $u^{1+\alpha}$ as a test function in problem (4.51). As in (4.55) and (4.56), we see that

$$\|u\|^{\alpha+p}_{\overline{p}^*(1+\alpha/p)} \leq C\frac{(p+\alpha)^p}{1+\alpha}\max\{1, \|u\|^{\beta_0+\alpha}_{\beta_0+\alpha}\},$$

where $C = \frac{8(b_2+1)(1+|\Omega|)^2(p-1)}{C_* p^p C_3}$. With the number α^* determined in Lemma 4.39, we construct the sequence $\{\alpha_m\}$ by $\alpha_0 := \alpha^*$ and $\alpha_{m+1} := \overline{p}^*\big(1+\frac{\alpha_m}{p}\big) - \beta_0$.

Proceeding as in the proof of Lemma 4.38, we prove the estimate

$$\|u\|_\infty \le C' \max\{1, \|u\|_{\beta_0+\alpha^*}\}^{(\overline{p}^*-p)(\beta_0+\alpha^*)/(\overline{p}^*\alpha^*)}$$

with a positive constant C' independent of ε and M. Applying Lemma 4.39, the conclusion ensues. □

We turn to problem (4.47). We introduce the notation

$$\mu_1(m) = \inf\left\{\int_\Omega |\nabla u|^q\,dx \,:\, u \in W_0^{1,q}(\Omega) \text{ and } \int_\Omega m|u|^q\,dx = 1\right\} \tag{4.58}$$

corresponding to a weight function m, and

$$A_q(t) := \frac{C_2}{q-1}\left(\frac{C_2}{C_4}\right)^{q-1} g_0(t)^q \quad \text{for } t \ge 0, \tag{4.59}$$

where g_0, C_2, C_4 are given in hypothesis H(A). It is worth noting that $A_q \equiv 1$ if $A(x,t,y) = |y|^{p-2}y + |y|^{q-2}y$ (i.e., the case of the (p,q)-Laplacian).

We formulate two conditions, corresponding to two different situations, permitting us to investigate the approximate problem (4.47), and eventually (4.43).

$\mathrm{H}_1(f)$ There exist $m \in L^\infty(\Omega)$ and positive constants b_0, b_0', r' with

$$|\{x \in \Omega : m(x) > 0\}| > 0, \quad b_0 > \mu_1(m)A_q(0), \quad r' > q \tag{4.60}$$

such that for each $\rho > 0$ there is $K_\rho > 0$ satisfying

$$f(x,t,\xi) \ge (b_0 m(x) - \rho)t^{q-1} - b_0'|\xi|^{r'-1} \tag{4.61}$$

for all $x \in \Omega$, $t \in [0, K_\rho]$ and $\xi \in \mathbb{R}^N$ with $|\xi| \le K_\rho$.

$\mathrm{H}_2(f)$ There exist $m \in L^\infty(\Omega)$ and positive constants b_0, b_0', r' with

$$|\{x \in \Omega : m(x) > 0\}| > 0, \quad b_0 > \lambda_{1,a_0}(m), \quad r' > q \tag{4.62}$$

(with a_0 as in $\mathrm{H}(A)_0$ and $\lambda_{1,a_0}(m)$ as in (4.44)) such that for each $\rho > 0$ there is $K_\rho > 0$ satisfying

$$f(x,t,\xi) \ge (b_0 m(x) - \rho)t^{q-1} - b_0'|\xi|^{r'-1} \tag{4.63}$$

for all $x \in \Omega$, $t \in [0, K_\rho]$ and $\xi \in \mathbb{R}^N$ with $|\xi| \le K_\rho$.

Example 4.41. The function f in Example 4.33 satisfies conditions $\mathrm{H}_1(f)$ and $\mathrm{H}_2(f)$ provided $1 < r_3 < q \le r_5$.

Remark 4.42. If $\mathrm{H}_1(f)$ or $\mathrm{H}_2(f)$ is verified, then condition $\mathrm{H}(\widetilde{f})$ in Section 2.3 holds.

First, we present an easy consequence.

Lemma 4.43. *Suppose* $H_1(f)$ *or* $H_2(f)$. *Then for each* $K > 0$ *there exists* $\lambda_0 > 0$ *such that*

$$f(x,t,\xi) \geq -\lambda_0 \left(t^{q-1} + |\xi|^{q-1}\right)$$
$$\text{for all } x \in \Omega,\ t \in [0, K] \text{ and } \xi \in \mathbb{R}^N \text{ with } |\xi| \leq K.$$

Proof. Fix $K > 0$. By $H_1(f)$ or $H_2(f)$, there exist $\delta > 0$ such that

$$f(x,t,\xi) \geq -(b_0 + 1)\|m\|_\infty t^{q-1} - b_0' \delta^{r'-q} |\xi|^{q-1}$$

for all $x \in \Omega$, $t \in [0, \delta]$, $|\xi| \leq \delta$. On the other hand, from (4.45) it follows that

$$\begin{aligned} f(x,t,\xi) &\geq -b_1(1 + t^{r_1-1} + |\xi|^{p+1}) \\ &\geq -\frac{b_1(1 + K^{r_1-1})}{\delta^{q-1}} t^{q-1} - b_1 K^{p-q+2} |\xi|^{q-1} \end{aligned}$$

provided $\delta \leq t \leq K$ and $|\xi| \leq K$. Similarly, if $0 \leq t \leq K$ and $\delta \leq |\xi| \leq K$, then we have

$$f(x,t,\xi) \geq -b_1(1 + t^{r_1-1} + |\xi|^{p+1}) \geq -\frac{b_1(1 + K^{r_1-1} + K^{p+1})}{\delta^{q-1}} |\xi|^{q-1}.$$

□

The desired estimates for problem (4.47) are now available.

Proposition 4.44. *Suppose* $H_1(f)$ *or* $H_2(f)$. *Then, for every* $\varepsilon \in (0, 1]$, *problem (4.47) has a positive solution* $u_\varepsilon \in \operatorname{int}(C_0^1(\overline{\Omega})_+)$. *Moreover, there exist* $D_2 > 0$ *and* $\gamma \in (0, 1)$ *such that* $u_\varepsilon \in C_0^{1,\gamma}(\overline{\Omega})$ *and* $\|u_\varepsilon\|_{C_0^{1,\gamma}(\overline{\Omega})} \leq D_2$ *for all* $\varepsilon \in (0, 1]$.

Proof. Let $u_{M,\varepsilon}$ be a nonnegative solution for (4.51) whose existence is guaranteed by Proposition 4.36. Choose $M_0 > D_1$, where D_1 is obtained in Proposition 4.40. Then $0 \leq u_{M,\varepsilon}(x) \leq D_1 < M_0$ for every $\varepsilon \in (0, 1]$ and $M \geq M_0$, so $u_{M,\varepsilon}$ is a solution for

$$-\operatorname{div} A(x, u, \nabla u) = f_M(x, u, \nabla u) + \varepsilon \quad \text{in } \Omega, \qquad u = 0 \quad \text{on } \partial\Omega. \tag{4.64}$$

Since f_M satisfies (4.45) with the same constant b_1 and $\|u_{M,\varepsilon}\|_\infty \leq D_1$, the regularity result in [129] and [130] applied to (4.64) provides constants $D_2 > 0$ and $\gamma \in (0, 1)$ such that $u_{M,\varepsilon} \in C_0^{1,\gamma}(\overline{\Omega})$ and $\|u_{M,\varepsilon}\|_{C_0^{1,\gamma}(\overline{\Omega})} \leq D_2$ for all $\varepsilon \in (0, 1]$ and $M \geq M_0$.

Taking M_1 with $M_1^{\beta} > \max\{D_2, M_0\}$, it holds

$$f_{M_1}(x, u_{M_1,\varepsilon}, \nabla u_{M_1,\varepsilon}) = f(x, u_{M_1,\varepsilon}, \nabla u_{M_1,\varepsilon}),$$

whence $u_{\varepsilon} := u_{M_1,\varepsilon} \in C_0^{1,\gamma}(\overline{\Omega})$ is a nonnegative solution of (4.47). Note that $u_{\varepsilon} \not\equiv 0$. Since Lemma 4.43 ensures that $\widetilde{f}(x,s,\xi) := f(x,s,\xi) + \varepsilon$ satisfies (2.49), Theorem 2.19 implies that $u_{\varepsilon} \in \operatorname{int}(C_0^1(\overline{\Omega})_+)$. □

Some preliminaries are still necessary.

Lemma 4.45. *Let $\varphi, u \in \operatorname{int}(C_0^1(\overline{\Omega})_+)$. Then*

$$\int_{\Omega} A(x,u,\nabla u) \cdot \nabla\left(\frac{\varphi^q}{u^{q-1}}\right) dx$$
$$\le \frac{C_1^q}{(p-1)C_3^{q-1}} \int_{\Omega} g_0(u)^q |\nabla u|^{p-q} |\nabla \varphi|^q \, dx + \int_{\Omega} A_q(u) |\nabla \varphi|^q \, dx,$$

with the function A_q in (4.59) and the constants C_1 and C_3 in hypothesis H(A).

Proof. Since $\varphi, u \in \operatorname{int}(C_0^1(\overline{\Omega})_+)$, there exist $\delta_1 > \delta_2 > 0$ such that $\delta_1 u \ge \varphi \ge \delta_2 u$ in $\overline{\Omega}$, hence u/φ, $\varphi/u \in L^{\infty}(\Omega)$. Therefore, by Proposition 4.32 (ii) and (iii), we find that

$$A(x,u,\nabla u) \cdot \nabla\left(\frac{\varphi^q}{u^{q-1}}\right)$$
$$= q\left(\frac{\varphi}{u}\right)^{q-1} A(x,u,\nabla u) \cdot \nabla \varphi - (q-1)\left(\frac{\varphi}{u}\right)^{q} A(x,u,\nabla u) \cdot \nabla u$$
$$\le g_0(u) \frac{qC_1}{p-1} \left(\frac{\varphi}{u}\right)^{q-1} |\nabla u|^{p-1} |\nabla \varphi| - \frac{(q-1)C_3}{p-1} \left(\frac{\varphi}{u}\right)^{q} |\nabla u|^p$$
$$+ g_0(u) \frac{qC_2}{q-1} \left(\frac{\varphi}{u}\right)^{q-1} |\nabla u|^{q-1} |\nabla \varphi| - C_4 \left(\frac{\varphi}{u}\right)^{q} |\nabla u|^q. \tag{4.65}$$

Young's inequality shows that

$$g_0(u) \frac{qC_1}{p-1} \left(\frac{\varphi}{u}\right)^{q-1} |\nabla u|^{p-1} |\nabla \varphi|$$
$$\le \frac{(q-1)C_3}{p-1} \left(\frac{\varphi}{u}\right)^{q} |\nabla u|^p + g_0(u)^q \frac{C_1^q}{(p-1)C_3^{q-1}} |\nabla u|^{p-q} |\nabla \varphi|^q. \tag{4.66}$$

Similarly, taking into account (4.59), we get

$$g_0(u) \frac{qC_2}{q-1} \left(\frac{\varphi}{u}\right)^{q-1} |\nabla u|^{q-1} |\nabla \varphi| \le C_4 \left(\frac{\varphi}{u}\right)^{q} |\nabla u|^q + A_q(u) |\nabla \varphi|^q. \tag{4.67}$$

Thanks to (4.65), (4.66), and (4.67), the conclusion follows. □

Lemma 4.46. *Let* $a_0 \in C(\overline{\Omega}, [0, +\infty))$ *and* $\varphi, u \in \mathrm{int}\,(C_0^1(\overline{\Omega})_+)$. *Then*

$$\int_\Omega a_0(x)|\nabla\varphi|^{q-2}\nabla\varphi \cdot \nabla\Big(\frac{\varphi^q - u^q}{\varphi^{q-1}}\Big)\,dx - \int_\Omega a_0(x)|\nabla u|^{q-2}\nabla u \cdot \nabla\Big(\frac{\varphi^q - u^q}{u^{q-1}}\Big)\,dx \geq 0.$$

The proof can be found in [226, Lemma 10].

Lemma 4.47. *Assume* $\mathrm{H}_1(f)$ *or* $\mathrm{H}_2(f)$. *Then there exists a constant* $D_4 > 0$ *such that*

$$\int_\Omega \Big(\frac{\varphi|\nabla u_\varepsilon|}{u_\varepsilon}\Big)^q dx \leq D_4 \int_\Omega (|\nabla\varphi|^q + \varphi^q)\,dx$$

for all $\varphi \in C_0^1(\overline{\Omega})_+$ *and* $u_\varepsilon \in \mathrm{int}\,(C_0^1(\overline{\Omega})_+)$ *given by Proposition 4.44 with* $\varepsilon \in (0, 1]$.

Proof. Fix any $\varphi \in C_0^1(\overline{\Omega})_+$. By Proposition 4.44, $\{u_\varepsilon : \varepsilon \in (0, 1]\}$ is bounded in $C_0^1(\overline{\Omega})$. Then Lemma 4.43 provides a positive constant λ_0 satisfying

$$f(x, u_\varepsilon, \nabla u_\varepsilon) \geq -\lambda_0(u_\varepsilon^{q-1} + |\nabla u_\varepsilon|^{q-1})$$

whenever $\varepsilon \in (0, 1]$. Note that $\varphi/u_\varepsilon \in L^\infty(\Omega)$ because $\varphi \in C_0^1(\overline{\Omega})_+$ and Proposition 4.44 guarantees that $u_\varepsilon \in \mathrm{int}\,(C_0^1(\overline{\Omega})_+)$. Using $\varphi^q u_\varepsilon^{1-q} \in W_0^{1,p}(\Omega)$ as a test function gives

$$\begin{aligned}
\int_\Omega A(x, u_\varepsilon, \nabla u_\varepsilon) \cdot \nabla\left(\frac{\varphi^q}{u_\varepsilon^{q-1}}\right) dx &= \int_\Omega \frac{f(x, u_\varepsilon, \nabla u_\varepsilon) + \varepsilon}{u_\varepsilon^{q-1}}\varphi^q\,dx \\
\geq -\lambda_0 \int_\Omega \varphi^q\,dx - \lambda_0 \int_\Omega \frac{|\nabla u_\varepsilon|^{q-1}}{u_\varepsilon^{q-1}}\varphi^q\,dx& \\
\geq -\lambda_0 \int_\Omega \varphi^q\,dx - \frac{C_4}{4}\int_\Omega \left(\frac{\varphi|\nabla u_\varepsilon|}{u_\varepsilon}\right)^q dx - D_5 \int_\Omega \varphi^q\,dx,& \qquad (4.68)
\end{aligned}$$

through Young's inequality and $D_5 = \lambda_0^q 4^{q-1} C_4^{1-q}$. On the other hand, by easy estimates involving Proposition 4.32 (ii) and (iii) (note that $1-q<0$), we obtain

$$\begin{aligned}
&\int_\Omega A(x, u_\varepsilon, \nabla u_\varepsilon) \cdot \nabla\left(\frac{\varphi^q}{u_\varepsilon^{q-1}}\right) dx \\
&\leq -C_4 \int_\Omega \left(\frac{\varphi|\nabla u_\varepsilon|}{u_\varepsilon}\right)^q dx - \frac{C_3(q-1)}{p-1}\int_\Omega \left(\frac{\varphi|\nabla u_\varepsilon|}{u_\varepsilon}\right)^q |\nabla u_\varepsilon|^{p-q}\,dx \\
&+ \frac{qC_2}{q-1}\int_\Omega g_0(u_\varepsilon)\left(\frac{\varphi|\nabla u_\varepsilon|}{u_\varepsilon}\right)^{q-1} |\nabla\varphi|\,dx
\end{aligned}$$

$$+\frac{qC_1}{p-1}\int_\Omega g_0(u_\varepsilon)\left(\frac{\varphi|\nabla u_\varepsilon|}{u_\varepsilon}\right)^{q-1}|\nabla u_\varepsilon|^{p-q}|\nabla\varphi|\,dx.$$

Applying Young's inequality to the last two terms yields

$$\begin{aligned}&\int_\Omega A(x,u_\varepsilon,\nabla u_\varepsilon)\cdot\nabla\left(\frac{\varphi^q}{u_\varepsilon^{q-1}}\right)dx\\ &\leq -\frac{C_4}{2}\int_\Omega\left(\frac{\varphi|\nabla u_\varepsilon|}{u_\varepsilon}\right)^q dx-\frac{C_3(q-1)}{2(p-1)}\int_\Omega\left(\frac{\varphi|\nabla u_\varepsilon|}{u_\varepsilon}\right)^q|\nabla u_\varepsilon|^{p-q}\,dx\\ &\quad+D_6\int_\Omega g_0(u_\varepsilon)^q|\nabla\varphi|^q\,dx+D_7\int_\Omega g_0(u_\varepsilon)^q|\nabla\varphi|^q|\nabla u_\varepsilon|^{p-q}\,dx,\end{aligned}\tag{4.69}$$

where D_6 and D_7 are positive constants independent of u_ε and φ. From (4.68) and (4.69), the existence of a positive constant D_8 independent of u_ε and φ such that

$$\begin{aligned}\int_\Omega\left(\frac{\varphi|\nabla u_\varepsilon|}{u_\varepsilon}\right)^q dx&\leq D_8\int_\Omega g_0(u_\varepsilon)^q|\nabla\varphi|^q(1+|\nabla u_\varepsilon|^{p-q})\,dx+D_8\int_\Omega\varphi^q\,dx\\ &\leq D_8G^q(1+D_3^{p-q})\int_\Omega|\nabla\varphi|^q\,dx+D_8\int_\Omega\varphi^q\,dx,\end{aligned}$$

where

$$G:=\sup\{g_0(t)\,:\,t\in(0,D_3]\}\quad\text{and}\quad D_3:=\sup\left\{\|u_\varepsilon\|_{C_0^1(\overline{\Omega})}\,:\,\varepsilon\in(0,1]\right\},$$

follows, which completes the proof. □

Lemma 4.48. *Assume that conditions* $\mathrm{H}_2(f)$ *and* $\mathrm{H}(A)_0$ *hold. If* $u_\varepsilon\to 0$ *in* $C_0^1(\overline{\Omega})$ *as* $\varepsilon\to 0^+$, *with* $u_\varepsilon\in\operatorname{int}(C_0^1(\overline{\Omega})_+)$ *solution of (4.47), then for every* $\varphi\in\operatorname{int}(C_0^1(\overline{\Omega})_+)$ *it is true that*

$$\lim_{\varepsilon\to 0^+}\int_\Omega A_0(x,u_\varepsilon,\nabla u_\varepsilon)\cdot\nabla\left(\frac{\varphi^q-u_\varepsilon^q}{u_\varepsilon^{q-1}}\right)dx=0,$$

where A_0 *is the function introduced in condition* $\mathrm{H}(A)_0$.

Proof. Fix any $\delta>0$. Since $u_\varepsilon\to 0$ in $C_0^1(\overline{\Omega})$, we may admit that $\|u_\varepsilon\|_{C_0^1(\overline{\Omega})}\leq s_0$ provided $\varepsilon>0$ is sufficiently small, where s_0 is the constant given in $\mathrm{H}(A)_0$. From assumption $\mathrm{H}(A)_0$ we have that

$$|A_0(x,u_\varepsilon,\nabla u_\varepsilon)|\leq\delta|\nabla u_\varepsilon|^{q-1}$$

for every $x\in\Omega$ and small enough $\varepsilon>0$. Therefore, by Lemma 4.47 and Hölder's inequality, we arrive at

$$
\begin{aligned}
&\left|\int_\Omega A_0(x,u_\varepsilon,\nabla u_\varepsilon)\cdot\nabla\left(\frac{\varphi^q}{u_\varepsilon^{q-1}}\right)dx\right|\\
&\le q\int_\Omega \frac{|A_0(x,u_\varepsilon,\nabla u_\varepsilon)|\,|\nabla\varphi|\varphi^{q-1}}{u_\varepsilon^{q-1}}dx\\
&\quad+(q-1)\int_\Omega \frac{|A_0(x,u_\varepsilon,\nabla u_\varepsilon)|\,|\nabla u_\varepsilon|\varphi^{q}}{u_\varepsilon^{q}}dx\\
&\le \delta q\|\varphi\|_{C_0^1(\overline{\Omega})}\int_\Omega\left(\frac{\varphi|\nabla u_\varepsilon|}{u_\varepsilon}\right)^{q-1}dx+\delta(q-1)\int_\Omega\left(\frac{\varphi|\nabla u_\varepsilon|}{u_\varepsilon}\right)^{q}dx\\
&\le \delta\|\varphi\|^q_{C_0^1(\overline{\Omega})}\left(q(2D_4)^{1-1/q}+2(q-1)D_4\right)|\Omega|,
\end{aligned}
$$

where D_4 is the positive constant obtained in Lemma 4.47. Moreover, it is clear that

$$
\left|\int_\Omega A_0(x,u_\varepsilon,\nabla u_\varepsilon)\cdot\nabla u_\varepsilon\,dx\right|\le\delta\|\nabla u_\varepsilon\|_q^q\le\delta s_0^q.
$$

Recalling that $\delta>0$ is arbitrary, the result is proven. □

The statement of the first main existence result for problem (4.43) is as follows.

Theorem 4.49. *Assume that conditions* H(A), H(f), *and* H$_1(f)$ *hold. Then problem (4.43) has a (positive) solution* $u\in\operatorname{int}(C_0^1(\overline{\Omega})_+)$.

Proof. By Proposition 4.44 there exist constants $D_2>0$ and $0<\gamma<1$ such that

$$
\|u_\varepsilon\|_{C_0^{1,\gamma}(\overline{\Omega})}\le D_2,\ \forall\varepsilon\in(0,1].
$$

Due to the compact embedding of $C_0^{1,\gamma}(\overline{\Omega})$ into $C_0^1(\overline{\Omega})$, there exist a sequence $\varepsilon_n\to0^+$ and a solution $u_0\in C_0^1(\overline{\Omega})_+$ of problem (4.43) such that $u_n:=u_{\varepsilon_n}\to u_0$ in $C_0^1(\overline{\Omega})$ as $n\to\infty$. If $u_0\neq0$ occurs, then $u_0\in\operatorname{int}(C_0^1(\overline{\Omega})_+)$ by the argument carried on in the proof of Proposition 4.44, and the conclusion of the theorem is achieved.

It remains to prove that $u_0\neq0$. To this end, we argue by contradiction supposing that $u_n\to0$ in $C_0^1(\overline{\Omega})$.

Let $\varphi\in\operatorname{int}(C_0^1(\overline{\Omega})_+)$ be an eigenfunction corresponding to the first positive eigenvalue $\mu_1(m)$ (see (4.58)). Thus φ is a positive solution of the Dirichlet problem

$$
\begin{cases}
-\Delta_q u=\mu_1(m)m(x)|u|^{q-2}u & \text{in }\Omega,\\
u=0 & \text{on }\partial\Omega.
\end{cases}
$$

Since this equation is $(q-1)$-homogeneous, we may assume that

$$\|\nabla\varphi\|_q^q = \mu_1(m)\int_\Omega m(x)\varphi^q\,dx = \mu_1(m).$$

In view of (4.60) we can choose $\rho > 0$ such that

$$\rho < \frac{b_0 - \mu_1(m)A_q(0)}{\|\varphi\|_q^q}. \tag{4.70}$$

Recalling that $u_n \to 0$ in $C_0^1(\overline{\Omega})$ and using Lemma 4.45, we have

$$\int_\Omega A(x,u_n,\nabla u_n)\cdot\nabla\left(\frac{\varphi^p}{u_n^{p-1}}\right)dx \le \int_\Omega A_q(u_n)|\nabla\varphi|^q\,dx + o(1). \tag{4.71}$$

On account of (4.61) there exists $K_0 > 0$ such that

$$f(x,s,\xi) \ge (b_0 m(x) - \rho)s^{q-1} - b_0'|\xi|^{r'-1} \tag{4.72}$$

whenever $x \in \Omega$, $0 \le s \le K_0$ and $|\xi| \le K_0$. Since $\|u_n\|_\infty \le K_0$ and $\|\nabla u_n\|_\infty \le K_0$ hold for sufficiently large n, by (4.72) we obtain

$$\begin{aligned}\int_\Omega A(x,u_n,\nabla u_n)\cdot\nabla\left(\frac{\varphi^q}{u_n^{q-1}}\right)dx &= \int_\Omega \frac{f(x,u_n,\nabla u_n)+\varepsilon_n}{u_n^{q-1}}\varphi^q\,dx \\ \ge b_0\int_\Omega m(x)\varphi^q\,dx - \rho\|\varphi\|_q^q - b_0'\int_\Omega \frac{|\nabla u_n|^{r'-1}\varphi^q}{u_n^{q-1}}\,dx.\end{aligned} \tag{4.73}$$

Hölder's inequality and Lemma 4.47 entail

$$\begin{aligned}\int_\Omega \frac{|\nabla u_n|^{r'-1}\varphi^q}{u_n^{q-1}}\,dx &= \int_\Omega \left(\frac{|\nabla u_n|\varphi}{u_n}\right)^{q-1}|\nabla u_n|^{r'-q}\varphi\,dx \\ &\le D_4^{(q-1)/q}\|u_n\|_{C^1(\overline{\Omega})}^{r'-q}(\|\nabla\varphi\|_q^q + \|\varphi\|_q^q)^{(q-1)/q}\|\varphi\|_q = o(1)\end{aligned} \tag{4.74}$$

(note that $r' - q > 0$ and $\|u_n\|_{C^1(\overline{\Omega})} = o(1)$). Finally, (4.71), (4.73), and (4.74) lead to

$$b_0 - \mu_1(m)A_q(0) \le \int_\Omega (A_q(u_n) - A_q(0))|\nabla\varphi|^q\,dx + \rho\|\varphi\|_q^q + o(1).$$

Letting $n \to \infty$ contradicts (4.70), which completes the proof. □

Now we are able to present the second main result of this section, where $\mathrm{H}(A)_0$ is assumed, and $\mathrm{H}_2(f)$ is assumed instead of $\mathrm{H}_1(f)$.

Theorem 4.50. *Assume that* $\mathrm{H}(A)$, $\mathrm{H}(A)_0$, $\mathrm{H}(f)$, *and* $\mathrm{H}_2(f)$ *are satisfied. Then problem (4.43) has a (positive) solution* $u \in \operatorname{int}(C_0^1(\overline{\Omega})_+)$.

Proof. At the beginning we proceed as in the proof of Theorem 4.49, finding a solution $u_0 \in C_0^1(\overline{\Omega})_+$ of problem (4.43). We note that if $u_0 \neq 0$, then we are done by following the argument in the proof of Proposition 4.44. So, we need to show that $u_0 \neq 0$. We argue by contradiction by supposing that $u_n \to 0$ in $C_0^1(\overline{\Omega})$, where $\{u_n\}$ is the sequence considered in the proof of Theorem 4.49.

Take a (positive) eigenfunction $\varphi_1 \in \operatorname{int}(C_0^1(\overline{\Omega})_+)$ corresponding to $\lambda_{1,a_0}(m)$, which can be obtained as minimizer of (4.44). We notice that

$$0 < \int_\Omega a_0(x)|\nabla\varphi_1|^q\,dx = \lambda_{1,a_0}(m)\int_\Omega m(x)\varphi_1^q\,dx,$$

thus $\int_\Omega m(x)\varphi_1^q\,dx > 0$. According to (4.62), we can choose $\rho > 0$ satisfying

$$\rho < \frac{b_0 - \lambda_{1,a_0}(m)}{\|\varphi_1\|_q^q}\int_\Omega m(x)\varphi_1^q\,dx. \tag{4.75}$$

Since $u_n \in \operatorname{int}(C_0^1(\overline{\Omega})_+)$ is a solution of problem (4.47) with $\varepsilon = \varepsilon_n$, Lemmas 4.46, 4.48, and assumption $\mathrm{H}(A)_0$ yield

$$\begin{aligned}
0 &\le \int_\Omega a_0(x)|\nabla\varphi_1|^{q-2}\nabla\varphi_1\cdot\nabla\left(\frac{\varphi_1^q - u_n^q}{\varphi_1^{q-1}}\right)dx \\
&\quad - \int_\Omega a_0(x)|\nabla u_n|^{q-2}\nabla u_n\cdot\nabla\left(\frac{\varphi_1^q - u_n^q}{u_n^{q-1}}\right)dx \\
&\le \lambda_{1,a_0}(m)\int_\Omega m(x)(\varphi_1^q - u_n^q)\,dx - \int_\Omega \frac{f(x,u_n,\nabla u_n)}{u_n^{q-1}}\varphi_1^q\,dx \\
&\quad + \int_\Omega A_0(x,u_n,\nabla u_n)\cdot\nabla\left(\frac{\varphi_1^q - u_n^q}{u_n^{q-1}}\right)dx \\
&\quad + \int_\Omega f(x,u_n,\nabla u_n)u_n\,dx + \varepsilon_n\int_\Omega u_n\,dx \\
&= -\int_\Omega \frac{f(x,u_n,\nabla u_n)}{u_n^{q-1}}\varphi_1^q\,dx + \lambda_{1,a_0}(m)\int_\Omega m(x)\varphi_1^q\,dx + o(1).
\end{aligned} \tag{4.76}$$

Based on the proof of Theorem 4.49 with φ_1 in place of φ (see (4.73) and (4.74)), through (4.63) we have

$$\int_\Omega \frac{f(x,u_n,\nabla u_n)}{u_n^{q-1}}\varphi_1^q\,dx \ge b_0\int_\Omega m(x)\varphi_1^q\,dx - \rho\|\varphi_1\|_q^q + o(1). \tag{4.77}$$

Combining (4.76) and (4.77), and then letting $n \to \infty$, it turns out that

$$0 \le \left(\lambda_{1,a_0}(m) - b_0\right)\int_\Omega m(x)\varphi_1^q\,dx + \rho\|\varphi_1\|_q^q.$$

This contradicts (4.75), completing the proof. □

4.4 NOTES

Section 4.1 considers elliptic problems driven by the p-Laplacian, which is the prototype of quasilinear differential operator, and exhibiting gradient dependence in the right-hand side of the equation. The semilinear and quasilinear elliptic equations with gradient dependence on the reaction term occupy a distinct place in the literature. This is due to the fact that the variational methods cannot be applied. They are usually studied by means of topological degree, subsolution–supersolution method, fixed point theory and approximation techniques. For instance, in [8], assuming that f is a C^1 function with growth given by

$$|f(x,s,\xi)| \leq a(s)(1+|\xi|^2),$$

for some increasing function a, the authors obtain a solution of problem (4.1) in an ordered interval of subsolution–supersolution. In this respect, we also mention [201] where existence results for the same problem (4.1), when $p = 2$, are obtained via subsolutions–supersolutions in the Sobolev space $W^{2,q}(\Omega)$ with $q > N$, provided $f(x,s,\xi)$ is Lipschitz continuous with respect to ξ. For the general theory of subsolutions–supersolutions for nonlinear elliptic problems depending on the gradient, we refer to [38]. Different other results based on the subsolution–supersolution method can be found in [82]. In [212], by combining Krasnoselskii's fixed point theorem in cones with blow up techniques, the existence of a positive solution of problem (4.1) is proved when $f(x,s,\xi)$ is a nonnegative function and has a suitable growth with respect to s and ξ. Recently, in [81] and [226], under different growth in the gradient, the existence of a positive solution is achieved via an approximation on finite dimensional subspaces. An approximation approach in a general functional setting with a p-sublinear growth condition in the gradient can be found in [27]. A different approach is developed in [68] where the authors prove the existence of a positive and a negative solution for problem (4.1), when $p = 2$, through an iterative method involving mountain pass technique and assuming that $f(x,s,\xi)$ satisfies Lipschitz conditions in s and ξ in a neighborhood of zero and has growth

$$|f(x,s,\xi)| \leq a_1(1+|s|^q), \text{ with } 1 < q < 2^* - 1.$$

We also mention that in [93] an existence result for a positive solution of an elliptic problem with dependence on the gradient with Neumann boundary condition is given. This reference [93] has much in common with [80].

Theorem 4.12 is the main result in Faraci–Motreanu–Puglisi [80]. It establishes the existence of a positive and a negative solution for problem (4.1) by combining subsolution–supersolution techniques with Schaefer's fixed point theorem. In addition, it is shown in Corollary 4.13 that these opposite constant-sign solutions can be chosen to be extremal. The existence of a nontrivial solution (necessarily nodal) between the extremal opposite constant-sign solutions is an open problem. The idea of the proof of Theorem 4.12 is to consider an auxiliary problem which can be studied through operator theory and subsolution–supersolution method in conjunction with a fixed point approach. It is worth mentioning that we obtain positive and negative solutions in spite of the fact that our assumptions imply that the trivial solution exists too, which is a relevant aspect for quasilinear elliptic problems with general gradient dependence.

Section 4.2 investigates gradient dependent elliptic problems driven by the (p,q)-Laplacian, where $1 < q < p$, which is an operator essentially different from the p-Laplacian. Theorems 4.20 and 4.21 belong to Motreanu–Motreanu–Moussaoui [164]. They study the location of nodal (that is, sign-changing) solutions for problem (4.13). Specifically, they prove that if assumptions $\mathrm{H}(f)_1$, $\mathrm{H}(f)_2$ and $f(x,0,0)=0$ for a.e. $x \in \Omega$ are fulfilled, then every nodal solution of problem (4.13) should be between two opposite constant-sign solutions. In particular, this provides the powerful fact that the existence of a nodal solution implies under the stated hypotheses that two opposite constant-sign solutions must exist. Moreover, this phenomenon occurs even unilaterally: if we only have $f(x,0,0) \geq 0$ (resp. $f(x,0,0) \leq 0$) for a.e. $x \in \Omega$, then every nodal solution to problem (4.13) is bounded above by a positive solution (resp. bounded below by a negative solution).

In Theorems 4.27 and 4.28 the existence of barrier solutions to problem (4.13), that is, the biggest solution and the smallest solution, is established. Taking into account what was said before about the location of nodal solutions, the biggest nontrivial solution is positive and the smallest nontrivial solution is negative, and they belong to $C_0^1(\overline{\Omega})$. Our approach for obtaining the barrier solutions relies on the method of subsolutions–supersolutions related to problem (4.13). We investigate separately the cases $\mu = 0$ and $\mu > 0$ because there are relevant differences in the construction of subsolutions and supersolutions corresponding to the two cases.

Hypothesis $\mathrm{H}(f)$ enables us to apply recent estimates due to Cianchi–Maz'ya [53] ensuring the global boundedness of the gradient of any solution u to (4.13). Owing to the homogeneous boundary condition, it follows that $u \in L^\infty(\Omega)$ and $\|u\|_\infty$ is bounded above by a constant independent of u. This is actually a substitute for the Moser iteration technique (see, e.g., [156, Theorem C]) and has to be regarded in conjunction with Lieberman's estimates [129, 130] for the regularity up to the boundary. In particular, every solution of problem (4.13) belongs to the space $C_0^1(\overline{\Omega})$.

The presentation of Section 4.3 follows closely Motreanu–Tanaka [183]. The generality of problem (4.43) is twofold:

(i) the left-hand side of the equation is expressed in divergence form through a possibly nonhomogeneous operator depending on the solution u and its gradient ∇u;

(ii) the nonlinearity in the right-hand side of the equation is a so-called convection term, which means that it depends on the solution u and its gradient ∇u.

Such a general problem is considered for the first time in [183]. Moreover, the growth allowed in H(f) improves the corresponding conditions in the preceding works. We briefly describe some related results. The existence of positive solutions for problem (4.43) with

$$A(x,u,\nabla u) = |\nabla u|^{p-2}\nabla u,$$

that is, div $A(x,u,\nabla u)$ stands the p-Laplacian Δ_p, and with some convection term $f(x,u,\nabla u)$ are studied in [80,212,242]. Specifically, in [80] the focus is on the case of $f(x,u,\nabla u)$ with $(p-1)$-sublinear growth in u and ∇u. In [212] the growth

$$\max\{0, |u|^{r_1} - C|\nabla u|^{r_2}\} \le |f(x,u,\nabla u)| \le C(1+|u|^{r_1}+|\nabla u|^{r_2})$$

with

$$p-1 < r_1 < \frac{N(p-1)}{N-p}, \quad p-1 < r_2 < \frac{pr_1}{r_1+1}, \quad p < N,$$

is assumed. In [242] the growth

$$|f(x,u,\nabla u)| \le C(1+|u|^{r_1}+|\nabla u|^{\frac{pr_1}{r_1+1}})$$

with

$$p-1 < r_1 < \frac{N(p-1)+p}{N-p}, \quad p-1 < r_2 < \frac{pr_1}{r_1+1}, \quad p < N,$$

is required. In particular, for [212] and [242] it is needed that $f(x,u,\nabla u)$ be p-sublinear in ∇u. In [226], the following possibly nonhomogeneous case

$$A(x,u,\nabla u) = a(x,|\nabla u|)\nabla u,$$

for a scalar function $a(x,t)$ with certain properties, has been considered. In [81] and [82], problem (4.43) has been investigated for the (p,q)-Laplacian, that is,

$$\text{div}\, A(x,u,\nabla u) = \Delta_p u + \Delta_q u$$

and for a convection term $f(x,u,\nabla u)$ which is $(p-1)$-sublinear in u and ∇u.

Theorems 4.49 and 4.50 can be found in Motreanu–Tanaka [183]. They can handle problem (4.43) with a substantially more general operator $A(x,u,\nabla u)$ and a convection term $f(x,u,\nabla u)$ that can be, for instance, p-superlinear in ∇u. Our approach is built on appropriate truncations and limiting process in order to use directly hypothesis H(f). Theorems 4.49 and 4.50 provide existence of positive solutions with additional regularity properties. The approach relies on the construction of approximate solutions via pseudomonotone operators, a priori estimates, nonlinear regularity theory up to the boundary, and the strong maximum principle stated in Theorem 2.19. Problems with general gradient dependence involving the $p(x)$-Laplacian operator are addressed in Zhang–Motreanu [241].

Chapter 5

Constant-Sign and Sign-Changing Solutions for Quasilinear Elliptic Problems

5.1 THE CASE OF DIRICHLET BOUNDARY CONDITION

The problem considered here is formulated as follows: find $u \in W_0^{1,p}(\Omega)$ and $\lambda > 0$ such that

$$\begin{cases} -\Delta_p u = \lambda |u|^{p-2} u - g(x, u) & \text{in } \Omega, \\ u = 0 & \text{on } \partial\Omega, \end{cases} \tag{5.1}$$

where $\Omega \subset \mathbb{R}^N$ is a bounded domain with a C^2 boundary $\partial\Omega$, $1 < p < +\infty$ and $g : \Omega \times \mathbb{R} \to \mathbb{R}$ is a given function. Problem (5.1) is stated with homogeneous Dirichlet boundary condition. It is driven by the negative p-Laplacian operator $-\Delta_p : W_0^{1,p}(\Omega) \to W^{-1,p'}(\Omega)$ (where $\frac{1}{p} + \frac{1}{p'} = 1$) defined by

$$\langle -\Delta_p u, v \rangle = \int_\Omega |\nabla u(x)|^{p-2} \nabla u(x) \cdot \nabla v(x) \, dx \quad \text{for all } u, v \in W_0^{1,p}(\Omega) \tag{5.2}$$

(see Proposition 1.56).

We suppose that the following hypotheses on the nonlinear term $g(x, s)$ of problem (5.1) hold:

H(g) $g : \Omega \times \mathbb{R} \to \mathbb{R}$ is a Carathéodory function (that is, $g(\cdot, s)$ is measurable for all $s \in \mathbb{R}$ and $g(x, \cdot)$ is continuous for almost all $x \in \Omega$) such that

(i) $\displaystyle\lim_{|s| \to +\infty} \frac{g(x, s)}{|s|^{p-2} s} = +\infty$ uniformly for a.e. $x \in \Omega$;

(ii) $\displaystyle\lim_{s \to 0} \frac{g(x, s)}{|s|^{p-2} s} = 0$ uniformly for a.e. $x \in \Omega$;

(iii) g is bounded on bounded sets.

A simple example of nonlinearity $g(x, \cdot)$, with $x \in \Omega$, whose growth is not polynomial, where all the assumptions in H(g) are fulfilled, is provided below.

Nonlinear Differential Problems with Smooth and Nonsmooth Constraints. DOI: 10.1016/B978-0-12-813386-6.00005-5

Example 5.1. The function $g : \Omega \times \mathbb{R} \to \mathbb{R}$ defined by

$$g(x,s) = \begin{cases} |s|^{p-2}se^{-s-1} & \text{if } s < -1, \\ \dfrac{|s|^p}{2}((s-1)\cos(s+1)+s+1) & \text{if } -1 \le s \le 1, \\ (1+(s-1)|x|)s^{p-1}e^{s-1} & \text{if } s > 1 \end{cases}$$

verifies the conditions required in H(g).

A fundamental tool in our approach is represented by the notions of subsolution and supersolution.

Definition 5.2. A function $\underline{u} \in W^{1,p}(\Omega)$ is called a subsolution of problem (5.1) if $\underline{u} \le 0$ on $\partial\Omega$, $g(\cdot, \underline{u}) \in L^q(\Omega)$ with some $1 < q < p^*$ and

$$\langle -\Delta_p \underline{u}, v\rangle \le \int_\Omega \big(\lambda|\underline{u}|^{p-2}\underline{u} - g(x,\underline{u})\big)v\,dx$$
$$\text{for all } v \in W_0^{1,p}(\Omega) \text{ with } v \ge 0 \text{ a.e. in } \Omega.$$

The definition of a supersolution $\overline{u} \in W^{1,p}(\Omega)$ is obtained by reversing the inequalities. A function $u \in W^{1,p}(\Omega)$ is a solution to problem (5.1) if it is simultaneously a subsolution and a supersolution.

First, we establish the existence of opposite constant-sign solutions to problem (5.1). To this end, as known from Proposition 1.56, the operator $-\Delta_p$ on $W_0^{1,p}(\Omega)$ is maximal monotone and coercive. Hence, by Theorem 1.55, there exists $e \in W_0^{1,p}(\Omega)$ such that $-\Delta_p e = 1$. The nonlinear regularity theory and nonlinear strong maximum principle imply $e \in \operatorname{int}(C_0^1(\overline{\Omega})_+)$. By λ_1 we denote the first eigenvalue of $-\Delta_p$ on $W_0^{1,p}(\Omega)$ and by $\hat{u}_1$ we denote the corresponding positive, L^p-normalized eigenfunction, so that $\hat{u}_1 \in \operatorname{int}(C_0^1(\overline{\Omega})_+)$ (see Proposition 1.57).

Lemma 5.3. *Assume* H(g) (i)–(iii) *and* $\lambda > \lambda_1$. *Then there exists a constant* $a_\lambda > 0$ *such that* $a_\lambda e$ *and* $-a_\lambda e$ *are supersolution and subsolution, respectively, of problem (5.1). In addition,* $-\varepsilon\hat{u}_1$ *is a supersolution and* $\varepsilon\hat{u}_1$ *is a subsolution of problem (5.1) provided* $\varepsilon > 0$ *is sufficiently small.*

Proof. Fix $\lambda > \lambda_1$. By condition H(g) (i) there exists $s_\lambda > 0$ such that

$$\frac{g(x,s)}{|s|^{p-2}s} > \lambda \;\text{ for a.a. } x \in \Omega \text{ and all } |s| > s_\lambda.$$

Assumption H(g) (iii) ensures the existence of a constant $c_\lambda > 0$ for which one has

$$|g(x,s) - \lambda|s|^{p-2}s| \le c_\lambda \quad \text{for a.a. } x \in \Omega \text{ and all } |s| \le s_\lambda.$$

Then we get

$$g(x,s) \geq \lambda s^{p-1} - c_\lambda \quad \text{for a.a. } x \in \Omega \text{ and all } s > 0.$$

This estimate shows that $c_\lambda^{\frac{1}{p-1}} e$ is a supersolution of problem (5.1) (see Definition 5.2). Similarly, because

$$g(x,s) \leq \lambda |s|^{p-2} s + c_\lambda \quad \text{for a.a. } x \in \Omega \text{ and all } s < 0,$$

it turns out that $-c_\lambda^{\frac{1}{p-1}} e$ is a subsolution of problem (5.1), thereby the first part of the lemma holds with $a_\lambda = c_\lambda^{\frac{1}{p-1}}$.

Assumption H(g) (ii) guarantees that a number $\delta_\lambda > 0$ exists such that

$$\frac{|g(x,s)|}{|s|^{p-1}} < \lambda - \lambda_1 \quad \text{for a.a. } x \in \Omega \text{ and all } 0 < |s| \leq \delta_\lambda. \tag{5.3}$$

If $\varepsilon \in (0, \frac{\delta_\lambda}{\|\hat{u}_1\|_\infty}]$, then it is straightforward to check that $\varepsilon \hat{u}_1$ is a subsolution and $-\varepsilon \hat{u}_1$ is a supersolution of problem (5.1). □

We first point out the existence of two solutions of problem (5.1) having opposite constant sign and being extremal.

Theorem 5.4. *Assume conditions* H(g) (i)–(iii). *Then for every* $\lambda > \lambda_1$ *there exist a smallest positive solution* $u_+ = u_+(\lambda) \in \operatorname{int}(C_0^1(\overline{\Omega})_+)$ *in the ordered interval* $[0, a_\lambda e]$ *and a biggest negative solution* $u_- = u_-(\lambda) \in -\operatorname{int}(C_0^1(\overline{\Omega})_+)$ *in the ordered interval* $[-a_\lambda e, 0]$, *with* $a_\lambda > 0$ *determined in Lemma 5.3.*

Proof. Fix $\lambda > \lambda_1$. Lemma 5.3 ensures that $\overline{u} = a_\lambda e \in \operatorname{int}(C_0^1(\overline{\Omega})_+)$ is a supersolution for problem (5.1) and $\underline{u} = \varepsilon \hat{u}_1 \in \operatorname{int}(C_0^1(\overline{\Omega})_+)$ is a subsolution for problem (5.1) if $\varepsilon > 0$ is small enough. For a possibly smaller $\varepsilon > 0$ we may assume that $\varepsilon \hat{u}_1 \leq a_\lambda e$. Then by the method of subsolutions–supersolutions we know that within the ordered interval $[\varepsilon \hat{u}_1, a_\lambda e]$ there is the smallest solution $u_\varepsilon = u_\varepsilon(\lambda) \in \operatorname{int}(C_0^1(\overline{\Omega})_+)$ of problem (5.1) (see [38]). Thus for every positive integer n sufficiently large there is the smallest solution $u_n \in \operatorname{int}(C_0^1(\overline{\Omega})_+)$ of problem (5.1) within $[\frac{1}{n}\hat{u}_1, a_\lambda e]$. In view of the minimality property of u_n, we have

$$u_n \downarrow u_- \quad \text{pointwise,} \tag{5.4}$$

with some function $u_+ : \Omega \to \mathbb{R}$ satisfying $0 \leq u_+ \leq a_\lambda e$.

We claim that

$$u_+ \quad \text{is a solution of problem (5.1).} \tag{5.5}$$

Taking into account that u_n solves (5.1), we have

$$\|\nabla u_n\|_p^p = \int_\Omega (\lambda u_n(x)^p - g(x, u_n(x)) u_n(x))\, dx.$$

By H(g) (iii) and since u_n belongs to the ordered interval $[0, a_\lambda e]$, it follows that the sequence $\{u_n\}$ is bounded in $W_0^{1,p}(\Omega)$. Then due to (5.4) we derive that $u_+ \in W_0^{1,p}(\Omega)$, as well as

$$u_n \rightharpoonup u_+ \text{ in } W_0^{1,p}(\Omega)\,, \quad u_n \to u_+ \text{ in } L^p(\Omega) \text{ and a.e. in } \Omega. \tag{5.6}$$

The fact that u_n is a solution of (5.1) results in

$$\langle -\Delta_p u_n, \varphi\rangle = \int_\Omega (\lambda |u_n|^{p-2} u_n - g(x, u_n))\varphi(x)\, dx \quad \text{for all } \varphi \in W_0^{1,p}(\Omega). \tag{5.7}$$

Inserting $\varphi = u_n - u_+$ in (5.7) yields

$$\langle -\Delta_p u_n, u_n - u_+\rangle = \int_\Omega (\lambda |u_n|^{p-2} u_n - g(x, u_n))(u_n(x) - u_+(x))\, dx.$$

On the basis of (5.6), H(g) (iii) and the uniform boundedness of the sequence $\{u_n\}$, the Lebesgue's dominated convergence theorem ensures

$$\lim_{n\to\infty} \langle -\Delta_p u_n, u_n - u_+\rangle = 0.$$

Then the $(S)_+$-property of $-\Delta_p$ on $W_0^{1,p}(\Omega)$ (see Proposition 1.56) implies

$$u_n \to u_+ \text{ in } W_0^{1,p}(\Omega) \text{ as } n \to \infty. \tag{5.8}$$

The strong convergence in (5.8), combined with H(g) (iii) and the uniform boundedness of $\{u_n\}$, allows passing to the limit in (5.7), which leads to (5.5).

We know from (5.4) that $u_+ \in L^\infty(\Omega)$. Then (5.5) and assumption H(g) (iii) entail $\Delta_p u_+ \in L^\infty(\Omega)$. The nonlinear regularity theory shows that $u_+ \in C^{1,\gamma}(\overline{\Omega})$ for some $\gamma \in (0, 1)$, so $u_+ \in C_0^1(\overline{\Omega})$. On the other hand, it turns out from (5.3) and hypothesis H(g) (iii) that a constant $\tilde{c}_\lambda > 0$ can be found such that

$$|g(x, s)| \le \tilde{c}_\lambda s^{p-1} \text{ for a.a. } x \in \Omega \text{ and all } 0 \le s \le a_\lambda \|e\|_\infty. \tag{5.9}$$

Then (5.5) and (5.9) show

$$\Delta_p u_+ \le (\lambda + \tilde{c}_\lambda) u_+^{p-1}. \tag{5.10}$$

Therefore Vázquez's strong maximum principle [229] can be applied. Specifically, one takes in its statement the function $\beta(s) = (\lambda + \tilde{c}_\lambda)s^{p-1}$ for all $s > 0$. It is possible because

$$\int_{0^+} \frac{1}{(s\beta(s))^{\frac{1}{p}}}\, ds = +\infty.$$

We are thus conducted that if $u \neq 0$, then $u > 0$ in Ω and $\frac{\partial u}{\partial \nu} < 0$ on $\partial\Omega$, which means according to (1.39) that

$$u_+ \in \operatorname{int}(C_0^1(\overline{\Omega})_+). \tag{5.11}$$

Arguing by contradiction, we suppose that (5.11) does not hold. In view of the above discussion, we must have $u_+ = 0$. Then (5.4) reads as

$$u_n(x) \downarrow 0 \text{ for all } x \in \Omega. \tag{5.12}$$

Setting

$$\tilde{u}_n = \frac{u_n}{\|\nabla u_n\|_p} \quad \text{for all } n,$$

we may suppose, along a relabeled subsequence, that

$$\tilde{u}_n \rightharpoonup \tilde{u} \text{ in } W_0^{1,p}(\Omega), \quad \tilde{u}_n \to \tilde{u} \text{ in } L^p(\Omega) \text{ and a.e. in } \Omega, \tag{5.13}$$

with some $\tilde{u} \in W_0^{1,p}(\Omega)$, and there is a function $w \in L^p(\Omega)_+$ such that

$$|\tilde{u}_n(x)| \leq w(x) \quad \text{for almost all } x \in \Omega. \tag{5.14}$$

Notice that (5.7) can be expressed as

$$\langle -\Delta_p \tilde{u}_n, \varphi \rangle = \lambda \int_\Omega \tilde{u}_n^{p-1} \varphi \, dx - \int_\Omega \frac{g(x, u_n)}{u_n^{p-1}} \tilde{u}_n^{p-1} \varphi \, dx \quad \text{for all } \varphi \in W_0^{1,p}(\Omega). \tag{5.15}$$

In particular, for $\varphi = \tilde{u}_n - \tilde{u}$ we obtain

$$\langle -\Delta_p \tilde{u}_n, \tilde{u}_n - \tilde{u} \rangle = \lambda \int_\Omega \tilde{u}_n^{p-1} (\tilde{u}_n - \tilde{u}) \, dx - \int_\Omega \frac{g(x, u_n)}{u_n^{p-1}} \tilde{u}_n^{p-1} (\tilde{u}_n - \tilde{u}) \, dx. \tag{5.16}$$

We see from (5.9) and (5.14) that

$$\frac{|g(x, u_n(x))|}{u_n(x)^{p-1}} \tilde{u}_n^{p-1} |\tilde{u}_n(x) - \tilde{u}(x)| \leq \tilde{c}_\lambda w(x)^{p-1} (w(x) + |\tilde{u}(x)|) \text{ for a.a. } x \in \Omega.$$

Owing to the fact that the right-hand side of the above inequality is in $L^1(\Omega)$, on the basis of (5.13) we can apply the Lebesgue's dominated convergence theorem to get

$$\lim_{n \to \infty} \int_\Omega \frac{g(x, u_n)}{u_n^{p-1}} \tilde{u}_n^{p-1} (\tilde{u}_n - \tilde{u}) \, dx = 0.$$

Then (5.16) and (5.13) give

$$\lim_{n\to\infty}\langle -\Delta_p \tilde{u}_n, \tilde{u}_n - \tilde{u}\rangle = 0,$$

which in view of the $(S)_+$-property of $-\Delta_p$ on $W_0^{1,p}(\Omega)$ as stated in Proposition 1.56 results in

$$\tilde{u}_n \to \tilde{u} \ \text{ in } W_0^{1,p}(\Omega) \text{ as } n \to \infty, \tag{5.17}$$

so in particular $\|\nabla\tilde{u}\|_p = 1$. By (5.15), (5.17), (5.12), and hypothesis H(g) (ii), we infer

$$\langle -\Delta_p \tilde{u}, \varphi\rangle = \lambda \int_\Omega \tilde{u}^{p-1}\varphi\, dx \quad \text{for all } \varphi \in W_0^{1,p}(\Omega). \tag{5.18}$$

In (5.18) it is thus expressed that $\tilde{u} \geq 0$ is an eigenfunction of $-\Delta_p$ on $W_0^{1,p}(\Omega)$ corresponding to the eigenvalue $\lambda > \lambda_1$. This is impossible according to Proposition 1.57 because $\tilde{u}$ must change sign. The reached contradiction proves that the claim in (5.11) holds true.

We now show

$$u_+ \text{ is the smallest positive solution of (5.1) within } [0, \overline{u}]. \tag{5.19}$$

Let $u \in W_0^{1,p}(\Omega)$ be a positive solution of (5.1) belonging to the ordered interval $[0, \overline{u}]$. Since $u \in L^\infty(\Omega)$, then (5.1) and hypothesis H(g) (ii) imply $\Delta_p u \in L^\infty(\Omega)$. The nonlinear regularity theory allows us to derive $u \in C_0^1(\overline{\Omega})$. Then (5.10) enables us to apply Vázquez's strong maximum principle [229], which leads to $u \in \mathrm{int}\,(C_0^1(\overline{\Omega})_+)$. We deduce $u \in [\frac{1}{n}\hat{u}_1, \overline{u}]$ for n sufficiently large. This in conjunction with the fact that u_n is the smallest solution of (5.1) in $[\frac{1}{n}\hat{u}_1, \overline{u}]$ ensures $u_n \leq u$ if n is large enough. Making use of (5.4), we obtain $u_+ \leq u$, thereby (5.19) is valid.

The proof of the existence of the biggest negative solution of (5.1) within the ordered interval $[-a_\lambda e, 0]$ can be carried out in a similar way. This completes the proof. □

The main result of this section as stated below provides an additional nontrivial solution to problem (5.4). The novelty is that it is nodal, which means that it changes sign in Ω. By λ_2 we denote the second eigenvalue of $-\Delta_p$ on $W_0^{1p}(\Omega)$.

Theorem 5.5. *Under hypotheses* H(g) (i)–(iii), *for every $\lambda > \lambda_2$ problem (5.4) admits at least a (positive) solution $u_+ = u_+(\lambda) \in \mathrm{int}\,(C_0^1(\overline{\Omega})_+)$, a (negative) solution $u_- = u_-(\lambda) \in -\mathrm{int}\,(C_0^1(\overline{\Omega})_+)$, and a nontrivial sign-changing solution $u_0 = u_0(\lambda) \in C_0^1(\overline{\Omega})$.*

Proof. Fix any $\lambda > \lambda_2$. As $\lambda_2 > \lambda_1$, Theorem 5.4 ensures the existence of the extremal solutions $u_+ \in \operatorname{int}(C_0^1(\overline{\Omega})_+)$ and $u_- \in -\operatorname{int}(C_0^1(\overline{\Omega})_+)$. We introduce on $\Omega \times \mathbb{R}$ the truncation functions

$$\tau_+(x,s) = \begin{cases} 0 & \text{if } s \le 0, \\ s & \text{if } 0 < s < u_+(x), \\ u_+(x) & \text{if } s \ge u_+(x), \end{cases}$$

$$\tau_-(x,s) = \begin{cases} u_-(x) & \text{if } s \le u_-(x), \\ s & \text{if } u_-(x) < s < 0, \\ 0 & \text{if } s \ge 0, \end{cases}$$

$$\tau_0(x,s) = \begin{cases} u_-(x) & \text{if } s \le u_-(x), \\ s & \text{if } u_-(x) < s < u_+(x), \\ u_+(x) & \text{if } s \ge u_+(x). \end{cases}$$

By means of them, the corresponding functionals E_+, E_-, $E_0 \in C^1(W_0^{1,p}(\Omega))$ are defined as follows for every $u \in W_0^{1,p}(\Omega)$:

$$\begin{aligned} E_+(u) &= \frac{1}{p}\|\nabla u\|_p^p - \int_\Omega \int_0^{u(x)} (\lambda \tau_+(x,s)^{p-1} - g(x,\tau_+(x,s))\,ds\,dx, \\ E_-(u) &= \frac{1}{p}\|\nabla u\|_p^p \\ &\quad - \int_\Omega \int_0^{u(x)} (\lambda|\tau_-(x,s)|^{p-2}\tau_-(x,s) - g(x,\tau_-(x,s)))\,ds\,dx, \\ E_0(u) &= \frac{1}{p}\|\nabla u\|_p^p \\ &\quad - \int_\Omega \int_0^{u(x)} (\lambda|\tau_0(x,s)|^{p-2}\tau_0(x,s) - g(x,\tau_0(x,s)))\,ds\,dx. \end{aligned}$$

The following location property is valid:

$$v \text{ is a critical point of } E_+ \implies 0 \le v(x) \le u_+(x) \text{ for a.a. } x \in \Omega. \tag{5.20}$$

Indeed, if $v \in W_0^{1,p}(\Omega)$ is a critical point of E_+, then

$$\begin{aligned} &\langle -\Delta_p v + \Delta_p u_+, (v-u_+)^+ \rangle \\ &= \int_\Omega \Big(\lambda\tau_+(x,v)^{p-1} - g(x,\tau_+(x,v)) - \lambda\tau_+(x,u_+)^{p-1} + g(x,\tau_+(x,u_+))\Big) \\ &\quad \times (v-u_+)^+\,dx \\ &= 0, \end{aligned}$$

which implies $v \le u_+$. In the same way, if $v \in W_0^{1,p}(\Omega)$ is a critical point of E_+, we have

$$\langle -\Delta_p v, -v^- \rangle = -\int_\Omega \Big(\lambda \tau_+(x,v)^{p-1} - g(x, \tau_+(x,v))\Big) v^- \, dx = 0,$$

so $v \ge 0$. Therefore (5.20) is verified.

Since the functional E_+ is coercive and sequentially weakly lower semicontinuous, there exists a global minimizer $v_+ \in W_0^{1,p}(\Omega)$ of E_+, that is,

$$E_+(v_+) = \inf_{W_0^{1,p}(\Omega)} E_+. \tag{5.21}$$

We note from (5.20) and (5.21) that v_+ is a solution of problem (5.1) and belongs to the order interval $[0, u_+]$. Moreover, v_+ is nontrivial which can be seen as follows. By (5.11) we can choose $t > 0$ sufficiently small to have $t\hat{u}_1 \le u_+$ in Ω and $\|t\hat{u}_1\|_\infty \le \delta_\lambda$. Then (5.3) ensures that

$$E_+(t\hat{u}_1) = \frac{\lambda_1}{p} t^p - \int_\Omega \int_0^{t\hat{u}_1(x)} (\lambda s^{p-1} - g(x,s))\, ds\, dx < 0,$$

and in view of (5.21) we get $v_+ \ne 0$. We have that $v_+ \in L^\infty(\Omega)$ and, because v_+ solves problem (5.1), we derive $\Delta_p v_+ \in L^\infty(\Omega)$. This allows the nonlinear regularity theory be invoked, guaranteeing that $v_+ \in C_0^1(\overline{\Omega})$. In addition, taking into account (5.9), the estimate $\Delta_p v_+ \le (\lambda + \tilde{c}_\lambda) v_+^{p-1}$ ensues (see also (5.10)). Then Vázquez's strong maximum principle [229] implies

$$v_+ \in \operatorname{int}(C_0^1(\overline{\Omega})_+). \tag{5.22}$$

Using (5.22) and the inequality $v_+ \le u_+$ in Ω, we see that the minimality property of u_+ established in Theorem 5.4 forces $v_+ = u_+$. By (5.21) we deduce that u_+ is a local minimizer of E_0 on $C_0^1(\overline{\Omega})$. In view of [156, Theorem 23] we may conclude that u_+ is a local minimizer of E_0 on the space $W_0^{1,p}(\Omega)$. Proceeding in the same way with the functional E_-, we show that u_- is a local minimizer of E_0 on $W_0^{1,p}(\Omega)$.

As for (5.20) we can verify

$$v \text{ is a critical point of } E_0 \implies u_-(x) \le v(x) \le u_+(x) \text{ for a.a. } x \in \Omega. \tag{5.23}$$

It is readily seen that (5.23) ensures that every critical point of E_0 is a solution to problem (5.1). Since the functional E_0 is coercive, sequentially weakly lower semicontinuous, with

$$\inf_{W_0^{1,p}(\Omega)} \varphi_0 < 0,$$

it is clear that E_0 has a global minimizer $v_0 \in W_0^{1,p}(\Omega)$ satisfying $v_0 \neq 0$. Then, as already observed, (5.23) implies that v_0 is a nontrivial solution of problem (5.1) belonging to the ordered interval $[u_-, u_+]$. We may admit that $v_0 = u_+$ or $v_0 = u_-$ because otherwise, due to the extremal properties of solutions u_+ and u_- as given by Theorem 5.4, v_0 would change sign and the result be achieved setting $u_0 = v_0$.

Without loss of generality, suppose $v_0 = u_+$. We may also assume that u_- is a strict local minimizer of E_0 because otherwise we would find (infinitely many) critical points of E_0 that are sign changing thanks to (5.23) and the extremal properties of the solutions u_-, u_+ obtained in Theorem 5.4. Then there exists $\rho \in (0, \|u_0 - v_0\|)$ such that

$$E_0(u_+) \leq E_0(u_-) < \inf\{E_0(u) : u \in \partial B_\rho(u_-)\}, \tag{5.24}$$

where $\partial B_\rho(u_-) = \{u \in W_0^{1,p}(\Omega) : \|u - u_-\| = \rho\}$. Owing to (5.24) and noticing that the Palais–Smale condition holds for the functional E_0, we may apply the mountain pass theorem (see Theorem 1.44) which provides a point $u_0 \in W_0^{1,p}(\Omega)$ satisfying $E_0'(u_0) = 0$ and

$$\inf\{E_0(u) : u \in \partial B_\rho(u_-)\} \leq E_C(u_0) = \inf_{\gamma \in \Gamma} \max_{t \in [-1,1]} E_0(\gamma(t)), \tag{5.25}$$

where

$$\Gamma = \{\gamma \in C([-1, 1], W_0^{1,p}(\Omega)) : \gamma(-1) = u_-,\ \gamma(1) = u_+\}.$$

From (5.24) and (5.25) it follows that $u_0 \neq u_-$ and $u_0 \neq u_+$.

We claim

$$E_0(u_0) < 0. \tag{5.26}$$

To this end it suffices to construct a path $\hat{\gamma} \in \Gamma$ such that

$$E_0(\hat{\gamma}(t)) < 0 \text{ for all } t \in [-1, 1]. \tag{5.27}$$

Let $S = W_0^{1,p}(\Omega) \cap \partial B_1^{L^p(\Omega)}$, where $\partial B_1^{L^p(\Omega)} = \{u \in L^p(\Omega) : \|u\|_p = 1\}$, and let $S_C = S \cap C_0^1(\overline{\Omega})$ be endowed with the topologies induced by $W_0^{1,p}(\Omega)$ and $C_0^1(\overline{\Omega})$, respectively. For later use we set

$$\Gamma_{0,C} = \{\gamma \in C([-1, 1], S_C) : \gamma(-1) = -\hat{u}_1,\ \gamma(1) = \hat{u}_1\}.$$

Assumption H(g) (ii) provides constants $\mu \in (0, \lambda - \lambda_2)$ and $\delta > 0$ such that

$$\frac{|g(x,s)|}{|s|^{p-1}} \leq \mu \text{ for a.a. } x \in \Omega \text{ and all } 0 < |s| \leq \delta. \tag{5.28}$$

Then we choose $\rho_0 \in (0, \lambda - \lambda_2 - \mu)$. From Proposition 1.58 we know that there exists a path $\gamma \in \Gamma_0$ such that

$$\max_{t\in[-1,1]} \|\nabla\gamma(t)\|_p^p < \lambda_2 + \frac{\rho_0}{2}.$$

Fix a number r with $0 < r \le (\lambda_2 + \rho_0)^{\frac{1}{p}} - (\lambda_2 + \frac{\rho_0}{2})^{\frac{1}{p}}$. The density of S_C in S implies that $\Gamma_{0,C}$ is dense in Γ_0, so there is $\gamma_0 \in \Gamma_{0,C}$ satisfying

$$\max_{t\in[-1,1]} \|\nabla\gamma(t) - \nabla\gamma_0(t)\|_p < r,$$

thus it turns out that

$$\max_{t\in[-1,1]} \|\nabla\gamma_0(t)\|_p^p < \lambda_2 + \rho_0. \tag{5.29}$$

The boundedness of the set $\gamma_0([-1,1])(\overline{\Omega})$ in $\mathbb{R}$ ensures the existence of an $\varepsilon_1 > 0$ such that

$$\varepsilon_1|u(x)| \le \delta \ \text{ for all } x \in \Omega \text{ and all } u \in \gamma_0([-1,1]). \tag{5.30}$$

Since $u_+, -u_- \in \operatorname{int}(C_0^1(\overline{\Omega})_+)$ as obtained in Theorem 5.4, for every $u \in \gamma_0([-1,1])$ and any bounded neighborhood V_u of u in $C_0^1(\overline{\Omega})$ we can find positive numbers h_u and j_u satisfying

$$u_+ - \frac{1}{h}v \in \operatorname{int}(C_0^1(\overline{\Omega})_+) \ \text{ and } \ -u_- + \frac{1}{j}v \in \operatorname{int}(C_0^1(\overline{\Omega})_+)$$

whenever $h \ge h_u$, $j \ge j_u$, $v \in V_u$. Then the compactness of $\gamma_0([-1,1])$ in $C_0^1(\overline{\Omega})$ allows us to determine a number $\varepsilon_0 > 0$ for which one has

$$u_-(x) \le \varepsilon u(x) \le u_+(x) \ \text{ for all } x \in \Omega,\ u \in \gamma_0([-1,1]),\ \varepsilon \in (0,\varepsilon_0). \tag{5.31}$$

We note that $\varepsilon\gamma_0$ is a continuous path in $C_0^1(\overline{\Omega})$ joining $-\varepsilon\hat{u}_1$ and $\varepsilon\hat{u}_1$ for each $0 < \varepsilon < \min\{\varepsilon_0, \varepsilon_1\}$. By (5.31), (5.29), (5.30), (5.28), and taking into account the choice of ρ_0, as well as that $\gamma_0([-1,1]) \subset \partial B_1^{L^p(\Omega)}$, we obtain

$$\begin{aligned}
E_0(\varepsilon\gamma_0(t)) &= \frac{\varepsilon^p}{p}\|\nabla\gamma_0(t)\|_p^p - \frac{\varepsilon^p}{p}\lambda + \int_\Omega\int_0^{\varepsilon\gamma_0(t)(x)} g(x,\tau_0(x,s))\,ds\,dx \\
&\le \frac{\varepsilon^p}{p}(\lambda_2 + \rho_0 - \lambda) + \int_\Omega\int_0^{\varepsilon\gamma_0(t)(x)} g(x,s)\,ds\,dx \\
&\le \frac{\varepsilon^p}{p}(\lambda_2 + \rho_0 - \lambda + \mu) < 0 \ \text{ for all } t \in [-1,1].
\end{aligned} \tag{5.32}$$

We next apply the second deformation lemma (see Theorem 1.45) to the C^1 functional $E_+ : W_0^{1,p}(\Omega) \to \mathbb{R}$. To this end we denote

$$c_+ = c_+(\lambda) = E_+(\varepsilon\hat{u}_1), \quad m_+ = m_+(\lambda) = E_+(u_+),$$
$$E_+^{c_+} = \{u \in W_0^{1,p}(\Omega) : E_+(u) \le c_+\}.$$

The minimality property of u_+ enables us to admit that $m_+ < c_+$. Again using the minimality of u_+ in conjunction with (5.20), we see that E_+ has no critical values in the interval $(m_+, c_+]$. Since the functional E_+ satisfies the Palais–Smale condition due to its coercivity, the second deformation lemma can be applied to E_+, yielding a continuous mapping $\eta \in C([0,1] \times E_+^{c_+}, E_+^{c_+})$ such that $\eta(0,u) = u$ and $\eta(1,u) = u_+$ for all $u \in E_+^{c_+}$, as well as $E_+(\eta(t,u)) \le E_+(u)$ whenever $t \in [0,1]$ and $u \in E_+^{c_+}$. We are thus led to introduce the path $\gamma_+ : [0,1] \to W_0^{1,p}(\Omega)$ by $\gamma_+(t) = \eta(t, \varepsilon\hat{u}_1)^+ = \max\{\eta(t,\varepsilon\hat{u}_1), 0\}$ for all $t \in [0,1]$. Clearly, γ_+ is continuous on $W_0^{1,p}(\Omega)$ and joins $\varepsilon\hat{u}_1$ and u_+. Moreover, the properties of the deformation η and (5.32) imply

$$E_0(\gamma_+(t)) = E_+(\gamma_+(t)) \le E_+(\varepsilon\hat{u}_1) < 0 \text{ for all } t \in [0,1]. \tag{5.33}$$

Now, applying the second deformation lemma to the functional E_-, we construct a continuous path $\gamma_- : [0,1] \to W_0^{1,p}(\Omega)$ between $-\varepsilon\hat{u}_1$ and u_- such that

$$E_0(\gamma_-(t)) < 0 \quad \text{for all } t \in [0,1]\,. \tag{5.34}$$

By the concatenation of the curves γ_-, $\varepsilon\gamma_0$, and γ_+ we obtain a path $\hat{\gamma} \in \Gamma$. On the basis of (5.34), (5.32), and (5.33), $\hat{\gamma}$ satisfies (5.27). Then, as noticed before, (5.26) holds, so $u_0 \ne 0$. Recalling (5.23), we deduce that u_0 is a nontrivial solution of (5.1) distinct from u_- and u_+, with $u_- \le u_0 \le u_+$, and verifying $u_0 \in C_0^1(\overline{\Omega})$ by the nonlinear regularity theory. The extremal properties of the constant sign solutions u_- and u_+ as described in Theorem 5.4 render u_0 to be sign changing. This completes the proof. □

5.2 THE CASE OF NEUMANN BOUNDARY CONDITION

Let $\Omega \subset \mathbb{R}^N$ be a bounded domain with a C^2 boundary $\partial\Omega$. Here we study the following quasilinear elliptic problem with Neumann boundary condition

$$\begin{cases} -\Delta_p u = \lambda |u|^{p-2}u - f(x,u) & \text{in } \Omega, \\ \frac{\partial u}{\partial \nu} = 0 & \text{on } \partial\Omega, \end{cases} \tag{5.35}$$

where λ is a positive parameter, ν stands for the outward unit normal to $\partial\Omega$, $\frac{\partial u}{\partial \nu}$ denotes the normal derivative of u on $\partial\Omega$, i.e., $\frac{\partial u}{\partial \nu} = \nabla u \cdot \nu$, $\Delta_p u := \operatorname{div}(|\nabla u|^{p-2}\nabla u)$ is the p-Laplacian operator with $1 < p < +\infty$, and f :

$\Omega \times \mathbb{R} \to \mathbb{R}$ is a Carathéodory function (i.e., measurable in $x \in \Omega$ for all $s \in \mathbb{R}$ and continuous in $s \in \mathbb{R}$ for a.e. $x \in \Omega$). The statement in (5.35) is the Neumann counterpart of the Dirichlet problem (5.1).

By a weak solution of problem (5.35) we mean any function $u \in W^{1,p}(\Omega)$ such that

$$\int_\Omega |\nabla u|^{p-2}\nabla u \cdot \nabla v \, dx \tag{5.36}$$
$$= \lambda \int_\Omega |u|^{p-2} u v \, dx - \int_\Omega f(x,u) v \, dx \quad \text{for all } v \in W^{1,p}(\Omega).$$

Here are the notions of subsolution and supersolution for problem (5.35).

Definition 5.6. The function $\underline{u} \in W^{1,p}(\Omega) \cap L^\infty(\Omega)$ is called a subsolution for problem (5.35) and the function $\overline{u} \in W^{1,p}(\Omega) \cap L^\infty(\Omega)$ is called a supersolution for problem (5.35) if

$$\int_\Omega \left(|\nabla \underline{u}|^{p-2}\nabla \underline{u} \cdot \nabla v - \lambda |\underline{u}|^{p-2}\underline{u} v + f(x,\underline{u}) v \right) dx \le 0$$

and

$$\int_\Omega \left(|\nabla \overline{u}|^{p-2}\nabla \overline{u} \cdot \nabla v - \lambda |\overline{u}|^{p-2}\overline{u} v + f(x,\overline{u}) v \right) dx \ge 0$$

for all $v \in W^{1,p}(\Omega)$ with $v \ge 0$ a.e. in Ω.

In the analysis of problem (5.35), we use the Banach space $C^1(\overline{\Omega})$ endowed with the ordered cone

$$C_+ = \left\{ u \in C^1(\overline{\Omega}) : \ u(x) \ge 0 \text{ for all } x \in \overline{\Omega} \right\}$$

whose interior is nonempty and given by

$$\operatorname{int} C_+ = \{ u \in C_+ : \ u(x) > 0 \text{ for all } x \in \overline{\Omega} \}.$$

We are first going to prove the existence of two opposite constant-sign solutions $u_+ \in \operatorname{int} C_+$ and $u_- \in -\operatorname{int} C_+$ for problem (5.35). Given $\lambda > 0$, let us assume that the Carathéodory function $f : \Omega \times \mathbb{R} \to \mathbb{R}$ satisfies:

H(f) **(i)** there is a positive constant k such that for almost all $x \in \Omega$, one has

$$\min\{f(x,k), \ -f(x,-k)\} \ge \lambda k^{p-1};$$

(ii) there are two real numbers c, α, with $c \le \alpha < \lambda$ such that the following inequalities hold uniformly for a.e. $x \in \Omega$:

$$c \le \liminf_{s \to 0} \frac{f(x,s)}{|s|^{p-2}s} \le \limsup_{s \to 0} \frac{f(x,s)}{|s|^{p-2}s} \le \alpha;$$

(iii) f is bounded on bounded sets.

Remark 5.7. From (ii) it follows that $f(x,0)=0$ a.e. in Ω, so $u\equiv 0$ is solution to (5.35).

For the sake of simplicity, in this section, the usual norm of the space $W^{1,p}(\Omega)$ as given in Definition 1.14 is denoted by $\|\cdot\|$.

Theorem 5.8. *Assume that hypothesis* H(f) *holds. Then problem (5.35) admits two solutions* $u_+\in \operatorname{int} C_+$ *and* $u_-\in -\operatorname{int} C_+$.

Proof. Hypothesis H(f) (i) entails

$$-\Delta_p(k)-\lambda k^{p-1}+f(x,k)\geq 0, \tag{5.37}$$

whereas hypothesis H(f) (ii) guarantees that there exists $\delta\in(0,k)$ such that

$$f(x,s)<\lambda s^{p-1} \tag{5.38}$$

for a.a. $x\in\Omega$ and for all $s\in(0,\delta)$. Then for any $\varepsilon\in(0,\delta)$, we obtain from (5.38) that

$$-\Delta_p(\varepsilon)-\lambda\varepsilon^{p-1}+f(x,\varepsilon)<0 \tag{5.39}$$

for a.a. $x\in\Omega$. Therefore, by (5.37) and (5.39), we note that $\underline{u}=\varepsilon$ and $\overline{u}=k$ are a subsolution and a supersolution for problem (5.35), respectively.

We introduce the following truncation of the identity on $\mathbb{R}$:

$$\tau(s)=\begin{cases} k & \text{if } s\geq k,\\ s & \text{if } \varepsilon<s<k,\\ \varepsilon & \text{if } s\leq\varepsilon,\end{cases}$$

and by means of it, the nonlinear operator $H: W^{1,p}(\Omega)\to W^{1,p}(\Omega)^*$ by

$$\begin{aligned}\langle H(u),v\rangle=&\int_\Omega\left(|\nabla u|^{p-2}\nabla u\cdot\nabla v+|u|^{p-2}uv\right)dx\\ &-(\lambda+1)\int_\Omega \tau^{p-1}(u)v\,dx+\int_\Omega f(x,\tau(u))v\,dx\end{aligned}$$

for all $u,v\in W^{1,p}(\Omega)$. We have

$$\left|\int_\Omega \tau^{p-1}(u)v\,dx\right|\leq k^{p-1}\int_\Omega |v|\,dx. \tag{5.40}$$

From hypothesis H(f) (iii) there is a constant $L\geq 0$ such that

$$\sup_{(x,s)\in\Omega\times\mathbb{R}}|f(x,\tau(s))|=\sup_{(x,t)\in\Omega\times[0,k]}|f(x,t)|\leq L. \tag{5.41}$$

Through (5.40) and (5.41), we are able to show the boundedness and coercivity of the operator H. In order to show that H is also pseudomonotone, let $\{u_n\}$ be a sequence in $W^{1,p}(\Omega)$ such that $u_n \rightharpoonup u$ in $W^{1,p}(\Omega)$ and

$$\limsup_{n\to\infty}\langle H(u_n), u_n - u\rangle \le 0. \tag{5.42}$$

Then, along a relabeled subsequence, $u_n \to u$ and $\tau(u_n) \to \tau(u)$ in $L^p(\Omega)$ and a.e. in Ω. Using the boundedness of τ, by the Lebesgue dominated convergence theorem, we obtain for every $w \in W^{1,p}(\Omega)$ that

$$\int_\Omega \tau^{p-1}(u)(u-w)\,dx = \lim_{n\to\infty}\int_\Omega \tau^{p-1}(u_n)(u_n-w)\,dx$$

and

$$\int_\Omega f(x,\tau(u))(u-w)\,dx = \lim_{n\to\infty}\int_\Omega f(x,\tau(u_n))(u-w)\,dx.$$

At this point, (5.42) and Proposition 1.56 enable us to conclude that H is a pseudomonotone operator.

Theorem 1.55 provides some $u \in W^{1,p}(\Omega)$ such that $H(u) = 0$, which reads as

$$\int_\Omega \left(|\nabla u|^{p-2}\nabla u\cdot\nabla v + |u|^{p-2}uv\right)dx - (\lambda+1)\int_\Omega \tau^{p-1}(u)v\,dx$$
$$+\int_\Omega f(x,\tau(u))v\,dx = 0 \quad \text{for all } v \in W^{1,p}(\Omega). \tag{5.43}$$

We claim that $\varepsilon \le u(x) \le k$ for a.e. $x \in \Omega$. Acting on (5.43) with $v = (u-k)^+$ gives

$$\int_{\{u\ge k\}}\left(|\nabla u|^p + u^{p-1}(u-k)\right)dx - (\lambda+1)\int_{\{u>k\}} k^{p-1}(u-k)\,dx$$
$$+\int_{\{u>k\}} f(x,k)(u-k)\,dx = 0.$$

From H(f) (i) we know that $f(x,k) - \lambda k^{p-1} \ge 0$ a.e. in Ω, which results in

$$0 \le \int_{\{u>k\}}\left(u^{p-1} - k^{p-1}\right)(u-k)\,dx \le 0,$$

so $u \le k$ a.e. in Ω. Next, acting on (5.43) with $v = (\varepsilon - u)^+$ leads to

$$-\int_{\{u\le\varepsilon\}}|\nabla u|^p\,dx + \int_{\{u<\varepsilon\}}|u|^{p-2}u(\varepsilon-u)\,dx$$
$$-(\lambda+1)\int_{\{u<\varepsilon\}}\varepsilon^{p-1}(\varepsilon-u)\,dx + \int_{\{u<\varepsilon\}} f(x,\varepsilon)(\varepsilon-u)\,dx = 0.$$

Using (5.38), we arrive at

$$0 \le \int_{\{u<\varepsilon\}} \left(|u|^{p-2}u - \varepsilon^{p-1}\right)(\varepsilon - u)\,dx \le 0,$$

which implies $|\{u < \varepsilon\}| = 0$, proving the claim.

Since, $\varepsilon \le u(x) \le k$ a.e. in Ω, (5.43) becomes (5.36), ensuring that u is a weak solution for (5.35). Noting that $u \in L^\infty(\Omega)$, H(f) (iii) guarantees that the function $g(x) = f(x, u(x))$ belongs to $L^\infty(\Omega)$, and [127, Theorem 4.9] ensures that $u \in C^{1,\beta}(\overline{\Omega})$, for some $\beta > 0$, so [169, Theorem 8.27] yields $u \in \operatorname{int} C_+$. Consequently, the desired conclusion holds true with $u = u_+$.

The existence of a (negative) solution $u_- \in -\operatorname{int} C_+$ is proven analogously. □

The next objective is to provide the existence of extremal constant-sign solutions of problem (5.35).

Lemma 5.9. *Assume that conditions* H(f) *hold. Then, for every* $\varepsilon < k$, *problem (5.35) possesses the smallest (positive) solution* $v_{\varepsilon,+}$ *in* $[\varepsilon, k]$ *and the biggest (negative) solution* $v_{\varepsilon,-}$ *in* $[-k, -\varepsilon]$.

Proof. Let $\mathcal{S}_\varepsilon$ denote the set of all solutions of (5.35) belonging to the ordered interval $[\varepsilon, k]$, which is nonempty according to Theorem 5.8. On the pattern of the Dirichlet problem, it can be shown that the set $\mathcal{S}_\varepsilon$ is downward directed and compact in $W^{1,p}(\Omega)$ (see [38, Corollary 3.7, Theorem 3.10]). Since $W^{1,p}(\Omega)$ is separable, there exists a sequence $\{v_n\} \subset \mathcal{S}_\varepsilon$ which is dense in $\mathcal{S}_\varepsilon$. Set $u_1 = v_1$. Proceeding inductively and exploiting that $\mathcal{S}_\varepsilon$ is downward directed, we can choose $u_{n+1} \in \mathcal{S}_\varepsilon$ such that $u_{n+1} \le \min\{u_n, v_n\}$. The compactness of $\mathcal{S}_\varepsilon$ and the monotonicity of $\{u_n\}$ entail that $u_n \to v_{\varepsilon,+} := \inf\limits_n u_n \in \mathcal{S}_\varepsilon$ in $W^{1,p}(\Omega)$. We observe that $v_{\varepsilon,+} \le v_n$ for all n. In view of the density of $\{v_n\}$ in $\mathcal{S}_\varepsilon$, we infer that $v_{\varepsilon,+}$ is the smallest solution in $[\varepsilon, k]$. The existence of the biggest solution $v_{\varepsilon,-}$ in $[-k, -\varepsilon]$ can be obtained in a similar way. □

Theorem 5.10. *If hypotheses* H(f) *hold, then problem (5.35) admits the smallest positive solution* $v_+ \in \operatorname{int} C_+$ *and the biggest negative solution* $v_- \in -\operatorname{int} C_+$.

Proof. We only show the existence of the smallest positive solution. The existence of the biggest negative solution can be shown along the same lines. From Lemma 5.9, we know that for any integer n there is the smallest solution $v_{n,+} \in [\frac{1}{n}, k]$. Clearly, one has

$$v_{n,+} \rightharpoonup v_+ \text{ in } W^{1,p}(\Omega),\ v_{n,+} \to v_+ \text{ in } L^p(\Omega) \text{ and } v_{n,+} \downarrow v_+ \text{ a.e. in } \Omega. \tag{5.44}$$

From (5.36) with $u = v_{n,+}$ and $v = v_{n,+} - v_+$, by using (5.44) we get that

$$\lim_{n\to\infty} \int_\Omega |\nabla v_{n,+}|^{p-2}\nabla v_{n,+} \cdot \nabla(v_{n,+} - v_+)\, dx = 0.$$

Then through Proposition 1.56 and (5.44) we arrive at the strong convergence $v_{n,+} \to v_+$ in $W^{1,p}(\Omega)$. Therefore v_+ is a solution for (5.35).

Now we show that $v_+ \neq 0$. Arguing by contradiction, suppose $v_+ = 0$. Set

$$\tilde{v}_n = \frac{v_{n,+}}{\|v_{n,+}\|} \quad \text{and} \quad h_n = \frac{f(x, v_{n,+})}{(v_{n,+})^{p-1}}\tilde{v}_n^{p-1}.$$

By H(f) (ii) and (5.44), up to a subsequence we may admit that

$$\tilde{v}_n \rightharpoonup \tilde{v} \text{ in } W^{1,p}(\Omega),\ \tilde{v}_n \to \tilde{v} \text{ in } L^p(\Omega),$$
$$\tilde{v}_n \to \tilde{v} \text{ a.e. in } \Omega,\ h_n \rightharpoonup h \text{ in } L^{\frac{p}{p-1}}(\Omega).$$

Acting on (5.36) for $u = v_{n,+}$ with $\tilde{v}_n - \tilde{v}$, we obtain

$$\int_\Omega |\nabla \tilde{v}_n|^{p-2}\nabla \tilde{v}_n \cdot \nabla(\tilde{v}_n - \tilde{v})\, dx = \int_\Omega (\lambda \tilde{v}_n^{p-1} - h_n)(\tilde{v}_n - \tilde{v})\, dx.$$

Combining with hypotheses H(f) (ii), (iii) and the Lebesgue dominated convergence theorem ensures

$$\lim_{n\to\infty} \int_\Omega |\nabla \tilde{v}_n|^{p-2}\nabla \tilde{v}_n \cdot \nabla(\tilde{v}_n - \tilde{v})\, dx = 0.$$

From the $(S)_+$-property of $-\Delta_p$ in Proposition 1.56, it turns out that $\tilde{v}_n \to \tilde{v}$ in $W^{1,p}(\Omega)$. Since $\|\tilde{v}\| = 1$, we have $\tilde{v} \neq 0$.

On the basis of assumption H(f) (ii), we note that $h = g\tilde{v}^{p-1}$, with $g \in L^\infty(\Omega)$, $\lambda - g(x) > 0$ for a.e. $x \in \Omega$, thus $\tilde{v}$ is a solution of

$$\int_\Omega |\nabla \tilde{v}|^{p-2}\nabla \tilde{v} \cdot \nabla v\, dx = \int_\Omega (\lambda - g(x))\, \tilde{v}^{p-1} v\, dx \quad \text{for all } v \in W^{1,p}(\Omega).$$

Consequently, $\tilde{v}$ is an eigenfunction associated with the eigenvalue 1 for the weighted problem

$$\begin{cases} -\Delta_p \tilde{v} = (\lambda - g(x))|\tilde{v}|^{p-2}\tilde{v} & \text{in } \Omega, \\ \frac{\partial \tilde{v}}{\partial \nu} = 0 & \text{on } \partial\Omega. \end{cases}$$

Recalling $\lambda - g(x) > 0$ for a.e. $x \in \Omega$, $\tilde{v}$ must change sign (see [169, Proposition 9.33] or Remark 3.7). This contradicts $\tilde{v} \geq 0$, $\tilde{v} \neq 0$, which proves that $v_+ \neq 0$. Then, as in the proof of Theorem 5.8, we deduce that $v_+ \in \operatorname{int} C_+$.

Let us verify that v_+ is the smallest positive solution of problem (5.35). Given a positive solution u of (5.35), by the strong maximum principle we know that $u \in \operatorname{int} C_+$, so $\frac{1}{n} \leq u$ on $\overline{\Omega}$ whenever n is sufficiently large, say $n \geq n_0$. Without loss of generality we may suppose that $u \leq k$ because otherwise we can choose in place of u a solution of (5.35) in the ordered interval $[\frac{1}{n_0}, \min\{u, k\}]$, which exists since $\min\{u, k\}$ is a supersolution and $1/n_0$ is a subsolution of (5.35). Knowing that $v_{n,+}$ is the smallest solution of (5.35) in $[\frac{1}{n}, k]$, it follows that $v_{n,+} \leq u$. Letting $n \to \infty$ enables us to find that $v_+ \leq u$, which completes the proof. □

Next we prove the existence of a third nontrivial solution to (5.35), which has the additional property to be nodal (i.e., sign changing). To this end it is necessary to strengthen our assumption on λ. In order to do this, we need to recall some basic facts about the spectrum of the negative p-Laplacian with Neumann boundary condition.

Consider the nonlinear eigenvalue problem

$$\begin{cases} -\Delta_p u = \lambda |u|^{p-2} u & \text{in } \Omega, \\ \frac{\partial u}{\partial \nu} = 0 & \text{on } \partial\Omega. \end{cases} \tag{5.45}$$

The first eigenvalue in (5.45) is $\lambda_1 = 0$. By $\hat{u}_1$ we denote its positive L^p-normalized eigenfunction, i.e. $\hat{u}_1(x) = |\Omega|^{-\frac{1}{p}}$ where $|\cdot|$ stands for the Lebesgue measure on $\mathbb{R}^N$. Since $\lambda_1 = 0$ is isolated and the spectrum $\sigma(p)$ of $-\Delta_p$ on $W^{1,p}(\Omega)$ is closed, then the second eigenvalue λ_2 satisfies

$$\lambda_2 = \inf\{\lambda : \lambda \in \sigma(p), \ \lambda > 0\} > 0.$$

The following minimax characterization of λ_2 is given in [16] (see also Corollary 3.11).

Proposition 5.11. *Let $S = \{u \in W^{1,p}(\Omega) : \|u\|_p = 1\}$ be endowed with the topology induced by $W^{1,p}(\Omega)$. Then*

$$\lambda_2 = \inf_{\widehat{\gamma} \in \widehat{\Gamma}} \max_{-1 \leq t \leq 1} \|\nabla \widehat{\gamma}(t)\|_p^p,$$

where $\widehat{\Gamma} = \left\{\widehat{\gamma} \in C([-1, 1], S) : \widehat{\gamma}(-1) = -\hat{u}_1, \ \widehat{\gamma}(1) = \hat{u}_1\right\}$.

In order to develop a variational approach, we set

$$F(x, s) = \int_0^s f(x, t)\, dt \ \text{ for all } (x, s) \in \Omega \times \mathbb{R}.$$

Under hypotheses H(f), $F : \Omega \times \mathbb{R} \to \mathbb{R}$ is a Carathéodory function such that $F(\cdot, 0) = 0$, the partial derivative $\frac{\partial F}{\partial s}$ of $F(x, s)$ exists and is a Carathéodory function satisfying $\frac{\partial F}{\partial s}(x, s) = f(x, s)$ for almost all $x \in \Omega$.

Using the extremal constant-sign solutions u_+ and u_- constructed in Theorem 5.10, we introduce the following functions on $\Omega \times \mathbb{R}$:

$$\tau_+(x,s)=\begin{cases}0 & \text{if } s\le 0,\\ s & \text{if } 0<s<v_+(x),\\ v_+(x) & \text{if } s\ge v_+(x),\end{cases}$$

$$\tau_-(x,s)=\begin{cases}v_-(x) & \text{if } s\le v_-(x),\\ s & \text{if } v_-(x)<s<0,\\ 0 & \text{if } s\ge 0,\end{cases}$$

$$\tau_0(x,s)=\begin{cases}v_-(x) & \text{if } s\le v_-(x),\\ s & \text{if } v_-(x)<s<v_+(x),\\ v_+(x) & \text{if } s\ge v_+(x),\end{cases}$$

$$f_+(x,s)=f(x,\tau_+(x,s)),\quad f_-(x,s)=f(x,\tau_-(x,s)),$$
$$f_0(x,s)=f(x,\tau_0(x,s)).$$

Corresponding to these functions, we define

$$F_+(x,s)=\int_0^s f_+(x,t)\,dt,\quad F_-(x,s)=\int_0^s f_-(x,t)\,dt,$$
$$F_0(x,s)=\int_0^s f_0(x,t)\,dt.$$

The energy functionals related to the three truncated problems obtained from (5.35) are:

$$E_+(u):=\frac{1}{p}\|u\|^p-(\lambda+1)\int_\Omega\int_0^{u(x)}\tau_+^{p-1}(x,s)\,ds\,dx+\int_\Omega F_+(x,u(x))\,dx,$$
$$E_-(u)=\frac{1}{p}\|u\|^p-(\lambda+1)\int_\Omega\int_0^{u(x)}\tau_-^{p-1}(x,s)\,ds\,dx+\int_\Omega F_-(x,u(x))\,dx,$$
$$E_0(u)=\frac{1}{p}\|u\|^p-(\lambda+1)\int_\Omega\int_0^{u(x)}\tau_0^{p-1}(x,s)\,ds\,dx+\int_\Omega F_0(x,u))\,dx.$$

The functionals $E_\pm$, E_0 are C^1 on $W^{1,p}(\Omega)$.

We are in a position to state the main result of this section, which is based on requiring $\lambda>\lambda_2$.

Theorem 5.12. *Assume $\lambda>\lambda_2$ and that hypothesis* H(f) *holds with $\alpha<\lambda-\lambda_2$. Then problem (5.35) has at least three nontrivial solutions: the smallest positive solution $v_+\in \operatorname{int} C_+$, the biggest negative solution $v_-\in -\operatorname{int} C_+$, and a nodal solution $u_0\in C^1(\overline{\Omega})$, with $v_-\le u_0\le v_+$.*

Proof. Theorem 5.10 provides the smallest positive solution $v_+\in \operatorname{int} C_+$ and the biggest negative solution $v_-\in -\operatorname{int} C_+$ of problem (5.35).

We claim that v_+ and v_- are global minimizers for E_+ and E_-, respectively, as well as local minimizers of E_0.

To this end, we note that any critical point $v \in W^{1,p}(\Omega)$ of E_+ satisfies

$$0 \le v(x) \le v_+(x), \text{ for a.a. } x \in \Omega. \tag{5.46}$$

Indeed, acting on $E_+'(v) = 0$ with $w = (v - v_+)^+$ yields

$$\int_{\{v \ge v_+\}} \left(|\nabla v|^{p-2} \nabla v \cdot \nabla (v - v_+) + v^{p-1}(v - v_+) \right) dx$$
$$-(\lambda + 1) \int_{\{v \ge v_+\}} (v_+)^{p-1}(v - v_+)\, dx + \int_{\{v \ge v_+\}} f(x, v_+)(v - v_+)\, dx = 0.$$

The fact that v_+ is a solution for (5.35) ensures

$$\int_{\{v \ge v_+\}} |\nabla v_+|^{p-2} \nabla v_+ \cdot \nabla (v - v_+)\, dx - \lambda \int_{\{v \ge v_+\}} (v_+)^{p-1}(v - v_+)\, dx$$
$$+ \int_{\{v \ge v_+\}} f(x, v_+)(v - v_+)\, dx = 0.$$

Subtracting the second equality from the first one results in

$$\int_{\{v \ge v_+\}} \left(|\nabla v|^{p-2} \nabla v - |\nabla v_+|^{p-2} \nabla v_+ \right) \cdot \nabla (v - v_+)\, dx$$
$$+ \int_{\{v \ge v_+\}} \left(v^{p-1} - (v_+)^{p-1} \right) (v - v_+)\, dx = 0,$$

which implies that $v(x) \le v_+(x)$ for a.a. $x \in \Omega$.

On the other hand, testing $E_+'(v) = 0$ with $w = -v^-$ has as effect

$$\int_{\{v \le 0\}} \left(|\nabla v|^p + |v|^{p-1} \right) dx - \int_{\{v \le 0\}} f_+(x, v) v^-\, dx = 0.$$

Since $f_+(x, s) = 0$ whenever $s \le 0$, we infer that $v^- = 0$ and so the validity of (5.46) is proven.

By hypothesis H(f) (ii), for each $\varepsilon > 0$ there exists $\delta > 0$ such that

$$F(x, s) \le \frac{\alpha + \varepsilon}{p} s^p \text{ for a.a. } x \in \Omega, \text{ all } s \in (0, \delta).$$

Selecting $0 < \delta < \min_{\overline{\Omega}} v_+$, one has that $F_+(x, s) = F(x, s)$ for all $s \in (0, \delta)$. Then for $t \in (0, \delta)$, by virtue of $\alpha < \lambda - \lambda_2$ we see that

$$E_+(t) \le (-\lambda + \alpha + \varepsilon) \frac{t^p}{p} |\Omega| < (-\lambda_2 + \varepsilon) \frac{t^p}{p} |\Omega|.$$

If $\varepsilon < \lambda_2$, we get $E_+(t) < 0$.

The functional E_+ has a global minimizer $z \in W^{1,p}(\Omega)$ because it is coercive and sequentially weakly lower semicontinuous. According to what was said before, it holds $E_+(z) < 0$, so $z \neq 0$. Owing to (5.46), it follows that $0 \leq z(x) \leq v_+(x)$ for a.e. $x \in \Omega$. Therefore z is a solution for Eq. (5.35). In view of the nonlinear regularity theory and strong maximum principle, we have that $z \in \operatorname{int} C_+$. Due to the minimality property of v_+, actually we obtain that $z = v_+$. Since E_+ and E_0 coincide on a $C^1(\overline{\Omega})$-neighborhood of v_+, we see that v_+ is a local $C^1(\overline{\Omega})$-minimizer of E_0, hence it is also a local minimizer of E_0 in $W^{1,p}(\Omega)$ (see [156, Proposition 24]).

Proceeding in a similar way with E_- in place of E_+, we obtain the corresponding result for v_-. Arguing exactly as for showing (5.46) we can establish that any critical point v of E_0 satisfies

$$v_-(x) \leq v(x) \leq v_+(x), \text{ for a.a. } x \in \Omega. \tag{5.47}$$

We may assume that v_+ and v_- are strict local minimizers of E_0 because otherwise handling (5.47) and the extremal properties of v_+ and v_- we find infinitely many sign-changing solutions for problem (5.35). Since E_0 is coercive, it satisfies the Palais–Smale condition. Therefore we can apply the mountain pass theorem (see, e.g., [207], [169, Theorem 5.40] or Theorem 1.44) to E_0 to obtain a critical point u_0 of E_0 satisfying

$$\max\{E_0(v_+), E_0(v_-)\} < E_0(u_0) = c := \inf_{\gamma \in \Gamma} \max_{-1 \leq t \leq 1} E_0(\gamma(t)), \tag{5.48}$$

with

$$\Gamma = \left\{\gamma \in C([-1,1], W^{1,p}(\Omega)) : \gamma(-1) = v_-,\ \gamma(1) = v_+\right\}.$$

We claim that

$$E_0(u_0) < 0 \tag{5.49}$$

To this end, we construct a path $\gamma \in \Gamma$ along which E_0 is negative. Towards this, we make use of the characterization of the second eigenvalue of the negative p-Laplacian given in Proposition 5.11. In addition to S endowed with the topology induced by $W^{1,p}(\Omega)$ (see Proposition 5.11), we consider $S_C = S \cap C^1(\overline{\Omega})$ endowed with the topology induced by $C^1(\overline{\Omega})$. Then, besides

$$\widehat{\Gamma} = \left\{\widehat{\gamma} \in C([-1,1], S) : \widehat{\gamma}(-1) = -\hat{u}_1,\ \widehat{\gamma}(1) = \hat{u}_1\right\},$$

we also take

$$\widehat{\Gamma}_C = \left\{\widehat{\gamma} \in C([-1,1], S_C) : \widehat{\gamma}(-1) = -\hat{u}_1,\ \widehat{\gamma}(1) = \hat{u}_1\right\}.$$

It is worthwhile to point out that $\widehat{\Gamma}_C$ is dense in $\widehat{\Gamma}$. Given any $\mu \in (0, \lambda - \lambda_2 - \alpha)$, hypothesis H($f$) (ii) yields $\delta > 0$ such that

$$F(x,s) < \frac{\alpha + \mu}{p}|s|^p \text{ for a.a. } x \in \Omega, \text{ all } |s| < \delta. \tag{5.50}$$

Then Proposition 5.11 and the density of $\widehat{\Gamma}_C$ in $\widehat{\Gamma}$ ensure that for every $\kappa \in (0, \lambda - \lambda_2 - \alpha - \mu)$ there exists a path $\gamma_0 \in \widehat{\Gamma}_C$ satisfying

$$\max_{-1 \le t \le 1} \|\nabla \gamma_0(t)\|_p^p < \lambda_2 + \kappa. \tag{5.51}$$

Using the compactness of $\gamma_0([-1, 1])$ in $C^1(\overline{\Omega})$, in conjunction with $v_+ \in \operatorname{int} C_+$ and $v_- \in -\operatorname{int} C_+$, we find that corresponding to $\delta > 0$ defined above there is an $\varepsilon > 0$ such that, for all $x \in \Omega$,

$$v_-(x) < \varepsilon u(x) < v_+(x) \text{ and } \varepsilon |u(x)| < \delta \quad \text{for all } u = \gamma_0(t) \in \gamma_0([-1, 1]). \tag{5.52}$$

On the basis of (5.50), (5.51), and (5.52), we infer that

$$\begin{aligned} E_0(\varepsilon \gamma_0(t)) &= \frac{\varepsilon^p}{p}(\|\nabla u\|_p^p - \lambda) + \int_\Omega F(x, \varepsilon u(x))\, dx \\ &\le \frac{\varepsilon^p}{p}(\lambda_2 + \kappa - \lambda + \alpha + \mu) < 0 \text{ for all } t \in [-1, 1]. \end{aligned} \tag{5.53}$$

Notice that $E_+(v_+) =: m_+ < c_+ := E_+(\varepsilon \hat{u}_1)$. Due to the minimality property of v_+ and (5.46), there are no critical values of E_+ in the interval $(m_+, c_+]$. As a consequence we can apply the second deformation lemma (see, e.g., [169, Theorem 5.34] or Theorem 1.45) to the functional E_+. We obtain a continuous mapping $\eta : [0, 1] \times E_+^{c_+} \to E_+^{c_+}$, where $E_+^{c_+} = E_+^{-1}((-\infty, c_+])$, such that

$$\begin{aligned} &\eta(0, u) = u, \ \eta(1, u) = v_+, \\ &E_+(\eta(t, u)) \le E_+(u) \quad \text{for all } t \in [0, 1], \text{ all } u \in E_+^{c_+}. \end{aligned}$$

We observe that the continuous path

$$\gamma_+(t) = \eta(t, \varepsilon \hat{u}_1)^+ \text{ for all } t \in [0, 1]$$

joins $\varepsilon \hat{u}_1$ and v_+, and for all $t \in [0, 1]$,

$$E_0(\gamma_+(t)) = E_+(\gamma_+(t)) = E_+(\eta(t, \varepsilon \hat{u}_1)) - \frac{\|\eta(t, \varepsilon \hat{u}_1)^-\|^p}{p} \le E_+(\varepsilon \hat{u}_1) < 0. \tag{5.54}$$

For the functional E_-, we can similarly construct a continuous path γ_- joining v_- and $-\varepsilon \hat{u}_1$ such that

$$E_0(\gamma_-(t)) < 0 \text{ for all } t \in [0, 1]. \tag{5.55}$$

Concatenating γ_-, $\varepsilon \gamma_0$ and γ_+ produces a path γ belonging to Γ. Taking into account (5.53), (5.54), and (5.55), we derive that $E_0(\gamma(t)) < 0$ for all $t \in [0, 1]$. Hence (5.49) holds true. It turns out that u_0 is a nontrivial critical point of E_0

and by (5.47) that u_0 solves (5.46). The L^∞-boundedness of u_0, together with H(f) (iii), enables us to apply [127, Theorem 4.9] to deduce that $u_0 \in C^1(\overline{\Omega})$.

It remains only to show that u_0 must change sign. Arguing indirectly, suppose that u_0 is of constant sign, say $u_0 \geq 0$. By (5.47) we have that $u_0 = v_+$, which comes in contradiction with the fact that $E_0(v_+) < E_0(u_0)$ as known from (5.48). Through the same argument we can show that $u_0 \leq 0$ cannot happen. This amounts to saying that u_0 changes sign, which completes the proof. □

5.3 PROBLEMS WITH MULTIVALUED TERMS

This section deals with the following inclusion problem involving the Dirichlet boundary condition:

$$\begin{cases} -\Delta_p u \in \lambda \partial j(x, u) & \text{in } \Omega, \\ u = 0 & \text{on } \partial\Omega. \end{cases} \tag{5.56}$$

Here $\Omega \subset \mathbb{R}^N$ ($N \geq 3$) is a bounded domain with a C^2 boundary $\partial\Omega$, $1 < p < +\infty$, $\lambda > 0$ is a parameter and Δ_p denotes the p-Laplacian operator

$$\Delta_p u := \operatorname{div}\left(|\nabla u|^{p-2}\nabla u\right) \quad \text{for all } u \in W_0^{1,p}(\Omega),$$

that is, (5.2) holds true.

As usual, p^* will denote the Sobolev critical exponent. The multivalued reaction term $\partial j(x, s)$ is the generalized gradient of a nonsmooth potential $s \mapsto j(x, s)$, which is subjected to the following conditions:

H(j) $j : \Omega \times \mathbb{R} \to \mathbb{R}$ is a Carathéodory function for which there exist constants $a_1 > 0$, $1 < q < p^*$, $0 < a_2 \leq a_3$ such that:

(i) $j(x, \cdot)$ is locally Lipschitz for a.a. $x \in \Omega$ and $j(\cdot, 0) \in L^1(\Omega)$;

(ii) $|\xi| \leq a_1\left(1 + |s|^{q-1}\right)$ for a.e. $x \in \Omega$ and for all $s \in \mathbb{R}$, $\xi \in \partial j(x, s)$;

(iii) $\limsup\limits_{|s| \to +\infty} \dfrac{j(x, s)}{|s|^p} \leq 0$ uniformly for a.e. $x \in \Omega$;

(iv) $a_2 \leq \liminf\limits_{s \to 0^+} \dfrac{\min \partial j(x, s)}{s^{p-1}} \leq \limsup\limits_{s \to 0^+} \dfrac{\max \partial j(x, s)}{s^{p-1}} \leq a_3$ uniformly for a.e. $x \in \Omega$;

(v) $a_2 \leq \liminf\limits_{s \to 0^-} \dfrac{\max \partial j(x, s)}{|s|^{p-2}s} \leq \limsup\limits_{s \to 0^-} \dfrac{\min \partial j(x, s)}{|s|^{p-2}s} \leq a_3$ uniformly for a.e. $x \in \Omega$.

The functions $(x, s) \mapsto \min \partial j(x, s)$, $\max \partial j(x, s)$ are well defined because $\partial j(x, s)$ is a compact interval. Moreover, they are measurable.

The functional space that fits problem (5.56) is the Sobolev space $W_0^{1,p}(\Omega)$ equipped with the norm $\|u\| := \|\nabla u\|_p$ (see Remark 1.18). In addition, we utilize $C_0^1(\overline{\Omega})$, which turns out to be an ordered Banach space with the order cone

$$C_0^1(\overline{\Omega})_+ := \left\{ u \in C_0^1(\overline{\Omega}) \; : \; u(x) \geq 0 \quad \text{for all } x \in \Omega \right\},$$

whose interior is nonempty and characterized in (1.39) as

$$\begin{aligned} &\operatorname{int}(C_0^1(\overline{\Omega})_+) \\ &= \left\{ u \in C_0^1(\overline{\Omega})_+ \; : \; u(x) > 0, \; \forall x \in \Omega \text{ and } \frac{\partial u}{\partial \nu}(x) < 0, \; \forall x \in \partial\Omega \right\}, \end{aligned}$$

where, as usual, $\nu(x)$ denotes the outward unit normal to $\partial\Omega$ at any point x.

Given $\lambda > 0$, we say that $u \in W_0^{1,p}(\Omega)$ is a (weak) solution of problem (5.56) if $\Delta_p u \in L^{\zeta}(\Omega)$ for some $\zeta > 1$ and

$$-\Delta_p u(x) \in \lambda \partial j(x, u(x)) \text{ a.e. in } \Omega.$$

From H(j) (iii) it follows that (5.56) admits the zero solution for all $\lambda > 0$. We are interested in finding nontrivial solutions.

Now we can state our main result on problem (5.56). Hereafter by λ_1 and λ_2 we denote the first two eigenvalues of $-\Delta_p$ on $W_0^{1,p}(\Omega)$ (see Propositions 1.57 and 1.58).

Theorem 5.13. *Assume that hypotheses* H(j) *hold. Then, for every* $\lambda > \lambda_2/a_2 + 1$, *problem (5.56) possesses at least three nontrivial solutions, the smallest positive, the biggest negative, and a nodal one, lying in* $C_0^1(\overline{\Omega})$.

We provide a simple example for which Theorem 5.13 applies. For simplicity, we drop the dependence of $j(x, s)$ with respect to $x \in \Omega$.

Example 5.14. The nonsmooth potential $j : \mathbb{R} \to \mathbb{R}$ defined for every $s \in \mathbb{R}$ by

$$j(s) = \begin{cases} \log(1 + |s|)|s|^{p-1} & \text{if } |s| \leq 1, \\ \log(2)|s|^q & \text{if } |s| > 1, \end{cases}$$

for $1 < q < p$, is locally Lipschitz and satisfies hypotheses H(j) with $a_1 = \log(2)$ and $a_2 = a_3 = 1$.

Theorem 5.13 is going to be proven through a nonsmooth variational approach. The nonsmooth energy functional $\varphi_\lambda : W_0^{1,p}(\Omega) \to \mathbb{R}$ associated to the inclusion (5.56) is defined by

$$\varphi_\lambda(u) = \frac{\|u\|^p}{p} - \lambda \int_\Omega j(x, u)\, dx \quad \text{for all } u \in W_0^{1,p}(\Omega).$$

Proposition 5.15. *The functional* $\varphi_\lambda : W_0^{1,p}(\Omega) \to \mathbb{R}$ *is locally Lipschitz. Moreover, if* $u \in W_0^{1,p}(\Omega)$ *is a critical point of* φ_λ, *then* $u \in C_0^1(\overline{\Omega})$ *and* u *solves (5.56).*

Proof. The function $u \mapsto \frac{\|u\|^p}{p}$ is of class C^1 whose differential is the p-Laplacian $-\Delta_p : W_0^{1,p}(\Omega) \to W^{-1,p'}(\Omega)$. Aubin–Clarke's theorem (see, e.g., [38, Theorem 2.181]) ensures that the function

$$u \mapsto \int_\Omega j(x,u)\,dx$$

is Lipschitz continuous on any bounded subset of $L^q(\Omega)$ and its generalized gradient is included in the set

$$N(u) := \left\{ w \in L^{q'}(\Omega) \ : \ w(x) \in \partial j(x, u(x)) \text{ for a.e. } x \in \Omega \right\}, \tag{5.57}$$

where $1 < q < p^*$ and $\frac{1}{q} + \frac{1}{q'} = 1$. Since, according to Theorem 1.20, $W_0^{1,p}(\Omega)$ is continuously embedded in $L^q(\Omega)$, the functional φ_λ turns out to be locally Lipschitz on $W_0^{1,p}(\Omega)$. We have

$$\partial\varphi_\lambda(u) \subset -\Delta_p(u) - \lambda N(u).$$

So, if $0 \in \partial\varphi_\lambda(u)$, then

$$-\Delta_p u = \lambda w \ \text{ in } W_0^{-1,p'}(\Omega) \tag{5.58}$$

with some $w \in N(u)$. Hence, we obtain that $\Delta_p u \in L^{q'}(\Omega)$ and u solves (5.56). Combining H(j) (ii) with (5.58) yields the estimate

$$-u\Delta_p u \le a_1(|u| + |u|^q) \ \text{ for a.e. } x \in \Omega,$$

which on account of [95, Theorem 1.5.5] implies $u \in L^\infty(\Omega)$. From H(j) (ii) and (5.58) it follows that $\Delta_p u \in L^\infty(\Omega)$. So, by the nonlinear regularity theory, the function u belongs to $C_0^1(\overline{\Omega})$. □

We complement Proposition 5.15 with a strong maximum principle-type result for problem (5.56).

Proposition 5.16. *If $u \in C_0^1(\overline{\Omega})_+$ is a solution of problem (5.56) satisfying $u \neq 0$, then $u \in \operatorname{int}(C_0^1(\overline{\Omega})_+)$.*

Proof. Fix $0 < \theta < a_2$. Through H(j) (iii) one can find $\delta > 0$ such that

$$\min \partial j(x,s) \ge \theta s^{p-1}$$

for a.e. $x \in \Omega$ and all $0 < s < \delta$. Since in (5.58) we now have $w \in L^\infty(\Omega)$, there exists a constant $c > 0$ such that

$$\Delta_p u \le c u^{p-1} \ \text{ a.e. in } \Omega.$$

Vázquez's maximum principle [229, Theorem 5] directly gives $u \in \operatorname{int}(C_0^1(\overline{\Omega})_+)$. □

Definition 5.17. A function $\underline{u} \in W^{1,p}(\Omega)$ is called a subsolution of problem (5.56) provided $\underline{u} \leq 0$ on $\partial\Omega$ and there exists $w \in N(\underline{u})$ (see (5.57) with some $1 < q < p^*$) such that

$$\langle -\Delta_p \underline{u}, v \rangle \leq \lambda \int_\Omega w v \, dx \quad \text{for all } v \in W_0^{1,p}(\Omega) \text{ with } v(x) \geq 0 \text{ a.e. in } \Omega.$$

The definition of a supersolution is obtained by reversing the inequalities involving $\underline{u}$.

The following comparison principle for differential inclusions in [41, Corollary 4.1] will be employed later on.

Proposition 5.18. *If $\underline{u}, \overline{u} \in W^{1,p}(\Omega)$ are a subsolution and a supersolution, respectively, of problem (5.56) and $\underline{u}(x) \leq \overline{u}(x)$ for a.a. $x \in \Omega$, then the set*

$$H := \left\{ u \in W_0^{1,p}(\Omega) \ : \ u \text{ solves (5.56) and } \underline{u}(x) \leq u(x) \leq \overline{u}(x) \text{ for a.e. } x \in \Omega \right\}$$

is nonempty, compact, and downward and upward directed with respect to the pointwise order. Moreover, H has the smallest and the biggest element.

We proceed by obtaining a minimal positive solution.

Theorem 5.19. *If $\lambda > \lambda_1/a_2$ then the inclusion problem (5.56) possesses the smallest positive solution $\hat{u}_\lambda \in \operatorname{int}(C_0^1(\overline{\Omega})_+)$.*

Proof. Fix $\lambda > \lambda_1/a_2$. We split the proof in several steps.

Claim 1. Problem (5.56) has a positive solution $\hat{u} \in \operatorname{int}(C_0^1(\overline{\Omega})_+)$.

Consider the truncation $j_+ : \Omega \times \mathbb{R} \to \mathbb{R}$ defined by

$$j_+(x,s) = j(x,s^+) \qquad \text{for all } (x,s) \in \Omega \times \mathbb{R},$$

where, as before, $s^+ := \max\{0, s\}$. Its generalized gradient with respect to s is equal to

$$\partial j_+(x,s) = \begin{cases} \{0\} & \text{if } s < 0, \\ \{\tau\xi \ : \ \xi \in \partial j(x,0), \ \tau \in [0,1]\} & \text{if } s = 0, \\ \partial j(x,s) & \text{if } s > 0. \end{cases}$$

The related functional $\varphi_\lambda^+ : W_0^{1,p}(\Omega) \to \mathbb{R}$ given by

$$\varphi_\lambda^+(u) = \frac{\|u\|^p}{p} - \lambda \int_\Omega j_+(x,u) \, dx \quad \text{for all } u \in W_0^{1,p}(\Omega)$$

is weakly sequentially lower semicontinuous and locally Lipschitz on $W_0^{1,p}(\Omega)$. We are going to prove that φ_λ^+ is coercive. Let $0 < \varepsilon < \lambda_1/(p\lambda)$. By H($j$) (iii) there exists $M > 0$ such that

$$j_+(x,s) < \varepsilon s^p \text{ for a.a. } x \in \Omega \text{ and all } s > M,$$

while H(j) (ii) and Proposition 1.40 provide

$$j_+(x,s) \le j(x,0) + a_1(M + M^q)$$
$$\text{for almost every } x \in \Omega \text{ and every } 0 \le s \le M.$$

Using the variational characterization of λ_1 in Proposition 1.57, we obtain

$$\begin{aligned}\varphi_\lambda^+(u) &\ge \frac{\|u\|^p}{p} - \lambda \int_{\{u \le M\}} \big(j(x,0) + a_1(M + M^q)\big)\, dx - \lambda\varepsilon \|u\|_p^p \\ &\ge \left(\frac{1}{p} - \frac{\varepsilon\lambda}{\lambda_1}\right) \|u\|^p - M',\end{aligned}$$

with a constant $M' > 0$. Since $p > 1$, the choice of ε allows us to conclude that φ_λ^+ is coercive.

Now the Weierstrass theorem (see, e.g., [152, Theorem 1.1]) yields $\hat{u} \in W_0^{1,p}(\Omega)$ fulfilling

$$\varphi_\lambda^+(\hat{u}) = \inf_{u \in W_0^{1,p}(\Omega)} \varphi_\lambda^+(u). \tag{5.59}$$

As in the proof of Proposition 5.15, we see that $\hat{u} \in C_0^1(\overline{\Omega})$ and there exists $\hat{w} \in L^{q'}(\Omega)$ with the properties

$$\hat{w}(x) \in \partial j_+(x, \hat{u}(x)) \quad \text{for a.e. } x \in \Omega$$

and

$$-\Delta_p \hat{u} = \lambda \hat{w} \text{ in } W^{-1,p'}(\Omega). \tag{5.60}$$

Acting on (5.60) with the test function $\hat{u}^-$ leads to

$$-\int_\Omega |\nabla \hat{u}^-|^p\, dx = \lambda \int_\Omega \hat{w}\, \hat{u}^-\, dx = 0$$

because $\hat{w}(x) = 0$ whenever $\hat{u}(x) < 0$, thus $\hat{u} \in C_0^1(\overline{\Omega})_+$.

Next we prove that $\hat{u} \neq 0$. Let us observe that due to our hypothesis we can choose $\theta < a_2$ with $\lambda > \lambda_1/\theta$. By assumption H($j$) (iii) there exists $\delta > 0$ such that

$$\min \partial j_+(x,s) > \theta s^{p-1} \text{ for a.e. } x \in \Omega \text{ and every } 0 < s < \delta.$$

Proposition 1.40 implies

$$j_+(x,s) \ge j(x,0) + \frac{\theta s^p}{p}.$$

Since the function $\hat{u}_1$ given by Proposition 1.57 is bounded, we can find $\tau > 0$ such that $\tau\hat{u}_1(x) < \delta$ for all $x \in \Omega$, which enables us to write

$$\varphi_\lambda^+(\tau\hat{u}_1) \leq \frac{\tau^p\|\hat{u}_1\|^p}{p} - \lambda \int_\Omega \left(j(x,0) + \frac{\theta\tau^p\hat{u}_1^p}{p} \right) dx$$
$$\leq \frac{\tau^p}{p}(\lambda_1 - \lambda\theta) + \varphi_\lambda^+(0) < \varphi_\lambda^+(0).$$

On account of (5.59) this renders $\hat{u} \neq 0$. Hence, we are in a position to invoke Proposition 5.16, from which we infer that $\hat{u} \in \mathrm{int}\,(C_0^1(\overline{\Omega})_+)$. In view of H$(j)$ (i), we derive that $\partial j_+(x,s) \subset \partial j(x,s)$ for a.e. $x \in \Omega$ and all $s > 0$, so $\hat{u}$ solves (5.56).

Claim 2. Problem (5.56) possesses the smallest positive solution $\hat{u}_\lambda \in \mathrm{int}\,(C_0^1(\overline{\Omega})_+)$.

Let $\varepsilon_0 > 0$ be such that $\varepsilon\hat{u}_1 < \hat{u}$ in Ω for all $\varepsilon \in (0, \varepsilon_0)$, which exists because $\hat{u} \in \mathrm{int}\,(C_0^1(\overline{\Omega})_+)$. Thanks to H$(j)$ (iii) and the inequality $\lambda > \lambda_1/a_2$, one has for $\varepsilon > 0$ sufficiently small that

$$-\Delta_p(\varepsilon\hat{u}_1) = \lambda_1\varepsilon^{p-1}\hat{u}_1^{p-1} \leq \lambda \min \partial j_+(x, \varepsilon\hat{u}_1) \text{ a.e. in } \Omega.$$

Therefore, according to Definition 5.17, $\varepsilon\hat{u}_1$ turns out to be a subsolution of problem (5.56). Since $\hat{u}$ is a solution of (5.56), it can be regarded as a supersolution. Proposition 5.18 guarantees that the ordered set

$$H_\varepsilon := \Big\{ u \in W_0^{1,p}(\Omega) \, : \varepsilon\hat{u}_1(x) \leq u(x) \leq \hat{u}(x) \text{ for a.e. } x \in \Omega, \text{ and } u \text{ solves (5.56)} \Big\}$$

is nonempty and contains a solution of (5.56) which is the smallest one with respect to the pointwise order. If, for n large enough, we denote by u_n, with n large enough, the smallest solution of (5.56) lying in $H_{1/n}$ then

$$-\Delta_p u_n = \lambda w_n \text{ in } W^{-1,p'}(\Omega) \tag{5.61}$$

with some $w_n \in N(u_n)$ (see (5.57)). By the minimality property of u_n, we get $u_n \geq u_{n+1}$. So, there exists $\hat{u}_\lambda : \Omega \to (0, +\infty)$ such that

$$\lim_{n\to\infty} u_n(x) = \hat{u}_\lambda(x) \text{ pointwise in } \Omega. \tag{5.62}$$

The sequence $\{u_n\}$ is bounded in $W_0^{1,p}(\Omega)$ because (5.61), H(j) (ii), and Proposition 1.40 entail

$$\|u_n\|^p = \langle -\Delta_p u_n, u_n \rangle = \lambda \int_\Omega w_n u_n \, dx \leq \lambda a_1 \int_\Omega \left(u_n + u_n^q \right) dx$$
$$\leq \lambda a_1 \int_\Omega \left(\hat{u} + \hat{u}^q \right) dx.$$

From (5.62), passing to a relabeled subsequence if necessary, it follows that $\hat{u}_\lambda \in W_0^{1,p}(\Omega)$, $u_n \rightharpoonup \hat{u}_\lambda$ in $W_0^{1,p}(\Omega)$, as well as $u_n \to \hat{u}_\lambda$ in $L^q(\Omega)$. Again through (5.61), H(j) (ii), and Proposition 1.40 we achieve

$$\begin{aligned}\langle -\Delta_p u_n, u_n - \hat{u}_\lambda \rangle &= \lambda \int_\Omega w_n (u_n - \hat{u}_\lambda)\, dx \\ &\leq \lambda a_1 \int_\Omega (1 + u_n^{q-1})(u_n - \hat{u}_\lambda)\, dx \leq M \|u_n - \hat{u}_\lambda\|_q,\end{aligned}$$

with some $M > 0$. This implies

$$\limsup_{n\to\infty} \langle -\Delta_p u_n, u_n - \hat{u}_\lambda \rangle \leq 0.$$

Since $-\Delta_p$ has the $(S)_+$-property (see Proposition 1.56), $u_n \to \hat{u}_\lambda$ in $W_0^{1,p}(\Omega)$.

By assumption H(j) (ii), the sequence $\{w_n\}$ turns out to be bounded in $W^{-1,p'}(\Omega)$. Thus, possibly along a relabeled subsequence, $w_n \rightharpoonup \hat{w}_\lambda$ in $W^{-1,p'}(\Omega)$. At this point, Proposition 1.39, in conjunction with (5.61), ensures that $\hat{w}_\lambda \in N(\hat{u}_\lambda)$ and

$$-\Delta_p \hat{u}_\lambda = \lambda \hat{w}_\lambda \text{ in } W^{-1,p'}(\Omega), \tag{5.63}$$

that is, $\hat{u}_\lambda$ solves problem (5.56). The reasoning developed in the proof of Proposition 5.15 demonstrates that $\hat{u}_\lambda \in C_0^1(\overline{\Omega})_+$.

Next we verify that $\hat{u}_\lambda \neq 0$. Arguing by contradiction, suppose $\hat{u}_\lambda = 0$. Notice that $v_n := u_n/\|u_n\|$ is a weak solution of the Dirichlet problem

$$\begin{cases} -\Delta_p v = \lambda \dfrac{w_n(x)}{u_n(x)^{p-1}} v^{p-1} & \text{in } \Omega, \\ v = 0 & \text{on } \partial\Omega. \end{cases} \tag{5.64}$$

The weight function in (5.64) belongs to $L^\infty(\Omega)$. Indeed, choosing $\theta \in (\lambda_1/\lambda, a_2)$, then (5.62) with $\hat{u}_\lambda = 0$ and H(j) (iii) show for any sufficiently large n that

$$\frac{w_n(x)}{u_n(x)^{p-1}} \geq \frac{\min \partial j(x, u_n(x))}{u_n(x)^{p-1}} > \theta \tag{5.65}$$

and

$$\frac{w_n(x)}{u_n(x)^{p-1}} \leq \frac{\max \partial j(x, u_n(x))}{u_n(x)^{p-1}} < 2a_3$$

a.e. in Ω.

Through (5.65) and the last part of Proposition 1.57 we obtain

$$\lambda_1\left(\frac{w_n}{u_n^{p-1}}\right) < \lambda_1(\theta) = \frac{\lambda_1}{\theta} < \lambda,$$

which expresses that $\lambda > 0$ is an eigenvalue of the weighted problem (5.64) greater than the first one. Hence, by Proposition 1.57, the corresponding eigenfunction v_n should be nodal. However, this comes in contradiction with the definition of v_n, whence $\hat{u}_\lambda \neq 0$.

On the basis of Proposition 5.16, it turns out that $\hat{u}_\lambda \in \mathrm{int}\,(C_0^1(\overline{\Omega})_+)$.

In order to prove that $\hat{u}_\lambda$ is the smallest among the positive solutions of (5.56), let $u \in W_0^{1,p}(\Omega)$ be a positive solution of (5.56). Applying again Proposition 5.16, we know that $u \in \mathrm{int}\,(C_0^1(\overline{\Omega})_+)$, whence

$$\frac{\hat{u}_1(x)}{n} \leq u(x) \quad \text{for all } x \in \Omega$$

provided n is large enough. The function $\frac{1}{n}\hat{u}_1$ is a subsolution and $\min\{u(x), \hat{u}(x)\}$ is a supersolution of (5.56) (see Definition 5.17). Consequently, by Proposition 5.18 there exists a solution $v \in W_0^{1,p}(\Omega)$ of (5.56) such that

$$\frac{\hat{u}_1(x)}{n} \leq v(x) \leq \min\{u(x), \hat{u}(x)\} \text{ for a.e. } x \in \Omega.$$

Then the minimality of u_n in $H_{1/n}$ entails

$$u_n(x) \leq v(x) \leq u(x) \text{ for a.e. } x \in \Omega.$$

Letting $n \to \infty$ provides $\hat{u}_\lambda \leq u$, which completes the proof. □

The proof of the below result parallels the proof of Theorem 5.19.

Theorem 5.20. *If $\lambda > \lambda_1/a_2$ then problem (5.56) possesses the biggest negative solution $\check{u}_\lambda \in -\mathrm{int}\,(C_0^1(\overline{\Omega})_+)$.*

Fix $\lambda > \lambda_1/a_2$ and consider the extremal solutions $\hat{u}_\lambda$ and $\check{u}_\lambda$ obtained in Theorems 5.19 and 5.20, respectively. Now we deal with the existence of solutions to problem (5.56) that lie in the interval $[\check{u}_\lambda, \hat{u}_\lambda]$, which will eventually lead to a nodal solution. To this end, we introduce a truncation-perturbation of the nonsmooth potential $j(x,s)$ as follows:

$$\tilde{j}(x,s) := \begin{cases} j(x,\check{u}_\lambda(x)) + \max \partial j(x,\check{u}_\lambda(x))(s - \check{u}_\lambda(x)) & \text{if } s < \check{u}_\lambda(x), \\ j(x,s) & \text{if } \check{u}_\lambda(x) \leq s \leq \hat{u}_\lambda(x), \\ j(x,\hat{u}_\lambda(x)) + \min \partial j(x,\hat{u}_\lambda(x))(s - \hat{u}_\lambda(x)) & \text{if } s > \hat{u}_\lambda(x). \end{cases}$$

It is straightforward to show that the function $\tilde{j} : \Omega \times \mathbb{R} \to \mathbb{R}$ satisfies the Carathéodory condition with $\tilde{j}(x,\cdot)$ locally Lipschitz, and hypotheses H(j) hold true for $\tilde{j}$ in place of j. Hence Proposition 5.15 guarantees that the functional $\tilde{\varphi}_\lambda : W_0^{1,p}(\Omega) \to \mathbb{R}$ defined by

$$\tilde{\varphi}_\lambda(u) = \frac{\|u\|^p}{p} - \lambda \int_\Omega \tilde{j}(x,u)\,dx \quad \text{for all } u \in W_0^{1,p}(\Omega)$$

is locally Lipschitz. Moreover, the below result is valid.

Proposition 5.21. *If u is a critical point of $\tilde{\varphi}_\lambda$, then $u \in C_0^1(\overline{\Omega})$, u solves (5.56) and has the location property $\check{u}_\lambda \le u \le \hat{u}_\lambda$ in Ω.*

Proof. According to Definition 1.41, we have $0 \in \partial\tilde{\varphi}_\lambda(u)$. As in the proof of Proposition 5.15, we get that $u \in C_0^1(\overline{\Omega})$ and there exists $w \in \tilde{N}(u)$, where

$$\tilde{N}(u) := \left\{ w \in L^{q'}(\Omega) \ : \ w(x) \in \partial\tilde{j}(x, u(x)) \text{ for a.e. } x \in \Omega \right\},$$

such that

$$-\Delta_p u = \lambda w \text{ in } W^{-1,p'}(\Omega). \tag{5.66}$$

Using (5.66), (5.63) with the test function $(u - \hat{u}_\lambda)^+ \in W_0^{1,p}(\Omega)$ and the definition of the potential $\tilde{j}$ yield

$$\begin{aligned}\langle -\Delta_p u + \Delta_p \hat{u}_\lambda, (u - \hat{u}_\lambda)^+ \rangle &= \lambda \int_\Omega (w - \hat{w}_\lambda)(u - \hat{u}_\lambda)^+ \, dx \\ &= \lambda \int_\Omega \left(\min \partial j(x, \hat{u}_\lambda) - \hat{w}_\lambda \right)(u - \hat{u}_\lambda)^+ \, dx \le 0.\end{aligned}$$

This is true because

$$\partial\tilde{j}(x, s) = \left\{ \min \partial j(x, \hat{u}_\lambda(x)) \right\} \text{ for a.e. } x \in \Omega \text{ and all } s > \hat{u}_\lambda(x),$$

whereas $\hat{w}_\lambda \in N(\hat{u}_\lambda)$. The strict monotonicity of $-\Delta_p$ implies $u \le \hat{u}_\lambda$. A similar argument provides $u \ge \check{u}_\lambda$, thus $\check{u} \le u \le \hat{u}$ in Ω. Taking into account that

$$\partial\tilde{j}(x, \hat{u}_\lambda(x)) \subset \partial j(x, \hat{u}_\lambda(x)) \quad \text{and} \quad \partial\tilde{j}(x, \check{u}_\lambda(x)) \subset \partial j(x, \check{u}_\lambda(x)),$$

we arrive at $w \in N(u)$. Therefore, by (5.66), u solves (5.56). □

The next result deals with the existence of a sign-changing solution, which in addition satisfies an important location property.

Theorem 5.22. *If $\lambda > \lambda_2/a_2 + 1$, then problem (5.56) possesses a nodal solution $\tilde{u}_\lambda \in C_0^1(\overline{\Omega})$ such that $\check{u}_\lambda(x) \le \tilde{u}_\lambda(x) \le \hat{u}_\lambda(x)$ for all $x \in \Omega$.*

Proof. Since $\lambda_2/a_2 + 1 > \lambda_1/a_2$, Theorems 5.19 and 5.20 provide the smallest positive solution $\hat{u}_\lambda \in \operatorname{int}(C_0^1(\overline{\Omega})_+)$ and the biggest negative solution $\check{u}_\lambda \in -\operatorname{int}(C_0^1(\overline{\Omega})_+)$ of (5.56). Set $\tilde{j}_+(x, s) := \tilde{j}(x, s^+)$. The functional $\tilde{\varphi}_\lambda^+ :$ $W_0^{1,p}(\Omega) \to \mathbb{R}$ given by

$$\tilde{\varphi}_\lambda^+(u) := \frac{\|u\|^p}{p} - \lambda \int_\Omega \tilde{j}_+(x, u) \, dx \quad \text{for all } u \in W_0^{1,p}(\Omega)$$

turns out to be locally Lipschitz, weakly sequentially lower semicontinuous, and coercive. So, as in the proof of Theorem 5.19, there exists $\tilde{u} \in W_0^{1,p}(\Omega)$ fulfilling

$$\tilde{\varphi}_\lambda^+(\tilde{u}) = \inf_{u \in W_0^{1,p}(\Omega)} \tilde{\varphi}_\lambda^+(u) < \tilde{\varphi}_\lambda^+(0). \tag{5.67}$$

We claim that $\tilde{u} = \hat{u}_\lambda$. By Proposition 5.16 we know that $\tilde{u} \in \operatorname{int}(C_0^1(\overline{\Omega})_+)$. Let us note that the restrictions to $C_0^1(\overline{\Omega})_+$ of the functionals $\tilde{\varphi}_\lambda$ and $\tilde{\varphi}_\lambda^+$ coincide. Hence, $\tilde{u}$ is a $C_0^1(\overline{\Omega})$-local minimizer for the functional $\tilde{\varphi}_\lambda$. Thanks to [175], $\tilde{u}$ is also a $W_0^{1,p}(\Omega)$-local minimizer for $\tilde{\varphi}_\lambda$, whence $0 \in \partial\tilde{\varphi}_\lambda(\tilde{u})$. Through Proposition 5.16 we get

$$0 < \tilde{u}(x) \leq \hat{u}_\lambda(x) \quad \text{for all } x \in \Omega.$$

The minimality of $\hat{u}_\lambda$ forces $\tilde{u} = \hat{u}_\lambda$, and the claim is verified.

Accordingly, (5.67) reads as

$$\tilde{\varphi}_\lambda(\hat{u}_\lambda) = \inf_{u \in W_0^{1,p}(\Omega)} \tilde{\varphi}_\lambda^+(u) < \tilde{\varphi}_\lambda(0). \tag{5.68}$$

Observe that $\hat{u}_\lambda$ turns out to be a local minimizer of $\tilde{\varphi}_\lambda$ because $\hat{u}_\lambda \in \operatorname{int}(C_0^1(\overline{\Omega})_+)$. The same holds true for $\check{u}_\lambda \in -\operatorname{int}(C_0^1(\overline{\Omega})_+)$, with

$$\tilde{\varphi}_\lambda(\check{u}_\lambda) < \tilde{\varphi}_\lambda(0). \tag{5.69}$$

We examine two situations that may occur.

Case 1. The functional $\tilde{\varphi}_\lambda$ has not a strict local minimum at $\hat{u}_\lambda$ or $\check{u}_\lambda$. Consequently, there exists another local minimizer $\tilde{u}_\lambda \in W_0^{1,p}(\Omega) \setminus \{\hat{u}_\lambda, \check{u}_\lambda\}$ satisfying

$$\tilde{\varphi}_\lambda(\tilde{u}_\lambda) = \tilde{\varphi}_\lambda(\hat{u}_\lambda) \quad \text{or} \quad \tilde{\varphi}_\lambda(\tilde{u}_\lambda) = \tilde{\varphi}_\lambda(\check{u}_\lambda).$$

Proposition 5.16 ensures that $\tilde{u}_\lambda \in C_0^1(\overline{\Omega})$, $\tilde{u}_\lambda$ solves (5.56), and

$$\check{u}_\lambda(x) \leq \tilde{u}_\lambda(x) \leq \hat{u}_\lambda(x) \quad \text{for all } x \in \Omega. \tag{5.70}$$

From (5.68) or (5.69) it clearly follows $\tilde{u}_\lambda \neq 0$.

Let us show that the solution $\tilde{u}_\lambda$ is nodal. On the contrary, without loss of generality suppose that $\tilde{u}_\lambda \in C_0^1(\overline{\Omega})_+$, then thanks to Proposition 5.16, $\tilde{u}_\lambda \in \operatorname{int}(C_0^1(\overline{\Omega})_+)$. In view of (5.70) and the minimality of $\hat{u}_\lambda$, this is impossible.

Case 2. The functional $\tilde{\varphi}_\lambda$ has a strict local minimum at both $\hat{u}_\lambda$ and $\check{u}_\lambda$. In this situation we can find $r \in (0, \|\hat{u}_\lambda - \check{u}_\lambda\|)$ such that

$$\max\{\tilde{\varphi}_\lambda(\hat{u}_\lambda), \tilde{\varphi}_\lambda(\check{u}_\lambda)\} < \sigma_r := \inf_{u \in \partial B_r(\hat{u}_\lambda)} \tilde{\varphi}_\lambda(u). \tag{5.71}$$

As in the proof of Theorem 5.19 we can show that $\tilde{\varphi}_\lambda$ is coercive. In order to prove that the nonsmooth condition (PS) in Definition 1.42 is fulfilled, let a sequence $\{u_n\}$ in $W_0^{1,p}(\Omega)$ be such that $\{\tilde{\varphi}_\lambda(u_n)\}$ is bounded and

$$(\tilde{\varphi}_\lambda)^0(u_n; v-u_n)+\varepsilon_n\|v-u_n\|\geq 0 \quad \text{for all } v\in W_0^{1,p}(\Omega), \text{ all } n\in\mathbb{N}, \tag{5.72}$$

with $\varepsilon_n \to 0^+$. By the coercivity of $\tilde{\varphi}_\lambda$, the sequence $\{u_n\}$ must be bounded. So, passing to a subsequence if necessary, we may assume that $u_n \rightharpoonup u$ in $W_0^{1,p}(\Omega)$ and $u_n \to u$ in $L^q(\Omega)$. Through (5.72) and hypothesis H(j) (ii) one has

$$\begin{aligned}0 &\leq \langle -\Delta_p u_n, u-u_n\rangle + \lambda\int_\Omega a_1(1+|u_n|^{q-1})|u-u_n|\,dx+\varepsilon_n\|u-u_n\|\\ &\leq \langle -\Delta_p u_n, u-u_n\rangle+\lambda M\|u-u_n\|_q+\varepsilon_n\|u-u_n\|,\end{aligned}$$

with a constant $M>0$. This amounts to saying that

$$\limsup_{n\to\infty}\langle -\Delta_p u_n, u_n-u\rangle\leq 0.$$

Since, by Proposition 1.56, $-\Delta_p$ has the $(S)_+$-property, $u_n \to u$ in $W_0^{1,p}(\Omega)$, proving the nonsmooth condition (PS) in Definition 1.42, as desired.

We are in a position to apply Theorem 1.44. Specifically, setting

$$\Gamma := \left\{\gamma\in C([-1,1], W_0^{1,p}(\Omega)) \;:\; \gamma(-1)=\check{u}_\lambda,\ \gamma(1)=\hat{u}_\lambda\right\} \text{ and}$$
$$\tilde{c} := \inf_{\gamma\in\Gamma}\max_{t\in[-1,1]}\tilde{\varphi}_\lambda(\gamma(t)),$$

there exists a critical point $\tilde{u}_\lambda\in W_0^{1,p}(\Omega)$ of $\tilde{\varphi}_\lambda$ such that

$$\tilde{\varphi}_\lambda(\tilde{u}_\lambda)=\tilde{c}\geq\sigma_r.$$

By (5.71) this makes clear that $\tilde{u}_\lambda\neq\hat{u}_\lambda$ and $\tilde{u}_\lambda\neq\check{u}_\lambda$. Proposition 5.21 ensures that $\tilde{u}_\lambda\in C_0^1(\overline{\Omega})$, $\tilde{u}_\lambda$ solves (5.56), and

$$\check{u}_\lambda(x)\leq\tilde{u}_\lambda(x)\leq\hat{u}_\lambda(x) \quad \text{for all } x\in\Omega.$$

The main point for proving that the solution $\tilde{u}_\lambda$ is nodal consists in the following.

Claim. $\tilde{u}_\lambda\neq 0$.

Choose $\theta\in(0,a_2)$ such that $\lambda>\lambda_2/\theta+1$. Proposition 1.58 provides a path $\gamma_0\in\Gamma_0$ satisfying

$$\|\gamma_0(t)\|^p<\lambda_2+\theta \quad \text{for all } t\in[-1,1]. \tag{5.73}$$

Without loss of generality, we may suppose $\gamma_0 \in C([-1,1], C_0^1(\overline{\Omega}))$ because $S \cap C_0^1(\overline{\Omega})$ is dense in S (endowed with the induced $W_0^{1,p}(\Omega)$-topology) while $\gamma_0([-1,1])$ is compact. Through H(j) (i), (iv), (v), and Proposition 1.40, we find $\delta > 0$ such that

$$j(x,s) > j(x,0) + \frac{\theta |s|^p}{p} \quad \text{for a.e. } x \in \Omega \text{ and every } |s| \leq \delta. \tag{5.74}$$

Finally, since $\gamma_0([-1,1])$ is compact in $C_0^1(\overline{\Omega})$ while $\hat{u}_\lambda, -\check{u}_\lambda \in \text{int}(C_0^1(\overline{\Omega})_+)$, there exists $\varepsilon > 0$ such that

$$\varepsilon|u(x)| < \min\{\delta,\ \hat{u}_\lambda(x),\ -\check{u}_\lambda(x)\} \quad \text{for all } x \in \Omega,\ \text{all } u \in \gamma_0([-1,1]). \tag{5.75}$$

Gathering (5.73)–(5.75) and recalling that $\|u\|_p = 1$ results in

$$\begin{aligned}
\tilde{\varphi}_\lambda(\varepsilon u) &= \frac{\varepsilon^p}{p}\|u\|^p - \lambda \int_\Omega \tilde{j}(x, \varepsilon u)\, dx \\
&\leq \frac{\varepsilon^p}{p}(\lambda_2 + \theta) - \lambda \int_\Omega \left(j(x,0) + \frac{\theta \varepsilon^p |u|^p}{p} \right) dx \\
&\leq \frac{\varepsilon^p}{p}(\lambda_2 + \theta - \lambda\theta) + \tilde{\varphi}_\lambda(0) \quad \text{for all } u \in \gamma_0([-1,1]).
\end{aligned}$$

By the choice of λ this entails

$$\tilde{\varphi}_\lambda(\varepsilon\gamma_0(t)) < \tilde{\varphi}_\lambda(0) \quad \text{for all } t \in [-1,1]. \tag{5.76}$$

Let us now apply the deformation assertion in Theorem 1.45 to the functional $\tilde{\varphi}_\lambda^+$ posing $a := \tilde{\varphi}_\lambda^+(\hat{u}_\lambda)$ and $b := \tilde{\varphi}_\lambda^+(\varepsilon\hat{u}_1)$. As in the proof of Theorem 5.19 we show

$$a < b < \tilde{\varphi}_\lambda^+(0).$$

One has $K_c(\tilde{\varphi}_\lambda^+) \subset \{\hat{u}_\lambda\}$ for any $c < \tilde{\varphi}_\lambda^+(0)$, where $K_c(\tilde{\varphi}_\lambda^+)$ stands for the set of critical points of $\tilde{\varphi}_\lambda^+$ with critical value c. Indeed, if $u \in K_c(\tilde{\varphi}_\lambda^+) \setminus \{\hat{u}_\lambda\}$, then thanks to Propositions 5.16 and 5.21 we have that $u \in \text{int}(C_0^1(\overline{\Omega})_+)$, u is a positive solution of (5.56), $u \leq \hat{u}_\lambda$, and $u \neq \hat{u}_\lambda$, against the minimality of $\hat{u}_\lambda$. Consequently, $K_a(\tilde{\varphi}_\lambda^+) = \{\hat{u}_\lambda\}$, while $K_c(\tilde{\varphi}_\lambda^+) = \emptyset$ for each $c \in (a,b]$.

Due to Theorem 1.45, there exists a continuous mapping $h : [0,1] \times Z \to Z$, where

$$Z := \left\{ u \in W_0^{1,p}(\Omega) \ :\ \tilde{\varphi}_\lambda^+(u) \leq b \right\},$$

with the properties $h(0,u) = u$, $h(1,u) = \hat{u}_\lambda$ for all $u \in Z$, and

$$\tilde{\varphi}_\lambda^+(h(t,u)) \leq \tilde{\varphi}_\lambda^+(u) \quad \text{for all } (t,u) \in [0,1] \times Z.$$

Then the path $\gamma_+ : [0, 1] \to W_0^{1,p}(\Omega)$ defined by

$$\gamma_+(t) := (h(t, \varepsilon \hat{u}_1))^+ \quad \text{for all } t \in [0, 1],$$

satisfies: $\gamma_+(0) = \varepsilon \hat{u}_1$, $\gamma_+(1) = \hat{u}_\lambda$, $\tilde{\varphi}_\lambda = \tilde{\varphi}_\lambda^+$ on the set $\gamma_+([0, 1])$, and

$$\tilde{\varphi}_\lambda(\gamma_+(t)) < \tilde{\varphi}_\lambda(0) \quad \text{for all } t \in [0, 1]. \tag{5.77}$$

Likewise, we construct a continuous path $\gamma_- : [0, 1] \to W_0^{1,p}(\Omega)$ such that $\gamma_-(0) = \check{u}_\lambda$, $\gamma_-(1) = -\varepsilon \hat{u}_1$, and

$$\tilde{\varphi}_\lambda(\gamma_-(t)) < \tilde{\varphi}_\lambda(0) \quad \text{for all } t \in [0, 1]. \tag{5.78}$$

Concatenating γ_-, $\varepsilon\gamma_0$, and γ_+ determines a path $\gamma \in \Gamma$. In view of (5.76)–(5.78), γ fulfills

$$\tilde{\varphi}_\lambda(\gamma(t)) < \tilde{\varphi}_\lambda(0) \quad \text{for all } t \in [-1, 1].$$

It turns out that $\tilde{c} < \tilde{\varphi}_\lambda(0)$ and, thus the claim $\tilde{u}_\lambda \neq 0$ ensues.

Finally, we prove that $\tilde{u}_\lambda$ is nodal. If not, $\tilde{u}_\lambda$ would be of constant sign. In the case where $\tilde{u}_\lambda \in C_0^1(\overline{\Omega})_+$, by Proposition 5.16 we deduce that

$$\tilde{u}_\lambda \in \operatorname{int}(C_0^1(\overline{\Omega})_+), \quad 0 < \tilde{u}_\lambda \leq \hat{u}_\lambda.$$

Then the minimality of $\hat{u}_\lambda$ ensures $\tilde{u}_\lambda = \hat{u}_\lambda$, which is impossible according to (5.71). A similar reasoning prevents to have $\tilde{u}_\lambda \notin -C_0^1(\overline{\Omega})_+$. Consequently, $\tilde{u}_\lambda$ must be nodal, which completes the proof. □

5.4 NOTES

Section 5.1 treats multiple solutions with sign information for a quasilinear elliptic problem governed by the p-Laplacian operator with $1 < p < +\infty$. There are many results studying the existence of multiple solutions for boundary value problems involving nonlinear elliptic equations with or without parameters. A large category of them concerns problems driven by the p-Laplacian or (p, q)-Laplacian operators. In these results, one imposes on the involved nonlinearities convenient assumptions describing the behavior at infinity and at zero. A prototype is represented by the classical work of Ambrosetti, Brézis, and Cerami [9], which investigates the semilinear case (i.e., the case $p = 2$) with a positive parameter $\lambda > 0$ and the concave–convex nonlinearities

$$s \mapsto \lambda |s|^{q-2}s + |s|^{r-2}s \ \text{ with } 1 < q < 2 < r < 2^*.$$

The result has been extended by Ambrosetti, García Azorero, and Peral [10] to the case of the p-Laplacian with $1 < p < +\infty$.

Among the papers devoted to the study of multiple solutions, a distinct class is formed by those which seek solutions with prescribed qualitative properties. For instance, a challenging goal is to obtain solutions of required sign. In this respect, we mention that, for quasilinear elliptic equations with Dirichlet boundary condition, the existence of multiple solutions with precise sign information is studied in [40,42,49,156]. Section 5.1 addresses exactly this topic. It is based on [40] showing the existence of multiple solutions to problem (5.1) for all values $\lambda > \lambda_2$, where λ_2 denotes the second eigenvalue of $-\Delta_p$ on $W_0^{1,p}(\Omega)$, and guaranteeing that for any such λ there exist at least three nontrivial solutions so that two are of opposite constant sign whereas the third is nodal (i.e., sign changing). This is the content of Theorem 5.5. In particular, it extends the main result in [11] obtained for $p = 2$ and, a fortiori, the corresponding results from [12] and [220], by dropping certain requirements on the nonlinearity g in problem (5.1). The most significant feature of Theorem 5.5 is that it sets forth the significant new information of the existence of a sign-changing solution of problem (5.1). The topic of searching a sign-changing solution between extremal constant-sign solutions originates in the work of Carl–Perera [49], where the study is related to the Fučík spectrum of the negative p-Laplacian on the space $W_0^{1,p}(\Omega)$ by exploiting subsolution–supersolution techniques to get both extremal constant-sign and nodal solutions for a problem without parameters.

It is also worth to point out that the approach here is completely different from the one in [11] where the Morse theory and Lyapunov–Schmidt reduction are used. Specifically, in the case of Theorem 5.5 the approach relies on the existence of extremal constant-sign solutions whose existence is presented in Theorem 5.4. Other main tools are the method of subsolutions–supersolutions, truncation, strong maximum principle and variational arguments. For the use of these techniques we also refer to results ensuring the existence of three nontrivial solutions in the works of Carl–Motreanu [43], Jin [114], and Motreanu–Motreanu–Papageorgiou [166] provided the relevant positive parameter λ is sufficiently small. We also refer to the monograph [200] for quasilinear elliptic problems involving p-Laplacian type operators. The investigation of multiple solutions for nonlinear elliptic problems with a more general differential operator, possibly nonhomogeneous, in place of the p-Laplacian has been recently considered in Miyajima–Motreanu–Tanaka [156], Motreanu–Tanaka [180–182]. We also mention that the nonlinear elliptic problems with nonhomogeneous Dirichlet boundary condition were treated through a nonsmooth variational approach in Goldshtein–Motreanu–Motreanu [106].

In the case of nonlinear Neumann problems, there are many results ensuring the existence of multiple solutions (see [2,23,28,156,168,209]), but few results involving the study of the sign of multiple solutions. We refer to [178,179,232, 240] for the special case of jumping nonlinearities.

The existence of multiple solutions for problem (5.35) with a complete study of their sign was done in [22], on which is based the presentation in Section 5.2. The main result is stated as Theorem 5.12, which provides two opposite constant-sign solutions and besides a sign-changing solution between them. Our approach relies on variational methods combined with subsolution–supersolution and truncation techniques. We emphasize that we do not impose any growth condition for the nonlinearity $f(x,u)$ in problem (5.35). The reason is that we are dealing with truncated functionals for which the subcritical growth condition is automatically valid. There are essential differences between the results and tools used for the Dirichlet and Neumann boundary value problems as can be seen from Theorems 5.5 and 5.12. The technical difficulties are overcome thanks to specific contributions on Neumann problems involving the p-Laplacian (see [2,23]).

Section 5.3 studies a differential inclusion problem of elliptic type driven by the p-Laplacian operator with a set-valued reaction term in the form of the generalized gradient of a locally Lipschitz function (see Section 1.2). Such problems cover discontinuous single-valued nonlinearities (see Chang [51]). The presence of the set-valued reaction term $\partial j(x,s)$ requires completely different ideas in order to handle truncations and verify the appropriate Palais–Smale condition. As required in hypotheses H(j) (iii), (iv), the locally Lipschitz potential $j(x,\cdot)$ is p-sublinear at infinity and p-superlinear at zero. It is worthwhile pointing out that in this multivalued setting, the variational methods and subsolutions–supersolutions can still be employed to achieve multiplicity results. The comparison principle for differential inclusions in Proposition 5.18 is taken from Carl–Motreanu [41, Corollary 4.1] (see also Carl–Le–Motreanu [38, Corollary 4.24]).

The main result is formulated as Theorem 5.13, which extends to a nonsmooth inclusion setting the results obtained in the C^1 case in [144]. More precisely, for problem (5.56) depending on a parameter $\lambda > 0$ it is proven the existence of a smallest positive, a biggest negative, and a nodal (i.e., sign-changing) solution provided $\lambda > 0$ is large enough. By minimizing a truncated energy functional we first find a positive solution of (5.56). Subsolution–supersolution arguments then provide a smallest positive solution $\hat{u}_\lambda \in C_0^1(\overline{\Omega})$. Analogously, a biggest negative solution $\check{u}_\lambda$ is obtained. A third solution $\tilde{u}_\lambda \in C_0^1(\overline{\Omega})$ is next obtained through the nonsmooth mountain pass-like theorem while the nonsmooth second deformation lemma ensures that it is not zero. Finally, due to the extremality of the constant-sign solutions $\hat{u}_\lambda$ and $\check{u}_\lambda$, the solution $\tilde{u}_\lambda$ must be nodal. Results of the same nature in a nonsmooth setting are given in Averna–Marano–Motreanu [18], Carl–Motreanu [41], Iannizzotto–Papageorgiou [113].

An open problem is to find a second sign-changing solution for the nonsmooth problem (5.56). In the smooth case, a result of this type can be found in [165].

Chapter 6

Nonlinear Elliptic Systems

6.1 ELLIPTIC SYSTEMS FULLY DEPENDING ON THE GRADIENT OF THE SOLUTION

In this section we deal with the following homogeneous Dirichlet boundary value problem formed by a system of nonlinear elliptic equations:

$$\begin{cases} -\Delta_{p_1} u_1 - \mu_1 \Delta_{q_1} u_1 = f_1(x, u_1, u_2, \nabla u_1, \nabla u_2) & \text{in } \Omega, \\ -\Delta_{p_2} u_2 - \mu_2 \Delta_{q_2} u_2 = f_2(x, u_1, u_2, \nabla u_1, \nabla u_2) & \text{in } \Omega, \\ u_1 = u_2 = 0 & \text{on } \partial\Omega. \end{cases} \tag{6.1}$$

Here $\Omega \subset \mathbb{R}^N$ is a bounded domain with the boundary $\partial\Omega$, $\mu_1, \mu_2 \geq 0$ are parameters, Δ_{p_i} and Δ_{q_i} are the p_i-Laplacian and q_i-Laplacian, respectively, with $1 < q_i < p_i < +\infty$, and $f_i : \Omega \times \mathbb{R} \times \mathbb{R} \times \mathbb{R}^N \times \mathbb{R}^N \to \mathbb{R}$ is a Carathéodory function, which means that $x \mapsto f(x, s_1, s_2, \xi_1, \xi_2)$ is measurable for all $(s_1, s_2, \xi_1, \xi_2) \in \mathbb{R} \times \mathbb{R} \times \mathbb{R}^N \times \mathbb{R}^N$ and $(s_1, s_2, \xi_1, \xi_2) \mapsto f(x, s_1, s_2, \xi_1, \xi_2)$ is continuous for a.a. $x \in \Omega$, $i = 1, 2$. We recall that the p_i-Laplacian is the operator $\Delta_{p_i} : W_0^{1,p_i}(\Omega) \to W^{-1,p_i'}(\Omega)$ given by

$$\Delta_{p_i} u = \operatorname{div}(|\nabla u|^{p_i - 2} \nabla u).$$

For simplicity we suppose that $N > \max\{p_1, p_2\}$, so the Sobolev critical exponents corresponding to p_1 and p_2 are $p_1^* = \frac{Np_1}{N-p_1}$ and $p_2^* = \frac{Np_2}{N-p_2}$, respectively. In the case where $N \leq \max\{p_1, p_2\}$ the treatment is simpler, and thus it is omitted. Accordingly, for every $r \in [1, +\infty]$ we denote by r' its Hölder conjugate, i.e., r' is determined by $\frac{1}{r} + \frac{1}{r'} = 1$.

In the sequel we denote by λ_{1,p_i} the first eigenvalue of the negative p_i-Laplacian ($i = 1, 2$). According to Proposition 1.57, it has the following variational characterization

$$\lambda_{1,p_i} = \inf_{u \in W_0^{1,p_i}(\Omega) \setminus \{0\}} \frac{\|\nabla u\|_{p_i}^{p_i}}{\|u\|_{p_i}^{p_i}},$$

hence

$$\|u\|_{p_i}^{p_i} \leq \lambda_{1,p_i}^{-1} \|\nabla u\|_{p_i}^{p_i} \quad \text{for all } u \in W_0^{1,p_i}(\Omega). \tag{6.2}$$

Nonlinear Differential Problems with Smooth and Nonsmooth Constraints. DOI: 10.1016/B978-0-12-813386-6.00006-7

We consider the following conditions on the functions f_1 and f_2.

H$(f_1, f_2)_1$ There exist constants $a_i \geq 0$, $b_i \geq 0$, $\alpha_i \in [0, p_i^* - 1)$, $\beta_i \in [0, \frac{p_i}{(p_i^*)'})$ and functions $\sigma_i \in L^{\gamma_i'}(\Omega)$, with $\gamma_i \in [1, p_i^*)$, such that

$$
\begin{aligned}
|f_1(x, s_1, s_2, \xi_1, \xi_2)| &\leq \sigma_1(x) + a_1(|s_1|^{\alpha_1} + |s_2|^{\frac{\alpha_1 p_2^*}{p_1^*}}) \\
&\quad + b_1(|\xi_1|^{\beta_1} + |\xi_2|^{\frac{\beta_1 p_2}{p_1}}), \\
|f_2(x, s_1, s_2, \xi_1, \xi_2)| &\leq \sigma_2(x) + a_2(|s_1|^{\frac{\alpha_2 p_1^*}{p_2^*}} + |s_2|^{\alpha_2}) \\
&\quad + b_2(|\xi_1|^{\frac{\beta_2 p_1}{p_2}} + |\xi_2|^{\beta_2})
\end{aligned}
$$

for a.a. $x \in \Omega$ and all $s_1, s_2 \in \mathbb{R}$.

H$(f_1, f_2)_2$ The following unilateral growth condition holds:

$$
\begin{aligned}
&f_1(x, s_1, s_2, \xi_1, \xi_2)s_1 + f_2(x, s_1, s_2, \xi_1, \xi_2)s_2 \\
&\leq \omega(x) + c(|s_1|^{p_1} + |s_2|^{p_2}) + d(|\xi_1|^{p_1} + |\xi_2|^{p_2})
\end{aligned}
$$

for a.a. $x \in \Omega$ and all $s_1, s_2 \in \mathbb{R}$, $\xi_1, \xi_2 \in \mathbb{R}^N$, with $\omega \in L^1(\Omega)$, and positive constants c and d satisfying $\max\left\{\lambda_{1,p_1}^{-1}, \lambda_{1,p_2}^{-1}\right\} c + d < 1$.

The solutions of system (6.1) are understood in the weak sense. Precisely, a weak solution of problem (6.1) is a pair $(u_1, u_2) \in W_0^{1,p_1}(\Omega) \times W_0^{1,p_2}(\Omega)$ such that

$$
\begin{aligned}
&\int_\Omega |\nabla u_1|^{p_1-2}\nabla u_1 \cdot \nabla v_1\, dx + \mu_1 \int_\Omega |\nabla u_1|^{q_1-2}\nabla u_1 \cdot \nabla v_1\, dx \\
&= \int_\Omega f_1(x, u_1, u_2, \nabla u_1, \nabla u_2)v_1\, dx, \qquad (6.3) \\
&\int_\Omega |\nabla u_2|^{p_2-2}\nabla u_2 \cdot \nabla v_2\, dx + \mu_2 \int_\Omega |\nabla u_2|^{q_2-2}\nabla u_2 \cdot \nabla v_2\, dx \\
&= \int_\Omega f_2(x, u_1, u_2, \nabla u_1, \nabla u_2)v_2\, dx \qquad (6.4)
\end{aligned}
$$

for all $(v_1, v_2) \in W_0^{1,p_1}(\Omega) \times W_0^{1,p_2}(\Omega)$. It is easy to check that, assuming H$(f_1, f_2)_1$, the integrals in (6.3) and (6.4) exist.

We also consider the related Dirichlet problems driven by the p_i-Laplacian operators ($i = 1, 2$):

$$
\begin{cases}
-\Delta_{p_1} u_1 = f_1(x, u_1, u_2, \nabla u_1, \nabla u_2) & \text{in } \Omega, \\
-\Delta_{p_2} u_2 = f_2(x, u_1, u_2, \nabla u_1, \nabla u_2) & \text{in } \Omega, \\
u_1 = u_2 = 0 & \text{on } \partial\Omega,
\end{cases} \qquad (6.5)
$$

$$\begin{cases} -\Delta_{p_1} u = f_1(x, u, 0, \nabla u, 0) & \text{in } \Omega, \\ u = 0 & \text{on } \partial\Omega, \end{cases} \tag{6.6}$$

$$\begin{cases} -\Delta_{p_2} u = f_2(x, 0, u, 0, \nabla u) & \text{in } \Omega, \\ u = 0 & \text{on } \partial\Omega. \end{cases} \tag{6.7}$$

We can state the following existence theorem for solutions of problem (6.1). We set $W = W_0^{1,p_1}(\Omega) \times W_0^{1,p_2}(\Omega)$.

Theorem 6.1. *Assume that the Carathéodory functions $f_i : \Omega \times \mathbb{R} \times \mathbb{R} \times \mathbb{R}^N \times \mathbb{R}^N \to \mathbb{R}$ $(i = 1, 2)$ satisfy hypotheses* $\mathrm{H}(f_1, f_2)_1$ *and* $\mathrm{H}(f_1, f_2)_2$. *Then problem (6.1) has a solution for all $\mu_1, \mu_2 \geq 0$.*

Proof. Fix $\mu_1, \mu_2 \geq 0$. We associate to problem (6.1) the nonlinear operator $A : W \to W^* = W^{-1,p_1'}(\Omega) \times W^{-1,p_2'}(\Omega)$ defined by

$$\begin{aligned} A(u_1, u_2) = (&-\Delta_{p_1} u_1 - \mu_1 \Delta_{q_1} u_1 - N_1(u_1, u_2), \\ &-\Delta_{p_2} u_2 - \mu_2 \Delta_{q_2} u_2 - N_2(u_1, u_2)), \end{aligned} \tag{6.8}$$

where $N_i : W \to W^{-1,p_i'}(\Omega)$ denotes the Nemytskii operator determined by f_i, that is, $N_i(u_1, u_2) = f_i(\cdot, u_1(\cdot), u_2(\cdot), \nabla u_1(\cdot), \nabla u_2(\cdot))$. It is straightforward to verify that due to hypothesis $\mathrm{H}(f_1, f_2)_1$ the operator A is well defined.

We first prove that the operator A in (6.8) is pseudomonotone. To this end, consider a sequence $\{(u_{1,n}, u_{2,n})\} \subset W$ weakly converging to (u_1, u_2) in W with

$$\limsup_{n\to\infty} \langle A(u_{1,n}, u_{2,n}), (u_{1,n} - u_1, u_{2,n} - u_2)\rangle \leq 0. \tag{6.9}$$

We claim that for $i = 1, 2$ we have

$$\lim_{n\to\infty} \int_\Omega f_i(x, u_{1,n}, u_{2,n}, \nabla u_{1,n}, \nabla u_{2,n})(u_{i,n} - u_i)\, dx = 0. \tag{6.10}$$

In order to show (6.10), on the basis of the growth condition in hypothesis $\mathrm{H}(f_1, f_2)_1$ it suffices to note that

$$\int_\Omega |\sigma_i||u_{i,n} - u_i|\, dx \leq \|\sigma_i\|_{\gamma_i'} \|u_{i,n} - u_i\|_{\gamma_i} \to 0,$$

$$\int_\Omega |u_{i,n}|^{\alpha_i} |u_{i,n} - u_i|\, dx \leq \|u_{i,n}\|_{p_i^*}^{\alpha_i} \|u_{i,n} - u_i\|_{\frac{p_i^*}{p_i^* - \alpha_i}} \to 0,$$

$$\int_\Omega |\nabla u_{i,n}|^{\beta_i} |u_{i,n} - u_i|\, dx \leq \|\nabla u_{i,n}\|_{p_i}^{\beta_i} \|u_{i,n} - u_i\|_{\frac{p_i}{p_i - \beta_i}} \to 0,$$

$$\int_\Omega |u_{j,n}|^{\frac{\alpha_i p_j^*}{p_i^*}} |u_{i,n} - u_i|\, dx \le \|u_{j,n}\|_{p_j^*}^{\frac{\alpha_i p_j^*}{p_i^*}} \|u_{i,n} - u_i\|_{\frac{p_i^*}{p_i^* - \alpha_i}} \to 0, \ \ i \ne j,$$

$$\int_\Omega |\nabla u_{j,n}|^{\frac{\beta_i p_j}{p_i}} |u_{i,n} - u_i|\, dx \le \|\nabla u_{j,n}\|_{p_j}^{\frac{\beta_i p_j}{p_i}} \|u_{i,n} - u_i\|_{\frac{p_i}{p_i - \beta_i}} \to 0, \ \ i \ne j.$$

These convergence relations are consequences of Hölder's inequality and Rellich–Kondrachov compactness embedding theorem (see Theorem 1.20). Therefore, in view of hypothesis $\mathrm{H}(f_1, f_2)_1$, (6.10) holds true.

Inserting (6.10) in (6.9) leads to

$$\begin{aligned}\limsup_{n\to\infty}\big(&\langle -\Delta_{p_1} u_{1,n} - \mu_1 \Delta_{q_1} u_{1,n}, u_{1,n} - u_1\rangle \\ &+ \langle -\Delta_{p_2} u_{2,n} - \mu_2 \Delta_{q_2} u_{2,n}, u_{2,n} - u_2\rangle\big) \le 0.\end{aligned} \tag{6.11}$$

We claim that

$$\limsup_{n\to\infty}\langle -\Delta_{p_i} u_{i,n} - \mu_i \Delta_{q_i} u_{i,n}, u_{i,n} - u_i\rangle \le 0, \ \ i = 1, 2. \tag{6.12}$$

If not, we may admit that along relabeled subsequences one has

$$\begin{aligned}&\lim_{n\to\infty}\langle -\Delta_{p_1} u_{1,n} - \mu_1 \Delta_{q_1} u_{1,n}, u_{1,n} - u_1\rangle > 0, \\ &\lim_{n\to\infty}\langle -\Delta_{p_2} u_{2,n} - \mu_2 \Delta_{q_2} u_{2,n}, u_{2,n} - u_2\rangle < 0.\end{aligned} \tag{6.13}$$

The second inequality in (6.13) enables us to utilize the $(S)_+$-property of the map $-\Delta_{p_2} - \mu_2 \Delta_{q_2}$ on $W_0^{1,p_2}(\Omega)$ (see [169, p. 39], also Proposition 1.56) to infer the strong convergence $u_{2,n} \to u_2$ in $W_0^{1,p_2}(\Omega)$. Then a contradiction arises between (6.11) and (6.13), thereby (6.12) holds true.

Then on the basis of (6.12) and invoking again the $(S)_+$-property of the map $-\Delta_{p_i} - \mu_i \Delta_{q_i}$ on $W_0^{1,p_i}(\Omega)$, we are able to infer the strong convergence $u_{i,n} \to u_i$ in $W_0^{1,p_i}(\Omega)$ for $i = 1, 2$. Once this strong convergence is achieved, it is a straightforward matter to prove that for each $(v_1, v_2) \in W$ we have

$$\lim_{n\to\infty}\langle A(u_{1,n}, u_{2,n}), (u_{1,n} - v_1, u_{2,n} - v_2)\rangle = \langle A(u_1, u_2), (u_1 - v_1, u_2 - v_2)\rangle.$$

The operator A is thus pseudomonotone.

Notice that the operator A is bounded, in the sense that it maps bounded sets into bounded sets. This is because the operator $-\Delta_{p_i} - \mu_i \Delta_{q_i}$ is bounded on $W_0^{1,p_i}(\Omega)$ for $i = 1, 2$, and assumption $\mathrm{H}(f_1, f_2)_1$ is supposed to be valid.

The next step in the proof is to show that the operator A in (6.8) is coercive, which means

$$\lim_{\|u\|\to+\infty}\frac{\langle Au, u\rangle}{\|u\|} = +\infty, \tag{6.14}$$

where for all $u=(u_1,u_2)\in W$ we set $\|u\|=\|\nabla u_1\|_{p_1}+\|\nabla u_2\|_{p_2}$. Indeed, arguing by means of hypothesis $\mathrm{H}(f_1,f_2)_2$ gives

$$\begin{aligned}\langle A(u_1,u_2),(u_1,u_2)\rangle &= \|\nabla u_1\|_{p_1}^{p_1}+\|\nabla u_2\|_{p_2}^{p_2}+\mu_1\|\nabla u_1\|_{q_1}^{q_1}+\mu_2\|\nabla u_2\|_{q_2}^{q_2}\\ &-\int_\Omega f_1(x,u_1,u_2,\nabla u_1,\nabla u_2)u_1\,dx-\int_\Omega f_2(x,u_1,u_2,\nabla u_1,\nabla u_2)u_2\,dx\\ &\geq \left(1-\max\left\{\lambda_{1,p_1}^{-1},\lambda_{1,p_2}^{-1}\right\}c-d\right)(\|\nabla u_1\|_{p_1}^{p_1}+\|\nabla u_2\|_{p_2}^{p_2})-\|\omega\|_1.\end{aligned}$$

Thanks to $1-\max\left\{\lambda_{1,p_1}^{-1},\lambda_{1,p_2}^{-1}\right\}c-d>0$ as posed in assumption $\mathrm{H}(f_1,f_2)_2$, it turns out that (6.14) is satisfied.

Since $A: W\to W^*$ introduced in (6.8) is a pseudomonotone, bounded, and coercive operator, we may apply the main theorem on pseudomonotone operators (see Theorem 1.55), which ensures that a solution of the equation $Au=0$ exists. According to (6.8), this implies that (u_1,u_2) is a weak solution of system (6.1). □

A natural topic is the uniqueness of solution for problem (6.1).

It is convenient to introduce the vector field $f:\Omega\times\mathbb{R}^2\times(\mathbb{R}^N)^2\to\mathbb{R}^2$ given by

$$\begin{aligned}&f(x,s,\xi)=(f_1(x,s,\xi),f_2(x,s,\xi))\\ &\quad\text{for a.e. } x\in\Omega,\ \text{every } s\in\mathbb{R}^2,\ \xi\in(\mathbb{R}^N)^2.\end{aligned}\tag{6.15}$$

A new assumption is now formulated.

($\mathcal{U}$) There exist constants $a,b\geq 0$ and a function $\tau=(\tau_1,\tau_2)\in L^1(\Omega,\mathbb{R}^2)$ such that $f(x,s,\cdot)-\tau(x)$ is linear on $(\mathbb{R}^N)^2$, and

$$\begin{aligned}&(f(x,s,\xi)-f(x,t,\xi))\cdot(s-t)\leq a|s-t|^2\\ &\quad\text{for a.e. } x\in\Omega,\ \text{every } s,t\in\mathbb{R}^2,\ \xi\in(\mathbb{R}^N)^2,\end{aligned}\tag{6.16}$$

$$\begin{aligned}&|f(x,s,\xi)-\tau(x)|\leq b|\xi|\\ &\quad\text{for a.e. } x\in\Omega,\ \text{every } (s,\xi)\in\mathbb{R}^2\times(\mathbb{R}^N)^2.\end{aligned}\tag{6.17}$$

Theorem 6.2. *Assume that conditions* $\mathrm{H}(f_1,f_2)_1$, $\mathrm{H}(f_1,f_2)_2$, *and* ($\mathcal{U}$) *are satisfied.*

- **(i)** *When* $p_1=p_2=2$ *and* $a\lambda_{1,2}^{-1}+b(2\lambda_{1,2}^{-1})^{\frac{1}{2}}<1$, *the solution of problem (6.1) is unique for every* $\mu_1,\mu_2\geq 0$.
- **(ii)** *When* $q_1=q_2=2$, *the solution of problem (6.1) is unique for every* $\mu_1,\mu_2>a\lambda_{1,2}^{-1}+b(2\lambda_{1,2}^{-1})^{\frac{1}{2}}$.

Proof. Theorem 6.1 guarantees the existence of solutions of problem (6.1). Let us admit that there are two solutions $u=(u_1,u_2),v=(v_1,v_2)\in W_0^{1,p_1}(\Omega)\times W_0^{1,p_2}(\Omega)$ for (6.1) with a fixed (μ_1,μ_2).

Testing the first equation in the system (6.1) with $u_1 - v_1 \in W_0^{1,p_1}(\Omega)$ and the second equation with $u_2 - v_2 \in W_0^{1,p_2}(\Omega)$ entails

$$\begin{aligned}
&\langle -\Delta_{p_1} u_1 + \Delta_{p_1} v_1, u_1 - v_1\rangle + \mu_1\langle -\Delta_{q_1} u_1 + \Delta_{q_1} v_1, u_1 - v_1\rangle \\
&+ \langle -\Delta_{p_2} u_2 + \Delta_{p_2} v_2, u_2 - v_2\rangle + \mu_2\langle -\Delta_{q_2} u_2 + \Delta_{q_2} v_2, u_2 - v_2\rangle \\
= &\int_\Omega (f(x,u,\nabla u) - f(x,v,\nabla u)) \cdot (u - v)\, dx \\
&+ \int_\Omega (f(x,v,\nabla u) - \tau(x) - f(x,v,\nabla v) + \tau(x)) \cdot (u - v)\, dx. \qquad (6.18)
\end{aligned}$$

(i) If $p_1 = p_2 = 2$, by (6.18), (6.15), (6.16), (6.17), the monotonicity of $-\Delta_{q_1}$ and $-\Delta_{q_2}$ in conjunction with Cauchy–Schwarz inequality, we get the estimate

$$\begin{aligned}
&\|\nabla(u_1 - v_1)\|_2^2 + \|\nabla(u_2 - v_2)\|_2^2 \le a(\|u_1 - v_1\|_2^2 + \|u_2 - v_2\|_2^2) \\
&+ \int_\Omega (f_1(x, v_1, v_2, (u_1 - v_1)\nabla(u_1 - v_1), (u_1 - v_1)\nabla(u_2 - v_2)) - \tau_1(x))\, dx \\
&+ \int_\Omega (f_2(x, v_1, v_2, (u_2 - v_2)\nabla(u_1 - v_1), (u_2 - v_2)\nabla(u_2 - v_2)) - \tau_2(x))\, dx \\
\le\, & a\lambda_{1,2}^{-1}(\|\nabla(u_1 - v_1)\|_2^2 + \|\nabla(u_2 - v_2)\|_2^2) \\
&+ b\int_\Omega (|u_1 - v_1| + |u_2 - v_2|)(|\nabla(u_1 - v_1)|^2 + |\nabla(u_2 - v_2)|^2)^{\frac{1}{2}}\, dx \\
\le\, & (a\lambda_{1,2}^{-1} + b(2\lambda_{1,2}^{-1})^{\frac{1}{2}})(\|\nabla(u_1 - v_1)\|_2^2 + \|\nabla(u_2 - v_2)\|_2^2).
\end{aligned}$$

Since $a\lambda_{1,2}^{-1} + b(2\lambda_{1,2}^{-1})^{\frac{1}{2}} < 1$, we derive that $u_i = v_i$ for $i = 1, 2$.

(ii) In the case of $q_1 = q_2 = 2$, we can proceed as in part (i) to obtain

$$\begin{aligned}
&\mu_1\|\nabla(u_1 - v_1)\|_2^2 + \mu_2\|\nabla(u_2 - v_2)\|_2^2 \\
&\le (a\lambda_{1,2}^{-1} + b(2\lambda_{1,2}^{-1})^{\frac{1}{2}})(\|\nabla(u_1 - v_1)\|_2^2 + \|\nabla(u_2 - v_2)\|_2^2).
\end{aligned}$$

The conclusion ensues by choosing $\mu_1, \mu_2 > a\lambda_{1,2}^{-1} + b(2\lambda_{1,2}^{-1})^{\frac{1}{2}}$. □

We point out a useful a priori estimate for the solutions of problem (6.1).

Lemma 6.3. *If the Carathéodory functions $f_i : \Omega \times \mathbb{R} \times \mathbb{R} \times \mathbb{R}^N \times \mathbb{R}^N \to \mathbb{R}$, $i = 1, 2$, satisfy assumptions $\mathrm{H}(f_1, f_2)_1$, $\mathrm{H}(f_1, f_2)_2$, then the set of solutions of problem (6.1) is bounded in $W_0^{1,p_1}(\Omega) \times W_0^{1,p_2}(\Omega)$ uniformly with respect to $\mu_1, \mu_2 \ge 0$.*

Proof. If $(u_1, u_2) \in W_0^{1,p_1}(\Omega) \times W_0^{1,p_2}(\Omega)$ is a solution of problem (6.1), by (6.3), (6.4), (6.2), and hypothesis $\mathrm{H}(f_1, f_2)_2$, we get the estimate

$$
\begin{aligned}
&\|\nabla u_1\|_{p_1}^{p_1} + \|\nabla u_2\|_{p_2}^{p_2} \\
&\le \int_\Omega (f_1(x, u_1, u_2, \nabla u_1, \nabla u_2)u_1 + f_2(x, u_1, u_2, \nabla u_1, \nabla u_2)u_2)\,dx \\
&\le \int_\Omega (|\omega(x)| + c(|u_1|^{p_1} + |u_2|^{p_2}) + d(|\nabla u_1|^{p_1} + |\nabla u_2|^{p_2}))\,dx \\
&\le \|\omega\|_1 + \left(\max\left\{\lambda_{1,p_1}^{-1}, \lambda_{1,p_2}^{-1}\right\} c + d\right)(\|\nabla u_1\|_{p_1}^{p_1} + \|\nabla u_2\|_{p_2}^{p_2}).
\end{aligned}
$$

Since it was supposed that $1 - \max\left\{\lambda_{1,p_1}^{-1}, \lambda_{1,p_2}^{-1}\right\} c - d > 0$, we obtain the bound

$$
\|\nabla u_1\|_{p_1}^{p_1} + \|\nabla u_2\|_{p_2}^{p_2} \le \frac{\|\omega\|_1}{1 - \max\left\{\lambda_{1,p_1}^{-1}, \lambda_{1,p_2}^{-1}\right\} c - d} =: c_0, \tag{6.19}
$$

with the constant c_0 independent of (μ_1, μ_2), which yields the desired conclusion. □

Using that μ_1, μ_2 are parameters, we can examine the asymptotic property involving the limit as $\mu_1 \to 0$ and $\mu_2 \to 0$.

Theorem 6.4. *Assume conditions* $\mathrm{H}(f_1, f_2)_1$ *and* $\mathrm{H}(f_1, f_2)_2$. *Given a sequence of parameters* $\{(\mu_{1,n}, \mu_{2,n})\}$ *with* $\mu_{1,n} \to 0^+$ *and* $\mu_{2,n} \to 0^+$, *and a sequence* $\{(u_{1,n}, u_{2,n})\}$ *of solutions of the corresponding problems (6.1) (with* $(\mu_1, \mu_2) = (\mu_{1,n}, \mu_{2,n})$*), one can find a relabeled subsequence of* $\{(u_{1,n}, u_{2,n})\}$ *such that* $u_{1,n} \to u_1$ *in* $W_0^{1,p_1}(\Omega)$ *and* $u_{2,n} \to u_2$ *in* $W_0^{1,p_2}(\Omega)$, *with* $(u_1, u_2) \in W_0^{1,p_1}(\Omega) \times W_0^{1,p_2}(\Omega)$ *weak solution of problem (6.5).*

Proof. The fact that $(u_{1,n}, u_{2,n})$ is a weak solution of problem (6.1) (with $(\mu_1, \mu_2) = (\mu_{1,n}, \mu_{2,n})$) permits applying Lemma 6.3 and deducing that the sequence $\{(u_{1,n}, u_{2,n})\}$ is bounded in $W_0^{1,p_1}(\Omega) \times W_0^{1,p_2}(\Omega)$. Then a relabeled subsequence can be found to have the weak convergence $u_{1,n} \rightharpoonup u_1$ in $W_0^{1,p_1}(\Omega)$ and $u_{2,n} \rightharpoonup u_2$ in $W_0^{1,p_2}(\Omega)$ for some $(u_1, u_2) \in W_0^{1,p_1}(\Omega) \times W_0^{1,p_2}(\Omega)$.

As in the proof of Theorem 6.1, hypothesis $\mathrm{H}(f_1, f_2)_1$ implies that (6.10) holds true. This essentially relies on Rellich–Kondrachov compactness embedding theorem. Then, using that $\mu_{1,n} \to 0^+$ and $\mu_{2,n} \to 0^+$, we note that the equations in (6.1) (with $(\mu_1, \mu_2) = (\mu_{1,n}, \mu_{2,n})$) yield in the limit

$$
\lim_{n\to\infty} \langle -\Delta_{p_i} u_{i,n}, u_{i,n} - u_i \rangle = 0 \quad \text{for } i = 1, 2.
$$

At this point we can refer to the $(S)_+$-property of the operators $-\Delta_{p_i} : W_0^{1,p_i}(\Omega) \to W^{-1,p_i'}(\Omega)$ $(i = 1, 2)$ in Proposition 1.56 for obtaining the strong convergence $u_{1,n} \to u_1$ in $W_0^{1,p_1}(\Omega)$ and $u_{2,n} \to u_2$ in $W_0^{1,p_2}(\Omega)$. Once the strong convergence is obtained, we can pass to the limit in the equations of (6.1) (with $(\mu_1, \mu_2) = (\mu_{1,n}, \mu_{2,n})$) because $-\Delta_{p_i} u_{i,n} \to -\Delta_{p_i} u_i$ and

$$\begin{aligned} &f_i(\cdot, u_{1,n}(\cdot), u_{2,n}(\cdot), \nabla u_{1,n}(\cdot), \nabla u_{2,n}(\cdot)) \\ &\to f_i(\cdot, u_1(\cdot), u_2(\cdot), \nabla u_1(\cdot), \nabla u_2(\cdot)) \text{ in } W^{-1,p_i'}(\Omega) \end{aligned} \tag{6.20}$$

for $i = 1, 2$. Then due to the fact that $\mu_{i,n} \to 0^+$, (6.20) and the boundedness of the operators $-\Delta_{q_i}$ for $i = 1, 2$, letting $n \to \infty$ in (6.1) (with $(\mu_1, \mu_2) = (\mu_{1,n}, \mu_{2,n})$) results in the fact that (u_1, u_2) becomes a weak solution of problem (6.5). This completes the proof. $\square$

Next we focus on the asymptotic study as $(\mu_1, \mu_2) \to (0, +\infty)$ (the situation where $(\mu_1, \mu_2) \to (+\infty, 0)$ is symmetric).

Theorem 6.5. *Assume* $\mathrm{H}(f_1, f_2)_1$ *and* $\mathrm{H}(f_1, f_2)_2$, *and in addition that* $\beta_1 \le \frac{q_2 p_1}{(p_1^*)' p_2}$. *For any sequence* $\{(\mu_{1,n}, \mu_{2,n})\}$ *of parameters with* $\mu_{1,n} \to 0^+$ *and* $\mu_{2,n} \to +\infty$, *and any sequence* $\{(u_{1,n}, u_{2,n})\}$ *of solutions of the corresponding problems (6.1) (with* $(\mu_1, \mu_2) = (\mu_{1,n}, \mu_{2,n})$*), one can find a relabeled subsequence* $\{(u_{1,n}, u_{2,n})\}$ *such that* $u_{1,n} \to u_1$ *strongly in* $W_0^{1,p_1}(\Omega)$ *and* $u_{2,n} \to 0$ *strongly in* $W_0^{1,q_2}(\Omega)$, *with* $u_1 \in W_0^{1,p_1}(\Omega)$ *weak solution of problem (6.6).*

Proof. By Lemma 6.3 we know that $\{(u_{1,n}, u_{2,n})\}$ is bounded in $W_0^{1,p_1}(\Omega) \times W_0^{1,p_2}(\Omega)$, thus it contains a relabeled subsequence with the weak convergence $u_{1,n} \rightharpoonup u_1$ in $W_0^{1,p_1}(\Omega)$ and $u_{2,n} \rightharpoonup u_2$ in $W_0^{1,p_2}(\Omega)$ for some $(u_1, u_2) \in W_0^{1,p_1}(\Omega) \times W_0^{1,p_2}(\Omega)$.

Actually, (6.1) (with $(\mu_1, \mu_2) = (\mu_{1,n}, \mu_{2,n})$) renders

$$\begin{cases} -\Delta_{p_1} u_{1,n} - \mu_{1,n} \Delta_{q_1} u_{1,n} = f_1(x, u_{1,n}, u_{2,n}, \nabla u_{1,n}, \nabla u_{2,n}) & \text{in } \Omega, \\ -\frac{1}{\mu_{2,n}} \Delta_{p_2} u_{2,n} - \Delta_{q_2} u_{2,n} = \frac{1}{\mu_{2,n}} f_2(x, u_{1,n}, u_{2,n}, \nabla u_{1,n}, \nabla u_{2,n}) & \text{in } \Omega, \\ u_{1,n} = u_{2,n} = 0 & \text{on } \partial\Omega. \end{cases} \tag{6.21}$$

Taking into account that $\mu_{1,n} \to 0$ and $\mu_{2,n} \to +\infty$, we infer from (6.21) that

$$\lim_{n\to\infty} \langle -\Delta_{p_1} u_{1,n}, u_{1,n} - u_1 \rangle = \lim_{n\to\infty} \langle -\Delta_{q_2} u_{2,n}, u_{2,n} - u_2 \rangle = 0.$$

Here it was also used the boundedness of the operators $-\Delta_{p_2}$, $-\Delta_{q_1}$ as well as (6.10). Through the $(S)_+$-property of the operators $-\Delta_{p_1} : W_0^{1,p_1}(\Omega) \to W^{-1,p_1'}(\Omega)$ and $-\Delta_{q_2} : W_0^{1,q_2}(\Omega) \to W^{-1,q_2'}(\Omega)$ in Proposition 1.56, it turns out that $u_{1,n} \to u_1$ in $W_0^{1,p_1}(\Omega)$ and $u_{2,n} \to u_2$ in $W_0^{1,q_2}(\Omega)$. Then, recalling that $\mu_{2,n} \to +\infty$, the second equation in (6.21) provides in the limit that $\Delta_{q_2} u_2 = 0$, thereby $u_2 = 0$. Now, recalling that $\mu_{1,n} \to 0$, we can pass to the limit as $n \to \infty$ in the first equation of (6.21) on the basis of $\mathrm{H}(f_1, f_2)_1$ with $\beta_1 \le \frac{q_2 p_1}{(p_1^*)' p_2}$. We conclude that u_1 is a solution of problem (6.6). $\square$

Finally, we consider the limiting process as $\mu_1 \to +\infty$ and $\mu_2 \to +\infty$.

Theorem 6.6. *Assume that conditions* $\mathrm{H}(f_1, f_2)_1$ *and* $\mathrm{H}(f_1, f_2)_2$ *hold. Given a sequence* $\{(\mu_{1,n}, \mu_{2,n})\}$ *with* $\mu_{1,n} \to +\infty$ *and* $\mu_{2,n} \to +\infty$, *every sequence* $\{(u_{1,n}, u_{2,n})\}$ *of solutions of the corresponding problems (6.1) (with* $(\mu_1, \mu_2) = (\mu_{1,n}, \mu_{2,n})$*) satisfies* $u_{1,n} \to 0$ *strongly in* $W_0^{1,q_1}(\Omega)$ *and* $u_{2,n} \to 0$ *strongly in* $W_0^{1,q_2}(\Omega)$.

Proof. By Lemma 6.3, the sequence $\{(u_{1,n}, u_{2,n})\}$ is bounded in $W_0^{1,p_1}(\Omega) \times W_0^{1,p_2}(\Omega)$, so we can extract a relabeled subsequence satisfying the weak convergence $u_{1,n} \rightharpoonup u_1$ in $W_0^{1,p_1}(\Omega)$ and $u_{2,n} \rightharpoonup u_2$ in $W_0^{1,p_2}(\Omega)$ with some $(u_1, u_2) \in W_0^{1,p_1}(\Omega) \times W_0^{1,p_2}(\Omega)$.

Writing (6.1) (with $(\mu_1, \mu_2) = (\mu_{1,n}, \mu_{2,n})$) in the form

$$\begin{cases} -\frac{1}{\mu_{1,n}}\Delta_{p_1} u_{1,n} - \Delta_{q_1} u_{1,n} = \frac{1}{\mu_{1,n}} f_1(x, u_{1,n}, u_{2,n}, \nabla u_{1,n}, \nabla u_{2,n}) & \text{in } \Omega, \\ -\frac{1}{\mu_{2,n}}\Delta_{p_2} u_{2,n} - \Delta_{q_2} u_{2,n} = \frac{1}{\mu_{2,n}} f_2(x, u_{1,n}, u_{2,n}, \nabla u_{1,n}, \nabla u_{2,n}) & \text{in } \Omega, \\ u_{1,n} = u_{2,n} = 0 & \text{on } \partial\Omega, \end{cases} \tag{6.22}$$

we derive on the same pattern as before that

$$\lim_{n\to\infty} \langle -\Delta_{q_1} u_{1,n}, u_{1,n} - u_1 \rangle = \lim_{n\to\infty} \langle -\Delta_{q_2} u_{2,n}, u_{2,n} - u_2 \rangle = 0.$$

The validity of these equalities is basically the consequence of (6.22), where $\mu_{1,n} \to +\infty$ and $\mu_{2,n} \to +\infty$. We are thus in a position to invoke the $(S)_+$-property of the operators $-\Delta_{q_1} : W_0^{1,q_1}(\Omega) \to W^{-1,q_1'}(\Omega)$ and $-\Delta_{q_2} : W_0^{1,q_2}(\Omega) \to W^{-1,q_2'}(\Omega)$ in Proposition 1.56, which entails that strongly $u_{1,n} \to u_1$ in $W_0^{1,q_1}(\Omega)$ and $u_{2,n} \to u_2$ in $W_0^{1,q_2}(\Omega)$. Then letting $n \to \infty$ in (6.22) leads to $\Delta_{q_1} u_1 = 0$ and $\Delta_{q_2} u_2 = 0$, which reads as $u_1 = u_2 = 0$. Since this occurs for every convergent subsequence of $\{(u_{1,n}, u_{2,n})\}$, we have the strong convergence for the whole sequence, which completes the proof. □

6.2 SUBSOLUTIONS–SUPERSOLUTIONS FOR NONLINEAR ELLIPTIC SYSTEMS

In this section, we investigate the system (6.1) of Section 6.1 through a different method, namely by means of subsolution–supersolution. In comparison with equations, the subsolution–supersolution method for systems has specific features. We suppose that $N > \max\{p_1, p_2\}$, which implies that the Sobolev critical exponent corresponding to p_i is $p_i^* = \frac{Np_i}{N-p_i}$, $i = 1, 2$. The case $N \leq \max\{p_1, p_2\}$ is basically simpler and can be carried out along the same lines. We also assume in this section that the bounded domain $\Omega \subset \mathbb{R}^N$ has a C^2 boundary $\partial\Omega$.

We say that $(\underline{u}_1, \underline{u}_2), (\overline{u}_1, \overline{u}_2) \in W^{1,p_1}(\Omega) \times W^{1,p_2}(\Omega)$ is a subsolution–supersolution of problem (6.1) if $\underline{u}_i \leq \overline{u}_i$ a.e. in Ω, $\underline{u}_i \leq 0 \leq \overline{u}_i$ on $\partial\Omega$ for $i = 1, 2$, and

$$\begin{aligned}
&\int_\Omega \Big(|\nabla \underline{u}_1|^{p_1-2} \nabla \underline{u}_1 \cdot \nabla v_1 + \mu_1 |\nabla \underline{u}_1|^{q_1-2} \nabla \underline{u}_1 \cdot \nabla v_1 \\
&\quad - f_1(x, \underline{u}_1, w_2, \nabla \underline{u}_1, \nabla w_2) v_1 \Big) \\
&\quad + \int_\Omega \Big(|\nabla \underline{u}_2|^{p_2-2} \nabla \underline{u}_2 \cdot \nabla v_2 + \mu_2 |\nabla \underline{u}_2|^{q_2-2} \nabla \underline{u}_2 \cdot \nabla v_2 \\
&\quad - f_2(x, w_1, \underline{u}_2, \nabla w_1, \nabla \underline{u}_2) v_2 \Big) \leq 0
\end{aligned}$$

and

$$\begin{aligned}
&\int_\Omega \Big(|\nabla \overline{u}_1|^{p_1-2} \nabla \overline{u}_1 \cdot \nabla v_1 + \mu_1 |\nabla \overline{u}_1|^{q_1-2} \nabla \overline{u}_1 \cdot \nabla v_1 \\
&\quad - f_1(x, \overline{u}_1, w_2, \nabla \overline{u}_1, \nabla w_2) v_1 \Big) \\
&\quad + \int_\Omega \Big(|\nabla \overline{u}_2|^{p_2-2} \nabla \overline{u}_1 \cdot \nabla v_2 + \mu_2 |\nabla \overline{u}_2|^{q_2-2} \nabla \overline{u}_1 \cdot \nabla v_2 \\
&\quad - f_2(x, w_1, \overline{u}_2, \nabla w_1, \nabla \overline{u}_2) v_2 \Big) \geq 0
\end{aligned}$$

for all $(v_1, v_2) \in W_0^{1,p_1}(\Omega) \times W_0^{1,p_2}(\Omega)$ with $v_1, v_2 \geq 0$ a.e. in Ω, and all $(w_1, w_2) \in W^{1,p_1}(\Omega) \times W^{1,p_2}(\Omega)$ such that $\underline{u}_i \leq w_i \leq \overline{u}_i$ a.e. in Ω, $i = 1, 2$.

In the sequel, we will require the existence of a subsolution–supersolution $(\underline{u}_1, \underline{u}_2), (\overline{u}_1, \overline{u}_2) \in W^{1,p_1}(\Omega) \times W^{1,p_2}(\Omega)$ of problem (6.1) satisfying:

H(f_1, f_2) There exist constants $b_i \geq 0$, $\beta_i \in [0, \frac{p_i}{(p_i^*)'})$ and functions $\sigma_i \in L^{\gamma_i'}(\Omega)$, with $\gamma_i \in [1, p_i^*)$, such that

$$\begin{aligned}
|f_1(x, s_1, s_2, \xi_1, \xi_2)| &\leq \sigma_1(x) + b_1(|\xi_1|^{\beta_1} + |\xi_2|^{\frac{\beta_1 p_2}{p_1}}), \\
|f_2(x, s_1, s_2, \xi_1, \xi_2)| &\leq \sigma_2(x) + b_2(|\xi_1|^{\frac{\beta_2 p_1}{p_2}} + |\xi_2|^{\beta_2})
\end{aligned}$$

for a.a. $x \in \Omega$, all $s = (s_1, s_2) \in [\underline{u}_1(x), \overline{u}_1(x)] \times [\underline{u}_2(x), \overline{u}_2(x)]$, $\xi_1, \xi_2 \in \mathbb{R}^N$, $i = 1, 2$.

Under H(f_1, f_2), the integrals in the definition of subsolution–supersolution exist.

Fix $\mu_1, \mu_2 \geq 0$ and according to H(f_1, f_2) let $(\underline{u}_1, \underline{u}_2), (\overline{u}_1, \overline{u}_2) \in W^{1,p_1}(\Omega) \times W^{1,p_2}(\Omega)$ be a subsolution–supersolution of problem (6.1). For $i = 1, 2$, we consider the truncation operators $T_i : W^{1,p_i}(\Omega) \to W^{1,p_i}(\Omega)$ defined by

$$(T_i u)(x) = \begin{cases} \overline{u}_i(x) & \text{if } u(x) > \overline{u}_i(x), \\ u(x) & \text{if } \underline{u}_i(x) \le u(x) \le \overline{u}_i(x), \\ \underline{u}_i(x) & \text{if } u(x) < \underline{u}_i(x). \end{cases}$$

The operators T_i are continuous and bounded.

With the constants β_i in $\mathrm{H}(f_1, f_2)$, we set for a.a. $x \in \Omega$, all $s \in \mathbb{R}$, $i = 1, 2$:

$$\pi_i(x, s) = \begin{cases} (s - \overline{u}_i(x))^{\frac{\beta_i}{p_i - \beta_i}} & \text{if } s > \overline{u}_i(x), \\ 0 & \text{if } \underline{u}_i(x) \le s \le \overline{u}_i(x), \\ -(\underline{u}_i(x) - s)^{\frac{\beta_i}{p_i - \beta_i}} & \text{if } s < \underline{u}_i(x). \end{cases} \tag{6.23}$$

It is clear that π_i is a Carathéodory function satisfying the estimate

$$|\pi_i(x, s)| \le \rho_i(x) + c_i |s|^{\frac{\beta_i}{p_i - \beta_i}} \tag{6.24}$$

for a.a. $x \in \Omega$, all $s \in \mathbb{R}$, with a constant $c_i \ge 0$ and a function $\rho_i \in L^{\frac{p_i}{\beta_i}}(\Omega)$. We have that $\rho_i \in L^{\frac{p_i}{\beta_i}}(\Omega)$ because $\underline{u}_i, \overline{u}_i \in W^{1,p_i}(\Omega)$, so $\underline{u}_i, \overline{u}_i \in L^{p_i^*}(\Omega)$ by the Sobolev embedding theorem (see Theorem 1.20), and $\beta_i < \frac{p_i}{(p_i^*)'}$ by hypothesis $\mathrm{H}(f_1, f_2)$. Moreover, a direct reasoning based on (6.23) shows the existence of positive constants $r_1^{(i)}$ and $r_2^{(i)}$ such that

$$\int_\Omega \pi_i(x, u) u \, dx \ge r_1^{(i)} \|u\|_{\frac{p_i}{p_i - \beta_i}}^{\frac{p_i}{p_i - \beta_i}} - r_2^{(i)} \quad \text{for all } u \in W^{1,p_i}(\Omega), \ i = 1, 2. \tag{6.25}$$

On the basis of (6.24), the corresponding Nemytskii operators $\Pi_i : W^{1,p_i}(\Omega) \to L^{\frac{p_i}{\beta_i}}(\Omega)$ given by $\Pi_i u(x) = \pi_i(x, u(x))$ are completely continuous due to the compact embedding of $W^{1,p_i}(\Omega)$ into $L^{p_i}(\Omega)$ (see Theorem 1.20).

For any $\lambda > 0$, consider the auxiliary problem

$$\begin{cases} -\Delta_{p_1} u_1 - \mu_1 \Delta_{q_1} u_1 + \lambda \Pi_1(u_1) \\ \quad = f_1(x, T_1 u_1, T_2 u_2, \nabla(T_1 u_1), \nabla(T_2 u_2)) & \text{in } \Omega, \\ -\Delta_{p_2} u_2 - \mu_2 \Delta_{q_2} u_2 + \lambda \Pi_2(u_2) \\ \quad = f_2(x, T_1 u_1, T_2 u_2, \nabla(T_1 u_1), \nabla(T_2 u_2)) & \text{in } \Omega, \\ u_1 = u_2 = 0 & \text{on } \partial\Omega. \end{cases} \tag{6.26}$$

Theorem 6.7. *Let* $(\underline{u}_1, \underline{u}_2), (\overline{u}_1, \overline{u}_2) \in W^{1,p_1}(\Omega) \times W^{1,p_2}(\Omega)$ *be a subsolution–supersolution of problem (6.1) such that condition* $\mathrm{H}(f_1, f_2)$ *is satisfied. Then, for every* $\lambda > 0$ *sufficiently large, problem (6.26) has a (weak) solution* $(u_1, u_2) \in W_0^{1,p_1}(\Omega) \times W_0^{1,p_2}(\Omega)$.

Proof. With the subsolution–supersolution $(\underline{u}_1, \underline{u}_2)$, $(\overline{u}_1, \overline{u}_2)$, we introduce the ordered intervals

$$[\underline{u}_i, \overline{u}_i] = \{u \in W^{1,p_i}(\Omega) : \underline{u}_i \leq u \leq \overline{u}_i \text{ a.e. in } \Omega\}, \; i = 1, 2. \tag{6.27}$$

By (6.27) and hypothesis H(f_1, f_2), the Nemytskii operator

$$\begin{aligned} N : [\underline{u}_1, \overline{u}_1] \times [\underline{u}_2, \overline{u}_2] &\subset W^{1,p_1}(\Omega) \times W^{1,p_2}(\Omega) \\ \to L^{\frac{p_1}{\beta_1}}(\Omega) \times L^{\frac{p_2}{\beta_2}}(\Omega) &\hookrightarrow W^{-1,p_1'}(\Omega) \times W^{-1,p_2'}(\Omega) \end{aligned}$$

defined through the functions f_i by

$$N(u_1, u_2) = (f_1(x, u_1, u_2, \nabla u_1, \nabla u_2), f_2(x, u_1, u_2, \nabla u_1, \nabla u_2)) \tag{6.28}$$

is bounded and completely continuous thanks to Rellich–Kondrachov compactness embedding theorem (see Theorem 1.20).

For our fixed $\lambda > 0$, we define the nonlinear operator $A : W_0^{1,p_1}(\Omega) \times W_0^{1,p_2}(\Omega) \to W^{-1,p_1'}(\Omega) \times W^{-1,p_2'}(\Omega)$ by

$$\begin{aligned} A(u_1, u_2) &= (A_1(u_1, u_2), A_2(u_1, u_2)) \\ &:= (-\Delta_{p_1} u_1 - \mu_1 \Delta_{q_1} u_1 + \lambda \Pi_1 u_1, -\Delta_{p_2} u_2 - \mu_2 \Delta_{q_2} u_2 + \lambda \Pi_2 u_2) \\ &\quad - N(T_1 u_1, T_2 u_2). \end{aligned}$$

According to (6.24), (6.27), (6.28) and hypothesis H(f_1, f_2), the operator A is well defined, bounded, and continuous.

Now we show that the operator A is pseudomonotone. Let $(u_{1,n}, u_{2,n}) \rightharpoonup (u_1, u_2)$ in $W_0^{1,p_1}(\Omega) \times W_0^{1,p_2}(\Omega)$ and

$$\limsup_{n \to \infty} \langle A(u_{1,n}, u_{2,n}), (u_{1,n} - u_1, u_{2,n} - u_2) \rangle \leq 0. \tag{6.29}$$

Since $\frac{p_i}{p_i - \beta_i} < p_i^*$ as known from assumption H(f_1, f_2), it follows the strong convergence $u_{i,n} \to u_i$ in $L^{\frac{p_i}{p_i - \beta_i}}(\Omega)$. Taking into account (6.24), we get

$$\lim_{n \to \infty} \int_\Omega \pi_i(x, u_{i,n}(x))(u_{i,n} - u_i)\, dx = 0, \; i = 1, 2. \tag{6.30}$$

Similarly, by Hölder's inequality and Rellich–Kondrachov compactness embedding theorem in Theorem 1.20, the weak convergence $(u_{1,n}, u_{2,n}) \rightharpoonup (u_1, u_2)$ in $W_0^{1,p_1}(\Omega) \times W_0^{1,p_2}(\Omega)$ enables us to derive that

$$\int_\Omega |\sigma_i| \, |u_{i,n} - u_i|\, dx \leq \|\sigma_i\|_{\gamma_i'} \|u_{i,n} - u_i\|_{\gamma_i} \to 0 \text{ as } n \to +\infty. \tag{6.31}$$

We claim that

$$\int_\Omega |\nabla(T_i u_{i,n})|^{\beta_i} |u_{i,n} - u_i| \, dx \to 0 \text{ as } n \to +\infty. \tag{6.32}$$

From the expression of the truncation operators, it turns out that

$$\begin{aligned}\int_\Omega |\nabla(T_i u_{i,n})|^{\beta_i} |u_{i,n} - u_i| \, dx &= \int_{\{u_{i,n} < \underline{u}_i\}} |\nabla \underline{u}_i|^{\beta_i} |u_{i,n} - u_i| \, dx \\ + \int_{\{\underline{u}_i \le u_{i,n} \le \overline{u}_i\}} &|\nabla u_{i,n}|^{\beta_i} |u_{i,n} - u_i| \, dx + \int_{\{u_{i,n} > \overline{u}_i\}} |\nabla \overline{u}_i|^{\beta_i} |u_{i,n} - u_i| \, dx.\end{aligned}$$

Recalling that $\frac{p_i}{p_i - \beta_i} < p_i^*$, we have

$$\int_{\{u_{i,n} < \underline{u}_i\}} |\nabla \underline{u}_i|^{\beta_i} |u_{i,n} - u_i| \, dx \le \|\nabla \underline{u}_i\|_{p_i}^{\beta_i} \|u_{i,n} - u_i\|_{\frac{p_i}{p_i - \beta_i}} \to 0,$$

$$\int_{\{\underline{u}_i \le u_{i,n} \le \overline{u}_i\}} |\nabla u_{i,n}|^{\beta_i} |u_{i,n} - u_i| \, dx \le \|\nabla u_{i,n}\|_{p_i}^{\beta_i} \|u_{i,n} - u_i\|_{\frac{p_i}{p_i - \beta_i}} \to 0,$$

$$\int_{\{u_{i,n} > \overline{u}_i\}} |\nabla \overline{u}_i|^{\beta_i} |u_{i,n} - u_i| \, dx \le \|\nabla \overline{u}_i\|_{p_i}^{\beta_i} \|u_{i,n} - u_i\|_{\frac{p_i}{p_i - \beta_i}} \to 0,$$

whence (6.32).

Similarly, for $i \neq j$, we can prove

$$\int_\Omega |\nabla(T_j u_{j,n})|^{\frac{\beta_i p_j}{p_i}} |u_{i,n} - u_i| \, dx \le \|\nabla(T_j u_{j,n})\|_{p_j}^{\frac{\beta_i p_j}{p_i}} \|u_{i,n} - u_i\|_{\frac{p_i}{p_i - \beta_i}} \to 0. \tag{6.33}$$

Combining (6.31), (6.32), and (6.33) with hypothesis H(f_1, f_2) yields

$$\lim_{n \to \infty} \int_\Omega f_i(x, T_1 u_{1,n}, T_2 u_{2,n}, \nabla(T_1 u_{1,n}), \nabla(T_2 u_{2,n}))(u_{i,n} - u_i) \, dx = 0, \quad i = 1, 2. \tag{6.34}$$

If we insert (6.30) and (6.34) into (6.29), we arrive at

$$\begin{aligned}\limsup_{n \to \infty} \big(&\langle -\Delta_{p_1} u_{1,n} - \mu_1 \Delta_{q_1} u_{1,n}, u_{1,n} - u_1 \rangle \\ &+ \langle -\Delta_{p_2} u_{2,n} - \mu_2 \Delta_{q_2} u_{2,n}, u_{2,n} - u_2 \rangle \big) \le 0.\end{aligned} \tag{6.35}$$

We show that (6.35) implies

$$\limsup_{n \to \infty} \langle -\Delta_{p_i} u_{i,n} - \mu_i \Delta_{q_i} u_{i,n}, u_{i,n} - u_i \rangle \le 0, \quad i = 1, 2. \tag{6.36}$$

Arguing by contradiction, assume that

$$\lim_{n\to\infty} \langle -\Delta_{p_1} u_{1,n} - \mu_1 \Delta_{q_1} u_{1,n}, u_{1,n} - u_1 \rangle > 0,$$
$$\lim_{n\to\infty} \langle -\Delta_{p_2} u_{2,n} - \mu_2 \Delta_{q_2} u_{2,n}, u_{2,n} - u_2 \rangle < 0.$$

The second inequality and the $(S)_+$-property of $-\Delta_{p_2} - \mu_2 \Delta_{q_2}$ on $W_0^{1,p_2}(\Omega)$ (see [169, p. 39]) yield $u_{2,n} \to u_2$ in $W_0^{1,p_2}(\Omega)$, which contradicts (6.35), thus proving (6.36).

By (6.36) and the $(S)_+$-property of $-\Delta_{p_i} - \mu_i \Delta_{q_i}$ on $W_0^{1,p_i}(\Omega)$, we derive the strong convergence $u_{i,n} \to u_i$ in $W_0^{1,p_i}(\Omega)$ as $n \to \infty$, $i = 1, 2$. Consequently, for every element $(v_1, v_2) \in W_0^{1,p_1}(\Omega) \times W_0^{1,p_2}(\Omega)$, one gets

$$\lim_{n\to\infty} \langle A(u_{1,n}, u_{2,n}), (u_{1,n} - v_1, u_{2,n} - v_2) \rangle = \langle A(u_1, u_2), (u_1 - v_1, u_2 - v_2) \rangle,$$

establishing that the operator A is pseudomonotone.

Next we show that the operator A is coercive, which means that for every sequence $\{(u_{1,n}, u_{2,n})\} \subset W_0^{1,p_1}(\Omega) \times W_0^{1,p_2}(\Omega)$ such that $\|(u_{1,n}, u_{2,n})\| \to +\infty$, we have

$$\lim_{n\to\infty} \frac{\langle A(u_{1,n}, u_{2,n}), (u_{1,n}, u_{2,n}) \rangle}{\|(u_{1,n}, u_{2,n})\|} = +\infty. \tag{6.37}$$

By (6.25) and hypothesis H(f_1, f_2), we note that

$$\begin{aligned}
&\langle A_1(u_{1,n}, u_{2,n}), u_{1,n} \rangle = \|\nabla u_{1,n}\|_{p_1}^{p_1} + \mu_1 \|\nabla u_{1,n}\|_{q_1}^{q_1} \\
&+ \lambda \int_\Omega \pi_1(x, u_{1,n}) u_{1,n} \, dx \\
&- \int_\Omega f_1(x, T_1 u_{1,n}, T_2 u_{2,n}, \nabla(T_1 u_{1,n}), \nabla(T_2 u_{2,n})) u_{1,n} \, dx \\
&\geq \|\nabla u_{1,n}\|_{p_1}^{p_1} + \lambda \left(r_1^{(1)} \|u_{1,n}\|_{\frac{p_1}{p_1 - \beta_1}}^{\frac{p_1}{p_1 - \beta_1}} - r_2^{(1)} \right) - \int_\Omega \sigma_1(x) |u_{1,n}(x)| \, dx \\
&- b_1 \int_\Omega |\nabla(T_1 u_{1,n})|^{\beta_1} |u_{1,n}| \, dx - b_1 \int_\Omega |\nabla(T_2 u_{2,n})|^{\frac{\beta_1 p_2}{p_1}} |u_{1,n}| \, dx.
\end{aligned}$$

Let us estimate separately the three integral terms. First, observe that

$$\int_\Omega \sigma_1(x) |u_{1,n}(x)| \, dx \leq \|\sigma_1\|_{\gamma_1'} \|u_{1,n}\|_{\gamma_1} \leq C_1 \|\nabla u_{1,n}\|_{p_1},$$

with a constant $C_1 > 0$. Second, using Young's inequality with any $\varepsilon > 0$, we get

$$\int_\Omega |\nabla(T_1 u_{1,n})|^{\beta_1} |u_{1,n}(x)| \, dx \leq \varepsilon \|\nabla(T_1 u_{1,n})\|_{p_1}^{p_1} + C_1(\varepsilon) \|u_{1,n}\|_{\frac{p_1}{p_1 - \beta_1}}^{\frac{p_1}{p_1 - \beta_1}}$$

$$\leq \varepsilon \|\nabla u_{1,n}\|_{p_1}^{p_1} + \varepsilon \|\nabla \underline{u}_1\|_{p_1}^{p_1} + \varepsilon \|\nabla \overline{u}_1\|_{p_1}^{p_1} + C_1(\varepsilon)\|u_{1,n}\|_{\frac{p_1}{p_1-\beta_1}}^{\frac{p_1}{p_1-\beta_1}}$$

and third,

$$\int_\Omega |\nabla(T_2 u_{2,n})|^{\frac{\beta_1 p_2}{p_1}} |u_{1,n}(x)|\,dx \leq \varepsilon \|\nabla(T_2 u_{2,n})\|_{p_2}^{p_2} + C_2(\varepsilon)\|u_{1,n}\|_{\frac{p_1}{p_1-\beta_1}}^{\frac{p_1}{p_1-\beta_1}}$$

$$\leq \varepsilon \|\nabla u_{2,n}\|_{p_2}^{p_2} + \varepsilon \|\nabla \underline{u}_2\|_{p_2}^{p_2} + \varepsilon \|\nabla \overline{u}_2\|_{p_2}^{p_2} + C_2(\varepsilon)\|u_{1,n}\|_{\frac{p_1}{p_1-\beta_1}}^{\frac{p_1}{p_1-\beta_1}}$$

for some constants $C_1(\varepsilon), C_2(\varepsilon) > 0$. These estimates render

$$\begin{aligned}\langle A_1(u_{1,n}, u_{2,n}), u_{1,n}\rangle \geq (1-\varepsilon)\|\nabla u_{1,n}\|_{p_1}^{p_1} - C_1\|\nabla u_{1,n}\|_{p_1} \\ + (\lambda r_1^{(1)} - C(\varepsilon))\|u_{1,n}\|_{\frac{p_1}{p_1-\beta_1}}^{\frac{p_1}{p_1-\beta_1}} - \varepsilon\|\nabla u_{2,n}\|_{p_2}^{p_2} - D(\varepsilon),\end{aligned}$$

with positive constants $C(\varepsilon)$, $D(\varepsilon)$. Similarly, we obtain

$$\begin{aligned}\langle A_2(u_{1,n}, u_{2,n}), u_{2,n}\rangle \geq (1-\varepsilon)\|\nabla u_{2,n}\|_{p_2}^{p_2} - C_1'\|\nabla u_{2,n}\|_{p_2} \\ + (\lambda r_1^{(2)} - C'(\varepsilon))\|u_{2,n}\|_{\frac{p_2}{p_2-\beta_2}}^{\frac{p_2}{p_2-\beta_2}} - \varepsilon\|\nabla u_{1,n}\|_{p_1}^{p_1} - D'(\varepsilon),\end{aligned}$$

with positive constants C_1', $C'(\varepsilon)$, $D'(\varepsilon)$.

From the last two inequalities, by choosing $\varepsilon > 0$ sufficiently small and $\lambda > 0$ sufficiently large, we notice that (6.37) holds true because $p_1, p_2 > 1$. Therefore the operator $A : W_0^{1,p_1}(\Omega) \times W_0^{1,p_2}(\Omega) \to W^{-1,p_1'}(\Omega) \times W^{-1,p_2'}(\Omega)$ is pseudomonotone, bounded, and coercive. Consequently, we may apply the main theorem on pseudomonotone operators as stated in Theorem 1.55, so a solution $(u_1, u_2) \in W_0^{1,p_1}(\Omega) \times W_0^{1,p_2}(\Omega)$ of the equation

$$A(u_1, u_2) = 0 \tag{6.38}$$

exists. Taking into account that Eq. (6.38) reads as system (6.26), we can thus conclude that the solution (u_1, u_2) of (6.38) is a solution of the auxiliary problem (6.26), which completes the proof. □

Here is our main result on the existence of a solution to problem (6.1) within the ordered rectangle determined by a subsolution–supersolution.

Theorem 6.8. *Let* $(\underline{u}_1, \underline{u}_2), (\overline{u}_1, \overline{u}_2) \in W^{1,p_1}(\Omega) \times W^{1,p_2}(\Omega)$ *be a subsolution–supersolution of problem (6.1) such that condition* H(f_1, f_2) *is fulfilled. Then there exists a solution* $(u_1, u_2) \in W_0^{1,p_1}(\Omega) \times W_0^{1,p_2}(\Omega)$ *of problem (6.1) satisfying the enclosure property* $\underline{u}_i \leq u_i \leq \overline{u}_i$ *a.e. in* Ω, $i = 1, 2$.

Proof. Assumption H(f_1, f_2) enables us to apply Theorem 6.7, which provides a solution $(u_1, u_2) = (u_1(\lambda), u_2(\lambda))$ to the auxiliary problem (6.26) for every $\lambda > 0$ sufficiently large. We claim that $(u_1, u_2) \in [\underline{u}_1, \overline{u}_1] \times [\underline{u}_2, \overline{u}_2]$ (see (6.27)).

Let us check that $u_1 \le \overline{u}_1$ a.e. in Ω. To this end, we test with the function $(u_1 - \overline{u}_1)^+ = \max\{u_1 - \overline{u}_1, 0\} \in W_0^{1,p_1}(\Omega)$ in problem (6.26) and in the definition of supersolution to problem (6.1), which gives

$$\begin{aligned}&\langle -\Delta_{p_1} u_1 - \mu_1 \Delta_{q_1} u_1, (u_1 - \overline{u}_1)^+\rangle + \lambda \int_\Omega \pi_1(x, u_1)(u_1 - \overline{u}_1)^+ \, dx \\ &= \int_\Omega f_1(x, T_1 u_1, T_2 u_2, \nabla(T_1 u_1), \nabla(T_2 u_2))(u_1 - \overline{u}_1)^+ \, dx \end{aligned} \tag{6.39}$$

and

$$\begin{aligned}&\langle -\Delta_{p_1} \overline{u}_1 - \mu_1 \Delta_{q_1} \overline{u}_1, (u_1 - \overline{u}_1)^+\rangle \\ &\ge \int_\Omega f_1(x, \overline{u}_1, w_2, \nabla \overline{u}_1, \nabla w_2)(u_1 - \overline{u}_1)^+ \, dx \end{aligned} \tag{6.40}$$

whenever $w_2 \in W^{1,p_2}(\Omega)$ with $\underline{u}_2 \le w_2 \le \overline{u}_2$ a.e. in Ω. Thanks to the definition of the truncation operator T_2, $\underline{u}_2 \le T_2 u_2 \le \overline{u}_2$, so we can insert $w_2 = T_2 u_2$ in (6.40), resulting in

$$\begin{aligned}&\langle -\Delta_{p_1} \overline{u}_1 - \mu_1 \Delta_{q_1} \overline{u}_1, (u_1 - \overline{u}_1)^+\rangle \\ &\ge \int_\Omega f_1(x, \overline{u}_1, T_2 u_2, \nabla \overline{u}_1, \nabla(T_2 u_2))(u_1 - \overline{u}_1)^+ \, dx. \end{aligned} \tag{6.41}$$

Subtracting (6.41) from (6.39) provides

$$\begin{aligned}&\langle -\Delta_{p_1} u_1 - \mu_1 \Delta_{q_1} u_1 - (-\Delta_{p_1} \overline{u}_1 - \mu_1 \Delta_{q_1} \overline{u}_1), (u_1 - \overline{u}_1)^+\rangle \\ &\quad + \lambda \int_\Omega \pi_1(x, u_1)(u_1 - \overline{u}_1)^+ \, dx \\ &\le \int_\Omega (f_1(x, T_1 u_1, T_2 u_2, \nabla(T_1 u_1), \nabla(T_2 u_2)) \\ &\quad - f_1(x, \overline{u}_1, T_2 u_2, \nabla \overline{u}_1, \nabla(T_2 u_2)))(u_1 - \overline{u}_1)^+ dx \\ &= 0, \end{aligned}$$

where the last equality holds because $T_1 u_1 = \overline{u}_1$ on the set $\{u_1 > \overline{u}_1\}$. This amounts to saying that

$$\begin{aligned}&\int_{\{u_1 > \overline{u}_1\}} (|\nabla u_1|^{p_1 - 2} \nabla u_1 - |\nabla \overline{u}_1|^{p_1 - 2} \nabla \overline{u}_1) \cdot (\nabla u_1 - \nabla \overline{u}_1) \, dx \\ &\quad + \mu_1 \int_{\{u_1 > \overline{u}_1\}} (|\nabla u_1|^{q_1 - 2} \nabla u_1 - |\nabla \overline{u}_1|^{q_1 - 2} \nabla \overline{u}_1) \cdot (\nabla u_1 - \nabla \overline{u}_1) \, dx \end{aligned}$$

$$+\lambda \int_{\{u_1>\overline{u}_1\}} (u_1-\overline{u}_1)^{\frac{p_1}{p_1-\beta_1}}\,dx \le 0.$$

Then, the inequalities

$$(|\zeta|^{p_1-2}\zeta-|\eta|^{p_1-2}\eta)\cdot(\zeta-\eta)>0,\ (|\zeta|^{q_1-2}\zeta-|\eta|^{q_1-2}\eta)\cdot(\zeta-\eta)>0 \tag{6.42}$$

for all $\zeta,\eta\in\mathbb{R}^N$ with $\zeta\neq\eta$, ensure that $u_1\le\overline{u}_1$ a.e. in Ω.

On the same pattern, making use of adequately chosen test functions, we can show that $\underline{u}_1\le u_1$ and $\underline{u}_2\le u_2\le\overline{u}_2$ a.e. in Ω. Hence the claim $(u_1,u_2)\in[\underline{u}_1,\overline{u}_1]\times[\underline{u}_2,\overline{u}_2]$ is verified.

Due to the claim, we have that $T_iu_i=u_i$ and $\Pi_iu_i=0$ for $i=1,2$. Therefore problem (6.26) reduces to (6.1), thus (u_1,u_2) solves problem (6.1). □

As an application of Theorem 6.8 we show the existence of at least one positive solution of problem (6.1) under the following hypothesis:

$\mathrm{H}(f_1,f_2)'$ For $i=1,2$, there exist $a_i\in L^{\alpha_i}(\Omega)$, with $\alpha_i>N$ and $a_i(x)>0$ on a set of positive measure, $b_i\ge0$, $\beta_i\in[0,\frac{p_i}{(p_i^*)'})$, $\sigma_i\in L^{\gamma_i'}(\Omega)$, with $\gamma_i\in[1,p_i^*)$, and a Carathéodory function $g_i:\Omega\times\mathbb{R}\times\mathbb{R}^N\to\mathbb{R}$ satisfying

$$0\le a_i(x)\le f_i(x,s_1,s_2,\xi_1,\xi_2)\le g_i(x,s_i,\xi_i)\le\sigma_i(x)+b_i|\xi_i|^{\beta_i}$$

for a.a. $x\in\Omega$, all $s_1,s_2>0$, $\xi_1,\xi_2\in\mathbb{R}^N$, with b_i,β_i,σ_i as in $\mathrm{H}(f_1,f_2)$, and such that the Dirichlet problem

$$\begin{cases}-\Delta_{p_i}u-\mu_i\Delta_{q_i}u=g_i(x,u,\nabla u) & \text{in } \Omega,\\ u=0 & \text{on } \partial\Omega\end{cases}$$

has a supersolution $\overline{u}_i\in W^{1,p_i}(\Omega)\cap C(\overline{\Omega})$.

Theorem 6.9. *Assume that condition* $\mathrm{H}(f_1,f_2)'$ *is satisfied. Then there exists a solution* $(u_1,u_2)\in W_0^{1,p_1}(\Omega)\times W_0^{1,p_2}(\Omega)$ *of problem (6.1), which is positive, meaning that* $u_i>0$ *a.e. in* Ω, $i=1,2$.

Proof. In view of assumption $\mathrm{H}(f_1,f_2)'$, for $i=1,2$ it holds that $a_i\in W^{-1,p_i'}(\Omega)$, hence the Dirichlet problem

$$\begin{cases}-\Delta_{p_i}u-\mu_i\Delta_{q_i}u=a_i & \text{in } \Omega,\\ u=0 & \text{on } \partial\Omega\end{cases}$$

possesses a unique solution $\underline{u}_i\in W_0^{1,p_i}(\Omega)$. The assumption $a_i\in L^{\alpha_i}(\Omega)$ allows us to invoke [53, Theorem 3.1] for deducing that $\underline{u}_i$ is bounded. Then, through the nonlinear regularity theory (see Proposition 2.12), it follows that $\underline{u}_i \in C^1(\overline{\Omega})$.

Let us prove that

$$\underline{u}_i > 0 \text{ in } \Omega,\ i = 1, 2. \tag{6.43}$$

Acting with the test function $-\underline{u}_i^-$ produces

$$\langle -\Delta_{p_i}\underline{u}_i - \mu_i \Delta_{q_i}\underline{u}_i, -\underline{u}_i^-\rangle = -\int_\Omega a_i(x)\underline{u}_i^-(x) \leq 0,$$

which yields

$$\int_\Omega (|\nabla \underline{u}_i^-|^{p_i} + \mu_i |\nabla \underline{u}_i^-|^{q_i})\, dx \leq 0,$$

thereby $\underline{u}_i \geq 0$ a.e. in Ω. Taking into account that $\underline{u}_i$ is nontrivial, the strong maximum principle in Theorem 2.19 implies (6.43).

We further claim that

$$\underline{u}_i \leq \overline{u}_i \text{ a.e. in } \Omega,\ i = 1, 2. \tag{6.44}$$

Arguing indirectly, assume, for instance, that the open set $\{\underline{u}_1 > \overline{u}_1\}$ is nonempty. Using hypothesis $\mathrm{H}(f_1, f_2)'$ and acting with $(\underline{u}_1 - \overline{u}_1)^+$ imply

$$\begin{aligned}
&\langle -\Delta_{p_1}\underline{u}_1 - \mu_1 \Delta_{q_1}\underline{u}_1, (\underline{u}_1 - \overline{u}_1)^+\rangle = \int_\Omega a_1(x)(\underline{u}_1 - \overline{u}_1)^+\, dx \\
&\leq \int_\Omega g_1(x, \overline{u}_1(x), \nabla \overline{u}_1(x))(\underline{u}_1 - \overline{u}_1)^+\, dx \\
&\leq \langle -\Delta_{p_1}\overline{u}_1 - \mu_1 \Delta_{q_1}\overline{u}_1, (\underline{u}_1 - \overline{u}_1)^+\rangle,
\end{aligned}$$

hence

$$\begin{aligned}
&\int_{\{\underline{u}_1 > \overline{u}_1\}} (|\nabla \underline{u}_1|^{p_1-2}\nabla \underline{u}_1 - |\nabla \overline{u}_1|^{p_1-2}\nabla \overline{u}_1) \cdot (\nabla \underline{u}_1 - \nabla \overline{u}_1)\, dx \\
&\quad + \mu_1 \int_{\{\underline{u}_1 > \overline{u}_1\}} (|\nabla \underline{u}_1|^{q_1-2}\nabla \underline{u}_1 - |\nabla \overline{u}_1|^{q_1-2}\nabla \overline{u}_1) \cdot (\nabla \underline{u}_1 - \nabla \overline{u}_1)\, dx \leq 0.
\end{aligned}$$

From the strict inequalities in (6.42), we infer that the function $\overline{u}_1 - \underline{u}_1$ is locally constant on $\{\underline{u}_1 > \overline{u}_1\}$. Taking a path $\gamma : [0, 1] \to \overline{\Omega}$ such that $\gamma(0) \in \{\underline{u}_1 > \overline{u}_1\}$ and $\gamma(1) \in \partial\Omega$, and noting that $\underline{u}_1 \leq \overline{u}_1$ on $\partial\Omega$, we find $t_0 \in (0, 1]$ such that $\gamma(t_0) \notin \{\underline{u}_1 > \overline{u}_1\}$ and $\gamma(t) \in \{\underline{u}_1 > \overline{u}_1\}$ for all $t \in [0, t_0)$. The continuity of $\underline{u}_1, \overline{u}_1$ and the fact that $\underline{u}_1 - \overline{u}_1$ is locally constant on $\{\underline{u}_1 > \overline{u}_1\}$ yield a contradiction. So (6.44) ensues.

Next we prove that $(\underline{u}_1, \underline{u}_2)$, $(\overline{u}_1, \overline{u}_2)$ is a subsolution–supersolution for problem (6.1). By hypothesis $H(f_1, f_2)'$ we derive that

$$\begin{aligned}\langle -\Delta_{p_1}\underline{u}_1 - \mu_1\Delta_{q_1}\underline{u}_1, v\rangle &= \int_\Omega a_1(x)v(x)\,dx \\ &\le \int_\Omega f_1(x, \underline{u}_1, w_2, \nabla\underline{u}_1, \nabla w_2)v\,dx\end{aligned}$$

and

$$\begin{aligned}\int_\Omega f_1(x, \overline{u}_1, w_2, \nabla\overline{u}_1, \nabla w_2)v\,dx &\le \int_\Omega g_1(x, \overline{u}_1, \nabla\overline{u}_1)v\,dx \\ &\le \langle -\Delta_{p_1}\overline{u}_1 - \mu_1\Delta_{q_1}\overline{u}_1, v\rangle\end{aligned}$$

for all $v \in W_0^{1,p_1}(\Omega)$ with $v \ge 0$ a.e. in Ω and all $w_2 \in W_0^{1,p_2}(\Omega)$ with $\underline{u}_2 \le w_2 \le \overline{u}_2$ a.e. in Ω (notice that due to (6.44), it makes sense to consider $w_2 \in W_0^{1,p_2}(\Omega)$ such that $\underline{u}_2 \le w_2 \le \overline{u}_2$ a.e. in Ω). Similarly, from hypothesis $H(f_1, f_2)'$ it is seen that

$$\begin{aligned}\langle -\Delta_{p_2}\underline{u}_2 - \mu_2\Delta_{q_2}\underline{u}_2, v\rangle &= \int_\Omega a_2(x)v(x)\,dx \\ &\le \int_\Omega f_2(x, w_1, \underline{u}_2, \nabla w_1, \nabla\underline{u}_2)v\,dx\end{aligned}$$

and

$$\begin{aligned}\int_\Omega f_2(x, w_1, \overline{u}_2, \nabla w_1, \nabla\overline{u}_2)v\,dx &\le \int_\Omega g_2(x, \overline{u}_2, \nabla\overline{u}_2)v\,dx \\ &\le \langle -\Delta_{p_2}\overline{u}_2 - \mu_2\Delta_{q_2}\overline{u}_2, v\rangle \qquad (6.45)\end{aligned}$$

for all $v \in W_0^{1,p_2}(\Omega)$ with $v \ge 0$ a.e. in Ω and all $w_1 \in W_0^{1,p_1}(\Omega)$ with $\underline{u}_1 \le w_1 \le \overline{u}_1$ a.e. in Ω. Altogether, $(\underline{u}_1, \underline{u}_2)$, $(\overline{u}_1, \overline{u}_2)$ is a subsolution–supersolution for (6.1).

Now we are in a position to apply Theorem 6.8, which ensures the existence of a solution $(u_1, u_2) \in W_0^{1,p_1}(\Omega) \times W_0^{1,p_2}(\Omega)$ of problem (6.1) satisfying $\underline{u}_i \le u_i \le \overline{u}_i$ a.e. in Ω, $i = 1, 2$. Then, in view of (6.43), we infer that $u_i > 0$ a.e. in Ω, $i = 1, 2$. □

Example 6.10. Hypothesis $H(f_1, f_2)'$ is fulfilled whenever there exist $a_i, a_i' \in L^{\alpha_i}(\Omega)$, with $\alpha_i > N$ and $a_i(x) > 0$ on a set of positive measure, such that

$$0 \le a_i(x) \le f_i(x, s_1, s_2, \xi_1, \xi_2) \le a_i'(x)$$

for a.a. $x \in \Omega$ and all $s_1, s_2 > 0$, $\xi_1, \xi_2 \in \mathbb{R}^N$.

6.3 ELLIPTIC SYSTEMS WITH VARIATIONAL STRUCTURE

We turn to the study of the following system of quasilinear elliptic equations with homogeneous Dirichlet boundary condition in the variational form

$$\begin{cases} -\Delta_{p_1} u_1 = f_{s_1}(x, u_1(x), u_2(x)) & \text{in } \Omega, \\ -\Delta_{p_2} u_2 = f_{s_2}(x, u_1(x), u_2(x)) & \text{in } \Omega, \\ u_1 = u_2 = 0, & \text{on } \partial\Omega, \end{cases} \tag{6.46}$$

on a bounded domain $\Omega \subset \mathbb{R}^N$ with a C^2 boundary $\partial\Omega$, where $p_1 > 1$ and $p_2 > 1$ are fixed numbers.

In (6.46) we have the p_i-Laplacian operator $\Delta_{p_i} u = \operatorname{div}(|\nabla u|^{p_i-2}\nabla u)$ on the space $W_0^{1,p_i}(\Omega)$ for $i = 1, 2$. The first eigenvalue of $-\Delta_{p_i}$ on $W_0^{1,p_i}(\Omega)$ as given in Proposition 1.57 is denoted by λ_{1,p_i}, while by $\hat{u}_{1,p_i}$ we denote the corresponding positive eigenfunction satisfying $\|\hat{u}_{1,p_i}\|_{p_i} = 1$. The second eigenvalue of $-\Delta_{p_i}$ on $W_0^{1,p_i}(\Omega)$ is denoted by λ_{2,p_i}.

The right-hand sides of the equations in (6.46) are expressed through the partial derivatives f_{s_1} and f_{s_2} of a function $f(x, s_1, s_2)$ with $f : \Omega \times \mathbb{R} \times \mathbb{R} \to \mathbb{R}$ that is assumed to be Carathéodory (i.e., measurable in $x \in \Omega$ and continuous in $(s_1, s_2) \in \mathbb{R}^2$), twice differentiable with respect to $(s_1, s_2) \in \mathbb{R}^2$ and the partial derivatives f_{s_1}, f_{s_2}, $f_{s_1s_1}$, $f_{s_1s_2}$, $f_{s_2s_2}$ are Carathéodory functions on $\Omega \times \mathbb{R}^2$, with f_{s_1}, f_{s_2} bounded on bounded sets. We suppose without any loss of generality that $f(x, 0, 0) = 0$ for a.e. $x \in \Omega$.

To the system (6.46) we associate the following elliptic equations with homogeneous Dirichlet boundary condition

$$\begin{cases} -\Delta_{p_1} u_1 = f_{s_1}(x, u_1(x), 0) & \text{in } \Omega, \\ u_1 = 0, & \text{on } \partial\Omega \end{cases} \tag{6.47}$$

and

$$\begin{cases} -\Delta_{p_2} u_2 = f_{s_2}(x, 0, u_2(x)) & \text{in } \Omega, \\ u_2 = 0, & \text{on } \partial\Omega. \end{cases} \tag{6.48}$$

A weak solution of (6.46) is any pair $(u_1, u_2) \in W_0^{1,p_1}(\Omega) \times W_0^{1,p_2}(\Omega)$ such that

$$\int_\Omega |\nabla u_1|^{p_1-2}\nabla u_1 \cdot \nabla v_1 \, dx = \int_\Omega f_{s_1}(x, u_1, u_2) v_1 \, dx,$$
$$\int_\Omega |\nabla u_2|^{p_2-2}\nabla u_2 \cdot \nabla v_2 \, dx = \int_\Omega f_{s_2}(x, u_1, u_2) v_2 \, dx$$

for all $(v_1, v_2) \in W_0^{1,p_1}(\Omega) \times W_0^{1,p_2}(\Omega)$. Moreover, we say that the solution (u_1, u_2) is positive (resp. negative) if u_1, u_2 are both positive (resp. negative).

The next definition introduces the notion of subsolution–supersolution (or trapping region) in the case of the system (6.46). It is a particular case of (6.1) considered in Sections 6.1 and 6.2.

Definition 6.11. We say that $(\underline{u}_1, \underline{u}_2)$, $(\overline{u}_1, \overline{u}_2) \in W^{1,p_1}(\Omega) \times W^{1,p_2}(\Omega)$ form a subsolution–supersolution for problem (6.46) if $\underline{u}_i \le \overline{u}_i$ a.e. in Ω, $\underline{u}_i \le 0 \le \overline{u}_i$ a.e. on $\partial\Omega$ for $i = 1, 2$,

$$\int_\Omega \left(|\nabla \underline{u}_1|^{p_1-2}\nabla \underline{u}_1 \cdot \nabla v_1 - f_{s_1}(x, \underline{u}_1, w_2)v_1\right) dx + \int_\Omega \left(|\nabla \underline{u}_2|^{p_2-2}\nabla \underline{u}_2 \cdot \nabla v_2 - f_{s_2}(x, w_1, \underline{u}_2)v_2\right) dx \le 0,$$

and

$$\int_\Omega \left(|\nabla \overline{u}_1|^{p_1-2}\nabla \overline{u}_1 \cdot \nabla v_1 - f_{s_1}(x, \overline{u}_1, w_2)v_1\right) dx + \int_\Omega \left(|\nabla \overline{u}_2|^{p_2-2}\nabla \overline{u}_2 \cdot \nabla v_2 - f_{s_2}(x, w_1, \overline{u}_2)v_2\right) dx \ge 0$$

for all $(v_1, v_2) \in W_0^{1,p_1}(\Omega) \times W_0^{1,p_2}(\Omega)$ such that $v_1, v_2 \ge 0$ a.e. in Ω, and all $(w_1, w_2) \in W^{1,p_1}(\Omega) \times W^{1,p_2}(\Omega)$ with $\underline{u}_i \le w_i \le \overline{u}_i$ for $i = 1, 2$.

We formulate the hypotheses on the function $f(x, s_1, s_2)$ in (6.46):

H$(f)_1$ There are constants $k_1 > 0$, $k_2 > 0$, $d_1 < 0$, $d_2 < 0$ such that

$$\begin{aligned}
&f_{s_1}(x, k_1, s_2) \le 0 \text{ for a.e. } x \in \Omega \text{ and every } s_2 \in [0, k_2],\\
&f_{s_1}(x, d_1, s_2) \ge 0 \text{ for a.e. } x \in \Omega \text{ and every } s_2 \in [d_2, 0],\\
&f_{s_2}(x, s_1, k_2) \le 0 \text{ for a.e. } x \in \Omega \text{ and every } s_1 \in [0, k_1],\\
&f_{s_2}(x, s_1, d_2) \ge 0 \text{ for a.e. } x \in \Omega \text{ and every } s_1 \in [d_1, 0].
\end{aligned}$$

H$(f)_2$ There are constants $c_1 > \lambda_{1,p_1}$ and $c_2 > \lambda_{1,p_2}$ such that

$$\begin{aligned}
&\liminf_{s_1\to 0^+} \frac{f_{s_1}(x, s_1, s_2)}{s_1^{p_1-1}} \ge c_1 \text{ uniformly for a.e. } x \in \Omega \text{ and every } 0 < s_2 \le k_2,\\
&\liminf_{s_1\to 0^-} \frac{f_{s_1}(x, s_1, s_2)}{|s_1|^{p_1-2}s_1} \ge c_1 \text{ uniformly for a.e. } x \in \Omega \text{ and every } d_2 \le s_2 < 0,\\
&\liminf_{s_2\to 0^+} \frac{f_{s_2}(x, s_1, s_2)}{s_2^{p_2-1}} \ge c_2 \text{ uniformly for a.e. } x \in \Omega \text{ and every } 0 < s_1 \le k_1,\\
&\liminf_{s_2\to 0^-} \frac{f_{s_2}(x, s_1, s_2)}{|s_2|^{p_2-2}s_2} \ge c_2 \text{ uniformly for a.e. } x \in \Omega \text{ and every } d_1 \le s_1 < 0.
\end{aligned}$$

We recall that the positive cone $C_0^1(\overline{\Omega})_+ \subset C_0^1(\overline{\Omega})$ has a nonempty interior characterized in (1.39).

Theorem 6.12. *Assume that hypotheses* $\mathrm{H}(f)_1$, $\mathrm{H}(f)_2$ *are satisfied. Then the following assertions hold:*

(a) *Problem (6.46) has a positive solution* $(u_1, u_2) \in \operatorname{int}(C_0^1(\overline{\Omega})_+) \times \operatorname{int}(C_0^1(\overline{\Omega})_+)$ *and a negative solution* $(v_1, v_2) \in \big(-\operatorname{int}(C_0^1(\overline{\Omega})_+)\big) \times \big(-\operatorname{int}(C_0^1(\overline{\Omega})_+)\big)$ *with* $u_i \leq k_i$ *and* $v_i \geq d_i$, $i = 1, 2$.

(b) *Problem (6.47) admits a positive solution* $u_1 \in \operatorname{int}(C_0^1(\overline{\Omega})_+)$ *with* $u_1 \leq k_1$ *and a negative solution* $v_1 \in -\operatorname{int}(C_0^1(\overline{\Omega})_+)$ *with* $v_1 \geq d_1$.

(c) *Problem (6.48) admits a positive solution* $u_2 \in \operatorname{int}(C_0^1(\overline{\Omega})_+)$ *with* $u_2 \leq k_2$ *and a negative solution* $v_2 \in -\operatorname{int}(C_0^1(\overline{\Omega})_+)$ *with* $v_2 \geq d_2$.

Proof. (a) Assumption $\mathrm{H}(f)_1$ implies that

$$\begin{cases} -\Delta_{p_1}(k_1) - f_{s_1}(x, k_1, s_2) \geq 0 & \text{for a.e. } x \in \Omega \text{ and every } s_2 \in [0, k_2], \\ -\Delta_{p_2}(k_2) - f_{s_2}(x, s_1, k_2) \geq 0 & \text{for a.e. } x \in \Omega \text{ and every } s_1 \in [0, k_1]. \end{cases} \tag{6.49}$$

Hypothesis $\mathrm{H}(f)_2$ guarantees that there exists $\delta \in (0, \min\{k_1, k_2\})$ such that

$$\begin{cases} f_{s_1}(x, s_1, s_2) > \lambda_{1,p_1} s_1^{p_1 - 1} & \text{for a.e. } x \in \Omega, \text{ every } s_1 \in (0, \delta), \, s_2 \in (0, k_2], \\ f_{s_2}(x, s_1, s_2) > \lambda_{1,p_2} s_2^{p_2 - 1} & \text{for a.e. } x \in \Omega, \text{ every } s_1 \in (0, k_1], \, s_2 \in (0, \delta). \end{cases} \tag{6.50}$$

Choose $\varepsilon > 0$ sufficiently small such that

$$\varepsilon \hat{u}_{1,p_i}(x) \leq \delta \quad \text{for all } x \in \Omega, \, i = 1, 2.$$

Then from (6.50) we infer that

$$\begin{cases} -\Delta_{p_1}(\varepsilon \hat{u}_{1,p_1}) - f_{s_1}(x, \varepsilon \hat{u}_{1,p_1}, s_2) \leq 0 & \text{for a.e. } x \in \Omega \text{ and every } s_2 \in [0, k_2], \\ -\Delta_{p_2}(\varepsilon \hat{u}_{1,p_2}) - f_{s_2}(x, s_1, \varepsilon \hat{u}_{1,p_2}) \leq 0 & \text{for a.e. } x \in \Omega \text{ and every } s_1 \in [0, k_1]. \end{cases} \tag{6.51}$$

By (6.49) and (6.51) we infer that $(\underline{u}_1, \underline{u}_2) := (\varepsilon \hat{u}_{1,p_1}, \varepsilon \hat{u}_{1,p_2})$ and $(\overline{u}_1, \overline{u}_2) := (k_1, k_2)$ form a pair of subsolution and supersolution in the sense of Definition 6.11 for problem (6.46) provided $\varepsilon > 0$ is sufficiently small. Now we apply the general result in Theorem 6.8 ensuring the existence of a positive solution $(u_1, u_2) \in \operatorname{int}(C_0^1(\overline{\Omega})_+) \times \operatorname{int}(C_0^1(\overline{\Omega})_+)$ of problem (6.46) satisfying $u_i \leq k_i$, $i = 1, 2$ (see also [38, Section 5.5]).

Following the same reasoning, we show that (d_1, d_2) and $(-\varepsilon \hat{u}_{1,p_1}, -\varepsilon \hat{u}_{1,p_2})$ constitute a pair of subsolution–supersolution in the sense of Definition 6.11 for problem (6.46) provided $\varepsilon > 0$ is sufficiently small. Consequently, as before, we obtain a negative solution $(v_1, v_2) \in \big(-\operatorname{int}(C_0^1(\overline{\Omega})_+)\big) \times \big(-\operatorname{int}(C_0^1(\overline{\Omega})_+)\big)$ with $v_i \geq d_i$, $i = 1, 2$.

(b) From (6.49) we derive

$$-\Delta_{p_1}(k_1) - f_{s_1}(x, k_1, 0) \geq 0 \text{ for a.e. } x \in \Omega,$$

whereas (6.51) implies

$$-\Delta_{p_1}(\varepsilon \hat{u}_{1,p_1}) - f_{s_1}(x, \varepsilon \hat{u}_{1,p_1}, 0) \leq 0 \text{ for a.e. } x \in \Omega.$$

Thus $\underline{u}_1 := \varepsilon \hat{u}_{1,p_1}$ and $\overline{u}_1 := k_1$ constitute a pair of subsolution and supersolution for problem (6.47) provided $\varepsilon > 0$ is sufficiently small. This results in the existence of a positive solution $u_1 \in \operatorname{int}(C_0^1(\overline{\Omega})_+)$ of (6.47) with $u_1 \leq k_1$.

On the same pattern, we construct an ordered pair of subsolution and supersolution for problem (6.47), which leads to a negative solution $v_1 \in -\operatorname{int}(C_0^1(\overline{\Omega})_+)$ of (6.47) with $v_1 \geq d_1$.

(c) The proof is similar to that of (b). □

Next we are concerned with the existence of a minimal positive solution and of a maximal negative solution of (6.46), (6.47), (6.48) in prescribed trapping regions away from zero.

Theorem 6.13. *Assume that hypotheses* $\mathrm{H}(f)_1$, $\mathrm{H}(f)_2$ *are satisfied. Then:*

(a) *Given a solution* (u_1, u_2) *of problem (6.46) in* $[\varepsilon \hat{u}_{1,p_1}, k_1] \times [\varepsilon \hat{u}_{1,p_2}, k_2]$ *for some* $\varepsilon > 0$, *there exists a minimal solution* $(u_1^\varepsilon, u_2^\varepsilon)$ *of (6.46) in* $[\varepsilon \hat{u}_{1,p_1}, k_1] \times [\varepsilon \hat{u}_{1,p_2}, k_2]$ *such that* $u_i^\varepsilon \leq u_i$, $i = 1, 2$. *Similarly, given a solution* (v_1, v_2) *of problem (6.46) in* $[d_1, -\varepsilon \hat{u}_{1,p_1}] \times [d_2, -\varepsilon \hat{u}_{1,p_2}]$ *for some* $\varepsilon > 0$, *there exists a maximal solution* $(v_1^\varepsilon, v_2^\varepsilon)$ *of (6.46) in* $[d_1, -\varepsilon \hat{u}_{1,p_1}] \times [d_2, -\varepsilon \hat{u}_{1,p_2}]$ *such that* $v_i^\varepsilon \geq v_i$, $i = 1, 2$.

(b) *There exist a minimal solution* u_1^ε *of (6.47) in* $[\varepsilon \hat{u}_{1,p_1}, k_1]$ *and a maximal solution* v_1^ε *of (6.47) in* $[d_1, -\varepsilon \hat{u}_{1,p_1}]$.

(c) *There exist a minimal solution* u_2^ε *of (6.48) in* $[\varepsilon \hat{u}_{1,p_2}, k_2]$ *and a maximal solution* v_2^ε *of (6.48) in* $[d_2, -\varepsilon \hat{u}_{1,p_2}]$.

Proof. (a) We provide the proof only for the first part of this assertion because the second part can be verified analogously. Theorem 6.12 ensures that the solution (u_1, u_2) exists. For a fixed $\varepsilon > 0$ let us denote by $\mathcal{S}_\varepsilon$ the set of all $(v_1, v_2) \in [\varepsilon \hat{u}_{1,p_1}, k_1] \times [\varepsilon \hat{u}_{1,p_2}, k_2]$ solving (6.46) and $v_i \leq u_i$, $i = 1, 2$. We prove that there is a minimal element of $\mathcal{S}_\varepsilon$, which will be done by applying Zorn's lemma. Consider a chain C in $\mathcal{S}_\varepsilon$. Then there is a sequence $\{(u_1^k, u_2^k)\}_{k \geq 1} \subset C$, with $u_i^{k+1} \leq u_i^k$, $i = 1, 2$, for all $k \geq 1$, such that

$$\inf C = \inf\{(u_1^k, u_2^k) : k \geq 1\}$$

(see, e.g., [112, Lemma 3.10]). Using that (u_1^k, u_2^k) is a solution of (6.46), we can easily establish the boundedness of the sequence $\{(u_1^k, u_2^k)\}_{k \geq 1}$ in $W_0^{1,p_1}(\Omega) \times W_0^{1,p_2}(\Omega)$. Along a subsequence, we may suppose that $u_i^k \rightharpoonup \hat{u}_i$ in $W_0^{1,p_i}(\Omega)$

and a.e. in Ω as $k \to \infty$, for $i = 1, 2$, thus $(\hat{u}_1, \hat{u}_2) \in [\varepsilon\hat{u}_{1,p_1}, k_1] \times [\varepsilon\hat{u}_{1,p_2}, k_2]$ and $\hat{u}_i \le u_i$, for $i = 1, 2$. Moreover, through the $(S)_+$-property of the operators $-\Delta_{p_1}$ and $-\Delta_{p_2}$ given in Proposition 1.56, we infer the strong convergence $u_i^k \to \hat{u}_i$ in $W_0^{1,p_i}(\Omega)$ as $k \to \infty$, for $i = 1, 2$. Consequently, $(\hat{u}_1, \hat{u}_2)$ is a solution of (6.46) that belongs to $\mathcal{S}_\varepsilon$ and $\inf C = (\hat{u}_1, \hat{u}_2) \in \mathcal{S}_\varepsilon$. Then Zorn's lemma can be applied, which yields the existence of $(u_1^\varepsilon, u_2^\varepsilon)$ as required.

Assertions (b) and (c) can be proven along the same lines. □

In order to obtain the existence of a minimal positive solution and of a maximal negative solution for (6.46), (6.47), and (6.48), we need a new hypothesis:

H$(f)_3$ There are constants $\alpha_1 \ge c_1$ and $\alpha_2 \ge c_2$ such that

$$\limsup_{s_1 \to 0^+} \frac{f_{s_1}(x, s_1, s_2)}{s_1^{p_1 - 1}} \le \alpha_1 \text{ uniformly for a.e. } x \in \Omega \text{ and every } 0 < s_2 \le k_2,$$

$$\limsup_{s_1 \to 0^-} \frac{f_{s_1}(x, s_1, s_2)}{|s_1|^{p_1 - 2} s_1} \le \alpha_1 \text{ uniformly for a.e. } x \in \Omega \text{ and every } d_2 \le s_2 < 0,$$

$$\limsup_{s_2 \to 0^+} \frac{f_{s_2}(x, s_1, s_2)}{s_2^{p_2 - 1}} \le \alpha_2 \text{ uniformly for a.e. } x \in \Omega \text{ and every } 0 < s_1 \le k_1,$$

$$\limsup_{s_2 \to 0^-} \frac{f_{s_2}(x, s_1, s_2)}{|s_2|^{p_2 - 2} s_2} \le \alpha_2 \text{ uniformly for a.e. } x \in \Omega \text{ and every } d_1 \le s_1 < 0.$$

Theorem 6.14. *Assume that hypotheses* H$(f)_1$–H$(f)_3$ *hold. Then:*

(a) *Problem (6.46) admits a positive solution* $(u_{1,+}, u_{2,+}) \in \operatorname{int}(C_0^1(\overline{\Omega})_+) \times \operatorname{int}(C_0^1(\overline{\Omega})_+)$ *with* $u_{i,+} \le k_i$, *for* $i = 1, 2$, *which is minimal among the positive solutions of (6.46), and a negative solution* $(u_{1,-}, u_{2,-}) \in \left(-\operatorname{int}(C_0^1(\overline{\Omega})_+)\right) \times \left(-\operatorname{int}(C_0^1(\overline{\Omega})_+)\right)$ *with* $u_{i,-} \ge d_i$, *for* $i = 1, 2$, *which is maximal among the negative solutions of (6.46).*

(b) *Problem (6.47) admits the smallest positive solution* $u_1 \in \operatorname{int}(C_0^1(\overline{\Omega})_+)$ *with* $u_1 \le k_1$ *and the biggest negative solution* $v_1 \in -\operatorname{int}(C_0^1(\overline{\Omega})_+)$ *with* $v_1 \ge d_1$.

(c) *Problem (6.48) admits the smallest positive solution* $u_2 \in \operatorname{int}(C_0^1(\overline{\Omega})_+)$ *with* $u_2 \le k_2$ *and the biggest negative solution* $v_2 \in -\operatorname{int}(C_0^1(\overline{\Omega})_+)$ *with* $v_2 \ge d_2$.

Proof. (a) We only prove the first part of this assertion because the second part can be proven through a similar reasoning. The successive application of Theorem 6.13 produces a sequence $\{(u_1^n, u_2^n)\}_{n \ge n_0}$ in $W_0^{1,p_1}(\Omega) \times W_0^{1,p_2}(\Omega)$, with n_0 sufficiently large, such that for every integer $n \ge n_0$ we have that (u_1^n, u_2^n) is a solution of problem (6.46) that is minimal (in the sense of minimal element in a partially ordered set) in the trapping region $[\frac{1}{n}\hat{u}_{1,p_1}, k_1] \times [\frac{1}{n}\hat{u}_{1,p_2}, k_2]$ and

$u_i^{n+1} \le u_i^n$, $i = 1, 2$. This enables us to get

$$u_i^n \rightharpoonup u_{i,+} \text{ in } W_0^{1,p_i}(\Omega) \text{ and } u_i^n \downarrow u_{i,+} \text{ pointwise in } \Omega \text{ for } i = 1, 2, \tag{6.52}$$

with some $(u_{1,+}, u_{2,+}) \in W_0^{1,p_1}(\Omega) \times W_0^{1,p_2}(\Omega)$. By the same reasoning as in the proof of Theorem 6.13 based on the $(S)_+$-property of the operators $-\Delta_{p_1}$ and $-\Delta_{p_2}$ given in Proposition 1.56, we can show that $(u_{1,+}, u_{2,+})$ is a solution for (6.46).

We claim that $u_{i,+} \neq 0$ for $i = 1, 2$. Arguing by contradiction, let us suppose that $u_{2,+} = 0$. For every $n \ge n_0$, we set

$$\tilde{u}_n = \frac{u_1^n}{\|u_1^n\|} \quad \text{and} \quad h_n = \frac{f_{s_1}(x, u_1^n, u_2^n)}{(u_1^n)^{p_1-1}} \tilde{u}_n^{p_1-1}.$$

Along a relabeled subsequence, on the basis of H$(f)_2$, H$(f)_3$ we may assume that

$$\tilde{u}_n \rightharpoonup \tilde{u} \text{ in } W_0^{1,p_1}(\Omega) \text{ a.e. in } \Omega, \text{ and } h_n \rightharpoonup h \text{ in } L^{\frac{p_1}{p_1-1}}(\Omega),$$

with some $\tilde{u} \in W_0^{1,p_i}(\Omega)$ and $h \in L^{\frac{p_1}{p_1-1}}(\Omega)$.

The fact that (u_1^n, u_2^n) is a solution of problem (6.46) entails

$$\int_\Omega |\nabla \tilde{u}_n|^{p_1-2} \nabla \tilde{u}_n \cdot \nabla(\tilde{u}_n - \tilde{u})\, dx = \int_\Omega h_n(\tilde{u}_n - \tilde{u})\, dx. \tag{6.53}$$

Then (6.53) gives

$$\lim_{n\to\infty} \int_\Omega |\nabla \tilde{u}_n|^{p_1-2} \nabla \tilde{u}_n \cdot \nabla(\tilde{u}_n - \tilde{u})\, dx = 0.$$

The $(S)_+$-property of the operator $-\Delta_{p_1}$ in Proposition 1.56 implies that $\tilde{u}_n \to \tilde{u}$ in $W_0^{1,p_1}(\Omega)$, which renders $\tilde{u} \neq 0$. Hypotheses H$(f)_2$ and H$(f)_3$ ensure that for any $\varepsilon > 0$ and a.a. $x \in \Omega$ there exists an integer $n(x)$ such that for every $n \ge n(x)$,

$$(c_1 - \varepsilon)\tilde{u}_n(x)^{p_1-1} \le h_n(x) \le (\alpha_1 + \varepsilon)\tilde{u}_n(x)^{p_1-1}.$$

Letting $n \to \infty$ and because $\varepsilon > 0$ is arbitrary, we arrive through Mazur's lemma at

$$c_1\tilde{u}(x)^{p_1-1} \le h(x) = g(x)\tilde{u}(x)^{p_1-1} \le \alpha_1\tilde{u}(x)^{p_1-1},$$

with $c_1 \le g(x) \le \alpha_1$ a.e. in Ω. Consequently, $\tilde{u}$ is an eigenfunction associated to the eigenvalue 1 for the weighted problem (with the weight $g(x) > 0$)

$$\begin{cases} -\Delta_{p_1}\tilde{u}(x) = g(x)\tilde{u}(x)^{p_1-1} & \text{in } \Omega, \\ \tilde{u} = 0 & \text{on } \partial\Omega. \end{cases} \tag{6.54}$$

Since the eigenfunction $\tilde{u}$ of (6.54) does not change sign, it follows that $1 = \lambda_1(g)$, that is, the first eigenvalue for

$$\begin{cases} -\Delta_{p_1}\tilde{u}(x) = \lambda g(x)\tilde{u}(x)^{p_1-1} & \text{in } \Omega, \\ \tilde{u} = 0 & \text{on } \partial\Omega \end{cases}$$

(see Proposition 1.57). On the other hand we know that $\lambda_1(g) \leq \lambda_1(c_1) < \lambda_1(\lambda_1) = 1$, which is a contradiction. Therefore $u_{i,+} \neq 0$ for $i = 1, 2$. Applying the strong maximum principle in Theorem 2.19 leads to $(u_{1,+}, u_{2,+}) \in \operatorname{int}(C_0^1(\overline{\Omega})_+) \times \operatorname{int}(C_0^1(\overline{\Omega})_+)$.

It remains to show that $(u_{1,+}, u_{2,+})$ is a minimal positive solution of problem (6.46). Let $(v_1, v_2) \in W_0^{1,p_1}(\Omega) \times W_0^{1,p_2}(\Omega)$ be a positive solution of (6.46) such that $v_1 \leq u_{1,+}$ and $v_2 \leq u_{2,+}$. By the strong maximum principle in Theorem 2.19, we know that $(v_1, v_2) \in \operatorname{int}(C_0^1(\overline{\Omega})_+) \times \operatorname{int}(C_0^1(\overline{\Omega})_+)$. This in conjunction with the construction of the solution $(u_{1,+}, u_{2,+})$ guarantees that

$$\frac{1}{n}\hat{u}_{i,p_i} \leq v_i \leq u_{i,+} \leq u_i^n \leq k_i, \quad i = 1, 2, \tag{6.55}$$

provided n is sufficiently large. Since (u_1^n, u_2^n) is a minimal solution of (6.46) in $[\frac{1}{n}\hat{u}_{1,p_1}, k_1] \times [\frac{1}{n}\hat{u}_{1,p_2}, k_2]$, from (6.55) we derive that $u_i^n \leq v_i$, $i = 1, 2$. By simple comparison with (6.55), it follows that $u_{1,+} = v_1$ and $u_{2,+} = v_2$, which completes the proof.

Assertions (b) and (c) can be proven along the same lines. □

Remark 6.15. The results of this section presented until now do not require a variational structure for problem (6.46). They hold for Carathéodory functions $g_1(x, s_1, s_2)$ and $g_2(x, s_1, s_2)$ in place of the partial derivatives $f_{s_1}(x, s_1, s_2)$ and $f_{s_2}(x, s_1, s_2)$, respectively, provided hypotheses $H(f)_1$–$H(f)_3$ are satisfied by g_1 and g_2.

Theorem 6.14 provides the existence of a minimal positive solution $(u_{1,+}, u_{2,+})$ of problem (6.46) with $u_{i,+} \leq k_i$ for $i = 1, 2$. Corresponding to it, we introduce the truncation $\tau_+ : \Omega \times \mathbb{R}^2 \to \mathbb{R}^2$ in the following way: if $x \in \Omega$ and $(s_1, s_2) \in \mathbb{R}^2$, we define $\tau_+(x, s_1, s_2)$ to be the projection of (s_1, s_2) on the closed convex subset $[0, u_{1,+}(x)] \times [0, u_{2,+}(x)]$ of $\mathbb{R}^2$. Similarly, using the maximal negative solution $(u_{1,-}, u_{2,-})$ of problem (6.46) with $u_{i,-} \geq d_i$ for $i = 1, 2$, which is provided by Theorem 6.14, we introduce the truncation $\tau_- : \Omega \times \mathbb{R}^2 \to \mathbb{R}^2$ by assigning to every $x \in \Omega$ and $(s_1, s_2) \in \mathbb{R}^2$ the projection $\tau_-(x, s_1, s_2)$ of (s_1, s_2) on the closed convex subset $[u_{1,-}(x), 0] \times [u_{2,-}(x), 0]$ of $\mathbb{R}^2$. Finally, we define the truncation $\tau_0 : \Omega \times \mathbb{R}^2 \to \mathbb{R}^2$ by letting $\tau_0(x, s_1, s_2)$ be the projection of (s_1, s_2) on the closed convex subset $[u_{1,-}(x), u_{1,+}(x)] \times [u_{2,-}(x), u_{2,+}(x)]$ of $\mathbb{R}^2$ for all $x \in \Omega$ and $(s_1, s_2) \in \mathbb{R}^2$.

On the basis of these truncations, let us define the following truncations-perturbations on $\Omega \times \mathbb{R}^2$ for $f(x, s_1, s_2)$ in (6.46):

$$\begin{aligned}
f_+(x, s_1, s_2) = {} & f(x, \tau_+(x, s_1, s_2)) \\
& + (s_1 - u_{1,+}(x))^+ f_{s_1}(x, \tau_+(x, s_1, s_2)) \\
& + (s_2 - u_{2,+}(x))^+ f_{s_2}(x, \tau_+(x, s_1, s_2)) \\
& - s_1^- f_{s_1}(x, \tau_+(x, s_1, s_2)) - s_2^- f_{s_2}(x, \tau_+(x, s_1, s_2)), \\
f_-(x, s_1, s_2) = {} & f(x, \tau_-(x, s_1, s_2)) \\
& - (s_1 - u_{1,-}(x))^- f_{s_1}(x, \tau_-(x, s_1, s_2)) \\
& - (s_2 - u_{2,-}(x))^- f_{s_2}(x, \tau_-(x, s_1, s_2)) \\
& + s_1^+ f_{s_1}(x, \tau_-(x, s_1, s_2)) + s_2^+ f_{s_2}(x, \tau_-(x, s_1, s_2)), \\
f_0(x, s_1, s_2) = {} & f(x, \tau_0(x, s_1, s_2)) \\
& - (s_1 - u_{1,-}(x))^- f_{s_1}(x, \tau_0(x, s_1, s_2)) \\
& - (s_2 - u_{2,-}(x))^- f_{s_2}(x, \tau_0(x, s_1, s_2)) \\
& + (s_1 - u_{1,+}(x))^+ f_{s_1}(x, \tau_0(x, s_1, s_2)) \\
& + (s_2 - u_{2,+}(x))^+ f_{s_2}(x, \tau_0(x, s_1, s_2)).
\end{aligned}$$

It is clear that $f_+, f_-, f_0 : \Omega \times \mathbb{R}^2 \to \mathbb{R}$ are Carathéodory functions, which are locally Lipschitz with respect to the variables $(s_1, s_2) \in \mathbb{R}^2$. Moreover, by means of a careful calculation we can show that their generalized gradients with respect to the variables $(s_1, s_2) \in \mathbb{R}^2$ reduce to the ordinary differential of f on the rectangles $[0, u_{1,+}(x)] \times [0, u_{2,+}(x)]$, $[u_{1,-}(x), 0] \times [u_{2,-}(x), 0]$ and $[u_{1,-}(x), u_{1,+}(x)] \times [u_{2,-}(x), u_{2,+}(x)]$, respectively. Precisely, denoting by f' the differential of f with respect to the variables $(s_1, s_2) \in \mathbb{R}^2$, we have that

$$\begin{aligned}
& \partial_{(s_1,s_2)} f_+(x, s_1, s_2) = \{f'(x, s_1, s_2)\} \\
& \text{for a.e. } x \in \Omega, \text{ every } (s_1, s_2) \in [0, u_{1,+}(x)] \times [0, u_{2,+}(x)],
\end{aligned} \tag{6.56}$$

$$\begin{aligned}
& \partial_{(s_1,s_2)} f_-(x, s_1, s_2) = \{f'(x, s_1, s_2)\} \\
& \text{for a.e. } x \in \Omega, \text{ every } (s_1, s_2) \in [u_{1,-}(x), 0] \times [u_{2,-}(x), 0],
\end{aligned} \tag{6.57}$$

$$\begin{aligned}
& \partial_{(s_1,s_2)} f_0(x, s_1, s_2) = \{f'(x, s_1, s_2)\} \\
& \text{for a.e. } x \in \Omega, \text{ every } (s_1, s_2) \in [u_{1,-}(x), u_{1,+}(x)] \times [u_{2,-}(x), u_{2,+}(x)].
\end{aligned} \tag{6.58}$$

These calculations are based on [55, Theorem 2.5.1].

Using the truncations-perturbations f_+, f_-, f_0, we introduce three functionals on $W_0^{1,p_1}(\Omega) \times W_0^{1,p_2}(\Omega)$ by:

$$E_+(u_1, u_2) = \frac{1}{p_1} \|\nabla u_1\|_{p_1}^{p_1} + \frac{1}{p_2} \|\nabla u_2\|_{p_2}^{p_2} - \int_\Omega f_+(x, u_1(x), u_2(x))\, dx,$$

$$E_-(u_1,u_2)=\frac{1}{p_1}\|\nabla u_1\|_{p_1}^{p_1}+\frac{1}{p_2}\|\nabla u_2\|_{p_2}^{p_2}-\int_\Omega f_-(x,u_1(x),u_2(x))\,dx,$$
$$E_0(u_1,u_2)=\frac{1}{p_1}\|\nabla u_1\|_{p_1}^{p_1}+\frac{1}{p_2}\|\nabla u_2\|_{p_2}^{p_2}-\int_\Omega f_0(x,u_1(x),u_2(x))\,dx.$$

The functionals E_+, E_-, E_0 are locally Lipschitz continuous on $W_0^{1,p_1}(\Omega)\times W_0^{1,p_2}(\Omega)$, so the notion of critical point as considered in Definition 1.41 makes sense for them. This means that (w_1,w_2) is a critical point of E_+ if

$$(0,0)\in\partial E_+(w_1,w_2). \tag{6.59}$$

We present a location result for the critical points of the functionals E_+, E_-, and E_0. To this end we need a new hypothesis:

H$(f)_4$ **(a)** for a.e. $x\in\Omega$, the function $f_{s_1}(x,s_1,\cdot)$ is nondecreasing on the interval $[d_2,k_2]$ whenever $s_1\in[d_1,k_1]$;
(b) for a.e. $x\in\Omega$, the function $f_{s_2}(x,\cdot,s_2)$ is nondecreasing on the interval $[d_1,k_1]$ whenever $s_2\in[d_2,k_2]$.

We illustrate the compatibility of hypotheses H$(f)_1$–H$(f)_4$ by an example.

Example 6.16. Hypotheses H$(f)_1$–H$(f)_4$ are satisfied for the system:

$$\begin{cases}-\Delta_{p_1}u_1=-\alpha(p_1+q_1)|u_1|^{p_1+q_1-2}u_1+\beta p_1(u_1^+)^{p_1-1}(u_2^+)^{p_2} & \\ \qquad +\gamma p_1|u_1|^{p_1-2}u_1 & \text{in } \Omega,\\ -\Delta_{p_2}u_2=-\alpha(p_2+q_2)|u_2|^{p_2+q_2-2}u_2+\beta p_2(u_2^+)^{p_2-1}(u_1^+)^{p_1} & \\ \qquad +\gamma p_2|u_2|^{p_2-2}u_2 & \text{in } \Omega,\\ u_1=u_2=0 & \text{on } \partial\Omega,\end{cases}$$

with constants $p_1\ge 2$, $p_2\ge 2$, $\alpha,\beta,q_1,q_2>0$, and $\gamma>\max\{\lambda_{1,p_1}/p_1,\lambda_{1,p_2}/p_2\}$. Here we have the potential

$$f(x,s_1,s_2)=-\alpha(|s_1|^{p_1+q_1}+|s_2|^{p_2+q_2})+\beta(s_1^+)^{p_1}(s_2^+)^{p_2}$$
$$+\gamma(|s_1|^{p_1}+|s_2|^{p_2}).$$

Actually, for any constants $k_1,k_2>0$ and $d_1,d_2<0$, hypotheses H$(f)_1$–H$(f)_4$ are fulfilled provided $\alpha>0$ is sufficiently large.

We can describe the location of the critical points of the functionals E_+, E_-, and E_0.

Lemma 6.17. *Assume* H$(f)_1$–H$(f)_4$. *Then:*

if $(v_1,v_2)\in W_0^{1,p_1}(\Omega)\times W_0^{1,p_2}(\Omega)$ *is a critical point of* E_+
then $0\le v_1(x)\le u_{1,+}(x)$ *and* $0\le v_2(x)\le u_{2,+}(x)$ *for a.e.* $x\in\Omega$; (6.60)

$$\begin{aligned}&\textit{if } (v_1, v_2) \in W_0^{1,p_1}(\Omega) \times W_0^{1,p_2}(\Omega) \textit{ is a critical point of } E_-\\ &\textit{then } u_{1,-}(x) \le v_1(x) \le 0 \textit{ and } u_{2,-}(x) \le v_2(x) \le 0 \textit{ for a.e. } x \in \Omega;\end{aligned} \tag{6.61}$$

$$\begin{aligned}&\textit{if } (v_1, v_2) \in W_0^{1,p_1}(\Omega) \times W_0^{1,p_2}(\Omega) \textit{ is a critical point of } E_0\\ &\textit{then } u_{1,-}(x) \le v_1(x) \le u_{1,+}(x) \textit{ and } u_{2,-}(x) \le v_2(x) \le u_{2,+}(x) \textit{ for a.e. } x \in \Omega.\end{aligned} \tag{6.62}$$

Proof. In order to check the claim in (6.60), let (v_1, v_2) be a critical point of E_+, i.e., (6.59) is fulfilled with $(w_1, w_2) = (v_1, v_2)$. This reads as

$$\begin{cases} -\Delta_{p_1} v_1 = g_1(x) & \text{in } \Omega,\\ -\Delta_{p_2} v_2 = g_2(x) & \text{in } \Omega,\\ v_1 = v_2 = 0 & \text{on } \partial\Omega, \end{cases} \tag{6.63}$$

with

$$(g_1(x), g_2(x)) \in \partial_{(s_1,s_2)} f_+(x, v_1(x), v_2(x)) \text{ for a.e. } x \in \Omega, \tag{6.64}$$

where [55, Theorem 2.7.5] has been applied.

Since $(u_{1,+}, u_{2,+})$ is a solution of (6.46), from (6.63) and (6.64) we deduce

$$\begin{aligned}&\langle -\Delta_{p_1} v_1 + \Delta_{p_1} u_{1,+}, (v_1 - u_{1,+})^+\rangle\\ &= \int_{\{v_1 > u_{1,+}\}} (g_1(x) - f_{s_1}(x, u_{1,+}(x), u_{2,+}(x)))(v_1(x) - u_{1,+}(x))\, dx\\ &= \int_{\{v_1 > u_{1,+};\ v_2 < 0\}} (f_{s_1}(x, u_{1,+}, 0) - f_{s_1}(x, u_{1,+}, u_{2,+}))(v_1 - u_{1,+})\, dx\\ &\quad + \int_{\{v_1 > u_{1,+};\ 0 \le v_2 \le u_{2,+}\}} (f_{s_1}(x, u_{1,+}, v_2) - f_{s_1}(x, u_{1,+}, u_{2,+}))(v_1 - u_{1,+})\, dx,\end{aligned}$$

where we essentially use that

$$\partial_{(s_1,s_2)} f_+(x, s_1, s_2) = \{f'(x, s_1, s_2)\} \text{ for a.e. } x \in \Omega.$$

$H(f)_4$ (a) implies that

$$\langle -\Delta_{p_1} v_1 + \Delta_{p_1} u_{1,+}, (v_1 - u_{1,+})^+\rangle \le 0.$$

On the basis of the strict monotonicity of the function $\xi \mapsto |\xi|^{p_1-2}\xi$ on $\mathbb{R}^N$, we derive that $v_1 \le u_{1,+}$. In the same way, through $H(f)_4$ (b), we prove that $v_2 \le u_{2,+}$.

By $H(f)_2$ and $H(f)_3$, we know that $f_{s_1}(x, 0, s_2) = 0$ for a.e. $x \in \Omega$, every $s_2 \in [0, k_2]$. Then from (6.63) and (6.64) it turns out that

$$\langle -\Delta_{p_1} v_1, -v_1^-\rangle = -\int_{\{v_1<0\}} g_1 v_1\, dx = -\int_{\{v_1<0;\ v_2<0\}} f_{s_1}(x, 0, 0) v_1\, dx$$

$$-\int_{\{v_1<0;\ 0\le v_2\le u_{2,+}\}} f_{s_1}(x,0,v_2)v_1\,dx$$
$$-\int_{\{v_1<0;\ v_2>u_{2,+}\}} f_{s_1}(x,0,u_{2,+})v_1\,dx=0,$$

which results in $v_1 \geq 0$. Similarly, we prove that $v_2 \geq 0$. Claim (6.60) is thus true.

Assertions (6.61) and (6.62) can be proven in the same way. □

The below lemma clarifies the relationship between the extremal constant-sign solutions and minimizers of the functionals E_+, E_-, and E_0.

Lemma 6.18. *Assume* $\mathrm{H}(f)_1$–$\mathrm{H}(f)_4$. *Then the minimal positive solution* $(u_{1,+}, u_{2,+})$ *of (6.46) is the unique global minimizer of* E_+ *and a local minimizer of* E_0. *Similarly, the maximal negative solution* $(u_{1,-}, u_{2,-})$ *of (6.46) is the unique global minimizer of* E_- *and a local minimizer of* E_0.

Proof. Notice that the locally Lipschitz functional $E_+ : W_0^{1,p_1}(\Omega) \times W_0^{1,p_2}(\Omega) \to \mathbb{R}$ is coercive and sequentially weakly lower semicontinuous. Therefore there exists a global minimizer $(w_1, w_2) \in W_0^{1,p_1}(\Omega) \times W_0^{1,p_2}(\Omega)$ of E_+, which is thus a critical point of E_+ in the sense of (6.59). Then, by (6.60), we know that

$$0 \leq w_i(x) \leq u_{i,+}(x) \leq k_i \text{ for a.e. } x \in \Omega,\ i = 1, 2. \tag{6.65}$$

Hypothesis $\mathrm{H}(f)_2$ ensures that for every $\varepsilon > 0$ there is $t > 0$ such that

$$E_+(t\hat{u}_{1,p_1}, t\hat{u}_{1,p_2}) \leq (\lambda_{1,p_1} - c_1 + \varepsilon)\frac{t^{p_1}}{p_1} + (\lambda_{1,p_2} - c_2 + \varepsilon)\frac{t^{p_2}}{p_2}.$$

Taking $\varepsilon > 0$ sufficiently small implies that $E_+(t\hat{u}_{1,p_1}, t\hat{u}_{1,p_2}) < 0$. This guarantees that $(w_1, w_2) \neq (0, 0)$ because

$$E_+(w_1, w_2) = \inf_{W_0^{1,p_1}(\Omega)\times W_0^{1,p_2}(\Omega)} E_+ < 0. \tag{6.66}$$

Actually,

$$w_i \neq 0 \text{ for } i = 1, 2. \tag{6.67}$$

Indeed, if for instance $w_2 = 0$, then $w_1 \neq 0$ is valid. On account of hypothesis $\mathrm{H}(f)_2$, we observe that there is $\delta > 0$ such that

$$f(x, s_1, s_2) - f(x, s_1, 0) > \lambda_{1,p_2}\frac{s_2^{p_2}}{p_2}$$

for a.e. $x \in \Omega$, every $0 < s_1 \leq k_1$ and $0 < s_2 \leq \delta$. Then, for $t > 0$ small enough, we find

$$\begin{aligned} E_+(w_1, t\hat{u}_{1,p_2}) &= E_+(w_1, 0) + \lambda_{1,p_2}\frac{t^{p_2}}{p_2} \\ &\quad - \int_\Omega (f(x, w_1(x), t\hat{u}_{1,p_2}(x)) - f(x, w_1(x), 0))dx \\ &< E_+(w_1, 0). \end{aligned}$$

This contradicts that $(w_1, 0)$ is a global minimizer of E_+, so (6.67) is verified.

Writing the explicit expression of (6.59) in the form of (6.63) and (6.64), we obtain on the basis of (6.60) and (6.56) that (w_1, w_2) is a solution of problem (6.46). Hypothesis $\mathrm{H}(f)_2$ and (6.65) allow us to apply the strong maximum principle in Theorem 2.19 to both equations in (6.46) with (w_1, w_2) in place of (u_1, u_2). Thanks to (6.67) and (6.65), this entails that $(w_1, w_2) \in \operatorname{int}(C_0^1(\overline{\Omega})_+) \times \operatorname{int}(C_0^1(\overline{\Omega})_+)$. Recalling (6.60) and that $(u_{1,+}, u_{2,+})$ is a minimal positive solution of (6.46), we find that $(w_1, w_2) = (u_{1,+}, u_{2,+})$. In this way we can infer that $(u_{1,+}, u_{2,+})$ is a local minimizer of E_0 on $C_0^1(\overline{\Omega}) \times C_0^1(\overline{\Omega})$ because the functions E_+ and E_0 coincide on $\operatorname{int}(C_0^1(\overline{\Omega})_+) \times \operatorname{int}(C_0^1(\overline{\Omega})_+)$. Then it follows that $(u_{1,+}, u_{2,+})$ is a local minimizer of E_0 on $W_0^{1,p_1}(\Omega) \times W_0^{1,p_2}(\Omega)$ due to the property of Sobolev versus Hölder local minimizers for quasilinear elliptic equations as shown in [91]. Arguing similarly with the functional E_- in place of E_+ and utilizing (6.57) instead of (6.56), we get the desired conclusion for $(u_{1,-}, u_{2,-})$. □

Here is a proposition revealing how Eqs. (6.47) and (6.48) are useful for studying system (6.46).

Proposition 6.19. *Assume* $\mathrm{H}(f)_1$–$\mathrm{H}(f)_4$. *Then there exists* $(u_+, v_+) \in \operatorname{int}(C_0^1(\overline{\Omega})_+) \times \operatorname{int}(C_0^1(\overline{\Omega})_+)$ *such that* u_+ *is a solution of (6.47),* v_+ *is a solution of (6.48),*

$$\begin{aligned} u_+ \leq u_{1,+}, \quad v_+ \leq u_{2,+}, \\ E_+(u_+, 0) = \inf E_+(\cdot, 0), \quad E_+(0, v_+) = \inf E_+(0, \cdot). \end{aligned}$$

Similarly, there exists $(u_-, v_-) \in \left(-\operatorname{int}(C_0^1(\overline{\Omega})_+)\right) \times \left(-\operatorname{int}(C_0^1(\overline{\Omega})_+)\right)$ *such that* u_- *is a solution of (6.47),* v_- *is a solution of (6.48),*

$$\begin{aligned} u_- \geq u_{1,-}, \quad v_- \geq u_{2,-}, \\ E_-(u_-, 0) = \inf E_-(\cdot, 0), \quad E_-(0, v_-) = \inf E_-(0, \cdot). \end{aligned}$$

Proof. Since $E_+(\cdot, 0) : W_0^{1,p_1}(\Omega) \to \mathbb{R}$ is coercive and sequentially weakly lower semicontinuous, there exists $u_+ \in W_0^{1,p_1}(\Omega)$ such that

$$E_+(u_+, 0) = \inf E_+(\cdot, 0).$$

This ensures that u_+ is a critical point of $E_+(\cdot, 0)$, which means that $0 \in \partial E_+(\cdot, 0)(u_+)$. By hypothesis $\mathrm{H}(f)_4$ we have

$$-\Delta_{p_1} u_{1,+} = f_{s_1}(x, u_{1,+}(x), u_{2,+}(x)) \geq f_{s_1}(x, u_{1,+}(x), 0),$$

so $u_{1,+}$ is a supersolution of (6.47). It follows that

$$\begin{aligned}&\langle -\Delta_{p_1} u_+ + \Delta_{p_1} u_{1,+}, (u_+ - u_{1,+})^+\rangle \\ &\leq \int_{\{u_+ > u_{1,+}\}} ((f_+)_{s_1}(x, u_+, 0) - f_{s_1}(x, u_{1,+}, u_{2,+}))(u_+ - u_{1,+})\, dx = 0.\end{aligned}$$

From this direct comparison we obtain that $0 \leq u_+(x) \leq u_{1,+}(x)$ for a.e. $x \in \Omega$ and $u_+ \neq 0$. This ensures that u_+ is solution of problem (6.47). Through the strong maximum principle (see Theorem 2.19) it follows that $u_+ \in \operatorname{int}(C_0^1(\overline{\Omega})_+)$. The assertions related to v_+, u_-, and v_- can be shown to be true along the same lines. □

For the proof of our main result we require the following subhomogeneous conditions:

$\mathrm{H}(f)_5$ For any $t \in [0, 1]$ we have

(i) $f_{s_1}(\cdot, ts_1, 0) \leq t^{p_1 - 1} f_{s_1}(\cdot, s_1, 0)$ for all $d_1 \leq s_1 \leq 0$,
(ii) $f_{s_1}(\cdot, ts_1, 0) \geq t^{p_1 - 1} f_{s_1}(\cdot, s_1, 0)$ for all $0 \leq s_1 \leq k_1$.

$\mathrm{H}(f)_6$ For any $t \in [0, 1]$ we have

(i) $f_{s_2}(\cdot, 0, ts_2) \leq t^{p_2 - 1} f_{s_2}(\cdot, 0, s_2)$ for all $d_2 \leq s_2 \leq 0$,
(ii) $f_{s_2}(\cdot, 0, ts_2) \geq t^{p_2 - 1} f_{s_2}(\cdot, 0, s_2)$ for all $0 \leq s_2 \leq k_2$.

Remark 6.20. One can check that $\mathrm{H}(f)_5$ and $\mathrm{H}(f)_6$ are verified for Example 6.16.

We are in a position to state the main result regarding problem (6.46).

Theorem 6.21. *Assume* $\mathrm{H}(f)_1$–$\mathrm{H}(f)_6$ *and that* $\mathrm{H}(f)_2$ *holds (in a stronger form) with constants* $c_1 > \lambda_{2,p_1}$ *and* $c_2 > \lambda_{2,p_2}$. *Then problem (6.46) admits at least three nontrivial solutions: a minimal positive solution*

$$(u_{1,+}, u_{2,+}) \in \operatorname{int}(C_0^1(\overline{\Omega})_+) \times \operatorname{int}(C_0^1(\overline{\Omega})_+),$$

a maximal negative solution

$$(u_{1,-}, u_{2,-}) \in (-\operatorname{int}(C_0^1(\overline{\Omega})_+)) \times (-\operatorname{int}(C_0^1(\overline{\Omega})_+)),$$

and a third solution $(u_{1,0}, u_{2,0}) \in C_0^1(\overline{\Omega}) \times C_0^1(\overline{\Omega})$ *such that*

$$u_{i,-} \leq u_{i,0} \leq u_{i,+} \ \text{ in } \Omega, \text{ for } i = 1, 2,$$

which is neither positive nor negative.

Proof. By Lemma 6.18 we know that $(u_{1,+}, u_{2,+})$ and $(u_{1,-}, u_{2,-})$ are local minimizers of the functional E_0. Due to (6.62), they are strict local minimizers being extremal positive and negative solutions of (6.46), respectively. Since the functional E_0 is coercive, it satisfies the nonsmooth Palais–Smale condition in Definition 1.42, so we can apply to the locally Lipschitz functional E_0 the nonsmooth version of the mountain pass theorem in Theorem 1.44. This yields a critical point $(u_{1,0}, u_{2,0}) \in W_0^{1,p_1}(\Omega) \times W_0^{1,p_2}(\Omega)$ of E_0, that is,

$$(0,0) \in \partial E_0(u_{1,0}, u_{2,0}), \tag{6.68}$$

such that

$$\max\{E_0(u_{1,+}, u_{2,+}), E_0(u_{1,-}, u_{2,-})\} < E_0(u_{1,0}, u_{2,0}) = \inf_{\gamma \in \Gamma} \max_{0 \le t \le 1} E_0(\gamma(t)), \tag{6.69}$$

with

$$\begin{aligned} \Gamma = \{\gamma \in C([0,1], W_0^{1,p_1}(\Omega) \times W_0^{1,p_2}(\Omega)) : \gamma(0) &= (u_{1,-}, u_{2,-}), \\ \gamma(1) &= (u_{1,+}, u_{2,+})\}. \end{aligned} \tag{6.70}$$

Exploiting the explicit expression of the generalized gradient $\partial E_0(u_{1,0}, u_{2,0})$, we infer from (6.68), (6.62), and (6.58) that $(u_{1,0}, u_{2,0})$ is a solution of (6.46). Moreover, using (6.69), it turns out that

$$(u_{1,0}, u_{2,0}) \neq (u_{1,+}, u_{2,+}) \quad \text{and} \quad (u_{1,0}, u_{2,0}) \neq (u_{1,-}, u_{2,-}). \tag{6.71}$$

We can obtain a path $\hat{\gamma} \in \Gamma$ with the property

$$E_0(\hat{\gamma}(t)) < 0$$

for all $t \in [0,1]$. The construction of such a path $\hat{\gamma}$ is performed on the pattern of the construction made in the proof of Theorem 5.5 (see the proof of [47, Theorem 4.1]). Then by (6.69) and (6.70) we have

$$E_0(u_{1,0}, u_{2,0}) < 0.$$

Therefore $(u_{1,0}, u_{2,0}) \in W_0^{1,p_1}(\Omega) \times W_0^{1,p_2}(\Omega)$ is a nontrivial solution of problem (6.46). The nonlinear regularity theory guarantees that $(u_{1,0}, u_{2,0}) \in C_0^1(\overline{\Omega}) \times C_0^1(\overline{\Omega})$. Finally, combining with (6.71), the extremal properties of constant-sign solutions $(u_{1,+}, u_{2,+})$ and $(u_{1,-}, u_{2,-})$ ensure that the nontrivial solution $(u_{1,0}, u_{2,0})$ is neither positive nor negative. This completes the proof. $\square$

6.4 NOTES

The general system (6.1) is studied in [184] from where the results in Section 6.1 are taken. The equation version of system (6.1) has been considered in [19]. The case when in (6.1) one takes $\mu_1 = \mu_2 = 0$, i.e., system (6.5), was investigated in [48] through the method of subsolution–supersolution (see also [38] for more particular situations). For the study of equations of type (6.6) and (6.7) exhibiting full gradient dependence, we refer to [68,80–82,212,226]. Equations with (p, q)-Laplacian but without gradient dependence in the right-hand side can be found in [143,191]. For the equations involving the p-Laplacian without gradient dependence in their right-hand side studied through variational and topological methods, we refer to [169]. See also the notes of Chapter 4.

A main feature of problem (6.1) is that the parameters μ_1, μ_2 enter the leading operators of the elliptic equations and not, as usually, in the right-hand sides of the elliptic equations with lower-order terms. The presence of four operators of type p-Laplacian in the principal parts of the equations, namely Δ_{p_i} and Δ_{q_i} for $i = 1, 2$, makes the problem highly nonlinear. Another important aspect of problem (6.1) is that the nonlinearities in the right-hand side depend on the solution and its gradient (which is often expressed as dealing with convection terms), which is rarely handled in the literature, especially in the system setting due to the interaction between different equations. The presence of convection terms is a major difficulty in the study mainly because such problems cannot be handled through variational methods due to the lack of variational structure.

Theorem 6.1 establishes the existence of weak solutions to problem (6.1) by using the theory of pseudomonotone operators. The pseudomonotone operators are thoroughly investigated in [38,169,228,238]. Next, a uniqueness result for problem (6.1) is proven in Theorem 6.2. For the case of equations, we refer to [19] for a uniqueness result regarding equations of type (6.6), (6.7), and to [188] in the situation where the functions f_i do not depend on the gradient of the solution. In Theorems 6.4, 6.5, and 6.6 we examine the asymptotic behavior of solutions to problem (6.1) as $(\mu_1, \mu_2) \to (0, 0)$, $(\mu_1, \mu_2) \to (+\infty, 0)$, $(\mu_1, \mu_2) \to (0, +\infty)$, and $(\mu_1, \mu_2) \to (+\infty, +\infty)$, respectively. Problems (6.5), (6.6), and (6.7) become limiting cases of problem (6.1), demonstrating a strong unifying effect for (6.1). These asymptotic properties are obtained through a priori estimates.

In Section 6.2 we study the Dirichlet boundary value problem (6.1) through the method of subsolution–supersolution. In this respect we follow the results obtained in Motreanu–Vetro–Vetro [185]. In comparison with the existence, uniqueness, and asymptotic results with respect to the parameters (μ_1, μ_2) given in Section 6.1, here we provide location and enclosure properties by subsolution–supersolution approach for the general system (6.1). We mention that the method of subsolution–supersolution for the case of (6.1) with $\mu_1 = \mu_2 = 0$ was investigated in [48], and without gradient dependence in the right-hand sides in [38]. Here we are able to provide a unifying approach. The method of subsolution–supersolution for system of equations differs substantially from what happens in the case of single equations. It is worth noticing that for systems one cannot define separately the notions of subsolution and supersolution, but the notion of subsolution–supersolution (or trapping region).

It is worth pointing out that assuming condition $H(f_1, f_2)$ of Section 6.2 we cannot apply the known results as available in [38,48,184], which hold under more restrictive hypotheses. Indeed, in [48] the driving differential operator is more specific, and for the nonlinearities in the reaction terms the following growth is supposed: there exist constants $c_i \geq 0$ and functions $\rho_i \in L^{p_i'}(\Omega)$ such that

$$|f_1(x, s_1, s_2, \xi_1, \xi_2)| \leq \rho_1(x) + c_1(|\xi_1|^{p_1-1} + |\xi_2|^{\frac{p_2}{p_1'}}),$$

$$|f_2(x, s_1, s_2, \xi_1, \xi_2)| \leq \rho_2(x) + c_2(|\xi_1|^{\frac{p_1}{p_2'}} + |\xi_2|^{p_2-1})$$

for a.a. $x \in \Omega$ and all $s = (s_1, s_2) \in [\underline{u}_1(x), \overline{u}_1(x)] \times [\underline{u}_2(x), \overline{u}_2(x)]$, $\xi_1, \xi_2 \in \mathbb{R}^N$, which is more restrictive than hypothesis $H(f_1, f_2)$ because

$$p_i - 1 = \frac{p_i}{p_i'} < \frac{p_i}{(p_i^*)'}$$

for $i = 1, 2$. In [38] the setting is even more restrictive. In [184] it is as in Section 6.1 where, due to a completely different approach, assumptions of another type are imposed, namely the generalized sign condition in Section 6.1.

Theorem 6.7 establishes under hypothesis $H(f_1, f_2)$ the existence and location of a solution to problem (6.1) within the ordered rectangle (or trapping region) $[\underline{u}_1, \overline{u}_1] \times [\underline{u}_2, \overline{u}_2]$ determined by a subsolution–supersolution. The proof relies on the study of an associated auxiliary problem as demonstrated in Theorem 6.8. In comparison with previous results, the main contribution here consists in the use of a cut-off function adapted to the general growth condition in assumption $H(f_1, f_2)$. In Theorem 6.9 the existence of a positive solution (u_1, u_2) of system (6.1) is shown, meaning that both components are positive. The essential point for obtaining a positive solution is the construction of an appropriate subsolution–supersolution of problem (6.1) permitting an application of Theorem 6.8. One can find in [184] criteria to get a negative solution, i.e., having both components negative, as well as hybrid solutions (u_1, u_2), in the sense that the components u_1 and u_2 are of opposite constant sign.

Section 6.3 is based on Carl–Motreanu [47]. The main result is stated in Theorem 6.21, which postulates the existence of three nontrivial solutions of system (6.46) in variational form. Two of these solutions are of opposite constant sign: one is positive in the sense that both components are positive and one is negative in the sense that both components are negative. They have extremal properties, which means that the former is minimal among the class of positive solutions, while the latter is maximal in the class of negative solutions.

Also there exists a third nontrivial solution of system (6.46) between the extremal opposite constant-sign solutions, which is neither positive nor negative. The complete proof of Theorem 6.21 can be found in [47].

The approach relies on subsolution–supersolution for systems and on variational methods. A significant aspect of Section 6.3 is that the smooth problem (6.46) is handled by means of a nonsmooth problem involving a locally Lipschitz potential, which is not differentiable. The nonsmooth problem is smooth on the relevant domain, which makes it possible to achieve solvability for the original problem (6.46). A crucial point is that a comparison argument permits checking that the critical points of the nonsmooth functional are solutions of system (6.46) within the prescribed trapping regions. Another relevant feature is that we use some associated equations, namely (6.47) and (6.48), in the study of problem (6.46). An open problem for nonlinear elliptic systems is to find solutions for which each component is sign-changing.

Chapter 7

Singular Quasilinear Elliptic Systems

7.1 THE CASE OF SINGULARITIES WITH RESPECT TO THE SOLUTION

Consider the system of quasilinear elliptic equations

$$\begin{cases} -\Delta_p u = f_1(u,v) & \text{in } \Omega, \\ -\Delta_q v = f_2(u,v) & \text{in } \Omega, \\ u, v > 0 & \text{in } \Omega, \\ u, v = 0 & \text{on } \partial\Omega \end{cases} \tag{7.1}$$

on a bounded domain $\Omega \subset \mathbb{R}^N$ with a $C^{1,\alpha}$ boundary $\partial\Omega$ for some $\alpha \in (0,1)$, for which we allow a singularity at zero. As usual, Δ_p and Δ_q stand for the p-Laplacian and q-Laplacian differential operators on $W_0^{1,p}(\Omega)$ and $W_0^{1,q}(\Omega)$, respectively. For avoiding repetitive discussions, we suppose that $1 < p, q \leq N$. The reaction terms in problem (7.1) are expressed by continuous functions $f_i : (0,+\infty) \times (0,+\infty) \to (0,+\infty)$, $i = 1,2$, for which we assume the growth conditions:

$$\begin{cases} m_1 s_1^{\alpha_1} s_2^{\beta_1} \leq f_1(s_1,s_2) \leq M_1 s_1^{\alpha_1} s_2^{\beta_1} \text{ for all } s_1, s_2 > 0, \text{ with } M_1, m_1 > 0, \\ \text{and } \alpha_1 \in \mathbb{R},\ \beta_1 < 0 \text{ such that } |\alpha_1| - \beta_1 < \min\{1, p-1\}, \end{cases} \tag{7.2}$$

$$\begin{cases} m_2 s_1^{\alpha_2} s_2^{\beta_2} \leq f_2(s_1,s_2) \leq M_2 s_1^{\alpha_2} s_2^{\beta_2} \text{ for all } s_1, s_2 > 0, \text{ with } M_2, m_2 > 0, \\ \text{and } \beta_2 \in \mathbb{R},\ \alpha_2 < 0 \text{ such that } |\beta_2| - \alpha_2 < \min\{1, q-1\}. \end{cases} \tag{7.3}$$

It is clear from conditions (7.2) and (7.3) that the system (7.1) possesses singularity at the origin.

In the analysis of problem (7.1) we use the Banach spaces $L^p(\Omega)$ and $W_0^{1,p}(\Omega)$ endowed with the usual norms $\|\cdot\|_p$ and $\|\cdot\| := \|\nabla \cdot\|_p$, respectively, as well as the analogous spaces $L^q(\Omega)$ and $W_0^{1,q}(\Omega)$. Actually, for reasons of nonlinear regularity theory, the natural spaces for solutions are the spaces $C_0^{1,\beta}(\overline{\Omega}) = \{u \in C^{1,\beta}(\overline{\Omega}) : u = 0 \text{ on } \partial\Omega\}$ with $\beta \in (0,1)$. The differential op-

Nonlinear Differential Problems with Smooth and Nonsmooth Constraints. DOI: 10.1016/B978-0-12-813386-6.00007-9

erators in problem (7.1) are the p-Laplacian $\Delta_p u = \operatorname{div}(|\nabla u|^{p-2}\nabla u)$ for all $u \in W_0^{1,p}(\Omega)$ and the q-Laplacian $\Delta_q u = \operatorname{div}(|\nabla u|^{q-2}\nabla u)$ for all $u \in W_0^{1,q}(\Omega)$. We denote by $\lambda_{1,p}$ and $\lambda_{1,q}$ the first eigenvalue of $-\Delta_p$ on $W_0^{1,p}(\Omega)$ and of $-\Delta_q$ on $W_0^{1,q}(\Omega)$, respectively (see Proposition 1.57). Corresponding to them, we denote by $\hat{u}_{1,p}$ the L^p-normalized positive eigenfunction of $-\Delta_p$ associated to $\lambda_{1,p}$, that is, $-\Delta_p \hat{u}_{1,p} = \lambda_{1,p}\hat{u}_{1,p}^{p-1}$ in Ω, $\hat{u}_{1,p} = 0$ on $\partial\Omega$, with $\|\hat{u}_{1,p}\|_p = 1$, and by $\hat{u}_{1,q}$ the L^q-normalized positive eigenfunction of $-\Delta_q$ associated to $\lambda_{1,q}$, that is, $-\Delta_q \hat{u}_{1,q} = \lambda_{1,q}\hat{u}_{1,q}^{q-1}$ in Ω, $\hat{u}_{1,q} = 0$ on $\partial\Omega$, with $\|\hat{u}_{1,q}\|_q = 1$. As known from the strong maximum principle (see Theorem 2.19), there exist positive constants l_1 and l_2 such that

$$l_1\hat{u}_{1,p}(x) \le \hat{u}_{1,q}(x) \le l_2\hat{u}_{1,p}(x) \ \text{ for all } x \in \Omega. \tag{7.4}$$

Recall that for any $1 < r < +\infty$ we denote $r' = \frac{r}{r-1}$.

For obtaining comparison principles and a priori estimates for problem (7.1), we need some technical tools. Let the functions w_1 and w_2 be defined as the solutions of the equations

$$\begin{cases} -\Delta_p w_1 = w_1^{\beta_1} & \text{in } \Omega, \\ w_1 > 0 & \text{in } \Omega, \\ w_1 = 0 & \text{on } \partial\Omega \end{cases} \quad \text{and} \quad \begin{cases} -\Delta_q w_2 = w_2^{\alpha_2} & \text{in } \Omega, \\ w_2 > 0 & \text{in } \Omega, \\ w_2 = 0 & \text{on } \partial\Omega. \end{cases} \tag{7.5}$$

From (7.5) and [101] we know that they satisfy

$$c_0\hat{u}_{1,p}(x) \le w_1(x) \le c_1\hat{u}_{1,p}(x) \ \text{ and } \ c_0'\hat{u}_{1,q}(x) \le w_2(x) \le c_1'\hat{u}_{1,q}(x) \ \text{ in } \Omega, \tag{7.6}$$

with positive constants c_0, c_1, c_0', c_1'. Consider now the functions z_1, z_2 defined by

$$-\Delta_p z_1 = \begin{cases} w_1^{\beta_1} & \text{in } \Omega \setminus \overline{\Omega}_\delta, \\ -w_1^{\beta_1} & \text{in } \Omega_\delta, \end{cases} \quad z_1 = 0 \ \text{ on } \partial\Omega \tag{7.7}$$

and

$$-\Delta_q z_2 = \begin{cases} w_2^{\alpha_2} & \text{in } \Omega \setminus \overline{\Omega}_\delta, \\ -w_2^{\alpha_2} & \text{in } \Omega_\delta, \end{cases} \quad z_2 = 0 \ \text{ on } \partial\Omega, \tag{7.8}$$

where

$$\Omega_\delta = \{x \in \Omega : d(x, \partial\Omega) < \delta\}$$

with a fixed $\delta > 0$ sufficiently small. Then (7.6) and the Hardy–Sobolev inequality in [5, Lemma 2.3] ensure that the right-hand side in (7.7) and (7.8) are elements of $W^{-1,p'}(\Omega)$ and $W^{-1,q'}(\Omega)$, respectively. According to Proposition 1.56 and Theorem 1.55, this guarantees the existence and uniqueness of z_1 and z_2 in (7.7) and (7.8). On the basis of (7.5), (7.6), (7.7), (7.8), the monotonicity of the operators $-\Delta_p$ and $-\Delta_q$, and [109, Corollary 3.1], we find the estimates

$$\frac{c_0}{2}\hat{u}_{1,p}(x) \le z_1(x) \le c_1\hat{u}_{1,p}(x), \quad \frac{c_0'}{2}\hat{u}_{1,q}(x) \le z_2(x) \le c_1'\hat{u}_{1,q}(x) \text{ in } \Omega. \tag{7.9}$$

Denote

$$R := \max\left\{\max_{x\in\overline{\Omega}} \hat{u}_{1,p}(x), \max_{x\in\overline{\Omega}} \hat{u}_{1,q}(x)\right\},$$

and fix a constant $\mu = \mu(\delta) > 0$ such that

$$\hat{u}_{1,p}(x),\ \hat{u}_{1,q}(x) \ge \mu \text{ in } \Omega \setminus \Omega_\delta.$$

Consider the Dirichlet problems

$$\begin{cases} -\Delta_p \xi_1 = \xi_1^{\gamma_1} & \text{in } \Omega, \\ \xi_1 > 0 & \text{in } \Omega, \\ \xi_1 = 0 & \text{on } \partial\Omega \end{cases} \quad \text{and} \quad \begin{cases} -\Delta_q \xi_2 = \xi_2^{\gamma_2} & \text{in } \Omega, \\ \xi_2 > 0 & \text{in } \Omega, \\ \xi_2 = 0 & \text{on } \partial\Omega, \end{cases} \tag{7.10}$$

with constants γ_1 and γ_2 satisfying

$$-1 < \gamma_i < \min\{0, \alpha_i + \beta_i\},\ i = 1, 2. \tag{7.11}$$

As known from [101], problems (7.10) possess unique solutions ξ_1 and ξ_2, and there exist constants $c_3 \ge c_2 > 0$ and $c_3' \ge c_2' > 0$ such that

$$c_2\hat{u}_{1,p}(x) \le \xi_1(x) \le c_3\hat{u}_{1,p}(x) \text{ and } c_2'\hat{u}_{1,q}(x) \le \xi_2(x) \le c_3'\hat{u}_{1,q}(x). \tag{7.12}$$

For a constant $C > 0$, set

$$(\underline{u}, \underline{v}) = C(z_1, z_2) \text{ and } (\overline{u}, \overline{v}) = \frac{1}{C}(\xi_1, \xi_2). \tag{7.13}$$

Useful comparison properties can be provided.

Proposition 7.1. *Assume that conditions (7.2) and (7.3) hold. Then, if $C > 0$ is small enough in (7.13), we obtain*

$$-\Delta_p \underline{u} \le \begin{cases} m_1 \underline{u}^{\alpha_1}\overline{v}^{\beta_1} & \text{if } \alpha_1 \ge 0, \\ m_1 \overline{u}^{\alpha_1}\overline{v}^{\beta_1} & \text{if } \alpha_1 < 0, \end{cases} \quad -\Delta_q \underline{v} \le \begin{cases} m_2 \overline{u}^{\alpha_2}\underline{v}^{\beta_2} & \text{if } \beta_2 \ge 0, \\ m_2 \overline{u}^{\alpha_2}\overline{v}^{\beta_2} & \text{if } \beta_2 < 0, \end{cases} \tag{7.14}$$

and

$$-\Delta_p\overline{u} \geq \begin{cases} M_1\overline{u}^{\alpha_1}\underline{v}^{\beta_1} & \text{if } \alpha_1 \geq 0, \\ M_1\underline{u}^{\alpha_1}\underline{v}^{\beta_1} & \text{if } \alpha_1 < 0, \end{cases} \quad -\Delta_q\overline{v} \geq \begin{cases} M_2\underline{u}^{\alpha_2}\overline{v}^{\beta_2} & \text{if } \beta_2 \geq 0, \\ M_2\underline{u}^{\alpha_2}\underline{v}^{\beta_2} & \text{if } \beta_2 < 0. \end{cases} \tag{7.15}$$

Proof. We provide a proof only for the first inequalities in (7.14) and (7.15) because the others can be established in the same way. By (7.13) and (7.7) we find

$$\underline{u}^{-\alpha_1}\overline{v}^{-\beta_1}(-\Delta_p\underline{u}) = -C^{p-1-\alpha_1+\beta_1}z_1^{-\alpha_1}\xi_2^{-\beta_1}w_1^{\beta_1} \leq 0 < m_1 \text{ in } \Omega_\delta,$$
$$\text{and } \quad \overline{u}^{-\alpha_1}\overline{v}^{-\beta_1}(-\Delta_p\underline{u}) = -C^{p-1+\alpha_1+\beta_1}\xi_1^{-\alpha_1}\xi_2^{-\beta_1}w_1^{\beta_1} \leq 0 < m_1 \text{ in } \Omega_\delta.$$

Now we turn to the corresponding estimates on $\Omega \setminus \overline{\Omega}_\delta$. If $\alpha_1 \geq 0$, by (7.13), (7.7), (7.6), (7.9), (7.12), and (7.2), it is seen that

$$\begin{aligned}
&\underline{u}^{-\alpha_1}\overline{v}^{-\beta_1}(-\Delta_p\underline{u}) \\
&= C^{p-1-\alpha_1+\beta_1}z_1^{-\alpha_1}\xi_2^{-\beta_1}w_1^{\beta_1} \\
&\leq C^{p-1-\alpha_1+\beta_1}\left(\frac{c_0}{2}\hat{u}_{1,p}\right)^{-\alpha_1}(c_3'\hat{u}_{1,q})^{-\beta_1}(c_0\hat{u}_{1,p})^{\beta_1} \\
&\leq C^{p-1-\alpha_1+\beta_1}2^{\alpha_1}c_0^{\beta_1-\alpha_1}(c_3'R)^{-\beta_1}\mu^{\beta_1-\alpha_1} \leq m_1 \text{ in } \Omega \setminus \overline{\Omega}_\delta
\end{aligned}$$

provided the constant $C > 0$ is sufficiently small. If $\alpha_1 < 0$, we deduce from (7.13), (7.7), (7.6), (7.12), and (7.2) that

$$\overline{u}^{-\alpha_1}\overline{v}^{-\beta_1}(-\Delta_p\underline{u}) \leq C^{p-1+\alpha_1+\beta_1}c_3^{-\alpha_1}\left(\frac{c_0\mu}{c_3'}\right)^{\beta_1}R^{-\alpha_1-\beta_1} \leq m_1 \text{ in } \Omega \setminus \overline{\Omega}_\delta$$

provided $C > 0$ is sufficiently small. Therefore the first part in (7.14) holds true.

Now we prove the first inequality in (7.15). If $\alpha_1 \geq 0$, we obtain from (7.13), (7.10), (7.12), (7.9), (7.4), and (7.11) that

$$\begin{aligned}
&\overline{u}^{-\alpha_1}\underline{v}^{-\beta_1}(-\Delta_p\overline{u}) \\
&\geq C^{-(p-1)+\alpha_1-\beta_1}(c_3\hat{u}_{1,p})^{\gamma_1-\alpha_1}\left(\frac{c_0'}{2}\hat{u}_{1,q}\right)^{-\beta_1} \\
&\geq C^{-(p-1)+\alpha_1-\beta_1}c_3^{\gamma_1-\alpha_1}\left(\frac{c_0'}{2}l_1\right)^{-\beta_1}\hat{u}_{1,p}^{\gamma_1-(\alpha_1+\beta_1)} \\
&\geq C^{-(p-1)+\alpha_1-\beta_1}c_3^{\gamma_1-\alpha_1}\left(\frac{c_0'}{2}l_1\right)^{-\beta_1}R^{\gamma_1-(\alpha_1+\beta_1)} \geq M_1 \text{ in } \Omega
\end{aligned}$$

provided $C > 0$ is sufficiently small. If $\alpha_1 < 0$, we note from (7.13), (7.10), (7.12), (7.9), (7.4), (7.11), and (7.2) that

$$\begin{aligned}\underline{u}^{-\alpha_1}\underline{v}^{-\beta_1}(-\Delta_p\overline{u}) &\geq C^{-(p-1)-\alpha_1-\beta_1}\left(\frac{c_0}{2}\right)^{-\alpha_1}\left(\frac{c_0'}{2}l_1\right)^{-\beta_1} c_3^{\gamma_1} R^{\gamma_1-(\alpha_1+\beta_1)} \\ &\geq M_1 \text{ in } \Omega\end{aligned}$$

provided the constant $C > 0$ is sufficiently small. This completes the proof of the first inequality in (7.15), whence the proof. □

At this point we are able to prove the existence and regularity of solutions of the singular system (7.1).

Theorem 7.2. *Under hypotheses (7.2) and (7.3), problem (7.1) has a (positive) solution* $(u, v) \in C_0^{1,\beta}(\overline{\Omega}) \times C_0^{1,\beta}(\overline{\Omega})$ *with some* $\beta \in (0, 1)$.

Proof. By using the functions $(\underline{u}, \underline{v})$ and $(\overline{u}, \overline{v})$ constructed in (7.13), we define the set

$$\mathcal{K} = \left\{(y_1, y_2) \in C(\overline{\Omega}) \times C(\overline{\Omega}) : \underline{u} \leq y_1 \leq \overline{u} \text{ and } \underline{v} \leq y_2 \leq \overline{v} \text{ in } \Omega\right\}.$$

A direct verification entails that $\mathcal{K}$ is a closed, bounded, and convex subset of $C(\overline{\Omega}) \times C(\overline{\Omega})$.

We define the mapping $\mathcal{T} : \mathcal{K} \to C(\overline{\Omega}) \times C(\overline{\Omega})$ by $\mathcal{T}(y_1, y_2) = (u, v)$, where (u, v) fulfills

$$\begin{cases} -\Delta_p u = f_1(y_1, y_2) & \text{in } \Omega, \\ -\Delta_q v = f_2(y_1, y_2) & \text{in } \Omega, \\ u, v = 0 & \text{on } \partial\Omega. \end{cases} \tag{7.16}$$

In order to confirm that the operator $\mathcal{T}$ is well defined, we show that problem (7.16) admits a unique solution (u, v). Indeed, given $(y_1, y_2) \in \mathcal{K}$, we infer from (7.2), (7.3), (7.13), (7.9), (7.12), and (7.4) the estimates

$$f_1(y_1, y_2) \leq M_1 y_1^{\alpha_1} y_2^{\beta_1} \leq \begin{cases} M_1 \overline{u}^{\alpha_1} \underline{v}^{\beta_1} & \text{if } \alpha_1 \geq 0 \\ M_1 \underline{u}^{\alpha_1} \underline{v}^{\beta_1} & \text{if } \alpha_1 < 0 \end{cases} \leq C_1 \hat{u}_{1,p}^{\alpha_1+\beta_1} \tag{7.17}$$

and

$$f_2(y_1, y_2) \leq M_2 y_1^{\alpha_2} y_2^{\beta_2} \leq \begin{cases} M_2 \underline{u}^{\alpha_2} \overline{v}^{\beta_2} & \text{if } \beta_2 \geq 0 \\ M_2 \underline{u}^{\alpha_2} \underline{v}^{\beta_2} & \text{if } \beta_2 < 0 \end{cases} \leq C_2 \hat{u}_{1,q}^{\alpha_2+\beta_2}, \tag{7.18}$$

with positive constants C_1, C_2. The estimates (7.17) and (7.18) combined with Hardy–Sobolev inequality (see [5, Lemma 2.3]) and hypotheses (7.2), (7.3) allow us to identify that $f_1(y_1, y_2) \in W^{-1,p'}(\Omega)$ and $f_2(y_1, y_2) \in W^{-1,q'}(\Omega)$.

Then it is sufficient to rely on Proposition 1.56 and Theorem 1.55 for obtaining the existence and uniqueness of solution (u, v) in (7.16).

The regularity theory up to the boundary as found in [129] if $\alpha_i + \beta_i \geq 0$ and in [101, Theorem B.1] if $\alpha_i + \beta_i < 0$ enables us to get that $(u, v) \in C_0^{1,\beta}(\overline{\Omega}) \times C_0^{1,\beta}(\overline{\Omega})$, with some $\beta \in (0, 1)$. Moreover, (7.17), (7.16), and the a priori estimates for singular problems (see [109, Lemma 3.1] or [101]) provide a constant $M > 0$ such that

$$\|u\|_{C_0^{1,\beta}(\overline{\Omega})}, \|v\|_{C_0^{1,\beta}(\overline{\Omega})} \leq M \tag{7.19}$$

whenever $(u, v) = \mathcal{T}(y_1, y_2)$ with $(y_1, y_2) \in \mathcal{K}$.

Let us prove the continuity of the operator $\mathcal{T}$. If $(y_{1,n}, y_{2,n}) \in \mathcal{K}$ is a sequence with $(y_{1,n}, y_{2,n}) \to (y_1, y_2)$ in $C(\overline{\Omega}) \times C(\overline{\Omega})$, denote $(u_n, v_n) = \mathcal{T}(y_{1,n}, y_{2,n})$, which explicitly means that

$$\int_\Omega |\nabla u_n|^{p-2}\nabla u_n \cdot \nabla\varphi \, dx = \int_\Omega f_1(y_{1,n}, y_{2,n})\varphi \, dx, \tag{7.20}$$

$$\int_\Omega |\nabla v_n|^{q-2}\nabla v_n \cdot \nabla\psi \, dx = \int_\Omega f_2(y_{1,n}, y_{2,n})\psi \, dx \tag{7.21}$$

for all $(\varphi, \psi) \in W_0^{1,p}(\Omega) \times W_0^{1,q}(\Omega)$. By (7.19) the sequences $\{u_n\}$ and $\{v_n\}$ are bounded in $W_0^{1,p}(\Omega)$ and $W_0^{1,q}(\Omega)$, respectively. Hence, along a relabeled subsequence, the weak convergence $(u_n, v_n) \rightharpoonup (u, v)$ with some $(u, v) \in W_0^{1,p}(\Omega) \times W_0^{1,q}(\Omega)$ holds. Inserting $\varphi = u_n - u$ in (7.20) and $\psi = v_n - v$ in (7.21) leads to

$$\lim_{n\to\infty} \langle -\Delta_p u_n, u_n - u\rangle = \lim_{n\to\infty} \langle -\Delta_q v_n, v_n - v\rangle = 0.$$

This conclusion is achieved via Lebesgue's dominated convergence theorem, which can be applied taking into account (7.17), (7.18), (7.19), and combining with Hardy–Sobolev inequality when $\alpha_i + \beta_i < 0$. Consequently, invoking Proposition 1.56, the $(S)_+$-property of $-\Delta_p$ on $W_0^{1,p}(\Omega)$ and of $-\Delta_q$ on $W_0^{1,q}(\Omega)$ ensures that $u_n \to u$ in $W_0^{1,p}(\Omega)$ and $v_n \to v$ in $W_0^{1,q}(\Omega)$. Due to (7.20) and (7.21), we obtain $(u, v) = \mathcal{T}(y_1, y_2)$. Furthermore, the boundedness of the sequence $\{(u_n, v_n)\}$ in $C_0^{1,\beta}(\overline{\Omega}) \times C_0^{1,\beta}(\overline{\Omega})$ as known from (7.19) yields $(u_n, v_n) \to (u, v)$ in $C(\overline{\Omega}) \times C(\overline{\Omega})$ because the embedding $C_0^{1,\beta}(\overline{\Omega}) \subset C(\overline{\Omega})$ is compact. We conclude that $\mathcal{T}$ is continuous.

Actually, we can establish that the nonlinear operator $\mathcal{T} : \mathcal{K} \to C(\overline{\Omega}) \times C(\overline{\Omega})$ is compact. Toward this, we point out that the estimates (7.19) and the compactness of the embedding $C_0^{1,\beta}(\overline{\Omega}) \subset C(\overline{\Omega})$ show that $\mathcal{T}(\mathcal{K})$ is a relatively compact subset of $C(\overline{\Omega}) \times C(\overline{\Omega})$. The compactness of the nonlinear map $\mathcal{T}$ follows.

Next we prove the inclusion $\mathcal{T}(\mathcal{K}) \subset \mathcal{K}$. Let $(y_1, y_2) \in \mathcal{K}$ and denote $(u, v) = \mathcal{T}(y_1, y_2)$. Making use of the definitions of $\mathcal{K}$ and $\mathcal{T}$ as given before, as well as of Proposition 7.1 and (7.17), results in

$$-\Delta_p u(x) = f_1(y_1(x), y_2(x)) \leq \begin{cases} M_1 \overline{u}(x)^{\alpha_1} \underline{v}(x)^{\beta_1} & \text{if } \alpha_1 \geq 0 \\ M_1 \underline{u}(x)^{\alpha_1} \underline{v}(x)^{\beta_1} & \text{if } \alpha_1 < 0 \end{cases}$$
$$\leq -\Delta_p \overline{u}(x) \text{ in } \Omega.$$

Along the same pattern, we are able to find

$$-\Delta_q v(x) = f_2(y_1(x), y_2(x)) \leq -\Delta_q \overline{v}(x) \text{ in } \Omega.$$

The same reasoning based on Proposition 7.1 and hypotheses (7.1), (7.2) permits getting

$$-\Delta_p u(x) = f_1(y_1(x), y_2(x)) \geq \begin{cases} m_1 \underline{u}(x)^{\alpha_1} \overline{v}(x)^{\beta_1} & \text{if } \alpha_1 \geq 0 \\ m_1 \overline{u}(x)^{\alpha_1} \overline{v}(x)^{\beta_1} & \text{if } \alpha_1 < 0 \end{cases}$$
$$\geq -\Delta_p \underline{u}(x) \text{ in } \Omega$$

and

$$-\Delta_q v(x) = f_2(y_1(x), y_2(x)) \geq -\Delta_q \underline{v}(x) \text{ in } \Omega.$$

Then the strict monotonicity of the operators $-\Delta_p$ and $-\Delta_q$ implies that $(u, v) \in \mathcal{K}$, which reads as $\mathcal{T}(\mathcal{K}) \subset \mathcal{K}$.

We have thus checked that all the requirements to apply Schauder's fixed point theorem (see, e.g., [237, p. 57]) with the set $\mathcal{K}$ and the map $\mathcal{T} : \mathcal{K} \to \mathcal{K}$ are fulfilled. This theorem ensures the existence of $(u, v) \in \mathcal{K}$ satisfying $(u, v) = \mathcal{T}(u, v)$. According to the definition of $\mathcal{T}$, we notice that (u, v) is a (positive) solution of problem (7.1). In addition, the fact that the solution (u, v) satisfies $(u, v) \in \mathcal{K}$ and that the growth conditions (7.2) and (7.3) hold enable us to apply the regularity theory up to the boundary (see [129] for nonsingular degenerate elliptic equations and [101,109] for singular degenerate elliptic equations) to infer that $(u, v) \in C_0^{1,\beta}(\overline{\Omega}) \times C_0^{1,\beta}(\overline{\Omega})$ with $\beta \in (0, 1)$. The proof is complete. □

7.2 THE CASE OF SINGULARITIES WITH RESPECT TO THE GRADIENT OF THE SOLUTION

In this section we focus on a class of singular systems where the singularity can also involve the gradient of the solution, which was not possible in the case of the setting in Section 7.1.

Let $\Omega \subset \mathbb{R}^N$ $(N \geq 2)$ be a bounded domain with C^2 boundary $\partial\Omega$. Given $1 < p, q < +\infty$, we consider the quasilinear elliptic problem

$$\begin{cases} -\Delta_p u = f(u, v, \nabla u, \nabla v) & \text{in } \Omega, \\ -\Delta_q v = g(u, v, \nabla u, \nabla v) & \text{in } \Omega, \\ u, v > 0 & \text{in } \Omega, \\ u, v = 0 & \text{on } \partial\Omega. \end{cases} \tag{7.22}$$

Here, as before, Δ_p and Δ_q stand for the p-Laplacian and q-Laplacian differential operators on $W_0^{1,p}(\Omega)$ and $W_0^{1,q}(\Omega)$, respectively. We stress the presence of the convection in the expression of problem (7.22) meaning the dependence on the solution and its gradient for the terms $f(u, v, \nabla u, \nabla v)$ and $g(u, v, \nabla u, \nabla v)$. Moreover, these terms can be singular both in the solution (u, v) and its gradient $(\nabla u, \nabla v)$.

We assume that $f, g : (0, +\infty) \times (0, +\infty) \times (\mathbb{R}^N \setminus \{0\}) \times (\mathbb{R}^N \setminus \{0\}) \to (0, +\infty)$ are continuous functions satisfying the growth conditions:

$$m_1(s_1 + |\xi_1|)^{\alpha_1}(s_2 + |\xi_2|)^{\beta_1} \leq f(s_1, s_2, \xi_1, \xi_2) \leq M_1(s_1 + |\xi_1|)^{\alpha_1}(s_2 + |\xi_2|)^{\beta_1} \tag{7.23}$$

and

$$m_2(s_1 + |\xi_1|)^{\alpha_2}(s_2 + |\xi_2|)^{\beta_2} \leq g(s_1, s_2, \xi_1, \xi_2) \leq M_2(s_1 + |\xi_1|)^{\alpha_2}(s_2 + |\xi_2|)^{\beta_2} \tag{7.24}$$

for all $s_1, s_2 > 0$, $(\xi_1, \xi_2) \in (\mathbb{R}^N \setminus \{0\}) \times (\mathbb{R}^N \setminus \{0\})$, with constants $m_1, m_2, M_1, M_2 > 0$, $\alpha_2, \beta_1 < 0$ and $\alpha_1, \beta_2 \in \mathbb{R}$ such that

$$\alpha_1\beta_2 > 0 \tag{7.25}$$

and

$$|\alpha_1| - \beta_1 < p - 1 \quad \text{and} \quad |\beta_2| - \alpha_2 < q - 1. \tag{7.26}$$

According to our assumptions, we notice that all the exponents α_1, α_2, β_1, β_2 can be negative. To give just an example, we can choose $\alpha_1 = \beta_2 = -\frac{1}{2}$, $\alpha_2 = \beta_1 = -\frac{1}{4}$ and $p, q \geq 2$. Another relevant feature of our problem is that the assumptions imposed on the functions f and g make the system (7.22) competitive. This is because $\alpha_2, \beta_1 < 0$, which generally prevents f and g from being increasing with respect to v and u, respectively.

We keep the notation of Section 7.1. In particular, we denote by $\lambda_{1,p}$ and $\lambda_{1,q}$ the first eigenvalues of $-\Delta_p$ on $W_0^{1,p}(\Omega)$ and of $-\Delta_q$ on $W_0^{1,q}(\Omega)$, respectively. In addition, let $\hat{u}_{1,p}$ be the L^p-normalized positive eigenfunction of $-\Delta_p$ corresponding to $\lambda_{1,p}$, that is, $-\Delta_p \hat{u}_{1,p} = \lambda_{1,p}\hat{u}_{1,p}^{p-1}$ in Ω, $\hat{u}_{1,p} = 0$ on $\partial\Omega$, with

$\|\hat{u}_{1,p}\|_p = 1$. Similarly, let $\hat{u}_{1,q}$ be the L^q-normalized positive eigenfunction of $-\Delta_q$ corresponding to $\lambda_{1,q}$, that is, $-\Delta_q \hat{u}_{1,q} = \lambda_{1,q} \hat{u}_{1,q}^{q-1}$ in Ω, $\hat{u}_{1,q} = 0$ on $\partial\Omega$, with $\|\hat{u}_{1,q}\|_q = 1$. For any $r \in (1, +\infty)$, we denote $r' = \frac{r}{r-1}$ (the conjugate exponent).

Set, as in Section 7.1,

$$R := \max \left\{ \max_{\Omega} \hat{u}_{1,p}, \max_{\Omega} \hat{u}_{1,q} \right\}. \tag{7.27}$$

We recall from Proposition 1.57 the fundamental property of $\hat{u}_{1,p}$ and $\hat{u}_{1,q}$ related to the normal derivatives on the boundary $\partial\Omega$, namely $\partial \hat{u}_{1,p}/\partial \nu < 0$ and $\partial \hat{u}_{1,q}/\partial \nu < 0$ on $\partial\Omega$, where ν stands for the outer unit normal. In Proposition 1.57 this property is contained in the relation $\hat{u}_{1,p}, \hat{u}_{1,q} \in \operatorname{int}(C_0^1(\overline{\Omega})_+)$. Consequently, we can fix numbers $\mu_1 > 0$ and $\mu_2 > 0$ such that

$$\hat{u}_{1,p} + \left|\nabla \hat{u}_{1,p}\right| \geq \mu_1 \text{ and } \hat{u}_{1,q} + \left|\nabla \hat{u}_{1,q}\right| \geq \mu_2 \text{ in } \overline{\Omega}. \tag{7.28}$$

This is certainly possible because, in view of $\hat{u}_{1,p}, \hat{u}_{1,q} \in \operatorname{int}(C_0^1(\overline{\Omega})_+)$, the properties $\hat{u}_{1,p}, \hat{u}_{1,q} \in C_0^1(\overline{\Omega})$, $\hat{u}_{1,p}, \hat{u}_{1,q} > 0$ in Ω and $\left|\nabla \hat{u}_{1,p}\right|, \left|\nabla \hat{u}_{1,q}\right| > 0$ on $\partial\Omega$ hold.

Let us also recall from [33, Lemma 1] that if $g \in L^\infty(\Omega)$ and if $u \in W_0^{1,r}(\Omega)$ ($r \in \{p, q\}$) is the weak solution of the equation

$$\begin{cases} -\Delta_r u = g & \text{in } \Omega, \\ u = 0 & \text{on } \partial\Omega, \end{cases} \tag{7.29}$$

then there exist a positive constant $K(r)$ depending only on r, N, and Ω, such that

$$\|\nabla u\|_{L^\infty(\Omega)} \leq K(r)(\|g\|_\infty)^{\frac{1}{r-1}}. \tag{7.30}$$

It is well known that the functions ξ_1 and ξ_2 defined as the solutions of the equations

$$\begin{cases} -\Delta_p \xi_1 = 1 & \text{in } \Omega, \\ \xi_1 = 0 & \text{on } \partial\Omega \end{cases} \quad \text{and} \quad \begin{cases} -\Delta_q \xi_2 = 1 & \text{in } \Omega, \\ \xi_2 = 0 & \text{on } \partial\Omega \end{cases} \tag{7.31}$$

belong to $C_0^{1,\alpha}(\overline{\Omega})$ for certain $\alpha \in (0, 1)$. Furthermore, there exist positive constants c_1 and c_2 such that

$$\xi_1(x) \geq c_1 \hat{u}_{1,p}(x) \text{ and } \xi_2(x) \geq c_2 \hat{u}_{1,q}(x) \text{ for all } x \in \Omega. \tag{7.32}$$

Denote

$$L_1 = \max\left\{ \|\xi_1\|_{C_0^{1,\alpha}(\overline{\Omega})}, K(p) \right\} \text{ and } L_2 = \max\left\{ \|\xi_2\|_{C_0^{1,\beta}(\overline{\Omega})}, K(q) \right\}, \tag{7.33}$$

$$\theta(\alpha_1, \beta_2) = \begin{cases} -1 & \text{if } \alpha_1, \beta_2 > 0, \\ 0 & \text{if } \alpha_1, \beta_2 < 0, \end{cases} \tag{7.34}$$

which makes sense in view of (7.25).

For any sufficiently large constant $C > 1$ that will be later selected, we set

$$(\underline{u}, \underline{v}) = (C^{-(q-1)^{\theta(\alpha_1, \beta_2)}} \hat{u}_{1,p}, C^{-(p-1)^{\theta(\alpha_1, \beta_2)}} \hat{u}_{1,q}) \tag{7.35}$$

and

$$(\overline{u}, \overline{v}) = (C^{(p-1)^{\theta(\alpha_1, \beta_2)}} \xi_1, C^{(q-1)^{\theta(\alpha_1, \beta_2)}} \xi_2). \tag{7.36}$$

Here $\theta(\alpha_1, \beta_2)$ is given in (7.34). From (7.35), (7.36), and (7.32) it turns out that if C is sufficiently large, for all $x \in \Omega$ one has

$$\begin{aligned}
\underline{u}(x) &= C^{-(q-1)^{\theta(\alpha_1, \beta_2)}} \hat{u}_{1,p}(x) \le C^{(p-1)^{\theta(\alpha_1, \beta_2)}} c_1 \hat{u}_{1,p}(x) \\
&\le C^{(p-1)^{\theta(\alpha_1, \beta_2)}} \xi_1(x) = \overline{u}(x), \\
\underline{v}(x) &= C^{-(p-1)^{\theta(\alpha_1, \beta_2)}} \hat{u}_{1,q}(x) \le C^{(q-1)^{\theta(\alpha_1, \beta_2)}} c_2 \hat{u}_{1,q}(x) \\
&\le C^{(q-1)^{\theta(\alpha_1, \beta_2)}} \xi_2(x) = \overline{v}(x).
\end{aligned}$$

The functions in (7.35) and (7.36) allow us to introduce the sets to be utilized for developing our fixed point argument. Let us set

$$\begin{aligned}
\mathcal{K}_{1,C} := \Big\{ y \in C_0^1(\overline{\Omega}) : \underline{u} \le y \le \overline{u} \text{ in } \Omega, \ \|\nabla y\|_\infty \le C^{(p-1)^{\theta(\alpha_1, \beta_2)}} L_1 \text{ in } \Omega \text{ and} \\
-\frac{\partial y}{\partial \nu} \ge C^{-(q-1)^{\theta(\alpha_1, \beta_2)}} |\nabla \hat{u}_{1,p}| \text{ on } \partial\Omega \Big\}, \\
\mathcal{K}_{2,C} := \Big\{ y \in C_0^1(\overline{\Omega}) : \underline{v} \le y \le \overline{v} \text{ in } \Omega, \ \|\nabla y\|_\infty \le C^{(q-1)^{\theta(\alpha_1, \beta_2)}} L_2 \text{ in } \Omega \text{ and} \\
-\frac{\partial y}{\partial \nu} \ge C^{-(p-1)^{\theta(\alpha_1, \beta_2)}} |\nabla \hat{u}_{1,q}| \text{ on } \partial\Omega \Big\}.
\end{aligned}$$

Combining the definitions of $\mathcal{K}_{1,C}$, $\mathcal{K}_{2,C}$ and formula (7.28), for every $(y_1, y_2) \in \mathcal{K}_{1,C} \times \mathcal{K}_{2,C}$ we have the estimates

$$y_1 + |\nabla y_1| \ge \underline{u} + |\nabla y_1| \ge C^{-(q-1)^{\theta(\alpha_1, \beta_2)}} \mu_0 \tag{7.37}$$

and

$$y_2 + |\nabla y_2| \ge \underline{v} + |\nabla y_2| \ge C^{-(p-1)^{\theta(\alpha_1, \beta_2)}} \mu_0, \tag{7.38}$$

for some $\mu_0 > 0$.

We need to establish comparison results for problem (7.22). First, we analyze the case $\alpha_1, \beta_2 > 0$.

Proposition 7.3. *Assume that conditions (7.23), (7.24), and (7.26) are satisfied with $\alpha_1, \beta_2 > 0$. Then, taking $C > 1$ large enough in (7.35) and (7.36), we have*

$$-\Delta_p \underline{u} \le m_1(\underline{u} + |\nabla y_1|)^{\alpha_1}(\overline{v} + L_2 C^{\frac{1}{q-1}})^{\beta_1} \quad in\ \Omega, \tag{7.39}$$

$$-\Delta_q \underline{v} \le m_2(\overline{u} + L_1 C^{\frac{1}{p-1}})^{\alpha_2}(\underline{v} + |\nabla y_2|)^{\beta_2} \quad in\ \Omega, \tag{7.40}$$

$$-\Delta_p \overline{u} \ge M_1(\overline{u} + L_1 C^{\frac{1}{p-1}})^{\alpha_1}(\underline{v} + |\nabla y_2|)^{\beta_1} \quad in\ \Omega, \tag{7.41}$$

$$-\Delta_q \overline{v} \ge M_2(\underline{u} + |\nabla y_1|)^{\alpha_2}(\overline{v} + L_2 C^{\frac{1}{q-1}})^{\beta_2} \quad in\ \Omega \tag{7.42}$$

for all $(y_1, y_2) \in \mathcal{K}_{1,C} \times \mathcal{K}_{2,C}$.

Proof. By (7.35), (7.36), (7.34), (7.33), (7.37), (7.27), and (7.26), we derive for every $(y_1, y_2) \in \mathcal{K}_{1,C} \times \mathcal{K}_{2,C}$ that

$$\begin{aligned}
&(\underline{u} + |\nabla y_1|)^{-\alpha_1}(\overline{v} + L_2 C^{\frac{1}{q-1}})^{-\beta_1}(-\Delta_p \underline{u}) \\
&= (\underline{u} + |\nabla y_1|)^{-\alpha_1}(C^{\frac{1}{q-1}}\xi_2 + L_2 C^{\frac{1}{q-1}})^{-\beta_1} C^{\frac{-(p-1)}{q-1}} \hat{u}_{1,p}^{p-1} \\
&\le C^{\frac{-(p-1)+\alpha_1}{q-1} - \frac{\beta_1}{q-1}} \mu_0^{-\alpha_1}(2L_2)^{-\beta_1} \hat{u}_{1,p}^{p-1} \\
&\le C^{\frac{\alpha_1 - \beta_1 - p + 1}{q-1}} \mu_0^{-\alpha_1}(2L_2)^{-\beta_1} R^{p-1} \le m_1 \quad \text{in } \Omega
\end{aligned}$$

provided that $C > 1$ is sufficiently large. This implies (7.39).

From (7.35), (7.36), (7.34), (7.33), (7.38), (7.27), and (7.26), for every $(y_1, y_2) \in \mathcal{K}_{1,C} \times \mathcal{K}_{2,C}$ we obtain

$$\begin{aligned}
&(\overline{u} + L_1 C^{\frac{1}{p-1}})^{-\alpha_2}(\underline{v} + |\nabla y_2|)^{-\beta_2}(-\Delta_q \underline{v}) \\
&= (C^{\frac{1}{p-1}}\xi_1 + L_1 C^{\frac{1}{p-1}})^{-\alpha_2}(\underline{v} + |\nabla y_2|)^{-\beta_2} C^{\frac{-q+1}{p-1}} \hat{u}_{1,q}^{q-1} \\
&\le C^{\frac{-\alpha_2 + \beta_2 - q + 1}{p-1}} (2L_1)^{-\alpha_2} \mu_0^{-\beta_2} R^{q-1} \le m_2 \quad \text{in } \Omega
\end{aligned}$$

provided that $C > 1$ is sufficiently large, which ensures (7.40).

By (7.36), (7.34), (7.31), (7.33), (7.38), (7.26), and $\beta_1 < 0$, for every $(y_1, y_2) \in \mathcal{K}_{1,C} \times \mathcal{K}_{2,C}$ it follows that

$$\begin{aligned}
&(\overline{u} + L_1 C^{\frac{1}{p-1}})^{-\alpha_1}(\underline{v} + |\nabla y_2|)^{-\beta_1}(-\Delta_p \overline{u}) \\
&= C^{\frac{p-1}{p-1}}(C^{\frac{1}{p-1}}\xi_1 + L_1 C^{\frac{1}{p-1}})^{-\alpha_1}(\underline{v} + |\nabla y_2|)^{-\beta_1} \\
&\ge C^{\frac{p-1+\beta_1-\alpha_1}{p-1}} (2L_1)^{-\alpha_1} \mu_0^{-\beta_1} \ge M_1 \quad \text{in } \Omega
\end{aligned}$$

provided that $C > 1$ is sufficiently large, which proves (7.41).

By (7.36), (7.34), (7.31), (7.33), (7.37), (7.26), and $\alpha_2 < 0$, for every $(y_1, y_2) \in \mathcal{K}_{1,C} \times \mathcal{K}_{2,C}$ we derive

$$
\begin{aligned}
&(\underline{u}+|\nabla y_1|)^{-\alpha_2}(\overline{v}+L_2C^{\frac{1}{q-1}})^{-\beta_2}(-\Delta_q\overline{v})\\
&=C^{\frac{q-1}{q-1}}(\underline{u}+|\nabla y_1|)^{-\alpha_2}(C^{\frac{1}{q-1}}\xi_2+L_2C^{\frac{1}{q-1}})^{-\beta_2}\\
&\geq C^{\frac{q-1+\alpha_2-\beta_2}{q-1}}\mu_0^{-\alpha_2}(2L_2)^{-\beta_2}\geq M_2 \quad \text{in } \Omega
\end{aligned}
$$

provided that $C>1$ is sufficiently large. So, we infer (7.42), which completes the proof. □

In the next proposition we discuss the case $\alpha_1, \beta_2<0$ that can also occur under condition (7.25).

Proposition 7.4. *Assume that conditions (7.23), (7.24), and (7.26) are satisfied with $\alpha_1, \beta_2<0$. Then, for $C>1$ large enough in (7.35) and (7.36), we have*

$$-\Delta_p\underline{u}\leq m_1(\overline{u}+L_1C)^{\alpha_1}(\overline{v}+L_2C)^{\beta_1} \quad \text{in } \Omega, \tag{7.43}$$

$$-\Delta_q\underline{v}\leq m_2(\overline{u}+L_1C)^{\alpha_2}(\overline{v}+L_2C)^{\beta_2} \quad \text{in } \Omega, \tag{7.44}$$

$$-\Delta_p\overline{u}\geq M_1(\underline{u}+|\nabla y_1|)^{\alpha_1}(\underline{v}+|\nabla y_2|)^{\beta_1} \quad \text{in } \Omega, \tag{7.45}$$

$$-\Delta_q\overline{v}\geq M_2(\underline{u}+|\nabla y_1|)^{\alpha_2}(\underline{v}+|\nabla y_2|)^{\beta_2} \quad \text{in } \Omega \tag{7.46}$$

for all $(y_1, y_2)\in\mathcal{K}_{1,C}\times\mathcal{K}_{2,C}$.

Proof. From (7.35), (7.36), (7.34), (7.33), (7.27), and (7.26), we find that

$$
\begin{aligned}
&(\overline{u}+L_1C)^{-\alpha_1}(\overline{v}+L_2C)^{-\beta_1}(-\Delta_p\underline{u})\\
&=(C\xi_1+L_1C)^{-\alpha_1}(C\xi_2+L_2C)^{-\beta_1}C^{-p+1}\hat{u}_{1,p}^{p-1}\\
&\leq C^{-p+1-\alpha_1-\beta_1}(2L_1)^{-\alpha_1}(2L_2)^{-\beta_1}\hat{u}_{1,p}^{p-1}\\
&\leq C^{-p+1-\alpha_1-\beta_1}2^{-\alpha_1-\beta_1}L_1^{-\alpha_1}L_2^{-\beta_1}R^{p-1}\leq m_1 \quad \text{in } \Omega
\end{aligned}
$$

provided that $C>1$ is sufficiently large, which proves (7.43).

Using (7.35), (7.36), (7.34), (7.33), (7.27), and (7.26), leads to

$$
\begin{aligned}
&(\overline{u}+L_1C)^{-\alpha_2}(\overline{v}+L_2C)^{-\beta_2}(-\Delta_q\underline{v})\\
&=(C\xi_1+L_1C)^{-\alpha_2}(C\xi_2+L_2C)^{-\beta_2}C^{-q+1}\hat{u}_{1,q}^{q-1}\\
&\leq C^{-q+1-\alpha_2-\beta_2}2^{-\alpha_2-\beta_2}L_1^{-\alpha_2}L_2^{-\beta_2}R^{q-1}\leq m_2 \quad \text{in } \Omega
\end{aligned}
$$

provided that $C>1$ is sufficiently large, hence (7.44) ensues.

According to (7.36), (7.31), (7.34), (7.37), (7.38), and (7.26), for every $(y_1, y_2)\in\mathcal{K}_{1,C}\times\mathcal{K}_{2,C}$ we obtain

$$
\begin{aligned}
&(\underline{u}+|\nabla y_1|)^{-\alpha_1}(\underline{v}+|\nabla y_2|)^{-\beta_1}(-\Delta_p\overline{u})\\
&=C^{p-1}(\underline{u}+|\nabla y_1|)^{-\alpha_1}(\underline{v}+|\nabla y_2|)^{-\beta_1}
\end{aligned}
$$

$$\geq C^{p-1+\alpha_1+\beta_1}\mu_0^{-\alpha_1-\beta_1} \geq M_1 \quad \text{in } \Omega$$

provided that $C > 1$ is sufficiently large. This shows that (7.45) is true.

By (7.36), (7.31), (7.34), (7.37), (7.38), and (7.26), for every $(y_1, y_2) \in \mathcal{K}_{1,C} \times \mathcal{K}_{2,C}$ we can derive

$$\begin{aligned}
&(\underline{u} + |\nabla y_1|)^{-\alpha_2}(\underline{v} + |\nabla y_2|)^{-\beta_2}(-\Delta_q \overline{v}) \\
&= C^{q-1}(\underline{u} + |\nabla y_1|)^{-\alpha_2}(\underline{v} + |\nabla y_2|)^{-\beta_2} \\
&\geq C^{q-1+\alpha_2+\beta_2}\mu_0^{-\alpha_2-\beta_2} \geq M_2 \text{ in } \Omega
\end{aligned}$$

provided that $C > 1$ is sufficiently large, so (7.46) holds. This completes the proof. □

We are now able to provide the existence of (positive) smooth solutions for the singular system (7.22).

Theorem 7.5. *Assume that conditions (7.23), (7.24), (7.25), and (7.26) hold. Then problem (7.22) has a positive, smooth solution (u, v) in $C_0^{1,\alpha}(\overline{\Omega}) \times C_0^{1,\alpha}(\overline{\Omega})$ for some $\alpha \in (0, 1)$.*

Proof. We use the sets $\mathcal{K}_{1,C}$ and $\mathcal{K}_{2,C}$ constructed before. Define the map $\mathcal{T} : \mathcal{K}_{1,C} \times \mathcal{K}_{2,C} \to C_0^1(\overline{\Omega}) \times C_0^1(\overline{\Omega})$ by $\mathcal{T}(y_1, y_2) = (u, v)$, where (u, v) is required to satisfy

$$\begin{cases}
-\Delta_p u = f(y_1, y_2, \nabla y_1, \nabla y_2) & \text{in } \Omega, \\
-\Delta_q v = g(y_1, y_2, \nabla y_1, \nabla y_2) & \text{in } \Omega, \\
u = v = 0 \text{ on } \partial\Omega.
\end{cases} \tag{7.47}$$

It is essential to point out that the operator $\mathcal{T}$ is well defined for $C > 1$ large enough because problem (7.47) has a unique smooth solution. Indeed, for $(y_1, y_2) \in \mathcal{K}_{1,C} \times \mathcal{K}_{2,C}$ with $C > 1$ sufficiently large, we derive from (7.23), (7.25), (7.34), (7.35), (7.36), (7.33), (7.37), (7.38), and (7.26) the estimate

$$\begin{aligned}
&f(y_1, y_2, \nabla y_1, \nabla y_2) \leq M_1(y_1 + |\nabla y_1|)^{\alpha_1}(y_2 + |\nabla y_2|)^{\beta_1} \\
&\leq M_1 \begin{cases} (\overline{u} + L_1 C^{\frac{1}{p-1}})^{\alpha_1}(y_2 + |\nabla y_2|)^{\beta_1} & \text{if } \alpha_1 > 0 \\ (y_1 + |\nabla y_1|)^{\alpha_1}(y_2 + |\nabla y_2|)^{\beta_1} & \text{if } \alpha_1 < 0 \end{cases} \\
&\leq M_1 \begin{cases} C^{\frac{\alpha_1-\beta_1}{p-1}}(2L_1)^{\alpha_1}\mu_0^{\beta_1} & \text{if } \alpha_1 > 0 \\ C^{-\alpha_1-\beta_1}\mu_0^{\alpha_1+\beta_1} & \text{if } \alpha_1 < 0 \end{cases} \\
&\leq \begin{cases} C & \text{if } \alpha_1 > 0 \\ C^{p-1} & \text{if } \alpha_1 < 0 \end{cases} = C^{(p-1)^{1+\theta(\alpha_1,\beta_2)}} \quad \text{in } \Omega.
\end{aligned} \tag{7.48}$$

Similarly, from (7.24), (7.25), (7.34), (7.35), (7.36), (7.33), (7.37), (7.38), and (7.26), for $(y_1, y_2) \in \mathcal{K}_{1,C} \times \mathcal{K}_{2,C}$ and taking the constant $C > 1$ sufficiently large, we have

$$\begin{aligned} g(y_1, y_2, \nabla y_1, \nabla y_2) &\le M_2 (y_1 + |\nabla y_1|)^{\alpha_2} (y_2 + |\nabla y_2|)^{\beta_2} \\ &\le M_2 \begin{cases} (y_1 + |\nabla y_1|)^{\alpha_2} (\overline{v} + L_2 C^{\frac{1}{q-1}})^{\beta_2} & \text{if } \beta_2 > 0 \\ (y_1 + |\nabla y_1|)^{\alpha_2} (y_2 + |\nabla y_2|)^{\beta_2} & \text{if } \beta_2 < 0 \end{cases} \\ &\le M_2 \begin{cases} C^{\frac{-\alpha_2+\beta_2}{q-1}} \mu_0^{\alpha_2} (2L_2)^{\beta_2} & \text{if } \beta_2 > 0 \\ C^{-\alpha_2-\beta_2} \mu_0^{\alpha_2+\beta_2} & \text{if } \beta_2 < 0 \end{cases} \\ &\le \begin{cases} C & \text{if } \beta_2 > 0 \\ C^{q-1} & \text{if } \beta_2 < 0 \end{cases} = C^{(q-1)^{1+\theta(\alpha_1,\beta_2)}} \quad \text{in } \Omega. \end{aligned} \tag{7.49}$$

In particular, the estimates obtained in (7.48) and (7.49) imply that $f(y_1, y_2, \nabla y_1, \nabla y_2) \in W^{-1,p'}(\Omega)$ and $g(y_1, y_2, \nabla y_1, \nabla y_2) \in W^{-1,q'}(\Omega)$. Then Proposition 1.56 and Theorem 1.55 ensure the existence and uniqueness of solution (u, v) in (7.47). Moreover, the nonlinear regularity theory up to the boundary in [129] guarantees that $(u, v) \in C_0^{1,\alpha}(\overline{\Omega}) \times C_0^{1,\alpha}(\overline{\Omega})$, with some $\alpha \in (0, 1)$. In addition, due to the estimates (7.48) and (7.49), there exists a constant $M > 0$ such that

$$\|u\|_{C_0^{1,\alpha}(\overline{\Omega})}, \|v\|_{C_0^{1,\alpha}(\overline{\Omega})} \le M \tag{7.50}$$

whenever $(u, v) = \mathcal{T}(y_1, y_2)$ with $(y_1, y_2) \in \mathcal{K}_{1,C} \times \mathcal{K}_{2,C}$.

The (weak) solutions of problem (7.22) coincide with the fixed points of the map $\mathcal{T}$ introduced in (7.47). It is straightforward to check that the set $\mathcal{K}_{1,C} \times \mathcal{K}_{2,C}$ is nonempty, closed, bounded, and convex in $C_0^1(\overline{\Omega}) \times C_0^1(\overline{\Omega})$.

In order to check that the map $\mathcal{T} : \mathcal{K}_{1,C} \times \mathcal{K}_{2,C} \to C_0^1(\overline{\Omega}) \times C_0^1(\overline{\Omega})$ is continuous and compact, let $(y_{1,n}, y_{2,n}) \in \mathcal{K}_{1,C} \times \mathcal{K}_{2,C}$ with $(y_{1,n}, y_{2,n}) \to (y_1, y_2)$ in $C_0^1(\overline{\Omega}) \times C_0^1(\overline{\Omega})$ as $n \to \infty$. Set $(u_n, v_n) = \mathcal{T}(y_{1,n}, y_{2,n})$, which reads as

$$\int_\Omega |\nabla u_n|^{p-2} \nabla u_n \cdot \nabla \varphi \, dx = \int_\Omega f(y_{1,n}, y_{2,n}, \nabla y_{1,n}, \nabla y_{2,n}) \varphi \, dx, \tag{7.51}$$

$$\int_\Omega |\nabla v_n|^{q-2} \nabla v_n \cdot \nabla \psi \, dx = \int_\Omega g(y_{1,n}, y_{2,n}, \nabla y_{1,n}, \nabla y_{2,n}) \psi \, dx \tag{7.52}$$

for all $(\varphi, \psi) \in W_0^{1,p}(\Omega) \times W_0^{1,q}(\Omega)$. Inserting $\varphi = u_n - u$ in (7.51) and $\psi = v_n - v$ in (7.52), we see from (7.48) and (7.49) that

$$\lim_{n\to\infty} \langle -\Delta_p u_n, u_n - u \rangle = \lim_{n\to\infty} \langle -\Delta_q v_n, v_n - v \rangle = 0.$$

Then the $(S)_+$-property of $-\Delta_p$ on $W_0^{1,p}(\Omega)$ and of $-\Delta_q$ on $W_0^{1,q}(\Omega)$, which was stated in Proposition 1.56, ensures that $u_n \to u$ in $W_0^{1,p}(\Omega)$ and $v_n \to v$

in $W_0^{1,q}(\Omega)$. Hence, we can pass to the limit in (7.51) and (7.52). This results in $(u, v) = \mathcal{T}(y_1, y_2)$. Furthermore, since the embedding $C_0^{1,\alpha}(\overline{\Omega}) \subset C_0^1(\overline{\Omega})$ is compact, the boundedness of the sequence $\{(u_n, v_n)\}$ in $C_0^{1,\alpha}(\overline{\Omega}) \times C_0^{1,\alpha}(\overline{\Omega})$ as known from (7.50) has the effect that up to a subsequence $(u_n, v_n) \to (u, v)$ in $C_0^1(\overline{\Omega}) \times C_0^1(\overline{\Omega})$. Therefore the map $\mathcal{T}$ is continuous. Again through (7.50) it follows that $\mathcal{T}(\mathcal{K}_{1,C} \times \mathcal{K}_{2,C})$ is bounded in $C_0^{1,\alpha}(\overline{\Omega}) \times C_0^{1,\alpha}(\overline{\Omega})$. Consequently, the compactness of the embedding $C_0^{1,\alpha}(\overline{\Omega}) \subset C_0^1(\overline{\Omega})$ ensures that $\mathcal{T}(\mathcal{K}_{1,C} \times \mathcal{K}_{2,C})$ is a relatively compact subset of $C_0^1(\overline{\Omega}) \times C_0^1(\overline{\Omega})$, which proves the claim.

A fundamental step in the proof is to show the invariance property

$$\mathcal{T}(\mathcal{K}_{1,C} \times \mathcal{K}_{2,C}) \subset \mathcal{K}_{1,C} \times \mathcal{K}_{2,C}.$$

For any $(y_1, y_2) \in \mathcal{K}_{1,C} \times \mathcal{K}_{2,C}$, we write as before $(u, v) = \mathcal{T}(y_1, y_2)$. By the definitions of $\mathcal{K}_{1,C}$, $\mathcal{K}_{2,C}$, $\mathcal{T}$ and in view of (7.23), (7.24), (7.47), (7.33), (7.34), Propositions 7.3 and 7.4 (specifically, (7.39)–(7.42) and (7.43)–(7.46)), it turns out that

$$\begin{aligned}
&-\Delta_p u = f(y_1, y_2, \nabla y_1, \nabla y_2) \leq M_1(y_1 + |\nabla y_1|)^{\alpha_1}(y_2 + |\nabla y_2|)^{\beta_1}\\
&\leq M_1 \begin{cases} (\overline{u} + L_1 C^{\frac{1}{p-1}})^{\alpha_1}(\underline{v} + |\nabla y_2|)^{\beta_1} & \text{if } \alpha_1 > 0 \\ (\underline{u} + |\nabla y_1|)^{\alpha_1}(\underline{v} + |\nabla y_2|)^{\beta_1} & \text{if } \alpha_1 < 0 \end{cases} \leq -\Delta_p \overline{u} \ \text{ in } \Omega,\\
&-\Delta_q v = g(y_1, y_2, \nabla y_1, \nabla y_2) \leq M_2(y_1 + |\nabla y_1|)^{\alpha_2}(y_2 + |\nabla y_2|)^{\beta_2}\\
&\leq M_2 \begin{cases} (\underline{u} + |\nabla y_1|)^{\alpha_2}(\overline{v} + L_2 C^{\frac{1}{q-1}})^{\beta_2} & \text{if } \beta_2 > 0 \\ (\underline{u} + |\nabla y_1|)^{\alpha_2}(\underline{v} + |\nabla y_2|)^{\beta_2} & \text{if } \beta_2 < 0 \end{cases} \leq -\Delta_q \overline{v} \ \text{ in } \Omega,\\
&-\Delta_p u = f(y_1, y_2, \nabla y_1, \nabla y_2) \geq m_1(y_1 + |\nabla y_1|)^{\alpha_1}(y_2 + |\nabla y_2|)^{\beta_1}\\
&\geq m_1 \begin{cases} (\underline{u} + |\nabla y_1|)^{\alpha_1}(\overline{v} + L_2 C^{\frac{1}{q-1}})^{\beta_1} & \text{if } \alpha_1 > 0 \\ (\overline{u} + L_1 C)^{\alpha_1}(\overline{v} + L_2 C)^{\beta_1} & \text{if } \alpha_1 < 0 \end{cases} \geq -\Delta_p \underline{u} \ \text{ in } \Omega,
\end{aligned} \tag{7.53}$$

$$\begin{aligned}
&-\Delta_q v = g(y_1, y_2, \nabla y_1, \nabla y_2) \geq m_2(y_1 + |\nabla y_1|)^{\alpha_2}(y_2 + |\nabla y_2|)^{\beta_2}\\
&\geq m_2 \begin{cases} (\overline{u} + L_1 C^{\frac{1}{p-1}})^{\alpha_2}(\underline{v} + |\nabla y_2|)^{\beta_2} & \text{if } \beta_2 > 0 \\ (\overline{u} + L_1 C)^{\alpha_2}(\overline{v} + L_2 C)^{\beta_2} & \text{if } \beta_2 < 0 \end{cases} \geq -\Delta_q \underline{v} \ \text{ in } \Omega.
\end{aligned} \tag{7.54}$$

Thus we get the inequality $-\Delta_p u \leq -\Delta_p \overline{u}$ in Ω. Acting with $(u - \overline{u})^+ \in W_0^{1,p}(\Omega)$ yields

$$\begin{aligned}
&\int_{\{u>\overline{u}\}} (|\nabla u|^{p-2}\nabla u - |\nabla \overline{u}|^{p-2}\nabla \overline{u}) \cdot (\nabla u - \nabla \overline{u})\, dx\\
&= \langle -\Delta_p u + \Delta_p \overline{u}, (u - \overline{u})^+ \rangle \leq 0.
\end{aligned}$$

The strict monotonicity of the map $\xi \mapsto |\xi|^{p-2}\xi$ on $\mathbb{R}^N$ implies that the set $\{u > \overline{u}\} = \{x \in \Omega : u(x) > \overline{u}(x)\}$ has zero Lebesgue measure. Therefore the inequality $u \leq \overline{u}$ a.e. in Ω is valid. Actually, proceeding in the same way, the estimates shown before entail

$$\underline{u} \leq u \leq \overline{u} \text{ and } \underline{v} \leq v \leq \overline{v} \text{ in } \Omega. \tag{7.55}$$

On the other hand, using (7.47), (7.48), (7.49), we infer through (7.29) and (7.30) that

$$\|\nabla u\|_\infty \leq K(p)C^{(p-1)^{\theta(\alpha_1, \beta_2)}} \text{ and } \|\nabla v\|_\infty \leq K(q)C^{(q-1)^{\theta(\alpha_1, \beta_2)}}.$$

In view of (7.33), we get

$$\|\nabla u\|_\infty \leq L_1 C^{(p-1)^{\theta(\alpha_1,\beta_2)}} \text{ and } \|\nabla v\|_\infty \leq L_2 C^{(q-1)^{\theta(\alpha_1,\beta_2)}}. \tag{7.56}$$

The nonlinear comparison principle (see, e.g., [61]) applied to (7.53) and (7.54) provides the inequalities

$$\frac{\partial u}{\partial \nu} \leq \frac{\partial \underline{u}}{\partial \nu} \text{ and } \frac{\partial v}{\partial \nu} \leq \frac{\partial \underline{v}}{\partial \nu} \text{ on } \partial\Omega. \tag{7.57}$$

The strong maximum principle applied to $\hat{u}_{1,p}$ and $\hat{u}_{1,q}$ yields that their normal derivatives are negative on the boundary $\partial\Omega$ (recall that $\hat{u}_{1,p}, \hat{u}_{1,q} \in \operatorname{int}(C_0^1(\overline{\Omega})_+)$ as known from Proposition 1.57). This fact, (7.57) and (7.35) imply

$$-\frac{\partial u}{\partial \nu} \geq -\frac{\partial \underline{u}}{\partial \nu} = |\nabla \underline{u}| = C^{-(q-1)^{\theta(\alpha_1, \beta_2)}} |\nabla \hat{u}_{1,p}| \text{ on } \partial\Omega, \tag{7.58}$$

$$-\frac{\partial v}{\partial \nu} \geq -\frac{\partial \underline{v}}{\partial \nu} = |\nabla \underline{v}| = C^{-(p-1)^{\theta(\alpha_1,\beta_2)}} |\nabla \hat{u}_{1,q}| \text{ on } \partial\Omega. \tag{7.59}$$

Gathering (7.55), (7.56), (7.58), and (7.59) enables us to conclude that $(u, v) \in \mathcal{K}_{1,C} \times \mathcal{K}_{2,C}$. Therefore the inclusion $\mathcal{T}(\mathcal{K}_{1,C} \times \mathcal{K}_{2,C}) \subset \mathcal{K}_{1,C} \times \mathcal{K}_{2,C}$ holds true.

We have checked all the conditions required to apply Schauder's fixed point theorem to the map $\mathcal{T} : \mathcal{K}_{1,C} \times \mathcal{K}_{2,C} \to \mathcal{K}_{1,C} \times \mathcal{K}_{2,C}$. We conclude that there exists a fixed point $(u, v) \in \mathcal{K}_{1,C} \times \mathcal{K}_{2,C}$, that is, $(u, v) = \mathcal{T}(u, v)$. Thanks to the definition of the map $\mathcal{T}$, it turns out that (u, v) is a (positive) smooth solution of problem (7.22). □

7.3 NOTES

Section 7.1 is based on Motreanu–Moussaoui [172] treating a singular quasilinear elliptic system where the singularities are exhibited by the vector field (f_1, f_2) and involve the solution (u, v). The singularities in problem (7.1) are located at 0 and arise through a competitive structure of the nonlinearities $f_1(u, v)$

and $f_2(u, v)$. The singularities occur because the exponents β_1 and α_2 are negative (see (7.2) and (7.3)). In particular, this prevents f_1 and f_2 from being increasing with respect to v and u, respectively. Consequently, the subsolution–supersolution method cannot be directly implemented in the case of the system (7.1). Additional assumptions are needed as it is apparent within an approach by the method of subsolution–supersolutions in Giacomoni–Schindler–Takáč [101]. A different existence result under another type of hypotheses and through adequate truncations can be found in Motreanu–Moussaoui [171]. We also mention the recent result in Alves–Moussaoui [7] where it is allowed to have certain convection terms. The semilinear case in problem (7.1) (i.e., $p = q = 2$) was handled in Ghergu [96] and Moussaoui–Khodja–Tas [190] using essentially the linearity of the principal part in the system. The complementary situation for the system (7.1) with respect to our setting is what is called the cooperative structure, which means to have positive exponents β_1 and α_2 in conditions (7.1) and (7.3). For this case we refer to [77,99,100,170].

Theorem 7.2 establishes the existence and regularity of (positive) solutions for problem (7.1). We emphasize that both components of the solution are positive. The proof relies on comparison arguments, which allow us to get a priori estimates. Using these estimates enables us to obtain the main result by applying the Schauder's fixed point theorem to a fixed point problem associated to the system (7.1). It is also worth mentioning that a careful inspection of the proof of Proposition 7.1 reveals that the constant $C > 0$ in the definition of the subsolution–supersolutions $(\underline{u}, \underline{v})$ and $(\overline{u}, \overline{v})$ in (7.13) can be precisely estimated.

Section 7.2 follows closely Motreanu–Moussaoui–Zhang [173], where for the first time the system (7.1) is studied with convection terms $f(u, v, \nabla u, \nabla v)$, $g(u, v, \nabla u, \nabla v)$ exhibiting singularities in both the solution and its gradient. Without loss of generality the singularity is located at zero. The singularity involving the gradient of the solution is a novelty in the literature. Singular systems with convection terms and singularities only in the solution were examined in the semilinear case, i.e., $p = q = 2$, by Alves–Carriao–Faria [4] and by Alves–Moussaoui [7]. For singular elliptic systems without gradient dependence in the lower order terms, we refer to [5,6,61,77,96,97,111,158,170–172].

The main result in Section 7.2 is Theorem 7.5 providing a positive solution with regularity information for the singular system (7.22) with singularities for the functions f and g in the right-hand side of the equations possibly occurring not only for u and v but also for their gradients ∇u and ∇v as it is apparent from conditions (7.23)–(7.26). The singularities in the gradient of the solutions produce serious technical difficulties because there are no available estimates for the gradient when it involves singularities. Relevant comparison arguments and a priori estimates for the singular system (7.22) are indicated in Proposition 7.3, which permits us in the proof of Theorem 7.5 to control the gradients ∇u and ∇v despite the fact that they can be singular. This enables us to apply the Schauder's fixed point theorem. The most challenging point in the fixed point approach is to identify the appropriate set where the Schauder's fixed point theorem should be applied taking into account the behavior of the gradient up to the boundary of the domain. It is worth pointing out a close analogy with the method of subsolutions–supersolutions in a singular framework. An open problem for nonlinear elliptic problems with singularities in the solutions and gradients of the solutions is to establish the existence of multiple positive solutions.

Chapter 8

Evolutionary Variational and Quasivariational Inequalities

8.1 EVOLUTIONARY VARIATIONAL INEQUALITIES

First, we present some notation. For a Banach space Z with the topological dual Z^*, the duality pairing between Z and Z^* will be denoted by $\langle\cdot,\cdot\rangle_Z$. By $\|\cdot\|_Z$ we denote the norm on Z. The domain and the graph of a multivalued map $F : Z \to 2^{Z^*}$ are given by $D(F) := \{x \in Z : F(x) \neq \emptyset\}$, and $G(F) := \{(x, w) : x \in D(F), w \in F(x)\}$, respectively. We denote the (effective) domain of a functional $\Phi : Z \to \mathbb{R} \cup \{+\infty\}$, which is $\not\equiv +\infty$, by $D(\Phi) := \{x \in Z : \Phi(x) < +\infty\}$. The strong and weak convergence are specified by $\to$ and $\rightharpoonup$, respectively.

We need multivalued versions of some definitions formulated in the single-valued setting of Section 1.3.

Definition 8.1. Given a Banach space X, let $F : X \to 2^{X^*}$ be a multivalued map.

(i) F is called monotone, if $\langle u - v, x - y\rangle_X \geq 0$ for every $(x, u), (y, v) \in G(F)$.

(ii) F is called maximal monotone, if F is monotone and $G(F)$ is not included in the graph of any other monotone map from X to X^*.

(iii) F is called generalized pseudomonotone if for any sequence $\{(x_n, w_n)\} \subset G(F)$ with $x_n \rightharpoonup x$ and $w_n \rightharpoonup w$ such that $\limsup_{n\to\infty}\langle w_n, x_n - x\rangle_X \leq 0$, we have $w \in F(x)$ and $\langle w_n, x_n\rangle_X \to \langle w, x\rangle_X$.

Definition 8.2. A multivalued map $F : X \to 2^{X^*}$ on a Banach space X is called pseudomonotone if it satisfies the conditions:

$(PM1)$ For each $x \in X$, the set $F(x)$ is nonempty, closed, and convex in X^*.

$(PM2)$ If $\{(x_n, w_n)\} \subset G(F)$ is a sequence such that $x_n \rightharpoonup x$ and

$$\limsup_{n\to\infty}\langle w_n, x_n - x\rangle_X \leq 0,$$

then for each $y \in X$ there exists $w(y) \in F(x)$ satisfying

$$\liminf_{n\to\infty}\langle w_n, x_n - y\rangle_X \geq \langle w(y), x - y\rangle_X.$$

Nonlinear Differential Problems with Smooth and Nonsmooth Constraints. DOI: 10.1016/B978-0-12-813386-6.00008-0

$(PM3)$ The map F is upper semicontinuous from each finite-dimensional subspace of X to the space X^* endowed with the weak topology.

Remark 8.3. It is shown in Kenmochi [117, Lemma 1.2] that, given a reflexive Banach space X, a map $F : X \to 2^{X^*}$ with $D(F) = X$ verifying conditions $(PM1)$ and $(PM2)$ of Definition 8.2 is pseudomonotone if it satisfies:

$(PM4)$ For each $x \in X$ and for each bounded subset B of X, there exists a constant $c(B, x)$ such that whenever $(z, u) \in G(F)$ with $z \in B$, one has $\langle u, z - x\rangle_X \geq c(B, x)$.

For instance, condition $(PM4)$ is fulfilled by any monotone map F with $D(F) = X$, as well as by bounded maps.

Remark 8.4. A pseudomonotone map $F : X \to 2^{X^*}$ on a reflexive Banach space X is generalized pseudomonotone. Conversely, a bounded, generalized pseudomonotone map $F : X \to 2^{X^*}$ on a reflexive Banach space X satisfying condition $(PM1)$ is pseudomonotone (see [38, Propositions 2.122, 2.123]).

We also quote the following result from Kenmochi [117, Proposition 4.1].

Proposition 8.5. *Let $F : X \to 2^{X^*}$ be a set-valued map on a reflexive Banach space X satisfying $(PM1)$, $(PM2)$, and $(PM4)$, let C be a nonempty, closed, convex, and bounded subset of X, let $\Phi : X \to \mathbb{R} \cup \{+\infty\}$ be lower semicontinuous, convex, with $C \cap D(\Phi) \neq \emptyset$, and let $f \in X^*$. Then there exists $x \in C \cap D(\Phi)$ such that for some $w \in F(x)$ we have*

$$\langle w - f, z - x\rangle_X \geq \Phi(x) - \Phi(z) \quad \textit{for every } z \in C.$$

Now we present general existence results for elliptic and evolutionary variational inequalities.

Theorem 8.6. *Consider reflexive Banach spaces X and Y with a compact linear map $i : X \to Y$, and a closed convex subset K of X with nonempty interior $\operatorname{int} K$. We assume that*

(A_1) *$L : D(L) \subset X \to X^*$ is a linear, maximal monotone operator.*
(A_2) *$A : X \to 2^{X^*}$ is a (possibly multivalued) bounded, pseudomonotone operator.*
(A_3) *$B : Y \to 2^{Y^*}$ is a (possibly multivalued) bounded operator, which has nonempty, convex, closed values, and whose graph is sequentially strongly–weakly closed (i.e., if $(y_n, b_n) \in G(B)$ with $y_n \to y$ in Y and $b_n \rightharpoonup b$ in Y^*, then $(y, b) \in G(B)$).*
(A_4) *$\Phi : K \to \mathbb{R} \cup \{+\infty\}$ is a convex, lower semicontinuous function, with $\operatorname{int} K \subset D(\Phi)$.*
(A_5) *$f \in X^*$, and when the set K is unbounded in X there exist $u_0 \in D(L) \cap \operatorname{int} K$ and an $r > 0$ satisfying*

$$\langle a, v - u_0\rangle_X + \langle b, iv - iu_0\rangle_Y > \langle f, v - u_0\rangle_X$$

for all $a \in A(v)$, $b \in B(iv)$, $v \in X$ with $\|v\|_X > r$.

Then there exists a solution of the following evolutionary variational inequality: find $u \in K \cap D(L) \cap D(\Phi)$ such that for some $a \in A(u)$ and $b \in B(iu)$ we have

$$\langle L(u)+a-f, v-u\rangle_X + \langle b, iv-iu\rangle_Y + \Phi(v)-\Phi(u) \geq 0 \quad \textit{for all } v \in K. \tag{8.1}$$

Proof. Define $\Psi : X \to \mathbb{R} \cup \{+\infty\}$ by

$$\Psi(x) = \begin{cases} \Phi(x) & \text{if } x \in K, \\ +\infty & \text{if } x \in X \setminus K. \end{cases}$$

From assumption (A_4) it follows that Ψ is convex, lower semicontinuous, and $\not\equiv +\infty$. Problem (8.1) is equivalent to the statement: find $u \in D(L) \cap D(\Psi)$ such that for some $a \in A(u)$ and $b \in B(iu)$ one has

$$\langle L(u)+a-f, v-u\rangle_X + \langle b, iv-iu\rangle_Y + \Psi(v)-\Psi(u) \geq 0 \quad \text{for all } v \in X. \tag{8.2}$$

In turn, (8.2) is equivalent to the inclusion problem: find $u \in D(L) \cap D(\partial\Psi)$ such that

$$f \in L(u) + A(u) + i^*B(iu) + \partial\Psi(u). \tag{8.3}$$

The notation $\partial\Psi$ stands for the subdifferential of Ψ in the sense of convex analysis. Indeed, admitting (8.2) yields

$$-(L(u)+a-f+i^*b) \in \partial\Psi(u),$$

thus (8.3) ensues. Conversely, assuming (8.3), inequality (8.2) follows by definition of the convex subdifferential $\partial\Psi(u)$. Therefore, in order to complete the proof, we need to show that (8.3) is solvable.

Knowing from hypothesis (A_4) that $\operatorname{int} K \subset D(\Psi)$, it turns out that

$$\operatorname{int} K \subset D(\partial\Psi) \subset K. \tag{8.4}$$

On the other hand, hypothesis (A_1) implies that

$$D(L) \text{ is dense in } X \tag{8.5}$$

as shown in [238, Theorem 32.L] (see also Proposition 1.52 (ii)). Then (8.5) gives

$$D(L) \cap \operatorname{int} K \neq \emptyset. \tag{8.6}$$

Due to (8.4) and (8.6), we are able to apply the sum theorem for maximal monotone operators obtaining

$$L + \partial\Psi : X \to 2^{X^*} \quad \text{is maximal monotone.} \tag{8.7}$$

We claim that

$$i^*Bi : X \to 2^{X^*} \quad \text{is pseudomonotone.} \tag{8.8}$$

Indeed, in view of assumption (A_3), the values of the operator i^*Bi are nonempty and convex. Let $x \in X$ and $\{b_n\} \subset B(ix)$ with $i^*b_n \to z$ in X^*. By (A_3), the sequence $\{b_n\}$ is bounded in Y^*, so along a relabeled subsequence we have $b_n \rightharpoonup b$ in Y^* with $b \in B(ix)$, so $z = i^*b$ showing that the multivalued mapping i^*Bi is closed valued. Furthermore, we note that if $x_n \rightharpoonup x$ in X and $w_n \rightharpoonup w$ in X^*, with $w_n \in i^*B(ix_n)$, then $ix_n \to ix$ in Y (because $i : X \to Y$ is compact) and $w_n = i^*b_n$ with $b_n \in B(ix_n)$. Since, by assumption (A_3), B is bounded, along a relabeled subsequence we get $b_n \rightharpoonup b$ in Y^*. Using once again (A_3), this results in $b \in B(ix)$. We infer that $i^*b_n \to i^*b$ in Y^* taking into account that the mapping i^* is compact, thus we can conclude that $w_n \to w = i^*b \in i^*B(ix)$. From this, recalling Definition 8.2 and taking into account (A_3) and Remark 8.3, we easily derive (8.8).

Since the sum of pseudomonotone operators is pseudomonotone, (A_2) and (8.8) imply

$$A + i^*Bi : X \to 2^{X^*} \quad \text{is pseudomonotone and bounded.} \tag{8.9}$$

At this point, on the basis of (8.7), (8.9), and assumption (A_5), the main theorem on pseudomonotone perturbations of maximal monotone mappings (see [238, Theorem 32.A and problem 32.4*]) ensures that the inclusion (8.3) holds true for some $u \in D(L) \cap D(\partial\Psi)$. The second inclusion in (8.4) provides $u \in K$. The equivalence between problems (8.1) and (8.3) completes the proof. □

Remark 8.7. The evolutionary character of the variational inequality (8.1), as well as subsequent problems, is determined by the linear, maximal monotone operator $L : D(L) \subset X \to X^*$, which generally is an unbounded linear operator (see Definition 1.46). The model case for the unbounded linear operator L is the time derivative, so incorporating in our setting evolutionary partial differential equations and inclusions.

Next we formulate an existence result for multivalued elliptic variational inequalities where, contrary to Theorem 8.6, the constraint set K can have an empty interior.

Theorem 8.8. *Given the reflexive Banach spaces X and Y with a compact linear map $i : X \to Y$, a nonempty closed convex subset K of X, and $f \in X^*$, assume conditions (A_2), (A_3), and*

(H_1) $\Phi : X \to \mathbb{R} \cup \{+\infty\}$ *is convex and lower semicontinuous;* $x_0 \in K \cap D(\Phi)$.
(H_2) *If K is unbounded then for every sequence $\{x_n\} \subset K \cap D(\Phi)$ with $\|x_n\|_X \to +\infty$ and $\frac{x_n}{\|x_n\|_X} \rightharpoonup x$ in X, for every $a_n \in A(x_n)$ and $b_n \in$*

$B(ix_n)$ one has

$$\liminf_{n\to\infty} \frac{1}{\|x_n\|_X} \left(\langle a_n + i^* b_n, x_n - x_0 \rangle_X + \Phi(x_n) \right) > \langle f, x \rangle_X. \tag{8.10}$$

Then there exists a solution of the following variational inequality: find $x \in K \cap D(\Phi)$ such that for some $a \in A(x)$ and for some $b \in B(ix)$ we have

$$\langle a - f, y - x \rangle_X + \langle b, iy - ix \rangle_Y + \Phi(y) - \Phi(x) \geq 0 \quad \textit{for all } y \in K. \tag{8.11}$$

Proof. For every integer n, we introduce $K_n := K \cap B_X(x_0, n)$, where $B_X(x_0, n)$ stands for the closed ball in X of center x_0 and of radius n. Here $x_0 \in K \cap D(\Phi)$ is the point postulated in assumption (H_1). Note that K_n is closed, convex, and bounded in X with $x_0 \in K_n$. By virtue of (A_2) and (A_3), the proof of Theorem 8.6 shows that the multivalued map $A + i^* Bi : X \to 2^{X^*}$ is bounded and pseudomonotone.

By hypothesis (H_1) we are thus allowed to apply Proposition 8.5, which ensures the existence of $x_n \in K_n \cap D(\Phi)$ such that for some $a_n \in A(x_n)$ and for some $b_n \in B(ix_n)$ we have

$$\langle a_n + i^* b_n - f, z - x_n \rangle_X \geq \Phi(x_n) - \Phi(z) \quad \text{for all } z \in K_n. \tag{8.12}$$

Now we claim that there exists an integer $k > 0$ with the property

$$\|x_k - x_0\|_X < k. \tag{8.13}$$

Arguing by contradiction, admit that for every positive integer n,

$$\|x_n - x_0\|_X = n.$$

Due to the reflexivity of X, along a subsequence we can suppose that $\frac{x_n}{\|x_n\|_X} \rightharpoonup x$ in X for some $x \in X$. By inserting $z = x_0$ in (8.12), we obtain

$$\langle a_n + i^* b_n, x_n - x_0 \rangle_X + \Phi(x_n) - \Phi(x_0) \leq \langle f, x_n - x_0 \rangle.$$

Letting $n \to \infty$ results in

$$\limsup_{n\to\infty} \left(\langle a_n + i^* b_n, \frac{x_n - x_0}{\|x_n\|_X} \rangle_X + \frac{\Phi(x_n)}{\|x_n\|_X} \right) \leq \langle f, x \rangle_X,$$

contradicting (8.10). Therefore assertion (8.13) is proven.

Let an arbitrary $y \in K$. By (8.13), for sufficiently small $t > 0$ we know that

$$\|x_k + t(y - x_k) - x_0\|_X < k.$$

With a sufficiently small $t > 0$, we can choose $z = x_k + t(y - x_k)$ in (8.12) for $n = k$. Using the convexity of Φ, we deduce

$$\langle a_k + i^* b_k - f, y - x_k \rangle_X \geq \Phi(x_k) - \Phi(y),$$

where $a_k \in A(x_k)$ and $b_k \in B(ix_k)$ are independent of y. We conclude that x_k solves (8.11), so the proof is complete. □

The following is an existence result for variational inequalities whose proof involves again pseudomonotone multivalued maps.

Theorem 8.9. *Given a reflexive Banach space X, a nonempty closed convex subset K of X, and $f \in X^*$, assume (H_1), and*

(C_1) *$A : X \to 2^{X^*}$ is a multivalued map satisfying conditions $(PM1)$, $(PM2)$, and $(PM4)$.*

(C_2) *If K is unbounded then for every $\{x_n\} \subset K \cap D(\Phi)$ with $\|x_n\|_X \to +\infty$ and $\frac{x_n}{\|x_n\|_X} \rightharpoonup x$ in X, for every $a_n \in A(x_n)$, we have*

$$\liminf_{n\to\infty} \frac{1}{\|x_n\|_X} \big(\langle a_n, x_n - x_0\rangle_X + \Phi(x_n)\big) > \langle f, x\rangle_X.$$

Then there exists a solution of the variational inequality: find $x \in K \cap D(\Phi)$ such that for some $a \in A(x)$ we have

$$\langle a - f, y - x\rangle_X + \Phi(y) - \Phi(x) \geq 0 \quad \textit{for all } y \in K.$$

The proof can be carried out through the arguments used in Theorem 8.8 with $B = 0$ and by using Remark 8.3.

The next result focuses on variational inequalities containing a multivalued operator A which is maximal monotone (see Definition 8.1 (ii)). Generally, we can have $D(A) \neq X$, so none of the above theorems is applicable. For the sake of simplicity, we suppose that $L = 0$ and $B = 0$.

Theorem 8.10. *Let K be a nonempty, closed, and convex subset of a reflexive Banach space X, let $f \in X^*$, and let $\Phi : X \to \mathbb{R} \cup \{+\infty\}$ be convex, lower semicontinuous. Assume the following conditions:*

(S_1) *$A : D(A) \subset X \to 2^{X^*}$ is a maximal monotone map, with*

$$0 \in K \cap \operatorname{int}(D(A)) \cap \operatorname{int}(D(\partial\Phi)),$$

$0 \in A(0)$, and the weak closure in X of $K \cap D(A) \cap D(\partial\Phi)$ is contained in $D(A)$.

(S_2) *For every sequence $\{x_n\} \subset K \cap D(A) \cap D(\partial\Phi)$ with $\|x_n\|_X \to +\infty$ and $\frac{x_n}{\|x_n\|_X} \rightharpoonup z$ in X, we have*

$$\liminf_{n\to\infty} \frac{\Phi(x_n)}{\|x_n\|_X} > \langle f, z\rangle_X. \tag{8.14}$$

(S_3) *For every $x \in K \cap D(A) \cap D(\Phi)$ there exists $a = a(x) \in A(x)$ such that if $z \in K \cap D(\Phi)$ there are a sequence $t_n \to 0^+$ as $n \to \infty$ with $x +$*

$t_n(z-x) \in D(A)$ and a sequence $\xi_n \in A((1-t_n)x + t_n z)$ satisfying

$$\liminf_{n\to\infty} \langle \xi_n, z-x\rangle_X \le \langle a, z-x\rangle_X.$$

Then there exists $x \in K \cap D(A) \cap D(\Phi)$ such that for $a = a(x) \in A(x)$ given in (S_3), we have

$$\langle a - f, z - x\rangle_X \ge \Phi(x) - \Phi(z) \quad \textit{for all } z \in K. \tag{8.15}$$

Proof. Since X is reflexive, we can consider a norm on X such that X and X^* become strictly convex. Define the multivalued map $T : X \to 2^{X^*}$ by $T := A + N_K + \partial\Phi$, where N_K denotes the subdifferential of the indicator function of K (equivalently, $N_K(x)$ is the cone of exterior normals to K at $x \in K$). From assumption (S_1) we know that T is a maximal monotone map (see, e.g., [238, Theorem 32.I]). Then a classical surjectivity result ensures that there exists $x_n \in D(T) = K \cap D(A) \cap D(\partial\Phi)$ such that $f \in T(x_n) + \epsilon_n J(x_n)$, where $J : X \to X^*$ stands for the duality mapping of X and the sequence $\{\epsilon_n\}$ satisfies $\epsilon_n \to 0^+$. Therefore, for every n, there exist $w_n \in A(x_n)$, $v_n \in N_K(x_n)$, and $u_n \in \partial\Phi(x_n)$ such that

$$\langle w_n + v_n + u_n + \epsilon_n J(x_n) - f, y - x_n\rangle_X = 0 \quad \text{for all } y \in X.$$

Thanks to the inequalities $\langle v_n, y - x_n\rangle_X \le 0$ and $\Phi(y) - \Phi(x_n) \ge \langle u_n, y - x_n\rangle_X$ whenever $y \in K$, it turns out that

$$\langle w_n + \epsilon_n J(x_n) - f, y - x_n\rangle_X \ge \Phi(x_n) - \Phi(y) \quad \text{for all } y \in K.$$

From this and the monotonicity of A we infer that

$$\begin{aligned} &\langle w + \epsilon_n J(y) - f, y - x_n\rangle_X \ge \Phi(x_n) - \Phi(y) \\ &\quad \text{for all } w \in A(y) \text{ and } y \in K \cap D(A). \end{aligned} \tag{8.16}$$

We claim that $\{x_n\}$ is bounded in X. If not, there would exist a relabeled subsequence such that $\|x_n\|_X \to +\infty$ as $n \to \infty$. In view of the reflexivity of X, we can find $z \in X$ such that up to a subsequence

$$\frac{x_n}{\|x_n\|_X} \rightharpoonup z \quad \text{in } X.$$

Setting $y = 0$ and $w = 0$ in (8.16), which is possible due to assumption (S_1), then dividing by $\|x_n\|_X$ and letting $n \to \infty$ gives

$$\limsup_{n\to\infty} \frac{\Phi(x_n)}{\|x_n\|_X} \le \langle f, z\rangle_X.$$

Since this contradicts assumption (S_2), the claim holds.

The reflexivity of X enables us to pass to a subsequence still denoted by $\{x_n\}$ converging weakly in X to some $x \in K$. Letting $n \to \infty$ in (8.16) yields

$$\langle w - f, y - x\rangle_X \geq \Phi(x) - \Phi(y) \quad \text{for all } w \in A(y) \text{ and } y \in K \cap D(A). \tag{8.17}$$

Setting $y = 0$ and $w = 0$ in (8.17) shows that $x \in D(\Phi)$, whereas from the last part of assumption (S_1) we know that $x \in D(A)$.

To $x \in K \cap D(A) \cap D(\Phi)$ there corresponds an element $a \in A(x)$ as stated in hypothesis (S_3). Let $z \in K \cap D(\Phi)$. We can find $t_n \to 0^+$ as $n \to \infty$ with $x + t_n(z - x) \in D(A)$ and a sequence $\xi_n \in A((1 - t_n)x + t_n z)$ that fulfills

$$\liminf_{n\to\infty} \langle \xi_n, z - x\rangle_X \leq \langle a, z - x\rangle_X. \tag{8.18}$$

Since $(1 - t_n)x + t_n z \in K \cap D(A)$, we can use $y = (1 - t_n)x + t_n z$ in (8.17), rendering through the convexity of Φ that

$$\langle w - f, z - x\rangle_X \geq \Phi(x) - \Phi(z) \quad \text{for all } w \in A((1 - t_n)x + t_n z).$$

Hence we obtain

$$\langle \xi_n - f, z - x\rangle_X \geq \Phi(x) - \Phi(z).$$

Then by means of (8.18) we get (8.15). Observe that (8.15) is valid for any $z \in K$ with $\Phi(z) = +\infty$, which completes the proof. □

In order to get more insight on Theorem 8.10, we present a result in the spirit of Minty's formulation for evolutionary variational inequalities.

Theorem 8.11. *Let K be a nonempty, closed, and convex subset of a reflexive Banach space X, let $L : D(L) \subset X \to X^*$ be linear, maximal monotone, let $A : D(A) \subset X \to 2^{X^*}$ be a maximal monotone map, let $\Phi : X \to \mathbb{R} \cup \{+\infty\}$ be a convex, lower semicontinuous functional $\not\equiv +\infty$, let $K \subset \operatorname{int}(D(A)) \cap \operatorname{int}(D(\partial\Phi))$, and let $f \in X^*$. Assume that there exists an element $x \in K \cap D(L)$ such that for every $z \in K \cap D(L)$ and for every $a_z \in A(z)$, we have*

$$\langle L(z) + a_z - f, z - x\rangle_X \geq \Phi(x) - \Phi(z). \tag{8.19}$$

Then there exists $a \in A(x)$ such that

$$\langle L(x) + a - f, z - x\rangle_X \geq \Phi(x) - \Phi(z) \quad \textit{for all } z \in K \cap D(L). \tag{8.20}$$

Proof. Since for every $z \in K \subset D(\partial\Phi)$, $z^* \in \partial\Phi(z)$, and $y \in X$, we have

$$\Phi(y) - \Phi(z) \geq \langle z^*, y - z\rangle_X,$$

it follows from (8.19) that for any $x \in K \cap D(L)$, $z \in K \cap D(L)$, $z^* \in \Phi(z)$, and $a_z \in A(z)$,

$$\langle L(z) + a_z + z^* - f, z - x\rangle_X \geq 0. \tag{8.21}$$

We define the multivalued map $T : X \to 2^{X^*}$ by $T := L + A + N_K + \partial\Phi$, where N_K denotes the subdifferential in the sense of convex analysis of the indicator function of K. It is known that T is a maximal monotone map with $D(T) = K \cap D(L)$ (see, e.g., [238, Theorem 32.1]). For any $u_z \in N_K(z)$ and $z \in K \cap D(L)$, we note from (8.21) that

$$\begin{aligned}&\langle L(z) + a_z + z^* + u_z - f, z - x\rangle_X \\ &= \langle L(z) + a_z + z^* - f, z - x\rangle_X + \langle u_z, z - x\rangle_X \geq 0,\end{aligned}$$

implying by the maximal monotonicity of T that $f \in (L + A + \partial\Phi + N_K)(x)$. Consequently, there exist $a \in A(x)$, $u \in N_K(x)$, and $x^* \in \partial\Phi(x)$ satisfying

$$\langle L(x) + a + u + x^* - f, z - x\rangle_X = 0 \quad \text{for all } z \in X,$$

which, by using the inequalities $\langle u, z - x\rangle_X \leq 0$ and $\Phi(z) - \Phi(x) \geq \langle x^*, z - x\rangle_X$ whenever $z \in K \cap D(L)$, confirms that (8.20) is true, and the proof is complete. □

Remark 8.12. Condition (S_3) in Theorem 8.8 represents a multivalued version of Minty's technique for variational inequalities driven by single-valued monotone, hemicontinuous operators. Complementing this, Theorem 8.11 provides a Minty formulation for evolutionary variational inequalities where it is shown that, by imposing the assumption $K \subset \text{int}(D(A)) \cap \text{int}(D(\partial\Phi))$, condition (S_3) can be dropped. We also note that the asymptotic recessive condition (8.14) in Theorem 8.9 is a requirement that does not involve A, in particular, it is not necessary to ask for any coercivity of A.

8.2 EVOLUTIONARY QUASIVARIATIONAL INEQUALITIES

For the sake of clarity we state the fixed point theorem of Kluge [123] that will be used in the sequel.

Theorem 8.13. *Let Z be a real reflexive Banach space and let $D \subset Z$ be nonempty, convex, bounded, and closed. Assume that $P : D \to 2^D$ is a set-valued map such that for every $u \in D$, the set $P(u)$ is nonempty, closed, and convex, and whose graph $G(P)$ is weakly closed. Then P has a fixed point, that is, $z \in P(z)$.*

Remark 8.14. The request in Theorem 8.13 on the set D to be bounded in Z can be replaced by requiring that the image $P(D)$ be bounded. To see this it is sufficient to apply Theorem 8.13 to the closed convex hull $\overline{\text{co}}(P(D))$ of $P(D)$ in place of D.

Let X and Y be reflexive Banach spaces with a compact linear map $i : X \to Y$, and let $L : D(L) \subset X \to X^*$ be a linear, maximal monotone map. Let $A : X \to 2^{X^*}$ and $B : Y \to 2^{Y^*}$ be multivalued maps, let K be a nonempty, closed, and convex subset of X, let $\Phi : X \to \mathbb{R} \cup \{+\infty\}$ be a functional $\not\equiv +\infty$, and let $C : K \to 2^K$ be a multivalued map such that for any $z \in K$, $C(z)$ is a closed, convex subset of K with nonempty interior in X, and let $f \in X^*$. With these data, we formulate the following quasivariational inequality: find $x \in C(x) \cap D(L) \cap D(\Phi)$ such that for some $a \in A(x)$ and $b \in B(ix)$ we have

$$\begin{aligned} &\langle L(x) + a - f, z - x\rangle_X + \langle b, iz - ix\rangle_Y + \Phi(z) - \Phi(x) \geq 0 \\ &\text{for all } z \in C(x) \cap D(L). \end{aligned} \tag{8.22}$$

The evolutionary feature of the inequality problem (8.22) comes from the unbounded linear operator L, which plays the role of time derivative in evolution equations and inclusions.

In order to focus on the evolutionary quasivariational inequality (8.22), we proceed through a parametric approach by means of evolutionary variational inequalities of type (8.1). Namely, for every $w \in K$, we consider the following parametric evolutionary variational inequality: find $x \in C(w) \cap D(L) \cap D(\Phi)$ such that for some $a \in A(x)$ and $b \in B(ix)$ we have

$$\begin{aligned} &\langle L(x) + a - f, z - x\rangle_X + \langle b, iz - ix\rangle_Y + \Phi(z) - \Phi(x) \geq 0 \\ &\text{for all } z \in C(w) \cap D(L). \end{aligned} \tag{8.23}$$

We define a multivalued map, called variational selection,

$$S : K \to 2^K \tag{8.24}$$

by assigning to each $y \in K$ the set $S(y)$ of all solutions of (8.23) with $w = y$. It is essential to note that if x is a fixed point of the multivalued map S in (8.24), that is, $x \in S(x)$, then x solves the evolutionary quasivariational inequality (8.22).

With the data in problem (8.22) we formulate the conditions:

(A_5') $f \in X^*$, and if the set K is unbounded in X, then there exist $r > 0$ and a point

$$u_0 \in D(L) \cap \operatorname{int}\Big(\bigcap_{w \in K} C(w) \Big) \tag{8.25}$$

satisfying

$$\begin{aligned} &\langle a, v - u_0\rangle_X + \langle b, iv - iu_0\rangle_Y \\ &> \max\{\langle f, v - u_0\rangle_X, \langle f - L(u_0), v - u_0\rangle_X\} \end{aligned}$$

for all $a \in A(v)$, $b \in B(iv)$, $v \in X$ with $\|v\|_X > r$.

(A_6) The operator $A + i^*Bi$ is monotone on $C(w)$, that is,

$$\langle a_v - a_u, v - u\rangle_X + \langle b_v - b_u, iv - iu\rangle_Y \geq 0 \quad \text{for all } u, v \in C(w),$$

$a_u \in A(u)$, $a_v \in A(v)$, $b_u \in B(iu)$, $b_v \in B(iv)$, $w \in K$.

(A_7) For every $x \in C(w) \cap D(L) \cap D(\Phi)$ there exist $A_x \in A(x)$ and $B_x \in B(ix)$ such that

$$\limsup_{t\to 0^+} \big(\langle a_t, z - x\rangle_X + \langle b_t, iz - ix\rangle_Y\big) \leq \langle A_x, z - x\rangle_X + \langle B_x, iz - ix\rangle_Y$$

whenever $z \in C(w) \cap D(L)$, $a_t \in A(x + t(z - x))$, $b_t \in B(ix + t(iz - ix))$.

(A_8) The map $C : K \to 2^K$ has the following Mosco-type continuity properties (see [159,160]):

(i) If $\{w_n\} \subset K$ and $u_n \in C(w_n) \cap D(L) \cap D(\Phi)$ satisfy $w_n \rightharpoonup w$ in X and $u_n \rightharpoonup u$ in X, with $u \in D(L)$ and $L(u_n) \rightharpoonup L(u)$ in X^*, then $u \in C(w)$.

(ii) For every sequence $\{w_n\} \subset K \cap D(L)$ with $w_n \rightharpoonup w$ in X and for every $v \in C(w) \cap D(L)$, there exist a subsequence $\{w_{n_k}\}$ of $\{w_n\}$ and a sequence $v_k \in C(w_{n_k}) \cap D(L)$ with $v_k \to v$ in X and $\Phi(v_k) \to \Phi(v)$.

Theorem 8.15. *Assume that conditions* (A_1)–(A_4), (A_5'), (A_6)–(A_8) *hold. Then there exists a solution of the quasivariational inequality* (8.22).

Proof. Conditions (A_1)–(A_5) are verified with $C(w)$ in place of K for each $w \in K$, so Theorem 8.6 with K substituted by $C(w)$ can be applied. This ensures that the parametric problem (8.23) is solvable for every $w \in K$, so the variational selection $S : K \to 2^K$ introduced in (8.24) has nonempty values. We split the proof in several parts.

Step 1. For every $w \in K$, problem (8.23) is equivalent to: find $x \in C(w) \cap D(L) \cap D(\Phi)$ such that for every $z \in C(w) \cap D(L)$, $a \in A(z)$ and $b \in B(iz)$, we have

$$\langle L(z) + a - f, z - x\rangle_X + \langle b, iz - ix\rangle_Y + \Phi(z) - \Phi(x) \geq 0. \tag{8.26}$$

Indeed, if $x \in C(w) \cap D(L) \cap D(\Phi)$ is a solution of (8.23), then it solves (8.5) on account of (A_1) and (A_6). Conversely, suppose that $x \in C(w) \cap D(L) \cap D(\Phi)$ is a solution of (8.26) and let $z \in C(w) \cap D(L)$. Since $C(w) \cap D(L)$ is a convex set, we can set $z_t := (1 - t)x + tz \in C(w) \cap D(L)$ for all $t \in (0, 1]$ in (8.26) obtaining

$$0 \leq \langle L(z_t) + a_t - f, z_t - x\rangle_X + \langle b_t, iz_t - ix\rangle_Y + \Phi(z_t) - \Phi(x)$$

whenever $a_t \in A(z_t)$ and $b_t \in B(iz_t)$. The linearity of L and the convexity of Φ as postulated in condition (A_4) imply

$$0 \leq \langle L(z) + (1 - t)L(x) + a_t - f, z - x\rangle_X + \langle b_t, iz - ix\rangle_Y + \Phi(z) - \Phi(x). \tag{8.27}$$

Letting $t \to 0$ in (8.27), by hypothesis (A_7) we find that x solves problem (8.23).

Step 2. The set $S(w)$ is convex and closed in X whenever $w \in K$.

Using Step 1, we first check that the solution set of (8.23) is convex in X. To this end, fix $w \in K$. Let u and v solve (8.26), so $u, v \in C(w) \cap D(L) \cap D(\Phi)$ such that for every $z \in C(w) \cap D(L)$ and for every $a \in A(z)$ and $b \in B(iz)$ we have

$$\langle L(z) + a - f, z - u \rangle_X + \langle b, iz - iu \rangle_Y + \Phi(z) - \Phi(u) \geq 0,$$
$$\langle L(z) + a - f, z - v \rangle_X + \langle b, iz - iv \rangle_Y + \Phi(z) - \Phi(v) \geq 0.$$

Let $t \in [0, 1]$. From the convexity of Φ, it is seen that

$$\langle L(z) + a - f, z - (1-t)u - tv \rangle_X + \Phi(z) - \Phi((1-t)u + tv) \geq 0,$$

which in view of Step 1 entails the convexity assertion.

Now let $\{u_n\} \subset S(w)$ with $u_n \to u$ in X. We infer that $u_n \in C(w) \cap D(L) \cap D(\Phi)$, and for some $a_n \in A(u_n)$ and $b_n \in B(iu_n)$ we have

$$\langle L(u_n) + a_n - f, v - u_n \rangle_X + \langle b_n, iv - iu_n \rangle_Y + \Phi(v) - \Phi(u_n) \geq 0 \quad (8.28)$$

for all $v \in C(w) \cap D(L)$. Since $C(w)$ is a closed subset of K, we get $u \in C(w)$. By hypotheses (A_2) and (A_3), the sequences $\{a_n\}$ and $\{b_n\}$ are bounded in X^* and Y^*, respectively. Consequently, along relabeled subsequences $a_n \rightharpoonup a$ in X^* and $b_n \rightharpoonup b$ in Y^*. Observing that $\lim_{n\to\infty} \langle a_n, u_n - u \rangle_X = 0$, we deduce from (A_2) that $a \in A(u)$ because then A is generalized pseudomonotone (see Remark 8.4). Similarly, in view of (8.8), we derive that $b \in B(iu)$.

Recalling that int $(C(w))$ is nonempty, we can choose a ball B in X with $B \subset C(w)$. Then (8.28), in conjunction with the density of $D(L)$ in X, guarantees that the sequence $\{L(u_n)\}$ is uniformly bounded below on the ball B, which ensures that $\{L(u_n)\}$ is bounded in X^*. Since L is linear, maximal monotone, so weakly graph closed, we have that $u \in D(L)$ and $L(u_n) \rightharpoonup L(u)$ in X^*. We also note from (8.28) that $u \in D(\Phi)$ thanks to the lower semicontinuity of Φ as required in (A_4). Altogether, we can pass to the limit in (8.28) as $n \to \infty$, proving that $u \in S(w)$.

Step 3. The image $S(K)$ of S is bounded in X.

Arguing by contradiction, we assume that there exists a sequence $\{u_n\} \subset S(K)$ satisfying

$$\|u_n\|_X \to +\infty \text{ as } n \to \infty. \quad (8.29)$$

Necessarily, by (8.29), K is unbounded. Let $w_n \in K$ verify $u_n \in S(w_n)$. Consequently, $u_n \in C(w_n) \cap D(L) \cap D(\Phi)$ and there exist $a_n \in A(u_n)$ and $b_n \in$

$B(iu_n)$ fulfilling (8.28). Due to (A_5') we can insert $v = u_0$ into (8.28), which by the monotonicity of L reads as

$$\begin{aligned}&\langle a_n, u_n - u_0\rangle_X + \langle b_n, iu_n - iu_0\rangle_Y + \Phi(u_n) - \Phi(u_0)\\ &\leq \langle f - L(u_0), u_n - u_0\rangle_X.\end{aligned}$$

Taking into account (8.29), we come to a contradiction with the inequality in hypothesis (A_5'). In view of this contradiction, Step 3 is valid.

Step 4. The graph $G(S)$ of S is sequentially weakly closed in $X \times X$.

Let $\{(w_n, u_n)\} \subset G(S)$ converge weakly to (w, u) in $X \times X$. The fact that $u_n \in S(w_n)$ means that $w_n \in K$, $u_n \in C(w_n) \cap D(L) \cap D(\Phi)$ and the inequality (8.28) holds for some $a_n \in A(u_n)$ and $b_n \in B(iu_n)$. Note that $w \in K$.

Owing to assertion (8.25) in (A_5'), we can choose a ball B in X such that $B \subset C(w_n)$ for all n. Then (8.28) and the density of $D(L)$ in X ensure that the sequence $\{L(u_n)\}$ is uniformly bounded below on the ball B, which guarantees that $\{L(u_n)\}$ is bounded in X^*. Since L, being linear and maximal monotone, is weakly graph closed, we obtain that $u \in D(L)$ and $L(u_n) \rightharpoonup L(u)$ in X^*. Then according to assumption (A_8) (i) we have that $u \in C(w)$. Moreover, by (A_4) and (8.28) we derive that $u \in D(\Phi)$.

Since $u \in C(w) \cap D(L) \cap D(\Phi)$, we can invoke assumption (A_8) (ii), obtaining a subsequence of $\{w_n\}$, denoted again by $\{w_n\}$, corresponding to which there exists a sequence $\{z_n\}$ with $z_n \in C(w_n) \cap D(L)$, $z_n \to u$ in X and $\Phi(z_n) \to \Phi(u)$. We can thus insert $v = z_n$ into (8.28), resulting in

$$\langle L(u_n) + a_n - f, z_n - u_n\rangle_X + \langle b_n, iz_n - iu_n\rangle_Y + \Phi(z_n) - \Phi(u_n) \geq 0,$$

which reads as

$$\begin{aligned}\langle a_n, u_n - u\rangle_X \leq &\langle a_n, z_n - u\rangle_X + \langle L(u_n) - f, z_n - u_n\rangle_X\\ &+ \langle b_n, iz_n - iu_n\rangle_Y + \Phi(z_n) - \Phi(u_n).\end{aligned} \tag{8.30}$$

On the other hand, the monotonicity of L, in conjunction with the facts that $u_{n_k} \rightharpoonup u$ in X and $L(u_{n_k}) \rightharpoonup L(u)$ in X^*, implies

$$\liminf_{n\to\infty} \langle L(u_n), u_n\rangle_X \geq \langle L(u), u\rangle_X. \tag{8.31}$$

Because the map $i : X \to Y$ is compact, we have that $iu_n \to iu$ in Y. We also notice from hypothesis (A_3) and (8.8) that along a relabeled sequence $b_n \rightharpoonup b$ in Y^* with $b \in B(iu)$. Thanks to the boundedness of the map A, up to a subsequence, we have $a_n \rightharpoonup a$ in X^*. So (8.30), (8.31), and (A_4) yield

$$\limsup_{n\to\infty} \langle a_n, u_n - u\rangle_X \leq 0.$$

This property, assumption (A_2) and Remark 8.4 entail

$$a \in A(u) \text{ and } \langle a_n, u_n \rangle_X \to \langle a, u \rangle_X. \tag{8.32}$$

Now let an arbitrary $v \in C(w) \cap D(L)$. On the basis of assumption (A_8) (ii), there exists $v_k \in C(w_{n_k}) \cap D(L)$ such that $v_k \to v$ in X and $\Phi(v_k) \to \Phi(v)$. By testing (8.28) with $v = v_k$ when $n = n_k$, we get

$$\langle L(u_{n_k}) + a_{n_k} - f, v_k - u_{n_k} \rangle_X + \langle b_{n_k}, iv_k - iu_{n_k} \rangle_Y + \Phi(v_k) - \Phi(u_{n_k}) \geq 0. \tag{8.33}$$

Letting $k \to \infty$ in (8.33) and taking into account (8.31) and (8.32) lead to

$$\langle L(u) + a - f, v - u \rangle_X + \langle b, iv - iu \rangle_Y + \Phi(v) - \Phi(u) \geq 0,$$

which proves that $(w, u) \in G(S)$. This confirms that $G(S)$ is sequentially weakly closed in $X \times X$.

Finally, we apply Theorem 8.13 by taking D as the closed convex hull of $S(K)$ in X and P as the restriction of S to this set D (see Remark 8.14). Steps 2–4 show that all the hypotheses in Theorem 8.13 are verified. Consequently, through Theorem 8.13 and Remark 8.14, we are able to obtain a fixed point of the multivalued map S, which represents a solution of the quasivariational inequality (8.22). This completes the proof. □

The following estimate from Alber–Notik [3] will be useful.

Lemma 8.16. *Let $(Z, \|\cdot\|)$ be a reflexive Banach space with its dual Z^*. Let $A : Z \to 2^{Z^*}$ be a monotone map and let $\bar{x} \in \operatorname{int}(D(A))$. Then there exists a constant $r = r(\bar{x}) > 0$ such that for every $(x, w) \in G(A)$ we have*

$$\langle w, x - \bar{x} \rangle \geq r\|w\| - (\|x - \bar{x}\| + r)c, \tag{8.34}$$

where $c := \sup\{\|w'\| : \|x' - \bar{x}\| \leq r \text{ and } w' \in A(x')\} < +\infty$.

We now resolve an evolutionary quasivariational inequality with a maximal monotone map A: find $x \in C(x) \cap D(L) \cap D(\Phi)$ such that for some $a \in A(x)$ we have

$$\langle L(x) + a - f, z - x \rangle_X + \Phi(z) - \Phi(x) \geq 0 \quad \text{for all } z \in C(x) \cap D(L). \tag{8.35}$$

As before we proceed through a parametric evolutionary variational inequality: find $x \in C(w) \cap D(L) \cap D(\Phi)$ such that for some $a \in A(x)$ we have

$$\langle L(x) + a - f, z - x \rangle_X + \Phi(z) - \Phi(x) \geq 0 \quad \text{for all } z \in C(w) \cap D(L). \tag{8.36}$$

Theorem 8.17. *Let K be a nonempty, closed, and convex subset of a reflexive Banach space X, let $f \in X^*$, and let $\Phi : X \to \mathbb{R} \cup \{+\infty\}$ be convex, lower semicontinuous. Assume (A_1), (A_8), and the conditions:*

(T_1) *$A : D(A) \subset X \to 2^{X^*}$ is a maximal monotone map, with*

$$0 \in \operatorname{int}\Big(\bigcap_{w \in K} C(w)\Big) \subset K \subset \operatorname{int}(D(A)) \cap \operatorname{int}(D(\Phi))$$

and $0 \in A(0)$.

(T_2) *For every sequence $\{x_n\} \subset K$ with $\|x_n\|_X \to +\infty$, and $\frac{x_n}{\|x_n\|_X} \rightharpoonup z$ in X, we have*

$$\liminf_{n \to \infty} \frac{\Phi(x_n)}{\|x_n\|_X} > \langle f, z \rangle_X. \tag{8.37}$$

Then the evolutionary quasivariational inequality (8.35) *has a nonempty solution set.*

Proof. The assumptions of Theorem 8.10 are fulfilled for every $C(w)$ in place of K. Specifically, condition (S_3) is true because A is hemicontinuous on K due to (T_1), whereas (T_1) and (T_2) imply (S_1) and (S_2), respectively. Then Theorem 8.10 ensures that the parametric evolutionary variational inequality (8.36) is solvable for every $w \in K$. Therefore, the associated variational selection $S : K \to 2^K$ (see (8.24)) has nonempty values for every $w \in K$.

We are going to show the existence of a fixed point of S. Using Theorem 8.11, it is seen that (8.36) is equivalent to the following evolutionary variational inequality: find $x \in C(w) \cap D(L)$ such that for every $a \in A(z)$, we have

$$\langle L(z) + a - f, z - x \rangle_X + \Phi(z) - \Phi(x) \geq 0 \quad \text{for all } z \in C(w) \cap D(L). \tag{8.38}$$

From (8.38) it is obvious that the solution set $S(w)$ is convex. It is also clear that it is closed. Here the essential fact is that A is locally bounded on K as guaranteed by assumption (T_1).

Let us show that the image $S(K)$ is bounded in X. By contradiction, suppose that there exists a sequence $\{u_n\} \subset S(K)$ such that $\|u_n\|_X \to +\infty$ as $n \to \infty$. Let $w_n \in K$ be such that $u_n \in S(w_n)$, that is $u_n \in C(w_n) \cap D(L)$ and there exists $a_n \in A(u_n)$ such that

$$\langle L(u_n) + a_n - f, z - u_n \rangle_X \geq \Phi(u_n) - \Phi(z) \quad \text{for all } z \in C(w_n) \cap D(L). \tag{8.39}$$

In view of assumption (T_1) we can set $z = 0$ in (8.39). Using the monotonicity of L and A, we arrive at

$$\Phi(u_n) - \Phi(0) \leq \langle f, u_n \rangle_X.$$

From here a contradiction to (8.37) in assumption (T_2) arises. Therefore $S(K)$ is bounded in X.

Next we prove that the graph $G(S)$ of S is sequentially weakly closed in $X \times X$. Let $\{(w_n, u_n)\} \subset G(S)$ converge weakly to (w, u) in $X \times X$. We know that $w_n \in K$ and $u_n \in C(w_n) \cap D(L)$ satisfies for some $a_n \in A(u_n)$ the inequality

$$\langle L(u_n) + a_n - f, v - u_n \rangle_X + \Phi(v) - \Phi(u_n) \geq 0 \quad \text{for all } v \in C(w_n) \cap D(L). \tag{8.40}$$

By assumption (T_1), we can choose a ball B in X such that $B \subset C(w_n)$ for all n. Then (8.40), the local boundedness of the monotone map A on K (see (T_1)) and the density of $D(L)$ in X ensure that the sequence $\{L(u_n)\}$ is bounded in X^*. Since L is weakly graph closed, it turns out that $u \in D(L)$ and $L(u_n) \rightharpoonup L(u)$ in X^*. Then assumption (A_8) (i) yields that $u \in C(w)$.

We claim that the sequence $\{a_n\}$ is bounded. Since $u \in C(w) \cap D(L)$, we find by assumption (A_8) (ii) a relabeled subsequence of $\{w_n\}$ and a sequence $\{z_n\}$ with $z_n \in C(w_n) \cap D(L)$, $z_n \to u$ in X and $\Phi(z_n) \to \Phi(u)$. Thus we can insert $v = z_n$ into (8.40), getting

$$\langle L(u_n) + a_n - f, z_n - u_n \rangle_X + \Phi(z_n) - \Phi(u_n) \geq 0.$$

Lemma 8.16 applied to $A - f$ implies that there are constants $c > 0$ and $r > 0$ as in the estimate (8.34) such that

$$\begin{aligned} r\|a_n - f\|_X &\leq \langle a_n - f, u_n - u \rangle_X + c(r + \|u_n - u\|_X) \\ &= \langle a_n - f, u_n - z_n \rangle_X + \langle a_n - f, z_n - u \rangle_X + c(r + \|u_n - u\|_X) \\ &\leq \|a_n - f\|_{X^*} \|z_n - u\|_X + c(r + \|u_n - u\|_X) \\ &\quad + \langle L(u_n), z_n - u_n \rangle_X + \Phi(z_n) - \Phi(u_n), \end{aligned}$$

thereby

$$\begin{aligned} &(r - \|z_n - u\|_X)\, \|a_n - f\|_{X^*} \\ &\leq c(r + \|u_n - u\|_X) + \langle L(u_n), z_n - u_n \rangle_X + \Phi(z_n) - \Phi(u_n). \end{aligned} \tag{8.41}$$

Since $z_n \to u$ and the right-hand side of the inequality (8.41) stays bounded, we deduce the boundedness of $\{a_n\}$ as claimed.

Let $v \in C(w) \cap D(L)$. By assumption (A_8) (ii), there exists $v_k \in C(w_{n_k}) \cap D(L)$ such that $v_k \to v$ in X and $\Phi(v_k) \to \Phi(v)$ as $k \to \infty$. From (8.40) with $v = v_k$ and $a_{n_k} \in A(u_{n_k})$ we infer that

$$\langle L(u_{n_k}) + a_{n_k} - f, v_k - u_{n_k} \rangle_X + \Phi(v_k) - \Phi(u_{n_k}) \geq 0. \tag{8.42}$$

Then, for any $a_v \in A(v)$, (8.42) enables us to find

$$
\begin{aligned}
&\langle L(v)+a_v, u_{n_k}-v\rangle_X \\
&\le \langle L(v)+a_v, u_{n_k}-v\rangle_X + \langle L(u_{n_k})+a_{n_k}-f, v_k-u_{n_k}\rangle_X + \Phi(v_k)-\Phi(u_{n_k}) \\
&= \langle L(u_{n_k})+a_{n_k}, v_k-v\rangle_X + \langle L(u_{n_k})+a_{n_k}-L(v)-a_v, v-u_{n_k}\rangle_X \\
&\quad + \langle f, u_{n_k}-v_k\rangle_X + \Phi(v_k)-\Phi(u_{n_k}) \\
&\le \langle L(u_{n_k})+a_{n_k}, v_k-v\rangle_X + \langle f, u_{n_k}-v_k\rangle_X + \Phi(v_k)-\Phi(u_{n_k}),
\end{aligned}
$$

where we used the monotonicity of L and A, too. In the limit we obtain

$$\langle L(v)+a_v, u-v\rangle_X \le \Phi(v)-\Phi(u)+\langle f, u-v\rangle.$$

We can now apply Theorem 8.13, choosing for D the closed convex hull of $S(K)$ in X and for P the restriction of S to this set D. As shown before all the requirements in Theorem 8.13 are fulfilled, which provides a fixed point of S, thus a solution of the quasivariational inequality (8.35). This completes the proof. □

The next part of this section concerns the evolutionary quasivariational inequality (8.22) with compact constraints $C(w)$, which will permit relaxing other data, in particular dropping the requirement on the linear map i to be compact. Suppose that X and Y are reflexive Banach spaces and a continuous linear map $i : X \to Y$ is given. Consider the maps $L : W := D(L) \subset X \to X^*$, $A : X \to 2^{X^*}$, and $B : Y \to 2^{Y^*}$ such that conditions (A_1), (A_2), and (A_3) are satisfied.

We notice that W endowed with the graph norm

$$\|x\|_W = \|x\|_X + \|L(x)\|_{X^*} \quad \text{for } x \in W$$

is a reflexive Banach space, and further assume that W is separable. Let K be a nonempty compact convex subset of W, let $\Phi : X \to \mathbb{R} \cup \{+\infty\}$ be a convex and lower semicontinuous function with $K \subset D(\partial\Phi)$, let $C : K \to 2^K$ be a multivalued map, and let $f \in X^*$. With these data, the quasivariational inequality (8.22) reads as: find $x \in C(x) \cap D(L)$ such that for some $a \in A(x)$ and $b \in B(ix)$ we have

$$\langle L(x)+a-f, z-x\rangle_X + \langle b, iz-ix\rangle_Y + \Phi(z)-\Phi(x) \ge 0 \quad \text{for all } z \in C(x). \tag{8.43}$$

An auxiliary result from Lunsford [140, Theorem 3.1] is necessary in the following.

Theorem 8.18. *Let X be a separable Banach space with its topological dual X^* and let K be a nonempty, compact, and convex subset of X. Let the multivalued maps $F : K \to 2^{X^*}$ and $G : K \to 2^K$ satisfy the conditions:*

(i) *The map* $M : K \to 2^K$ *defined by*

$$M(x) := \left\{ z \in K : \inf_{w \in F(x)} \langle w, x - z \rangle_X \le 0 \right\}$$

has a closed graph.

(ii) G *is lower semicontinuous and has closed graph and nonempty convex values.*

Then there exists a fixed point $x \in G(x)$ *such that*

$$\inf_{w \in F(x)} \langle w, x - z \rangle_X \le 0 \quad \text{for all } z \in G(x).$$

If, in addition, $F(x)$ *is weakly compact, then there exists* $w \in F(x)$ *such that*

$$\langle w, x - z \rangle_X \le 0 \quad \text{for all } z \in G(x).$$

The statement of our result on problem (8.43) is formulated below.

Theorem 8.19. *Under assumptions* (A_1), (A_2), (A_3), *if* $\partial\Phi$ *is bounded on* K *and the multivalued map* $C : K \to 2^K$ *is lower semicontinuous with closed graph and nonempty convex values, then the evolutionary quasivariational inequality* (8.43) *has a nonempty solution set.*

Proof. By the definition of the convex subdifferential $\partial\Phi$, it is sufficient to find $x \in C(x)$ such that for some $a \in A(x)$, $b \in B(ix)$, and $c \in \partial\Phi(x)$ one has

$$\langle L(x) + a + c - f, z - x \rangle_X + \langle b, iz - ix \rangle_Y \ge 0 \quad \text{for all } z \in C(x).$$

Since $K \subset D(\partial\Phi)$, we can introduce for every $x \in K$ the set

$$M(x) := \left\{ z \in K : \inf_{\substack{a \in A(x) \\ b \in B(ix) \\ c \in \partial\Phi(x)}} \left(\langle L(x) + a + c - f, x - z \rangle_X + \langle b, ix - iz \rangle_Y \right) \le 0 \right\}. \tag{8.44}$$

Notice that $x \in M(x)$ for every $x \in K$.

We claim that the multivalued map $M : K \to 2^K$ defined in (8.44) has a closed graph in the topology induced by W. Indeed, let $(x_n, z_n) \in G(M)$ be a sequence such that $x_n \to x$ and $z_n \to z$ in W. It is seen from (8.44) that there exist $a_n \in A(x_n)$, $b_n \in B(ix_n)$, $c_n \in \partial\Phi(x_n)$ such that

$$\langle L(x_n) + a_n + c_n - f, x_n - z_n \rangle_X + \langle b_n, ix_n - iz_n \rangle_Y \le \frac{1}{n}. \tag{8.45}$$

The fact that $x_n \to x$ in X, $z_n \to z$ in X, and $L(x_n) \to L(x)$ in X^* ensures

$$\langle L(x_n), x_n - z_n \rangle_X \to \langle L(x), x - z \rangle_X \quad \text{as } n \to \infty. \tag{8.46}$$

It turns out from assumptions (A_2) and (A_3) that along relabeled subsequences one has $a_n \rightharpoonup a$ in X^* and $b_n \rightharpoonup b$ in Y^* for some $a \in A(x)$ and $b \in B(ix)$. Moreover, the boundedness of $\partial\Phi$ on K renders that along a relabeled subsequence one has $c_n \rightharpoonup c$ in X^*, with $c \in \partial\Phi(x)$ because $\partial\Phi$ is maximal monotone. Then, taking into account (8.45) and (8.46), we infer that

$$\begin{aligned}&\langle L(x)+a+c-f, x-z\rangle_X + \langle b, ix-iz\rangle_Y \\ &\le \limsup_{n\to\infty}\big(\langle L(x_n)+a_n+c_n-f, x_n-z_n\rangle_X + \langle b_n, x_n-z_n\rangle_X\big) \le 0,\end{aligned}$$

which entails $z \in M(x)$, so our claim holds true. Hence hypothesis (i) in Theorem 8.18 is verified.

Since we assumed that $C : K \to 2^K$ is lower semicontinuous with closed graph and nonempty convex values, hypothesis (ii) in Theorem 8.18 holds also true. Therefore we are able to apply Theorem 8.18 to $X = W$, $F = (L + A + i^*Bi + \partial\Phi)|_K - f$, $G = C$, and our set K. Our hypotheses ensure that the set $F(x)$ is bounded, convex, and closed in X^*, so weakly compact in W^*. The final assertion of Theorem 8.18 enables us to complete the proof. □

Remark 8.20. Sufficient conditions to meet the properties required for the multivalued map $C : K \to 2^K$ in Theorem 8.19 can be found in [140].

We end the section with two applications related to Theorem 8.15. In order to avoid technical details and emphasize the main ideas, we take the set $K = X$.

The first application deals with an elliptic quasivariational inequality in the form of a hemivariational inequality. Given a bounded domain Ω in $\mathbb{R}^N$ and a number $p \in [2, +\infty)$, we consider the reflexive Banach spaces $W_0^{1,p}(\Omega)$ and $L^p(\Omega)$ endowed with their usual norms. By Theorem 1.20 we know that the inclusion $W_0^{1,p}(\Omega) \subset L^p(\Omega)$ is a dense compact embedding. The negative (Dirichlet) p-Laplacian $-\Delta_p : W_0^{1,p}(\Omega) \to W^{-1,p'}(\Omega)$ (see Proposition 1.56) satisfies

$$\langle -\Delta_p u + \Delta_p v, u - v\rangle_{W_0^{1,p}(\Omega)} \ge c(p)\|u-v\|^p_{W_0^{1,p}(\Omega)} \quad \text{for all } u \in W_0^{1,p}(\Omega), \tag{8.47}$$

with some constant $c(p) > 0$. It has a first eigenvalue $\lambda_1 > 0$ (see Proposition 1.57). Let $f \in W^{-1,p'}(\Omega)$ and consider a convex, continuous function $\Phi : W_0^{1,p}(\Omega) \to \mathbb{R}$. Next we fix a function $J : L^p(\Omega) \to \mathbb{R}$ which is Lipschitz continuous on the bounded sets in $L^p(\Omega)$ and whose generalized gradient $\partial J : L^p(\Omega) \to 2^{L^{p'}(\Omega)}$ fulfills $0 \in \partial J(0)$ and

$$\begin{aligned}&\langle \xi - \eta, u - v\rangle_{L^p(\Omega)} \ge -c_0\|u-v\|^p_{L^p(\Omega)} \\ &\text{for all } u, v \in L^p(\Omega),\ \xi \in \partial J(u),\ \eta \in \partial J(v),\end{aligned} \tag{8.48}$$

with a constant $c_0 < \lambda_1 c(p)$.

Gathering all these data, we formulate the following quasivariational inequality: given $\rho > 0$, find $u \in W_0^{1,p}(\Omega)$ satisfying

$$\|u\|_{W_0^{1,p}(\Omega)} \le \rho + \|u\|_{L^p(\Omega)} \tag{8.49}$$

such that for some $b \in \partial J(u)$ we have

$$\langle -\Delta_p u - f, v - u\rangle_{W_0^{1,p}(\Omega)} + \langle b, v - u\rangle_{L^p(\Omega)} + \Phi(v) - \Phi(u) \ge 0 \tag{8.50}$$

for all $v \in W_0^{1,p}(\Omega)$ with $\|v\|_{W_0^{1,p}(\Omega)} \le \rho + \|u\|_{L^p(\Omega)}$.

Theorem 8.21. *If the locally Lipschitz function $J : L^p(\Omega) \to \mathbb{R}$ verifies (8.48), then for any $\rho > 0$, problem (8.49)–(8.50) possesses at least one solution.*

Proof. Fix a number $\rho > 0$ and set $X = W_0^{1,p}(\Omega)$ and $Y = L^p(\Omega)$. Our goal is to apply Theorem 8.15 to $K = W_0^{1,p}(\Omega)$, $L = 0$, $A = -\Delta_p$, $B = \partial J$, and a convex, lower semicontinuous function $\Phi : W_0^{1,p}(\Omega) \to \mathbb{R}$. It is clear that conditions (A_1) and (A_4) are satisfied. It is known from (8.47) that $-\Delta_p$ is uniformly monotone, so taking into account Proposition 1.56 it is pseudomonotone. Therefore assumption (A_2) is verified.

In order to check assumption (A_3), let $b_n \in \partial J(y_n)$ with $y_n \to y$ in $L^p(\Omega)$ and $b_n \rightharpoonup b$ in $L^{p'}(\Omega)$. Then Proposition 1.39 (ii) ensures that $b \in \partial J(y)$. Furthermore, ∂J has nonempty, convex, closed values, and it is a bounded operator because J is supposed to be Lipschitz continuous on the bounded sets. This follows from Proposition 1.40. Altogether, we can conclude that assumption (A_3) is fulfilled.

We define the multivalued map $C : W_0^{1,p}(\Omega) \to 2^{W_0^{1,p}(\Omega)}$ by

$$C(w) := \left\{ v \in W_0^{1,p}(\Omega) : \|v\|_{W_0^{1,p}(\Omega)} \le \rho + \|w\|_{L^p(\Omega)} \right\}$$
$$\text{for all } w \in W_0^{1,p}(\Omega).$$

By (8.47) and (8.48), the following estimate is valid

$$\begin{aligned} &\langle -\Delta_p u - f, u\rangle_{W_0^{1,p}(\Omega)} + \langle b, u\rangle_{L^p(\Omega)} \\ &\ge (c(p) - \lambda_1^{-1} c_0)\|u\|^p_{W_0^{1,p}(\Omega)} - \|f\|_{W^{-1,p'}(\Omega)} \|u\|_{W_0^{1,p}(\Omega)} \end{aligned}$$

for every $u \in W_0^{1,p}(\Omega)$ and $b \in \partial J(u)$. Since $c_0 < \lambda_1 c(p)$, we derive that hypothesis (A_5') is satisfied with $u_0 = 0$. Through an estimate of the same type based again on (8.47) and (8.48) we find that hypothesis (A_6) holds true.

Given $u, z \in C(w)$ with $w \in W_0^{1,p}(\Omega)$, from Proposition 1.39 it turns out that

$$\limsup_{t \to 0^+} \langle b_t, z - u\rangle_{L^p(\Omega)} \in \{\langle b, z - u\rangle_{L^p(\Omega)} : b \in \partial J(u)\}$$

whenever $b_t \in \partial J(ix + t(z-x))$ for $t > 0$ small. This enables us to obtain

$$\limsup_{t\to 0^+} \langle b_t, z-u\rangle_{L^p(\Omega)} \le \max\{\langle b, z-u\rangle_{L^p(\Omega)} : b \in \partial J(u)\}$$
$$= \langle B_u, z-u\rangle_{L^p(\Omega)},$$

for some $B_u \in \partial J(u)$, because the generalized gradient $\partial J(u)$ is weakly* compact in $L^{p'}(\Omega)$. Combining with the continuity of $-\Delta_p$, this proves condition (A_7).

For checking condition (A_8), let $\{w_n\} \subset W_0^{1,p}(\Omega)$ and $u_n \in C(w_n)$ satisfy $w_n \rightharpoonup w$ and $u_n \rightharpoonup u$ in $W_0^{1,p}(\Omega)$. It is thus known that $\|u_n\|_{W_0^{1,p}(\Omega)} \le \rho + \|w_n\|_{L^p(\Omega)}$ and along a relabeled subsequence $w_n \to w$ in $L^p(\Omega)$ by applying Theorem 1.20. In the limit we get $\|u\|_{W_0^{1,p}(\Omega)} \le \rho + \|w\|_{L^p(\Omega)}$, so $u \in C(w)$, which yields condition (A_8) (i).

Let $\{w_n\} \subset W_0^{1,p}(\Omega)$ with $w_n \rightharpoonup w$ in $W_0^{1,p}(\Omega)$ and $v \in C(w)$. We set

$$v_n := \frac{\rho + \|w_n\|_{L^p(\Omega)}}{\rho + \|w\|_{L^p(\Omega)}} v.$$

Along a relabeled subsequence we have $v_n \to v$ in $W_0^{1,p}(\Omega)$ and $v_n \in C(w_n)$ because $v \in C(w)$ and thus

$$\|v_n\|_{W_0^{1,p}(\Omega)} = \frac{\rho + \|w_n\|_{L^p(\Omega)}}{\rho + \|w\|_{L^p(\Omega)}} \|v\|_{W_0^{1,p}(\Omega)} \le \rho + \|w_n\|_{L^p(\Omega)},$$

which shows that condition (A_8) (ii) holds, too.

Since all the hypotheses of Theorem 8.15 are verified, we are in a position to apply it to problem (8.49)–(8.50), which leads to the desired conclusion. □

We pass to our second application which concerns an evolutionary quasivariational inequality. Let a bounded domain Ω in $\mathbb{R}^N$ and numbers $p \in [2, +\infty)$ and $\tau > 0$. We set $X := L^p((0,\tau); W_0^{1,p}(\Omega))$, which is a reflexive Banach space under the usual norm and has the dual $X^* = L^{p'}((0,\tau); W^{-1,p'}(\Omega))$, where $1 < p < +\infty$ and $1/p + 1/p' = 1$. Let $f \in X^*$, and let us introduce the map $A : X \to 2^{X^*}$ by

$$A(u)(t) = -\Delta_p(u(t)) \quad \text{for all } u \in X,\ t \in [0,\tau]. \tag{8.51}$$

Fix a (possibly nonlinear) compact operator $T : X \to X$. We state the evolutionary quasivariational inequality: given a number $\rho > 0$, find $u \in X$ with $u' = \frac{du}{dt} \in X^*$ and $u(0) = 0$ such that

$$\|u\|_X \le \rho + \|Tu\|_X \tag{8.52}$$

and

$$\langle u' + A(u) - f, v - u\rangle_X + \Phi(v) - \Phi(u) \geq 0 \tag{8.53}$$

for all $v \in X$ with $\|v\|_X \leq \rho + \|Tu\|_X$.

Theorem 8.22. *For any $\rho > 0$, problem (8.52)–(8.53) possesses at least one solution.*

Proof. Fix $\rho > 0$. The time derivative is an unbounded linear operator $L : D(L) \subset X \to X^*$ defined by $L(u) = u'$ for all $u \in D(L)$, with the domain

$$D(L) = \{u \in X : u' \in X^*,\ u(0) = 0\}.$$

It is a maximal monotone operator (see, e.g., [38, Lemma 2.149]), so assumption (A_1) in Theorem 8.6 is satisfied. The map $A : X \to 2^{X^*}$ introduced in (8.51) is monotone, hemicontinuous, and bounded (see, e.g., [238, p. 878]), so pseudomonotone, which implies assumption (A_2) in Theorem 8.6. Hypothesis (A_3) is verified with $B = 0$.

Choose $K = X$, which makes hypothesis (A_4) be satisfied. Let us define the multivalued map $C : X \to 2^X$ by

$$C(w) = \{v \in X : \|v\|_X \leq \rho + \|Tw\|_X\} \quad \text{for all } w \in X.$$

On account of (8.47), the operator A introduced in (8.51) is coercive, which renders assumption (A_5') true, for instance, with $u_0 = 0$. Since the operator A is monotone, whereas $B = 0$, condition (A_6) is valid, while condition (A_7) holds because the operator A is hemicontinuous.

In order to check condition (A_8) (i), we consider sequences $\{w_n\} \subset X$ and $u_n \in C(w_n)$ satisfying $w_n \rightharpoonup w$ and $u_n \rightharpoonup u$ in X. We have that $\|u_n\|_X \leq \rho + \|Tw_n\|_X$ and, thanks to the compactness of the mapping T, we can pass to a relabeled subsequence such that $Tw_n \to Tw$ in X. In the limit it results in $\|u\|_X \leq \rho + \|Tw\|_X$, that is, $u \in C(w)$, thus (A_8) (i) holds. For proving condition (A_8) (ii), we consider $\{w_n\} \subset X$ with $w_n \rightharpoonup w$ in X and $v \in C(w)$. Setting

$$v_n := \frac{\rho + \|Tw_n\|_X}{\rho + \|Tw\|_X} v$$

and using the compactness of the map T, as well as that $v \in C(w)$, it is straightforward to show that along a relabeled subsequence $v_n \to v$ in X and $v_n \in C(w_n)$. Since the function Φ is continuous, condition (A_8) (ii) is satisfied.

Therefore all the hypotheses of Theorem 8.15 hold true, which enables us to infer through Theorem 8.15 the solvability of problem (8.52)–(8.53). □

8.3 NOTES

This chapter is conducted following Khan–Motreanu [118] and is devoted to evolutionary variational and quasivariational inequalities. The evolutionary feature of these problems is introduced through an unbounded, linear, maximal operator L, which is interpreted as abstract time derivative. Quasivariational inequalities not only subsume variational inequalities and nonlinear partial differential inequations, but also provide a unified framework for general boundary value problems with nonstandard unilateral boundary conditions. For different developments and applications we refer to [20,102,104,105,119,124,189,193]. Roughly speaking, for existence and approximation regarding quasivariational inequalities it is required that a variational inequality and a fixed point problem be solved simultaneously.

Section 8.1 sets forth several general results on elliptic and evolutionary variational inequalities incorporating numerous boundary value problems. We mention Theorem 8.6, extending different existence results for evolutionary variational inequalities, as well as variational–hemivariational inequalities as, for instance, [117, Theorem 4.1] and [132, Theorem 3.1]. Specifically, for variational–hemivariational inequalities, if $g : Y \to \mathbb{R}$ is a locally Lipschitz function, denoting its generalized gradient by $\partial g : Y \to 2^{Y^*}$, we can choose $B = \partial g$. Recall that if the function g is convex, the generalized gradient ∂g becomes the subdifferential of g in the sense of convex analysis. Notice that Theorem 8.6 can involve different types of pseudomonotone operators.

In Theorems 8.8 and 8.9, under different hypotheses, the assumption on the constraint set K to have a nonempty interior is dropped. Theorem 8.9 gives a new existence result for elliptic variational inequalities with generalized pseudomonotone maps. The recessive assumption (H_2) extends all the classical coercivity conditions. Theorem 8.10 manages elliptic variational inequalities driven by maximal monotone operators covering new situations. Theorem 8.11 is based on a multivalued version of Minty's technique for variational inequalities governed by single-valued monotone operators. Another trait of our results is that the imposed assumptions exploit the interplay between the properties of the operators involved in the problem and the geometry of the sets of constraints, for instance, their exterior normal cones.

Section 8.2 deals with quasivariational inequalities which incorporate evolutionary inequality problems thanks to the presence of a possibly unbounded linear operator L whose prototype is the time derivative $L(x) = x'$ on a space such as $X = L^p((0,\tau), V)$, with $1 < p < +\infty$, $\tau > 0$, and a reflexive Banach space V. For this evolutionary setting we refer to Zeidler [238] (see also [38]). The specific feature of the quasivariational inequalities treated here is that the constraint set $C(x)$ depends on the solution x. This dependence poses serious challenges in extending the properties from variational inequalities to quasivariational inequalities.

The quasivariational inequality (8.22) represents a general statement covering significant elliptic and evolutionary problems. In this direction we mention just a simple but relevant example, which in fact lies at the beginning of the theory. Namely, if the map A is single-valued with $D(A) = X$, $\Phi = 0$, $B = 0$, and $L = 0$, then (8.22) reduces to the quasivariational inequality given by Bensoussan–Lions [26]: find $x \in C(x)$ such that

$$\langle A(x) - f, z - x \rangle_X \geq 0 \quad \text{for every } z \in C(x).$$

If, additionally, $C(x) = K$ for every $x \in K$, then it recovers the standard statement of variational inequality (see [122]): find $x \in K$ such that

$$\langle A(x) - f, z - x \rangle_X \geq 0 \quad \text{for every } z \in K.$$

A powerful technique for solving quasivariational inequalities is by finding fixed points of the associated variational selection which is defined in (8.24). This procedure is implemented for resolving the general quasivariational inequality stated in problem (8.22) thus treating in a unifying way elliptic and evolutionary problems in Theorem 8.15. The proof is based on Mosco-type continuity properties and Kluge's fixed point theorem for multivalued maps. In Theorem 8.17 we give another existence result for quasivariational inequalities where this time the focus is on maximal monotone maps. The main tool in the proof is an extension of Minty's technique for evolutionary variational inequalities provided in Theorem 8.11. The object of Theorem 8.19 is the study of evolutionary quasivariational inequalities with compact constraints and taking as underlying space the domain of the linear, maximal monotone operator L endowed with the graph norm. Applications of the abstract results, specifically of Theorem 8.15, are given in Theorems 8.21 and 8.22. An open study in the field of variational and quasivariational inequalities is the qualitative theory of solutions.

Chapter 9

Control Problems for Evolutionary Differential Inclusions

9.1 SEMILINEAR EVOLUTIONARY INCLUSIONS WITH CONTROL PARAMETER

We start by mentioning certain needed preliminaries about the semilinear evolutionary inclusion on a complex separable Hilbert space H with the scalar product $\langle \cdot, \cdot \rangle_H$ and the associated norm $\| \cdot \|_H$:

$$\begin{cases} x'(t) \in A(t)x(t) + Bu(t) + G(t, x(t)), & t \in J = [0, b], \\ x(0) = x_0, \end{cases} \tag{9.1}$$

where $\{A(t) : t \in J\}$ is a family of linear closed densely defined operators $A(t)$ on H (see Definition 1.48) such that the domain of $A(t)$ does not depend on t, B is a bounded linear operator from a Hilbert space U (the admissible controls set) into H, $G : J \times H \to 2^H$ is a multivalued map, and $x_0 \in H$. The control function u belongs to $L^2(J, U)$. For notation convenience the resolvent of $A(t)$ whenever it exists is denoted by $R(\lambda, A(t))$ (see Section 1.3 in Chapter 1). We can regard (9.1) as a control problem where the control function u is in $L^2(J, U)$. We may consider the control problem on a Banach space, but for simplicity we deal with a Hilbert space.

We assume the following conditions taken from [198, p. 150]:

(A1) the operators $A(t)$ $(t \in J)$ have a common domain D which is dense in H;

(A2) for each $t \in J$, the resolvent $R(\lambda, A(t))$ of $A(t)$ exists for all $\lambda \in \mathbb{C}$ with $\operatorname{Re} \lambda \leq 0$ and there exists a constant $C_1 > 0$ independent of t and λ such that

$$\|R(\lambda, A(t))\| \leq \frac{C_1}{|\lambda| + 1};$$

(A3) there exist constants $\beta \in (0, 1]$ and $C_2 > 0$ such that whenever $t, s, \tau \in J$,

$$\|\big(A(t) - A(s)\big)A^{-1}(\tau)\| \leq C_2 |t - s|^{\beta};$$

(A4) for each $t \in J$ there exists $\lambda \in \rho(A(t))$ (the resolvent set of $A(t)$) such that the resolvent $R(\lambda, A(t))$ is a compact operator.

Nonlinear Differential Problems with Smooth and Nonsmooth Constraints. DOI: 10.1016/B978-0-12-813386-6.00009-2

Definition 9.1. ([198, p. 129]) A two-parameter family of bounded linear operators $U(t,s)$, $0 \le s \le t \le b$, on a Banach space E is called an evolution system if the following conditions are satisfied:

(i) $U(s,s) = I$, $U(t,r)U(r,s) = U(t,s)$ for all $0 \le s \le r \le t \le b$;
(ii) $(t,s) \mapsto U(t,s)$ is strongly continuous on the set $0 \le s \le t \le b$.

Here I denotes the identity map of H.

Lemma 9.2 ([198, p. 150]). *Under assumptions* (A1)–(A3), *there exists a unique evolution system $U(t,s)$ on the set $0 \le s \le t \le b$ in the sense of Definition 9.1 satisfying:*

(i) *There is a constant $M > 0$ such that $\|U(t,s)\| \le M$ for $0 \le s \le t \le b$;*
(ii) *For $0 \le s < t \le b$, one has $U(t,s) : H \to D$, and $t \mapsto U(t,s)$ is strongly differentiable with the derivative $\frac{\partial}{\partial t}U(t,s) \in \mathcal{L}(H,H)$ strongly continuous on the set $0 \le s < t \le b$. Moreover, for $0 \le s < t \le b$,*

$$\frac{\partial}{\partial t}U(t,s) + A(t)U(t,s) = 0, \qquad \left\|\frac{\partial}{\partial t}U(t,s)\right\| = \|A(t)U(t,s)\| \le \frac{M}{t-s},$$
$$\|A(t)U(t,s)A^{-1}(s)\| \le M;$$

(iii) *For every $y \in D$ and $t \in (0,b]$, $U(t,s)y$ is differentiable with respect to s on $0 \le s \le t \le b$ and*

$$\frac{\partial}{\partial s}U(t,s)y = U(t,s)A(s)y.$$

Lemma 9.3 ([86]). *Let $\{A(t) : t \in J\}$ satisfy conditions* (A1)–(A4). *If $\{U(t,s) : 0 \le s \le t \le b\}$ is the evolution system generated by $\{A(t) : t \in J\}$ as given in Lemma 9.2, then $\{U(t,s) : 0 \le s \le t \le b\}$ is a compact operator whenever $t - s > 0$.*

Taking advantage of what was said above, we introduce the following concept of solution for the differential system (9.1).

Definition 9.4. Given $u \in L^2(J,U)$, a function $x \in C(J,H)$ is called a mild solution of system (9.1) if there exists $f \in L^2(J,H)$ such that $f(t) \in G(t,x(t))$ for a.e. $t \in J$ and

$$x(t) = U(t,0)x_0 + \int_0^t U(t,s)Bu(s)\,ds + \int_0^t U(t,s)f(s)\,ds \quad \text{for all } t \in J.$$

We also consider the following semilinear evolution hemivariational inequality:

$$\begin{cases} \langle -x'(t) + A(t)x(t) + Bu(t), v\rangle_H + F^0(t,x(t);v) \ge 0 \\ \quad \text{for all } t \in J, \text{ all } v \in H, \\ x(0) = x_0, \end{cases} \tag{9.2}$$

where $F^0(t,\cdot;\cdot)$ stands for the generalized directional derivative of a locally Lipschitz function $F(t,\cdot): H \to \mathbb{R}$ (see Section 1.2 for the definition of $F^0(t,\cdot;\cdot)$).

Let us note that Definition 9.4 provides the notion of mild solution for problem (9.2), too. Indeed, according to the definition of the generalized gradient in Section 1.2, problem (9.2) is equivalent to the differential inclusion

$$\begin{cases} x'(t) - A(t)x(t) - Bu(t) \in \partial F(t, x(t)), & t \in J, \\ x(0) = x_0, \end{cases}$$

where we denote $\partial F(t,x) = \partial F(t,\cdot)(x)$. Hence Definition 9.4 becomes as follows.

Definition 9.5. Given $u \in L^2(J,U)$, a function $x \in C(J,H)$ is called a mild solution of the hemivariational inequality (9.2) if there exists $f \in L^2(J,H)$ such that $f(t) \in \partial F(t,x(t))$ for a.e. $t \in J$ and

$$x(t) = U(t,0)x_0 + \int_0^t U(t,s)Bu(s)\,ds + \int_0^t U(t,s)f(s)\,ds \quad \text{for all } t \in J.$$

The set $K_b(G) = \{x(b) \in H : x(\cdot)$ is a mild solution of the system (9.1) corresponding to a control $u \in L^2(J,U)$ with initial value $x_0 \in H\}$ represents the reachable set of the system (9.1). For the general theory of linear control systems we refer to [63].

Definition 9.6. The system (9.1) is said to be approximately controllable on the interval $J = [0,b]$ if $\overline{K_b(G)} = H$, where $\overline{K_b(G)}$ denotes the closure of $K_b(G)$ in H. In particular, the corresponding linear system is approximately controllable on J if $\overline{K_b(0)} = H$.

We impose the following hypotheses on the multivalued map $G: J \times H \to 2^H$ in problem (9.1):

H(G) **(i)** G is jointly measurable on $J \times H$ and has nonempty, convex, and weakly compact values;

(ii) for a.e. $t \in J$, $G(t,\cdot)$ has a strongly-weakly closed graph, i.e., if $x_n \to x$ in H and $h_n \rightharpoonup h$ in H with $h_n \in G(t,x_n)$, then $h \in G(t,x)$;

(iii) there exist a function $a \in L^2(J)$ and a constant $c > 0$ such that

$$\|G(t,x)\|_H := \sup\{\|\xi\|_H : \xi \in G(t,x)\} \le a(t) + c\|x\|_H$$

for a.a. $t \in J$ and all $x \in H$.

In order to establish the existence of mild solutions for problem (9.1), we introduce the multivalued map $\mathcal{N}: L^2(J,H) \to 2^{L^2(J,H)}$ by

$$\mathcal{N}(x) = \{w \in L^2(J,H) : w(t) \in G(t,x(t)) \quad \text{for a.e. } t \in J, \text{ whenever } x \in L^2(J,H)\}.$$

As in [155, Lemma 5.3] we can prove the result.

Lemma 9.7. *If hypotheses* H(G) *hold, then for each* $x \in L^2(J, H)$ *the set* $\mathcal{N}(x)$ *is nonempty, convex, and weakly compact.*

Applying [154, Lemma 11], we also get a useful closure statement.

Lemma 9.8. *If conditions* H(G) *hold, then the multivalued operator* $\mathcal{N}$ *defined above fulfills the property: if* $x_n \to x$ *in* $L^2(J, H)$, $w_n \rightharpoonup w$ *in* $L^2(J, H)$, *and* $w_n \in \mathcal{N}(x_n)$, *then* $w \in \mathcal{N}(x)$.

A key tool in the sequel is a fixed point theorem for multivalued maps that we quote from [107].

Theorem 9.9. *If* E *is a Banach space,* $\Omega \subset E$ *is nonempty, closed, and convex with* $0 \in \Omega$, *and* $\Gamma : \Omega \to 2^{\Omega}$ *is an upper semicontinuous multifunction that maps bounded sets into relatively compact sets and has nonempty, compact, convex values, then one of the following statements is true:*

(a) *the set* $V = \{x \in \Omega : x \in \lambda\Gamma(x),\ \lambda \in (0, 1)\}$ *is unbounded;*
(b) Γ *has a fixed point, i.e., there exists* $x \in \Omega$ *such that* $x \in \Gamma(x)$.

Now we are able to provide the existence of mild solutions to problem (9.1).

Theorem 9.10. *Assume that the hypotheses* (A1)–(A4) *and* H(G) *hold. Then, for each* $u \in L^2(J, U)$, *problem (9.1) has a mild solution on* J *in the sense of Definition 9.4.*

Proof. By Lemma 9.7, the multivalued map $\Gamma : C(J, H) \to 2^{C(J,H)}$ defined by

$$\Gamma(x) = \Big\{h \in C(J, H) : \\ h(t) = U(t, 0)x_0 + \int_0^t U(t, s)\big(f(s) + Bu(s)\big)\, ds, \quad f \in \mathcal{N}(x)\Big\}$$

has nonempty values. In order to show that Γ has a fixed point, we check that Γ fulfills the conditions of Theorem 9.9. For any $x \in C(J, H)$, $\Gamma(x)$ is convex due to the convexity of $\mathcal{N}(x)$. We subdivide the rest of the proof into five steps.

Step 1. Γ is bounded, that is it maps bounded sets into bounded sets in $C(J, H)$.

Given $r > 0$, we set $B_r := \{x \in C(J, H) : \|x\|_{C(J,H)} \le r\}$. For every $\varphi \in \Gamma(x)$ with $x \in B_r$, there exists $f \in \mathcal{N}(x)$ such that

$$\varphi(t) = U(t, 0)x_0 + \int_0^t U(t, s)Bu(s)\, ds + \int_0^t U(t, s)f(s)\, ds, \quad t \in J. \tag{9.3}$$

Using Lemma 9.2 (i), H(G) (iii), and Hölder's inequality, we obtain for all $t \in J$ that

$$\|\varphi(t)\|_H \le \|U(t, 0)x_0\|_H + \int_0^t \|U(t, s)Bu(s)\|_H\, ds + \int_0^t \|U(t, s)f(s)\|_H\, ds$$

$$\le M\Big(\|x_0\|_H + \int_0^t \big(\|B\|\,\|u(s)\|_U + a(s) + c\|x(s)\|_H\big)\,ds\Big)$$
$$\le M\Big(\|x_0\|_H + \|B\|\,\|u\|_{L^2(J,U)}\sqrt{b} + \|a\|_{L^2(J)}\sqrt{b} + crb\Big),$$

which proves the claim.

Step 2. $\Gamma(B_r) := \bigcup\{\Gamma(x) : x \in B_r\}$ is equicontinuous for every $r > 0$.

For each $x \in B_r$ and $\varphi \in \Gamma(x)$, there exists $f \in \mathcal{N}(x)$ such that (9.3) holds. By assumption H(G) (iii), we note for $0 < \tau_1 < \tau_2 \le b$ and $\delta > 0$ small enough that

$$\begin{aligned}
\|\varphi(\tau_2) - \varphi(\tau_1)\|_H \ &\le\ \|U(\tau_2, 0)x_0 - U(\tau_1, 0)x_0\|_H \\
&+ \int_0^{\tau_1} \|\big(U(\tau_2, s) - U(\tau_1, s)\big)\big(Bu(s) + f(s)\big)\|_H\,ds \\
&+ \int_{\tau_1}^{\tau_2} \|U(\tau_2, s)\big(Bu(s) + f(s)\big)\|_H\,ds \\
\le \|\big(U(\tau_2, 0) &- U(\tau_1, 0)\big)x_0\|_H \\
&+ \sup_{s\in[0,\tau_1-\delta]} \|U(\tau_2, s) - U(\tau_1, s)\|\,\big((\|B\|\,\|u\|_{L^2(J,U)} + \|a\|_{L^2(J)})\sqrt{b} + crb\big) \\
&+ M\,\big((\|B\|\,\|u\|_{L^2(J,U)} + \|a\|_{L^2(J)})(2\sqrt{\delta} + \sqrt{\tau_2 - \tau_1}) + cr(2\delta + \tau_2 - \tau_1)\big).
\end{aligned}$$

Lemma 9.3 ensures the compactness of $U(t, s)$ for $t - s > 0$. It is seen that $\|\varphi(\tau_2) - \varphi(\tau_1)\|_H \to 0$ uniformly with respect to $x \in B_r$ as $\tau_2 \to \tau_1$ and $\delta \to 0$. In addition, if $\tau_1 = 0$ and $0 < \tau_2 \le b$, it is easy to prove that $\|\varphi(\tau_2) - x_0\|_H \to 0$ uniformly with respect to $x \in B_r$ as $\tau_2 \to 0$. Altogether, we find that Step 2 holds true.

Step 3. Γ is completely continuous.

Fix any $r > 0$. We need to show that $\Gamma(B_r)$ is relatively compact in $C(J, H)$. Taking into account Steps 1 and 2 and making use of Ascoli–Arzelà theorem, we must prove that for every $t \in (0, b]$ the set $\Pi(t) := \{\varphi(t) : \varphi \in \Gamma(B_r)\}$ is relatively compact in H.

Given $x \in B_r$, $\varphi \in \Gamma(x)$ and $t \in (0, b]$, choose $\delta \in (0, t)$ and define

$$\begin{aligned}
\varphi^\delta(t) &:= U(t, 0)x_0 + \int_0^{t-\delta} U(t, s)\big(Bu(s) + f(s)\big)\,ds \\
&= U(t, 0)x_0 + U(t, t - \delta)\int_0^{t-\delta} U(t - \delta, s)\big(Bu(s) + f(s)\big)\,ds.
\end{aligned}$$

Lemma 9.3 ensures the compactness of $U(t, s)$ for $t - s > 0$. This, in conjunction with the boundedness of $\int_0^{t-\delta} U(t - \delta, s)(Bu(s) + f(s))\,ds$, implies that the set $\Pi^\delta(t) := \{\varphi^\delta(t) : \varphi \in \Gamma(B_r)\}$ is relatively compact in H.

Moreover, for every $\varphi \in \Gamma(B_r)$, it turns out that

$$\|\varphi(t) - \varphi^\delta(t)\|_H \leq M\big(\|B\| \, \|u\|_{L^2(J,U)}\sqrt{\delta} + \|a\|_{L^2(J)}\sqrt{\delta} + cr\delta\big) \\ \to 0 \text{ as } \delta \to 0,$$

with a constant $M > 0$. Therefore the set $\Pi(t)$ is totally bounded, so in view of Step 2, it is relatively compact in H.

Step 4. Γ has a closed graph.

Let $x_n \to x_*$ in $C(J, H)$ and $\varphi_n \to \varphi_*$ in $C(J, H)$, with $\varphi_n \in \Gamma(x_n)$. The fact that $\varphi_n \in \Gamma(x_n)$ means that there exists $f_n \in \mathcal{N}(x_n)$ such that

$$\varphi_n(t) = U(t, 0)x_0 + \int_0^t U(t, s)Bu(s)\,ds + \int_0^t U(t, s)f_n(s)\,ds. \tag{9.4}$$

By assumption H(G) (iii), we can show that the sequence $\{f_n\}$ is bounded in $L^2(J, H)$. Hence, passing to a subsequence if necessary, we may suppose that

$$f_n \rightharpoonup f_* \text{ in } L^2(J, H). \tag{9.5}$$

From (9.4), (9.5) and the compactness of the operator $U(t, s)$ for $t - s > 0$ as stated in Lemma 9.3, we get

$$\varphi_n(t) \to U(t, 0)x_0 + \int_0^t U(t, s)Bu(s)\,ds + \int_0^t U(t, s)f_*(s)\,ds. \tag{9.6}$$

Since $\varphi_n \to \varphi_*$ in $C(J, H)$ and $f_n \in \mathcal{N}(x_n)$, from Lemma 9.8 and (9.6), we obtain $f_* \in \mathcal{N}(x_*)$, hence $\varphi_* \in \Gamma(x_*)$.

Step 5. A priori estimate and conclusion.

Step 4 guarantees that the graph of Γ is closed. Then it follows from [155, Proposition 3.3.12(2)] that Γ is upper semicontinuous. Thus from Steps 1–4 we know that the multivalued map Γ satisfies all the hypotheses of Theorem 9.9. It remains to decide on the alternative of Theorem 9.9. To this end we claim that the set

$$V := \{x \in C(J, H) : x \in \lambda\Gamma(x),\ 0 < \lambda < 1\}$$

is bounded.

Let $x \in V$ and $\lambda \in (0, 1)$ with $x \in \lambda\Gamma(x)$. Then there exists $f \in \mathcal{N}(x)$ such that

$$x(t) = \lambda U(t, 0)x_0 + \lambda \int_0^t U(t, s)Bu(s)\,ds + \lambda \int_0^t U(t, s)f(s)\,ds.$$

By assumption H(G) (iii) we derive the estimate

$$\|x(t)\|_H \leq M\|x_0\|_H + M\|B\| \int_0^t \|u(s)\|_U \, ds + M \int_0^t \big(a(s) + c\|x(s)\|_H\big) \, ds$$
$$\leq \rho + Mc \int_0^t \|x(s)\|_H \, ds \quad \text{for all } t \in J,$$

where $\rho = M\left(\|x_0\|_H + \left(\|B\| \, \|u\|_{L^2(J,U)} + \|a\|_{L^2(J)}\right)\sqrt{b}\right)$, and with a constant $M > 0$. Then Gronwall's inequality yields $\|x(t)\|_H \leq \rho e^{Mct}$, showing that the set V is bounded. Now Theorem 9.9 provides a fixed point for Γ, thus a mild solution of problem (9.1). □

On the basis of Theorem 9.10 we can conduct the study of the approximate controllability for the control problem (9.1). For this we rely on the associated linear control system

$$\begin{cases} x'(t) = A(t)x(t) + Bu(t), \quad t \in J = [0, b], \\ x(0) = x_0. \end{cases} \tag{9.7}$$

The controllability map corresponding to (9.7) is the nonnegative bounded linear operator on H defined by

$$\Gamma_0^b = \int_0^b U(b, s)BB^*U^*(b, s) \, ds.$$

Here B^* and $U^*(t, s)$ denote the adjoint operators of B and $U(t, s)$, respectively. Therefore, the inverse of $\varepsilon I + \Gamma_0^b$ exists for any $\varepsilon > 0$, so the resolvent

$$R(\varepsilon, -\Gamma_0^b) = (\varepsilon I + \Gamma_0^b)^{-1}$$

is well defined; see also Section 1.3.

Lemma 9.11. *Given $x_1 \in H$ and $\varepsilon > 0$, there exists a unique (optimal) control $u^\varepsilon(\cdot) \in L^2(J, U)$ at which the functional $S : L^2(J, U) \to \mathbb{R}$ defined by*

$$S(u) = \|x^u(b) - x_1\|_H^2 + \varepsilon \int_0^b \|u(t)\|_U^2 \, dt, \tag{9.8}$$

where $x^u(\cdot)$ is the unique mild solution of (9.7), attains its minimum value and

$$u^\varepsilon(t) = B^*U^*(b, t)R(\varepsilon, -\Gamma_0^b)(x_1 - U(b, 0)x_0), \tag{9.9}$$
$$x^{u^\varepsilon}(b) - x_1 = \varepsilon R(\varepsilon, -\Gamma_0^b)(U(b, 0)x_0 - x_1). \tag{9.10}$$

Proof. The existence and uniqueness of an optimal control is the consequence of a general theorem on linear regulator problems (see [62]). We will establish Eqs. (9.9) and (9.10). The functional S in (9.8) is convex and differentiable. It has a minimizer u^ε which is characterized by $S'(u^\varepsilon)=0$, that is, for all $v \in L^2(J,U)$, one has

$$\begin{aligned}
0 &= \int_0^b \langle S'(u^\varepsilon), v\rangle_U \, dt = \lim_{h\to 0} \frac{S(u^\varepsilon + hv) - S(u^\varepsilon)}{h} \\
&= \lim_{h\to 0} \frac{1}{h}\Big(\langle x^{u^\varepsilon + hv}(b) - x_1, x^{u^\varepsilon + hv}(b) - x_1\rangle_H - \langle x^{u^\varepsilon}(b) - x_1, x^{u^\varepsilon}(b) - x_1\rangle_H \\
&\quad + \varepsilon \int_0^b \big(\langle u^\varepsilon(t) + hv(t), u^\varepsilon(t) + hv(t)\rangle_U - \langle u^\varepsilon(t), u^\varepsilon(t)\rangle_U\big)\, dt\Big) \\
&= 2\int_0^b \big(\langle x^{u^\varepsilon}(b) - x_1, U(b,t)Bv(t)\rangle_H + \varepsilon\langle u^\varepsilon(t), v(t)\rangle_U\big)\, dt \\
&= 2\int_0^b \big(\langle B^*U^*(b,t)(x^{u^\varepsilon}(b) - x_1) + \varepsilon u^\varepsilon(t), v(t)\rangle_U\big)\, dt.
\end{aligned}$$

Since $v \in L^2(J,U)$ is arbitrary, we get

$$u^\varepsilon(t) = -\frac{1}{\varepsilon} B^*U^*(b,t)(x^{u^\varepsilon}(b) - x_1) \text{ a.e. on } J. \tag{9.11}$$

Then, in view of (9.11), we have

$$\begin{aligned}
x^{u^\varepsilon}(b) &= U(b,0)x_0 + \int_0^b U(b,s)Bu^\varepsilon(s)\, ds \\
&= U(b,0)x_0 - \frac{1}{\varepsilon}\int_0^b U(b,s)BB^*U^*(b,s)(x^{u^\varepsilon}(b) - x_1)\, ds \\
&= U(b,0)x_0 - \frac{1}{\varepsilon}\Gamma_0^b(x^{u^\varepsilon}(b) - x_1),
\end{aligned}$$

implying

$$(\varepsilon I + \Gamma_0^b)x^{u^\varepsilon}(b) = \varepsilon U(b,0)x_0 + \Gamma_0^b x_1 = \varepsilon U(b,0)x_0 + (\varepsilon I + \Gamma_0^b)x_1 - \varepsilon x_1.$$

We obtain (9.10) because

$$x^{u^\varepsilon}(b) - x_1 = \varepsilon(\varepsilon I + \Gamma_0^b)^{-1}(U(b,0)x_0 - x_1) = \varepsilon R(\varepsilon, -\Gamma_0^b)(U(b,0)x_0 - x_1).$$

From (9.11) and (9.10) we derive

$$\begin{aligned}
u^\varepsilon(t) &= -\frac{1}{\varepsilon}B^*U^*(b,t)(x^{u^\varepsilon}(b) - x_1) \\
&= -\frac{1}{\varepsilon}B^*U^*(b,t)\varepsilon(\varepsilon I + \Gamma_0^b)^{-1}(U(b,0)x_0 - x_1) \\
&= B^*U^*(b,t)R(\varepsilon, -\Gamma_0^b)(x_1 - U(b,0)x_0).
\end{aligned}$$

Therefore (9.9) holds, which completes the proof. □

Through Lemma 9.11, [24, Theorem 2] can be adapted from the time-invariant case to our time depending system (9.7).

Lemma 9.12. *The linear control system (9.7) is approximately controllable on J in the sense of Definition 9.6 if and only if $\varepsilon R(\varepsilon, -\Gamma_0^b) \to 0$ as $\varepsilon \to 0^+$ in $L(H, H)$.*

At this point, we develop a fixed point approach. From Lemma 9.7 we know that for any $x \in C(J, H) \subset L^2(J, H)$ there holds $\mathcal{N}(x) \neq \emptyset$. Hence, for every $\varepsilon > 0$, we can define the multivalued map $\Gamma_\varepsilon : C(J, H) \to 2^{C(J,H)}$ with nonempty values as follows:

$$\Gamma_\varepsilon(x) = \Big\{h \in C(J, H) : \\ h(t) = U(t, 0)x_0 + \int_0^t U(t, s)\big(Bu_\varepsilon(s) + f(s)\big)\, ds, \quad f \in \mathcal{N}(x)\Big\},$$

where

$$u_\varepsilon(t) = B^*U^*(b, t)R(\varepsilon, -\Gamma_0^b)\left(x_1 - U(b, 0)x_0 - \int_0^b U(b, s)f(s)\, ds\right).$$

It is worth noting that for $G = 0$, we get (9.9). We seek fixed points for Γ_ε.

Theorem 9.13. *Suppose that conditions* (A1)–(A4) *and* H(G) *hold. Then the multivalued map $\Gamma_\varepsilon : C(J, H) \to 2^{C(J,H)}$ possesses a fixed point.*

Proof. The proof is carried out by applying Theorem 9.9 to the multivalued map Γ_ε. First, we note that the assumptions guarantee the applicability of Theorem 9.10. Using that $\mathcal{N}(x)$ is convex, it is clear that $\Gamma_\varepsilon(x)$ is convex for each $x \in C(J, H)$. As for Theorem 9.10, we subdivide the proof into steps.

Step 1. Γ_ε is bounded.

We must show that Γ_ε maps $B_R = \{x \in C(J, H) : \|x\|_{C(J,H)} \leq R\}$ into a bounded set in $C(J, H)$ for every $R > 0$. Let $x \in B_R$ and $\varphi \in \Gamma_\varepsilon(x)$. Then there exists $f \in \mathcal{N}(x)$ such that

$$\varphi(t) = U(t, 0)x_0 + \int_0^t U(t, s)Bu_\varepsilon(s)\, ds + \int_0^t U(t, s)f(s)\, ds, \quad t \in J. \quad (9.12)$$

Notice from H(G) (iii) that

$$\|u_\varepsilon(t)\|_U \leq \frac{M\|B\|}{\varepsilon}\left(\|x_1\|_H + M\|x_0\|_H + M\|\phi\|_{L^2(J)}\sqrt{b}\right) =: \Lambda,$$

with a constant $M > 0$ and $\phi \in L^2(J)$ given by

$$\phi(t) = \sup\{\|\xi\|_H : \xi \in G(t, x(t)),\ x \in B_R\} \text{ for a.e. } t \in J.$$

In view of (9.12), this entails that the assertion of Step 1 is valid.

Step 2. $\Gamma_\varepsilon(B_R) := \bigcup\{\Gamma_\varepsilon(x) : x \in B_R\}$ is equicontinuous for every $R > 0$.

As in the proof of Theorem 9.10, for $x \in B_R$ and $\varphi \in \Gamma_\varepsilon(x)$, there exists $f \in \mathcal{N}(x)$ satisfying (9.12). Using the estimate of $\|u_\varepsilon(t)\|_U$ in Step 1, we obtain the equicontinuity property on the pattern of Step 2 in the proof of Theorem 9.13.

Step 3. Γ_ε is completely continuous.

Let $\delta > 0$. For each $x \in B_R$ and $t > \delta$, we introduce

$$\begin{aligned}\varphi^\delta(t) &:= U(t,0)x_0 + \int_0^{t-\delta} U(t,s)\big(Bu_\varepsilon(s) + f(s)\big)\,ds\\ &= U(t,0)x_0 + U(t,t-\delta)\int_0^{t-\delta} U(t-\delta,s)\big(Bu_\varepsilon(s) + f(s)\big)\,ds,\end{aligned}$$

where f is related to x through the definition of Γ_ε. The estimate for $\|u_\varepsilon(t)\|_U$ in Step 1 and the compactness of $U(t,s)$ for $t-s>0$ imply that the set $\Pi^\delta(t) := \{\varphi^\delta(t) : \varphi \in \Gamma_\varepsilon(B_R)\}$ is relatively compact in H for every $t > \delta$. We also observe that for every $\varphi \in \Gamma_\varepsilon(B_R)$ we have

$$\|\varphi(t) - \varphi^\delta(t)\|_H \le M\int_{t-\delta}^t \big(\|B\|\Lambda + \phi(s)\big)\,ds \le M\big(\|B\|\Lambda\delta + \|\phi\|_{L^2(J)}\sqrt{\delta}\big),$$

with a constant $M > 0$ and $\phi \in L^2(J)$ in Step 1. We infer that the set $\Pi(t) := \{\varphi(t) : \varphi \in \Gamma_\varepsilon(B_R)\}$ is totally bounded, so relatively compact in H. By Ascoli–Arzelà theorem and Step 2, we conclude that Γ_ε is completely continuous.

Step 4. Γ_ε is upper semicontinuous.

We claim that Γ_ε has a closed graph. Let $x_n \to x_*$ and $\varphi_n \to \varphi_*$ in $C(J,H)$, with $\varphi_n \in \Gamma_\varepsilon(x_n)$. Thus there exists $f_n \in \mathcal{N}(x_n)$ such that

$$\begin{aligned}\varphi_n(t) = U(t,0)x_0 + \int_0^t U(t,s)\bigg(&BB^*U^*(b,s)R(\varepsilon,-\Gamma_0^b)\Big(x_1 - U(b,0)x_0\\ &- \int_0^b U(b,\tau)f_n(\tau)\,d\tau\Big) + f_n(s)\bigg)\,ds. \end{aligned}\tag{9.13}$$

Assumption H(G) (iii) ensures that $\{f_n\}_{n\ge1} \subset L^2(J,H)$ is bounded, hence passing to a subsequence if necessary,

$$f_n \rightharpoonup f_* \text{ in } L^2(J,H). \tag{9.14}$$

By (9.13), (9.14), and the compactness of the operator $U(t,s)$ for $t-s>0$, we get

$$\varphi_n(t) \to U(t,0)x_0 + \int_0^t U(t,s)\bigg(BB^*U^*(b,s)R(\varepsilon,-\Gamma_0^b)\Big(x_1 - U(b,0)x_0$$

$$-\int_0^b U(b,\tau)f_*(\tau)\,d\tau\Big)+f_*(s)\Big)\,ds \text{ as } n\to\infty. \tag{9.15}$$

Since $\varphi_n \to \varphi_*$ in $C(J,H)$ and $f_n \in \mathcal{N}(x_n)$, Lemma 9.8 and (9.15) yield $f_* \in \mathcal{N}(x_*)$, so $\varphi_* \in \Gamma_\varepsilon(x_*)$, which expresses that Γ_ε has a closed graph. Finally, [155, Proposition 3.3.12(2)] entails that Γ_ε is upper semicontinuous.

Step 5. A priori estimate.

We show that the set $\Omega := \{x \in C(J,H) : x \in \lambda\Gamma_\varepsilon(x),\ 0<\lambda<1\}$ is bounded in $C(J,H)$. Given $x \in \Omega$ and $\lambda \in (0,1)$ with $x \in \lambda\Gamma_\varepsilon(x)$, there exists $f \in \mathcal{N}(x)$ such that

$$x(t) = \lambda U(t,0)x_0 + \lambda\int_0^t U(t,s)\big(Bu_\varepsilon(s)+f(s)\big)\,ds.$$

From H(G) (iii), we deduce the estimate

$$\begin{aligned}\|x(t)\|_H &\le \|U(t,0)x_0\|_H + \Big\|\int_0^t U(t,s)Bu_\varepsilon(s)\,ds\Big\|_H \\ &\quad + \Big\|\int_0^t U(t,s)f(s)\,ds\Big\|_H \\ &\le M\|x_0\|_H + M\|\phi\|_{L^2(J)}\sqrt{b} \\ &\quad + \frac{M^2\|B\|^2 b}{\varepsilon}\Big(\|x_1\|_H + M\|x_0\|_H + M\|\phi\|_{L^2(J)}\sqrt{b}\Big),\end{aligned}$$

with a constant $M>0$, which proves Step 5.

By Steps 1–4 we know that the multivalued map Γ_ε fulfills all the conditions of Theorem 9.9, whereas Step 5 decides on the alternative in the conclusion of Theorem 9.10. Therefore Γ_ε has a fixed point, which completes the proof. □

Now we are in a position to provide the main result of approximate controllability.

Theorem 9.14. *Assume that the hypotheses of Theorem 9.14 are fulfilled and there exists a function $\phi \in L^2(J)$ such that*

$$\|G(t,x)\|_H := \sup\{\|\xi\|_H : \xi \in G(t,x)\} \le \phi(t) \text{ for a.a. } t\in J \text{ and all } x \in H.$$

If the associated linear control system (9.7) is approximately controllable on J, then the system (9.1) is approximately controllable on J in the sense of Definition 9.6.

Proof. For every $\varepsilon>0$, Theorem 9.13 provides a fixed point $x^\varepsilon \in C(J,H)$ of the multivalued operator Γ_ε, so x^ε is a mild solution of (9.1) corresponding to the control u_ε assigned with Γ_ε. In other words, there exists $f^\varepsilon \in \mathcal{N}(x^\varepsilon)$ such that

$$x^\varepsilon(t) = U(t,0)x_0 + \int_0^t U(t,s)\Big(BB^*U^*(b,s)R(\varepsilon, -\Gamma_0^b)(x_1 - U(b,0)x_0 - \int_0^b U(b,\tau)f^\varepsilon(\tau)\,d\tau) + f^\varepsilon(s)\Big)\,ds. \tag{9.16}$$

Since $I - \Gamma_0^b R(\varepsilon, -\Gamma_0^b) = \varepsilon R(\varepsilon, -\Gamma_0^b)$, we get from (9.16) and the definitions of the operators Γ_0^b and $R(\varepsilon, -\Gamma_0^b)$ that

$$x^\varepsilon(b) = x_1 - \varepsilon R(\varepsilon, -\Gamma_0^b)E(f^\varepsilon),$$

with

$$E(f^\varepsilon) := x_1 - U(b,0)x_0 - \int_0^b U(b,\tau)f^\varepsilon(\tau)\,d\tau.$$

By virtue of our assumption that $\|G(t,x)\|_H \le \phi(t)$ it follows that

$$\int_0^b \|f^\varepsilon(s)\|_H\,ds \le \|\phi\|_{L^2(J)}\sqrt{b}.$$

Consequently, the net $\{f^\varepsilon\}$ is bounded in $L^2(J,H)$, thus there is a subsequence, still denoted by $\{f^\varepsilon\}$, converging weakly to f in $L^2(J,H)$.

Set

$$z := x_1 - U(b,0)x_0 - \int_0^b U(b,\tau)f(\tau)\,d\tau.$$

Due to the compactness in Lemma 9.3, it is readily seen that

$$\|E(f^\varepsilon) - z\|_H \le \int_0^b \|U(b,\tau)\big(f^\varepsilon(\tau) - f(\tau)\big)\|_H\,d\tau \to 0 \tag{9.17}$$

as $\varepsilon \to 0$. From Lemma 9.12 that can be applied because the system (9.7) is approximately controllable and on the account of (9.17), it turns out that

$$\begin{aligned}\|x^\varepsilon(b) - x_1\|_H &= \|\varepsilon R(\varepsilon, \Gamma_0^b)E(f^\varepsilon)\|_H \\ &\le \|\varepsilon R(\varepsilon, \Gamma_0^b)(z)\|_H + \|\varepsilon R(\varepsilon, \Gamma_0^b)\big(E(f^\varepsilon) - z\big)\|_H \\ &\le \|\varepsilon R(\varepsilon, \Gamma_0^b)(z)\|_H + \|E(f^\varepsilon) - z\|_H \to 0 \text{ as } \varepsilon \to 0.\end{aligned}$$

Thanks to the fact that $x_1 \in H$ is arbitrary, we conclude that the multivalued system (9.1) is approximately controllable on J. □

A relevant special case is the control hemivariational inequality (9.2). Here the mild solutions are understood in the sense of Definition 9.5. The hypotheses on the function $F : J \times H \to \mathbb{R}$ in (9.2) are as follows:

H(F) **(i)** the function $t \mapsto F(t,x)$ is measurable for all $x \in H$;
(ii) the function $x \mapsto F(t,x)$ is locally Lipschitz for a.e. $t \in J$;
(iii) there exist a function $a \in L^2(J)$ and a constant $c > 0$ such that

$$\|\partial F(t,x)\|_H := \sup\{\|f\|_H : f \in \partial F(t,x)\} \le a(t) + c\|x\|_H$$

for a.a. $t \in J$ and all $x \in H$.

Corollary 9.15. *Assume that hypotheses* (A1)–(A4) *and* H(F) *are satisfied. If the associated linear control system (9.7) is approximately controllable on J, then the control hemivariational inequality (9.2) is approximately controllable on J as described in Definition 9.6 for the particular situation of (9.2).*

Proof. Using Proposition 1.39 it is easy to see that hypotheses H(F) (i)–(iii) imply conditions H(G) for $G(t,x) = \partial F(t,x)$, so Theorem 9.14 can be applied. In view of Definitions 9.5 and 9.6, the stated conclusion is achieved. □

We end this section with an application to a heat equation. For $J = [0,b]$ and $\Omega = (0,1) \subset \mathbb{R}$, consider the following initial–boundary value problem:

$$\begin{cases} \dfrac{\partial}{\partial t}x(t,y) = \dfrac{\partial^2}{\partial y^2}x(t,y) + a(t)x(t,y) + Bu(t,y) + \varphi(t,y), & t \in J, \quad y \in \Omega, \\ x(t,0) = x(t,1) = 0, \quad t \in J, \\ x(0,y) = x_0(y), \quad y \in \Omega. \end{cases} \tag{9.18}$$

The physical interpretation is that $x(t,y)$ stands for the temperature at the point $y \in \Omega$ and time $t \in J$. On the data in (9.18) we suppose that $a : J \to (0,+\infty)$ is continuous, $x_0 \in L^2((0,1))$, $\varphi = \psi + \chi$, with

$$-\psi(t,y) \in \partial F(t, x(t,y)), \quad (t,y) \in J \times \Omega,$$

for a measurable function F provided $F(t,\cdot)$ is locally Lipschitz on $\mathbb{R}$, so its generalized gradient denoted ∂F is well defined. A simple example of a function F satisfying hypotheses H(F) which is relevant for our purposes is $F(t,\eta) = F(\eta) = \max\{h_1(\eta), h_2(\eta)\}$, where $h_i : \mathbb{R} \to \mathbb{R}$ $(i = 1,2)$ are convex quadratic functions.

In order to match the setting of Corollary 9.15, take $H = L^2((0,1))$ and, for every $t \in [0,b]$, the operator $A(t)$ defined by $A(t)x = x'' + a(t)x$ with the common domain

$$D := \{x \in H : x, x' \text{ are absolutely continuous}, \ x'' \in H, \ x(0) = x(1) = 0\}.$$

The family of operators $\{A(t) : t \in [0, b]\}$ verifies conditions (A1)–(A4) and in view of Lemma 9.2 it generates a compact evolution system $U(t, s)$ given by

$$U(t, s) = T(t - s)e^{-\int_s^t a(\tau)\, d\tau},$$

where $T(t)$ for $t > 0$ is the compact semigroup generated by the operator $Ax = x''$ for $x \in D$ (see [87]). The eigenvalues of A are $-n^2\pi^2$ with the corresponding eigenvectors $e_n(y) = \sqrt{2}\sin(n\pi y)$, $n \in \mathbb{N}$. Thus, for $x \in D$, we have

$$A(t)x = \sum_{n=1}^{\infty}(-n^2 + a(t))\langle x, e_n\rangle_H e_n,$$

while for each $x \in H$, it holds

$$U(t, s)x = U^*(t, s)x = \sum_{n=1}^{\infty} e^{-n^2\pi^2(t-s)-\int_s^t a(\tau)\, d\tau}\langle x, e_n\rangle_H e_n.$$

The control space in problem (9.18) is chosen to be the Hilbert space

$$U := \{u : u = \sum_{n=2}^{\infty} u_n e_n \text{ with } \sum_{n=2}^{\infty} u_n^2 < +\infty\},$$

endowed with the norm $\|u\|_U = (\sum_{n=2}^{\infty} u_n^2)^{1/2}$, whereas the mapping $B \in \mathcal{L}(U, H)$ in (9.18) is given by

$$Bu = 2u_2e_1 + \sum_{n=2}^{\infty} u_n e_n \quad \text{for all } u = \sum_{n=2}^{\infty} u_n e_n \in U.$$

It follows that

$$B^*u = (2u_1 + u_2)e_2 + \sum_{n=3}^{\infty} u_n e_n \quad \text{for all } u = \sum_{n=1}^{\infty} u_n e_n \in H,$$

which gives

$$B^*U^*(t, s)x = e^{-\int_s^t a(\tau)\, d\tau}\Big((2x_1e^{-\pi^2(t-s)} + x_2e^{-4\pi^2(t-s)})e_2 + \sum_{n=3}^{\infty} e^{-n^2\pi^2(t-s)}x_n e_n\Big)$$

for all $x = \sum_{n=1}^{\infty} x_n e_n \in H$. We see that, if $B^*U^*(t, s)x = 0$ for some $t \in J$, then $x = 0$. This results in the fact that the linear control system corresponding to (9.18) is approximately controllable on J (see [63, Theorem 4.1.7] and [213] for results of this type). Therefore the hypotheses of Corollary 9.15 are satisfied, so problem (9.18) is approximately controllable.

9.2 EVOLUTIONARY PROBLEMS DRIVEN BY VARIATIONAL INEQUALITIES

The problem considered in this section consists of an evolution equation driven by a variational inequality

$$\begin{cases} \dot{x}(t) = Ax(t) + f(t, x(t), u(t)), & t \in [0, T], \\ u(t) \in S(K, g(t, x(t), \cdot), \phi), & t \in [0, T], \\ x(0) = x_0, \end{cases} \tag{9.19}$$

where $S(K, g(t, x(t), \cdot), \phi)$ stands for the solution set of the variational inequality: find $u : [0, T] \to K$ such that

$$\langle g(t, x(t), u(t)), v - u(t)\rangle + \phi(v) - \phi(u(t)) \geq 0 \quad \text{for all } v \in K. \tag{9.20}$$

Here, with real Banach spaces E and E_1, K is a convex subset of E_1, $A : D(A) \subset E \to E$ is the infinitesimal generator of a C_0-semigroup e^{At} in E, $\phi : E_1 \to (-\infty, +\infty]$ is a convex lower semicontinuous $\not\equiv +\infty$ functional, and $f : [0, T] \times E \times E_1 \to E$ and $g : [0, T] \times E \times K \to E_1^*$ are given mappings.

We introduce a notion of solution for problem (9.19) in a suitable mild sense.

Definition 9.16. A pair of functions (x, u), with $x \in C([0, T], E)$ and $u : [0, T] \to K$ measurable, is said to be a mild solution of problem (9.19) if

$$x(t) = e^{At}x_0 + \int_0^t e^{A(t-s)} f(s, x(s), u(s))\, ds, \quad t \in [0, T],$$

and $u(s) \in S(K, g(s, x(s), \cdot), \phi)$, $s \in [0, T]$. If (x, u) is a mild solution of (9.19), then x is called the mild trajectory and u the variational control.

We prepare our results on problem (9.19) by providing needed information on the variational inequality (9.20). We recall a useful classical result.

Lemma 9.17 (Ky Fan [79]). *Let K be a nonempty subset of a Hausdorff topological vector space E_1 and let $G : K \to 2^{E_1}$ be a set-valued mapping with the properties:*

(a) *G is a KKM mapping, that is, for any $\{v_1, v_2, \ldots, v_n\} \subset K$, one has that its convex hull $co\{v_1, v_2, \ldots, v_n\}$ is contained in $\bigcup_{i=1}^{n} G(v_i)$;*

(b) *$G(v)$ is closed in E_1 for every $v \in K$;*

(c) *$G(v_0)$ is compact in E_1 for some $v_0 \in K$.*

Then $\bigcap_{v \in K} G(v) \neq \emptyset$.

Given a reflexive Banach space E_1, a nonempty subset $K \subseteq E_1$, a mapping $Q : K \to E_1^*$, and a function $\phi : E_1 \to \mathbb{R} \cup \{+\infty\}$ which is $\not\equiv +\infty$, we consider the following variational inequality: find $u \in K$ such that

$$\langle Q(u), v - u\rangle + \phi(v) - \phi(u) \geq 0 \quad \text{for all } v \in K. \tag{9.21}$$

Theorem 9.18. *Assume that:*

(i) $Q : K \to E_1^*$ *is monotone on* K*, that is,*

$$\langle Q(v) - Q(u), v - u\rangle \geq 0 \quad \textit{for all } u, v \in K,$$

and satisfies

$$\liminf_{\lambda \to 0^+} \langle Q(\lambda u + (1-\lambda)v), v - u\rangle \leq \langle Q(v), v - u\rangle \quad \textit{for all } u, v \in K;$$

(ii) $\phi : E_1 \to \mathbb{R} \cup \{+\infty\}$ *is convex, lower semicontinuous, and* $\not\equiv +\infty$*;*

(iii) if the set K *is unbounded in* E_1*, there exist* $u_0 \in K$ *and* $r > 0$ *such that*

$$\langle Q(v), v - u_0\rangle + \phi(v) - \phi(u_0) > 0$$

for all $v \in K$ *with* $\|v\|_{E_1} > r$.

Then $u \in K$ *is a solution of (9.21) if and only if it fulfills*

$$\langle Q(v), v - u\rangle + \phi(v) - \phi(u) \geq 0 \quad \textit{for all } v \in K. \tag{9.22}$$

Moreover, the solution set of (9.21) is nonempty, convex and closed in E_1.

Proof. Clearly, if $u \in K$ is a solution of (9.21), it is also a solution of (9.22) by the monotonicity of Q in assumption (i).

Conversely, assume that $u \in K$ is a solution of (9.22). Using the convexity of the set K, one has $u_\lambda := \lambda v + (1-\lambda)u \in K$ for all $\lambda \in (0, 1)$ and all $v \in K$, so

$$\langle Q(u_\lambda), u_\lambda - u\rangle + \phi(u_\lambda) - \phi(u) \geq 0.$$

Then by assumption (ii), there holds

$$\langle Q(u_\lambda), v - u\rangle + \phi(v) - \phi(u) \geq 0. \tag{9.23}$$

Letting $\lambda \to 0$ in (9.22) and making use of hypotheses (i) and (ii) imply that $u \in K$ solves problem (9.21), proving the required equivalence.

In order to establish the properties of the solution set of (9.21), we first suppose that the set K is bounded in E_1. We introduce the multivalued map $G : K \to 2^K$ by

$$G(v) := \left\{u \in K : \langle Q(v), v - u\rangle + \phi(v) - \phi(u) \geq 0\right\} \text{ for all } v \in K.$$

We notice that $G(v)$ is convex and $v \in G(v)$ whenever $v \in K$. The set $G(v)$ is weakly closed in E_1 for all $v \in K$. Indeed, taking a sequence $\{u_n\} \subset G(v)$ with $u_n \rightharpoonup u$ in E_1, we have

$$\langle Q(v), v - u_n\rangle + \phi(v) - \phi(u_n) \geq 0. \tag{9.24}$$

Letting $n \to \infty$ in (9.24), on the basis of assumption (ii), results in

$$\langle Q(v), v-u\rangle + \phi(v) - \phi(u) \geq 0,$$

which means that $u \in G(v)$.

Now we show that G is a KKM mapping. Arguing by contradiction, assume that there exist a finite subset $\{v_1, v_2, \ldots, v_n\} \subset K$ and $u_0 = \sum_{i=1}^{n} \lambda_i v_i$ with $\lambda_i \in [0,1]$ and $\sum_{i=1}^{n} \lambda_i = 1$ satisfying $u_0 \notin \bigcup_{i=1}^{n} G(v_i)$, or equivalently

$$\langle Q(v_i), v_i - u_0\rangle + \phi(v_i) - \phi(u_0) < 0 \text{ for all } i \in \{1,2,\ldots,n\}. \tag{9.25}$$

By hypothesis (i), for each $i \in \{1,\ldots,n\}$ there holds

$$\langle Q(u_0) - Q(v_i), v_i - u_0\rangle \leq 0. \tag{9.26}$$

Then (9.25) and (9.26) lead to

$$\langle Q(u_0), v_i - u_0\rangle + \phi(v_i) - \phi(u_0) < 0 \text{ for all } i \in \{1,\ldots,n\},$$

from which the following contradiction arises:

$$\begin{aligned} 0 &= \langle Q(u_0), u_0 - u_0\rangle + \phi(u_0) - \phi(u_0) \\ &\leq \sum_{i=1}^{n} \lambda_i \big(\langle Q(u_0), v_i - u_0\rangle + \phi(v_i) - \phi(u_0)\big) < 0. \end{aligned}$$

Therefore G is a KKM mapping.

Since K is bounded, closed, and convex in the reflexive Banach space E_1, it follows that $G(v)$ is weakly compact in E_1 for each $v \in K$. We are in a position to apply Lemma 9.17 ensuring that $\bigcap_{v \in K} G(v) \neq \emptyset$. Hence the solution set of problem (9.21) is nonempty.

Next we suppose that the set K is unbounded in E_1. For every integer $n \geq 1$, using the element $u_0 \in K$ given in assumption (iii) we set $K_n := \{x \in K : \|x - u_0\|_{E_1} \leq n\}$, which is a bounded, closed and convex subset of E_1. According to the previous case, we can find $u_n \in K_n$ such that

$$\langle Q(u_n), v - u_n\rangle + \phi(v) - \phi(u_n) \geq 0 \quad \text{for all } v \in K_n. \tag{9.27}$$

We claim that there exists an integer $k \geq 1$ for which one has

$$\|u_k - u_0\|_{E_1} < k. \tag{9.28}$$

Arguing by contradiction, assume that $\|u_n - u_0\|_{E_1} = n$ for every $n \geq 1$. Inserting $v = u_0$ in (9.27), we observe that

$$\langle Q(u_n), u_n - u_0\rangle + \phi(u_n) - \phi(u_0) \leq 0,$$

which contradicts assumption (iii) provided n is sufficiently large. Hence (9.28) holds.

Let $x \in K$. By (9.28), for sufficiently small $t > 0$ we have that $\|u_k + t(x - u_k) - x_0\|_{E_1} < k$. This enables us to set $v = u_k + t(x - u_k)$ in (9.27) for $n = k$, which through the convexity of Φ yields

$$\langle Q(u_k), x - u_k \rangle + \phi(x) - \phi(u_k) \geq 0.$$

Consequently, $u = u_k$ is a solution of problem (9.21).

The equivalence between problems (9.21) and (9.22), in conjunction with assumption (ii), shows that the solution set of (9.21) is closed and convex in E_1, which completes the proof. □

Corollary 9.19. *Let K be a nonempty, compact, and convex subset of a real Banach space E_1. Assume that $Q : K \to E_1^*$ and $\phi : E_1 \to \mathbb{R}$ verify conditions (i), (ii) in Theorem 9.18. Then the conclusion of Theorem 9.18 is valid.*

Before continuing, some prerequisites are necessary concerning the measurability of multivalued mappings.

Definition 9.20. Let E and E_1 be Banach spaces and let an interval $I \subset \mathbb{R}$. We say that $F : I \times E \to 2^{E_1}$ with nonempty values is superpositionally measurable if, for every measurable multivalued mapping $Q : I \to 2^E$ with nonempty compact values, the composition $\Phi : I \to 2^{E_1}$ given by $\Phi(t) = F(t, Q(t))$ is measurable.

Theorem 9.21 (see [238]). *If $F : I \times E \to 2^{E_1}$ with nonempty compact values satisfies the Carathéodory condition or F is upper or lower semicontinuous, then F is superpositionally measurable in the sense of Definition 9.20.*

Theorem 9.22 ([115, Theorem 1.3.5]). *Let E and E_1 be Banach spaces. Assume that, for some $0 < T < +\infty$, the multivalued mapping $G : [0, T] \times E \to 2^{E_1}$ with nonempty compact values satisfies:*

(a) *for every $x \in E$, $G(\cdot, x) : [0, T] \to 2^{E_1}$ has a strongly measurable selection;*

(b) *for a.e. $t \in [0, T]$, $G(t, \cdot) : E \to 2^{E_1}$ is upper semicontinuous.*

Then, for every strongly measurable function $q : [0, T] \to E$, there exists a strongly measurable selection $g : [0, T] \to E_1$ of the composition $M : [0, T] \to 2^{E_1}$, $M(t) = G(t, q(t))$ for a.e. $t \in [0, T]$.

We focus on certain properties of the solution set for a class of variational inequalities related to problem (9.19).

Theorem 9.23. *Assume that K is a nonempty, compact, and convex subset of a real Banach space E_1. Assume that E is a real separable Banach space and, for some $T > 0$, the mapping $g : [0, T] \times E \times K \to E_1^*$ is such that $g(\cdot, \cdot, u)$ is* *continuous from* $[0, T] \times E$ *to* E_1^* *endowed with the weak* topology whenever* $u \in K$. *In addition, we assume that for every* $(t, x) \in [0, T] \times E$ *the mappings* $Q := g(t, x, \cdot)$ *and* $\phi : E_1 \to \mathbb{R} \cup \{+\infty\}$ *satisfy hypotheses* (i), (ii) *in Theorem 9.18. Then the multivalued mapping* $U : [0, T] \times E \to 2^K$ *defined by*

$$U(t, x) := \{u \in K : \langle g(t, x, u), v - u\rangle + \phi(v) - \phi(u) \geq 0 \quad \text{for all } v \in K\} \tag{9.29}$$

for all $(t, x) \in [0, T] \times E$ *has nonempty values and fulfills:*

(U_1) *U is upper semicontinuous;*
(U_2) *U is superpositionally measurable.*

Proof. Corollary 9.19 guarantees that for every $t \in [0, T]$ and $x \in E$ the set $U(t, x)$ is nonempty, so the multivalued mapping $U : [0, T] \times E \to 2^K$ has nonempty values. In order to prove assertion (U_1), we need to show that for each closed subset $C \subset K$, the set $U^-(C) := \{(t, x) \in [0, T] \times E : U(t, x) \cap C \neq \emptyset\}$ is closed in $\mathbb{R} \times E$. To this end, let a sequence $\{(t_n, x_n)\} \subset U^-(C)$ satisfy $(t_n, x_n) \to (t, x)$ in $\mathbb{R} \times E$. Thus we can choose $u_n \in U(t_n, x_n) \cap C$, and (9.29) and Corollary 9.19 render

$$\langle g(t_n, x_n, v), v - u_n\rangle + \phi(v) - \phi(u_n) \geq 0 \text{ for all } v \in K. \tag{9.30}$$

By the compactness of K, passing to a relabeled subsequence, we may assume that $u_n \to u$ in E_1 for some $u \in C$. Letting $n \to \infty$ in (9.30), from our assumptions we obtain

$$\langle g(t, x, v), v - u\rangle + \phi(v) - \phi(u) \geq 0 \quad \text{for all } v \in K.$$

This reads as $u \in U(t, x) \cap C$, whence $(t, x) \in U^-(C)$, thereby (U_1).

Finally, assertion (U_1) allows us to apply Theorem 9.21, which ensures that (U_2) holds true. This completes the proof. □

In order to study problem (9.19) with the set of constraints $K \subset E_1$ we impose the following assumptions on the mapping $f : [0, T] \times E \times E_1 \to E$ in (9.19):

H(f) **(i)** for every $(t, x) \in [0, T] \times E$ and every convex set $D \subset K$, the set $f(t, x, D)$ is convex in E;
(ii) there exists $\psi \in L^1([0, T])$ such that

$$\|f(t, x, u)\|_E \leq \psi(t)(1 + \|x\|_E) \text{ for all } (t, x, u) \in [0, T] \times E \times K;$$

(iii) $f(\cdot, x, u) : [0, T] \to E$ is measurable for every $(x, u) \in E \times E_1$;
(iv) $f(t, \cdot, \cdot) : E \times E_1 \to E$ is continuous for a.e. $t \in [0, T]$;
(v) there exists $k \in L^1([0, T])$ such that, for a.e. $t \in [0, T]$, every $x_0, x_1 \in E$ and $u \in K$,

$$\|f(t, x_0, u) - f(t, x_1, u)\|_E \leq k(t)\|x_0 - x_1\|_E.$$

Remark 9.24. A special case of condition H(f) (i) is when $f(t,x,\cdot): K \to E_1$ is affine for all $(t,x) \in [0,T] \times E$. In Pang–Stewart [196] and Liu–Loi–Obukhovskii [136] a function f of the form $f(t,x,u) = g(t,x) + B(t,x)u$ is utilized.

Next we briefly discuss the important notion of measure of noncompactness that is needed in the sequel.

Definition 9.25. Let E be a Banach space and let $(\mathfrak{A}, \leq)$ be a partially ordered set. A mapping $\beta : 2^E \to \mathfrak{A}$ is called a measure of noncompactness (MNC, for short) in E if $\beta(\overline{co}\Omega) = \beta(\Omega)$ for every $\Omega \in 2^E$.

An important example of MNC in the sense of Definition 9.25 is the Hausdorff MNC χ defined by diameters of sets:

$$\chi(\Omega) := \inf\big\{\epsilon > 0 : \exists \Omega_i \in 2^E \text{ with } \operatorname{diam}(\Omega_i) \leq \epsilon,\ i = 1,\dots,n, \text{ and } \Omega \subseteq \bigcup_{i=1}^{n} \Omega_i\big\}.$$

We also use the monotone nonsingular MNC in the space $C([0,T],E)$ introduced in [115, Example 2.1.4]. Fixing a constant $L > 0$, for each nonempty bounded set $\Omega \subset C([0,T],E)$ it is equal to

$$\nu(\Omega) := \max_{\omega \in \Delta(\Omega)} \{\gamma(\omega), \operatorname{mod}_C(\omega)\}, \tag{9.31}$$

where $\Delta(\Omega)$ denotes the collection of all countable subsets of Ω, while

$$\gamma(\omega) := \sup_{t \in [0,T]} e^{-Lt} \chi(\omega(t)), \quad \operatorname{mod}_C(\omega) := \limsup_{\delta \to 0} \max_{x \in \omega} \max_{|t_1 - t_2| \leq \delta} \|x(t_1) - x(t_2)\|_E.$$

In (9.31) the maximum is viewed in the sense of the order on $\mathfrak{A} = \mathbb{R}^2$.

Let β_1 and β_2 be real MNCs (that is, $\mathfrak{A} = [0, +\infty]$) in E_1 and E_2, respectively. For $k \geq 0$, the multivalued map $F : X \subset E_1 \to K(E_2)$ is said to be (k, β_1, β_2)-bounded if

$$\beta_2(F(\Omega)) \leq k\beta_1(\Omega) \text{ for all } \Omega \subseteq X.$$

The next criterion of (k, β_1, β_2)-boundedness is quoted from [115, Proposition 2.2.2].

Theorem 9.26. *Let χ_1 and χ_2 be the Hausdorff MNCs in E_1 and E_2, respectively. Suppose that $X \subset E_1$ and $F : X \times E_1 \to 2^{E_2}$ with nonempty compact values satisfy the conditions:*

(i) *there exists $k \in \mathbb{R}_+$ such that for every $x \in X$, $F(x,\cdot) : E_1 \to K(E_2)$ is k-Lipschitz with respect to the Hausdorff metric H, that is,*

$$H(F(x,y_1), F(x,y_2)) \leq k\|y_1 - y_2\| \text{ for all } y_1, y_2 \in E_1;$$

(ii) *the set* $F(\Omega \times \{y\})$ *is relatively compact in* E_2 *for every bounded subset* $\Omega \subset X$ *and* $y \in E_1$.

Then $G : X \to K(E_2)$ *defined as* $G(x) = F(x, x)$ *is* (k, χ_1, χ_2)*-bounded.*

The following definition will be useful in the forthcoming multivalued fixed point approach.

Definition 9.27. Let X be a closed subset of a Banach space E and let β be a real MNC in E. A multivalued map $F : X \to 2^E$ with nonempty compact values is said to be β-condensing if there is $k \in [0, 1)$ such that $\beta(F(\Omega)) \leq k\beta(\Omega)$ for every $\Omega \subset X$.

Relevant properties of the multivalued mapping $F : [0, T] \times E \to 2^E$ given by

$$F(t, x) = f(t, x, U(t, x)),$$

with U introduced in (9.29), are listed below.

Lemma 9.28. *Let* E *and* E_1 *be real Banach spaces, with* E *separable, and let* K *be a nonempty compact and convex subset of* E_1. *Assume that the hypotheses of Theorem 9.23, as well as conditions* H(f)*, are satisfied. Then:*

(i) *for all* $(t, x) \in [0, T] \times E$, $F(t, x)$ *is nonempty, compact, and convex in* E;
(ii) *for every* $x \in E$, $F(\cdot, x)$ *has a strongly measurable selection;*
(iii) *for a.e.* $t \in [0, T]$, $F(t, \cdot)$ *is upper semicontinuous;*
(iv) *for every bounded subset* $D \subset E$, *there exists* $l \in L^1([0, T])$ *such that*

$$\chi(F(t, D)) \leq l(t)\chi(D) \quad \text{for a.e. } t \in [0, T],$$

where χ *is the Hausdorff MNC in* E.

Proof. (i) Corollary 9.19 and Theorem 9.23 ensure that the multivalued mapping U takes nonempty, compact and convex values in K. Then hypotheses H(f) (i), (iv) imply the validity of assertion (i).

(ii) By virtue of H(f) (iii)–(iv), we have that $f(\cdot, x, \cdot) : [0, T] \times E_1 \to E$ satisfies the Carathéodory condition. Notice that $U(t, x)$ is nonempty, compact in E_1 for all $(t, x) \in [0, T] \times E$. Invoking assertion (U_1) of Theorem 9.23, together with Theorem 9.21, we infer that $F(\cdot, x) = f(\cdot, x, U(\cdot, x))$ is measurable for every $x \in E$, so it is strongly measurable because E is separable (see Definition 1.1). Therefore, taking into account (i), $F(\cdot, x)$ has a strongly measurable selection whenever $x \in E$ (see [115]).

(iii) For any $t \in [0, T]$, as known from property (U_2) in Theorem 9.23, the multivalued mapping $F_1 = (\cdot, U(t, \cdot)) : E \to E \times K$ is upper semicontinuous, whereas hypothesis H(f) (iv) guarantees that the mapping $F_2 = f(t, \cdot, \cdot) : E \times E_1 \to E$ is continuous. Then, since $F(t, \cdot) = F_2 \circ F_1$, we can infer that $F(t, \cdot)$ is upper semicontinuous.

(iv) Fixing $t \in [0, T]$, in view of Theorem 9.23, the multivalued mapping $G : E \times E \to 2^E$ given by $G(x, y) = f(t, y, U(t, x))$ for all $(x, y) \in E \times E$ has nonempty, compact, convex values. For each bounded subset $D \subset E$, $G(D, y) = f(t, y, U(t, D))$ is relatively compact in E whenever $y \in E$ due to the facts that $U(t, D) \subset K$ and K is compact. Hence the requirement (ii) in Theorem 9.26 is verified.

Let $x, y', y'' \in E$. If $z' \in G(x, y')$, there is $u \in U(t, x)$ such that $z' = f(t, y', u)$, and so $z'' = f(t, y'', u) \in G(x, y'')$. Assumption H($f$) (v) enables us to get

$$\|z' - z''\|_E = \|f(t, y', u) - f(t, y'', u)\|_E \leq k(t)\|y' - y''\|_E.$$

In this way, the requirement (i) in Theorem 9.26 is fulfilled. Consequently, we can apply Theorem 9.26, obtaining the desired conclusion with $l = k \in L^1((0, T))$. □

From parts (ii) and (iii) of Lemma 9.28 we see that the assumptions of Theorem 9.22 with F in place of G are fulfilled. Hence $F(\cdot, q(\cdot))$ has strongly measurable selections for every $q \in C([0, T], E)$. So, it is meaningful to define the set-valued mapping $P_F : C([0, T], E) \to 2^{L^1([0,T],E)}$ by

$$P_F(q) := \{g : g \text{ is strongly measurable and } g(t) \in F(t, q(t)) \text{ for a.e. } t \in [0, T]\}$$

for all $q \in C([0, T], E)$. In view of part of Lemma 9.28 (i), we can introduce the multivalued map $\Gamma : C([0, T], E) \to 2^{C([0,T],E)}$ with nonempty, closed, convex values by

$$\Gamma x := \left\{ y \in C([0, T], E) : y(t) = e^{At} x_0 + \int_0^t e^{A(t-s)} h(s)\, ds,\ h \in P_F(x) \right\}. \tag{9.32}$$

Here $A : D(A) \subset E \to E$ is the infinitesimal generator of the C_0-semigroup e^{At} in E as given in problem (9.19).

Theorem 9.29. *Under the hypotheses of Lemma 9.28, the multivalued map Γ in (9.32) is upper semicontinuous and ν-condensing in the sense of Definition 9.27 on every closed bounded subset of $C([0, T], E)$, with ν constructed in (9.31).*

Proof. The conclusion is achieved by combining Lemma 4.2.1, Corollary 5.1.2, Theorem 5.1.3, and Theorem 5.1.4 of [115] with Lemma 9.28. □

The proof of our main result relies on the following fixed point theorems that we cite from [115].

Theorem 9.30. *If S is a nonempty, closed, convex subset of a Banach space E and $F : S \to 2^S$ is a closed, β-condensing multivalued map in the sense of Definition 9.27, with nonempty compact convex values and a nonsingular MNC β in E, then the set of fixed points of F is nonempty.*

Theorem 9.31. *Let C be a nonempty closed subset of a Banach space E and let $F : C \to 2^E$ be a closed multivalued mapping, with nonempty compact values, which is β-condensing on every bounded subset of C, with a monotone MNC β in E. If the set of fixed points of F is bounded, then it is compact in E.*

We are now able to present our main result regarding problem (9.19).

Theorem 9.32. *Under the hypotheses of Lemma 9.28, the solution set of problem (9.19) in the sense of Definition 9.16 is nonempty and the set of all corresponding mild trajectories of (9.19) is compact in $C([0, T], E)$.*

Proof. Consider the evolutionary differential inclusion associated to problem (9.19):

$$\begin{cases} \dot{x}(t) \in Ax(t) + F(t, x(t)), & t \in [0, T], \\ x(0) = x_0, \end{cases} \tag{9.33}$$

where, as before, $F(t, x) = f(t, x, U(t, x))$ with $U(t, x)$ defined in (9.29).

Step 1. The solution set of (9.33) is nonempty.

By (9.32) it is equivalent to show that the set of fixed points of Γ is nonempty.

Fix a positive constant L sufficiently large such that

$$M \int_0^t e^{-L(t-s)} \psi(s) ds < 1 \quad \text{for all } t \in [0, T], \tag{9.34}$$

with $\psi \in L^1([0, T])$ in assumption H(f) (ii) and $M := \max_{t\in[0,T]} \|e^{At}\|$. According to (9.34), we may choose $r > 0$ such that

$$M\big(\|x_0\|_E + \|\psi\|_{L^1([0,T])}\big) + Mr \int_0^t e^{-L(t-s)} \psi(s) ds \le r \quad \text{for all } t \in [0, T]. \tag{9.35}$$

For the equivalent norm on the space $C([0, T], E)$ given by

$$\|x\|_* := \max_{t\in[0,T]} e^{-Lt} \|x(t)\|_E,$$

we consider the closed ball

$$\overline{B_r}(0) := \big\{x \in C([0, T], E) : \|x\|_* \le r\big\}.$$

We prove that Γ maps $\overline{B_r}(0)$ into $\overline{B_r}(0)$. Indeed, let $y \in \Gamma x$ with $x \in \overline{B_r}(0)$. From (9.32), there is $h \in P_F(x)$ such that

$$y(t) = e^{At}x_0 + \int_0^t e^{A(t-s)}h(s)ds$$

whenever $t \in [0, T]$. Then H(f) (ii) yields the estimate

$$\begin{aligned} e^{-Lt}\|y(t)\|_E &= e^{-Lt}\| e^{At}x_0 + \int_0^t e^{A(t-s)}h(s)ds \|_E \\ &\le M(\|x_0\|_E + \|\psi\|_{L^1([0,T])}) + M\|x\|_* \int_0^t e^{-L(t-s)}\psi(s)ds. \end{aligned}$$

Taking into account that $x \in \overline{B_r}(0)$ and (9.35), we find

$$e^{-Lt}\|y(t)\|_E \le M(\|x_0\|_E + \|\psi\|_{L^1([0,T])}) + Mr\int_0^t e^{-L(t-s)}\psi(s)ds \le r,$$

which means that $\|y\|_* \le r$.

Through Theorem 9.29, we apply Theorem 9.30 with $S = \overline{B_r}(0)$ and $F = \Gamma$ obtaining that the set of fixed points of Γ is nonempty, so Step 1 holds true.

Step 2. The solution set of (9.33) is compact in $C([0, T], E)$.

If $x \in C([0, T], E)$ is solution of (9.33), then with certain $h \in P_F(x)$, from the growth condition H(f) (ii) we have for any $t \in [0, T]$ that

$$\begin{aligned} \|x(t)\|_E &\le \|e^{At}\| \, \|x_0\|_E + \int_0^t \|e^{A(t-s)}\| \, \|h(s)\|_E ds, \\ &\le M\left(\|x_0\|_E + \|\psi\|_{L^1([0,T])} + \int_0^t \psi(s)\|x(s)\|_E ds \right) \end{aligned}$$

with $M > 0$ introduced in Step 1. Through Gronwall's inequality, we get the estimate

$$\|x(t)\|_E \le M(\|x_0\|_E + \|\psi\|_{L^1([0,T])})e^{M\|\psi\|_{L^1([0,T])}}$$

showing that the set of fixed points of Γ is bounded in $C([0, T], E)$. Due to Theorem 9.29, we can invoke Theorem 9.31 with $F = \Gamma$ in its statement. We deduce that the solution set of problem (9.33), which coincides with the set of fixed points of Γ, is compact in $C([0, T], E)$. Consequently, the set of all mild trajectories of problem (9.19) is compact in $C([0, T], E)$.

Step 3. The solution set of (9.19) is nonempty.

Assertion (U_2) of Theorem 9.23 guarantees that multivalued map $U : [0, T] \times E \to 2^K$ is superpositionally measurable as formulated in Definition 9.20. This allows us to apply Filippov implicit function lemma (see

[85,115]) to conclude that for every trajectory $x \in C([0,T],E)$ of (9.33) there exists a measurable selection $u(t) \in U(t,x(t))$ such that $\dot{x}(t) = Ax(t) + f(t,x(t),u(t))$ for $t \in [0,T]$, which ensures that (x,u) is a mild solution of problem (9.19) in the sense of Definition 9.25. The proof is complete. □

9.3 NOTES

Section 9.1 focuses on the approximate controllability for control systems and inclusions, which is a property frequently studied (see [24,63,137,142,213]). Our basic reference is Liu–Li–Motreanu [135] from which the main results are extracted. Specifically, we investigate the approximate controllability of the control inclusion problem (9.1), which contains as a particular case the control hemivariational inequality (9.2).

Hemivariational inequalities have been recognized as an interesting subject in mathematics, constituting also one of the best tools to describe the mechanical problems with nonsmooth and nonconvex energy potentials. For existence of solutions for hemivariational inequalities under different hypotheses, we refer to [29,37,39,58,133,167,174,177,192,194,199,208]. Motivated by the wide applications in engineering, physics, and finance, some works focused on the optimal control problems for hemivariational inequalities as [110,153,197].

A powerful method to study the controllability properties of nonlinear systems is to use the fixed point theory (see, e.g., [24,137,142,213]). The first paper in this direction for hemivariational inequalities is Liu–Li [134]. For investigating the control problems (9.1) and (9.2) we also develop a fixed point approach.

The first main result of Section 9.1 is stated as Theorem 9.10, providing the existence of solutions in a mild sense of the control evolutionary inclusion problem (9.1), which is more general than the inequality problem (9.2). Next, to (9.1) and (9.2) we associate a linear control system (9.7). The main controllability result presented in Theorem 9.13 establishes that the approximate controllability of the associated linear system (9.7) implies the approximate controllability of the control evolutionary inclusion problem (9.1), and in particular of the control hemivariational inequality (9.2). It extends the criterion of approximate controllability in [213, Theorem 4.5] and [134, Theorem 4.4] from the time-invariant case to a control inclusion problem with time-depending operators $A(t)$. Explicit expressions for the involved approximate controls and corresponding states are indicated. Theorem 9.14 sets forth an application of the approximate controllability result to a heat equation.

The object of Section 9.2 is a nonstandard problem stated in (9.19) which consists in mixing an evolutionary differential equation with a variational inequality depending mutually on their solutions. The exposition of this section is carried based on the results of Liu–Zeng–Motreanu [139].

Aspects related to problem (9.19) have been examined until now only in a finite-dimensional context or with $A = 0$ and other particularities (see [108,128, 136,196]). In those works an ordinary differential equation is parameterized by a variable required to solve a variational inequality in the state variable of the differential equation. Such problems taking into account simultaneously both dynamics and constraints in the form of inequalities arise in electrical circuits with ideal diodes, Coulomb friction problems for contacting bodies, economical dynamics, dynamic traffic networks, control systems (see [52,108,128,135,136,195,196]). Mixing partial differential equations and inequality problems with constraints is an open subject.

The crucial point in problem (9.19) is that it is described by a partial differential equation instead of an ordinary differential equation. Hence we can take, for example, the operator A to be the Laplacian (see (1.34)), subject to an infinite-dimensional variational inequality (9.20). This incorporates large classes of problems and models. In order to investigate problem (9.19), we first prove Theorem 9.18, which includes many important situations and extends the results in Costea–Rădulescu [57] and Liu–Zeng [138]. Assertion (9.22) is an extension of Minty's technique, whereas assumption (iii) in Theorem 9.18 expresses a generalized coercivity condition. Related results can be found in [118].

Theorem 9.23 shows essential properties of measurability and upper semicontinuity for the solution set of the general variational inequality associated to (9.19). Then, based on these properties, the main result of Section 9.2 is established in Theorem 9.32, stating that the solution set of problem (9.19) is nonempty and compact. Various analytical and topological tools such as KKM theorem, monotonicity, semigroup theory, measure of noncompactness, Filippov implicit function lemma, fixed points for set-valued mappings are used in the proof.

Bibliography

[1] Adams R. Sobolev spaces. Pure Appl Math. New York–London: Academic Press; 1975.

[2] Aizicovici S, Papageorgiou NS, Staicu V. The spectrum and an index formula for the Neumann p-Laplacian and multiple solutions for problems with a crossing nonlinearity. Discrete Contin Dyn Syst Ser A 2009;25(2):431–56.

[3] Alber YI, Notik AI. Perturbed unstable variational inequalities with unbounded operators on approximately given sets. Set-Valued Anal 1993;1(4):393–402.

[4] Alves CO, Carrião PC, Faria LFO. Existence of solutions to singular elliptic equations with convection terms via the Galerkin method. Electron J Differential Equations 2010;(12). 12 pp.

[5] Alves CO, Corrêa FJSA. On the existence of positive solution for a class of singular systems involving quasilinear operators. Appl Math Comput 2007;185(1):727–36.

[6] Alves CO, Corrêa FJSA, Gonçalves JVA. Existence of solutions for some classes of singular Hamiltonian systems. Adv Nonlinear Stud 2005;5(2):265–78.

[7] Alves CO, Moussaoui A. Existence of solutions for a class of singular elliptic systems with convection term. Asymptot Anal 2014;90(3–4):237–48.

[8] Amann H, Crandall MG. On some existence theorems for semi-linear elliptic equations. Indiana Univ Math J 1978;27(5):779–90.

[9] Ambrosetti A, Brezis H, Cerami G. Combined effects of concave and convex nonlinearities in some elliptic problems. J Funct Anal 1994;122(2):519–43.

[10] Ambrosetti A, García Azorero J, Peral I. Multiplicity results for some nonlinear elliptic equations. J Funct Anal 1996;137(1):219–42.

[11] Ambrosetti A, Lupo D. On a class of nonlinear Dirichlet problems with multiple solutions. Nonlinear Anal 1984;8(10):1145–50.

[12] Ambrosetti A, Mancini G. Sharp nonuniqueness results for some nonlinear problems. Nonlinear Anal 1979;3(5):635–45.

[13] Ambrosetti A, Prodi G. A primer of nonlinear analysis. Cambridge Stud Adv Math, vol. 34. Cambridge: Cambridge University Press; 1993.

[14] Anane A. Simplicité et isolation de la première valeur propre du p-laplacien avec poids. C R Acad Sci Paris Sér I Math 1987;305(16):725–8.

[15] Arias M, Campos J. Radial Fučik spectrum of the Laplace operator. J Math Anal Appl 1995;190(3):654–66.

[16] Arias M, Campos J, Gossez J-P. On the antimaximum principle and the Fučik spectrum for the Neumann p-Laplacian. Differential Integral Equations 2000;13(1–3):217–26.

[17] Arias M, Campos J, Cuesta M, Gossez J-P. An asymmetric Neumann problem with weights. Ann Inst H Poincaré Anal Non Linéaire 2008;25(2):267–80.

[18] Averna D, Marano SA, Motreanu D. Multiple solutions for a Dirichlet problem with p-Laplacian and set-valued nonlinearity. Bull Aust Math Soc 2008;77(2):285–303.

[19] Averna D, Motreanu D, Tornatore E. Existence and asymptotic properties for quasilinear elliptic equations with gradient dependence. Appl Math Lett 2016;61:102–7.

[20] Baiocchi C, Capelo A. Variational and quasivariational inequalities. Applications to free boundary problems. New York: John Wiley & Sons, Inc.; 1984.

[21] Barbu V. Nonlinear semigroups and differential equations in Banach spaces. Leiden: Noordhoff International Publishing; 1976.
[22] Barletta G, Candito P, Motreanu D. Constant sign and sign-changing solutions for quasilinear elliptic equations with Neumann boundary condition. J Convex Anal 2014;21(1):53–66.
[23] Barletta G, Papageorgiou NS. A multiplicity theorem for the Neumann p-Laplacian with an asymmetric nonsmooth potential. J Global Optim 2007;39(3):365–92.
[24] Bashirov AE, Mahmudov NI. On concepts of controllability for deterministic and stochastic systems. SIAM J Control Optim 1999;37(6):1808–21.
[25] Benouhiba N, Belyacine Z. On the solutions of the (p, q)-Laplacian problem at resonance. Nonlinear Anal 2013;77:74–81.
[26] Bensoussan A, Lions J-L. Nouvelles méthodes en contrôle impulsionnel. Appl Math Optim 1974/1975;1(4):289–312.
[27] Boccardo L, Gallouët T, Orsina L. Existence and nonexistence of solutions for some nonlinear elliptic equations. J Anal Math 1997;73:203–23.
[28] Bonanno G, Candito P. Three solutions to a Neumann problem for elliptic equations involving the p-Laplacian. Arch Math (Basel) 2003;80(4):424–9.
[29] Bonanno G, Motreanu D, Winkert P. Boundary value problems with nonsmooth potential, constraints and parameters. Dynam Systems Appl 2013;22(2–3):385–96.
[30] Brezis H. Opérateurs maximaux monotones et semi-groupes de contractions dans les espaces de Hilbert. North-Holl Math Stud, vol. 5. Amsterdam–London: North-Holland Publishing Co.; 1973.
[31] Brezis H. Functional analysis, Sobolev spaces and partial differential equations. Universitext. New York: Springer; 2011.
[32] Brezis H, Cazenave T. Course notes.
[33] Bueno H, Ercole G. A quasilinear problem with fast growing gradient. Appl Math Lett 2013;26(4):520–3.
[34] Các NP. On nontrivial solutions of a Dirichlet problem whose jumping nonlinearity crosses a multiple eigenvalue. J Differential Equations 1989;80(2):379–404.
[35] Caldiroli P, Montecchiari P. Homoclinic orbits for second order Hamiltonian systems with potential changing sign. Comm Appl Nonlinear Anal 1994;1(1):97–129.
[36] Carl S, Heikkilä S. Fixed point theory in ordered sets and applications. From differential and integral equations to game theory. New York: Springer; 2011.
[37] Carl S, Le VK. Multi-valued parabolic variational inequalities and related variational–hemivariational inequalities. Adv Nonlinear Stud 2014;14(3):631–59.
[38] Carl S, Le VK, Motreanu D. Nonsmooth variational problems and their inequalities. Comparison principles and applications. Springer Monogr Math. New York: Springer; 2007.
[39] Carl S, Motreanu D. Extremal solutions of quasilinear parabolic inclusions with generalized Clarke's gradient. J Differential Equations 2003;191(1):206–33.
[40] Carl S, Motreanu D. Constant-sign and sign-changing solutions for nonlinear eigenvalue problems. Nonlinear Anal 2008;68(9):2668–76.
[41] Carl S, Motreanu D. General comparison principle for quasilinear elliptic inclusions. Nonlinear Anal 2009;70(2):1105–12.
[42] Carl S, Motreanu D. Multiple and sign-changing solutions for the multivalued p-Laplacian equation. Math Nachr 2010;283(7):965–81.
[43] Carl S, Motreanu D. Constant-sign and sign-changing solutions of a nonlinear eigenvalue problem involving the p-Laplacian. Differential Integral Equations 2007;20(3):309–24.
[44] Carl S, Motreanu D. Sign-changing and extremal constant-sign solutions of nonlinear elliptic problems with supercritical nonlinearities. Comm Appl Nonlinear Anal 2007;14(4):85–100.
[45] Carl S, Motreanu D. Sign-changing solutions for nonlinear elliptic problems depending on parameters. Int J Differ Equ 2010:536236.
[46] Carl S, Motreanu D. Multiple solutions of nonlinear elliptic variational problems. In: Nonlinear analysis research trends. New York: Nova Sci. Publ.; 2009. p. 235–55.

[47] Carl S, Motreanu D. Multiple solutions for elliptic systems via trapping regions and related nonsmooth potentials. Appl Anal 2015;94(8):1594–613.
[48] Carl S, Motreanu D. Extremal solutions for nonvariational quasilinear elliptic systems via expanding trapping regions. Monatsh Math 2017;182(4):801–21.
[49] Carl S, Perera K. Sign-changing and multiple solutions for the p-Laplacian. Abstr Appl Anal 2002;7(12):613–25.
[50] Carriao PC, Miyagaki OH. Existence of homoclinic solutions for a class of time-dependent Hamiltonian systems. J Math Anal Appl 1999;230(1):157–72.
[51] Chang KC. Variational methods for nondifferentiable functionals and their applications to partial differential equations. J Math Anal Appl 1981;80(1):102–29.
[52] Chen X, Wang Z. Convergence of regularized time-stepping methods for differential variational inequalities. SIAM J Optim 2013;23(3):1647–71.
[53] Cianchi A, Maz'ya V. Global gradient estimates in elliptic problems under minimal data and domain regularity. Commun Pure Appl Anal 2015;14(1):285–311.
[54] Cingolani S, Degiovanni M. Nontrivial solutions for p-Laplace equations with right-hand side having p-linear growth at infinity. Comm Partial Differential Equations 2015;30(7–9):1191–203.
[55] Clarke FH. Optimization and nonsmooth analysis. New York: John Wiley & Sons Inc.; 1983.
[56] Costea N, Csirik M, Varga C. Linking-type results in nonsmooth critical point theory and applications. Set-Valued Var Anal 2017;25(2):333–56.
[57] Costea N, Rădulescu V. Inequality problems of quasi-hemivariational type involving set-valued operators and a nonlinear term. J Global Optim 2017;52(4):743–56.
[58] Costea N, Varga C. Systems of nonlinear hemivariational inequalities and applications. Topol Methods Nonlinear Anal 2013;41(1):39–65.
[59] Coti Zelati V, Rabinowitz PH. Homoclinic orbits for second order Hamiltonian systems possessing superquadratic potentials. J Amer Math Soc 1991;4(4):693–727.
[60] Cuesta M, de Figueiredo D, Gossez JP. The beginning of the Fučik spectrum for the p-Laplacian. J Differential Equations 1999;159(1):212–38.
[61] Cuesta M, Takáč P. Nonlinear eigenvalue problems for degenerate elliptic systems. Differential Integral Equations 2010;23(11–12):1117–38.
[62] Curtain RF, Pritchard AJ. Infinite dimensional linear systems theory. Lecture Notes in Control and Inform Sci, vol. 8. New York: Springer-Verlag; 1978.
[63] Curtain RF, Zwart H. An introduction to infinite dimensional linear systems theory. Texts Appl Math, vol. 21. New York: Springer-Verlag; 1995.
[64] Damascelli L. Comparison theorems for some quasilinear degenerate elliptic operators and applications to symmetry and monotonicity results. Ann Inst H Poincaré Anal Non Linéaire 1998;15(4):493–516.
[65] Dancer EN. On the Dirichlet problem for weakly non-linear elliptic partial differential equations. Proc Roy Soc Edinburgh Sect A 1976/1977;76(4):283–300.
[66] Dancer EN. Generic domain dependence for nonsmooth equations and the open set problem for jumping nonlinearities. Topol Methods Nonlinear Anal 1993;1(1):139–50.
[67] Dancer EN, Perera K. Some remarks on the Fucík spectrum of the p-Laplacian and critical groups. J Math Anal Appl 2001;254(1):164–77.
[68] De Figueiredo D, Girardi M, Matzeu M. Semilinear elliptic equations with dependence on the gradient via mountain-pass techniques. Differential Integral Equations 2004;17(1–2):119–26.
[69] De Figueiredo D, Gossez J-P. On the first curve of the Fučik spectrum of an elliptic operator. Differential Integral Equations 1994;7(5–6):1285–302.
[70] Deng S-G. Positive solutions for Robin problem involving the $p(x)$-Laplacian. J Math Anal Appl 2009;360(2):548–60.
[71] Deimling K. Nonlinear functional analysis. Berlin: Springer-Verlag; 1985.
[72] DiBenedetto E. C^{1+a} local regularity of weak solutions of degenerate elliptic equations. Nonlinear Anal 1983;7(8):827–50.

[73] Diestel J. Sequences and series in Banach spaces. Grad Texts in Math, vol. 92. New York: Springer-Verlag; 1984.
[74] Drábek P. Solvability and bifurcations of nonlinear equations. Harlow: Longman Scientific & Technical; 1982.
[75] Drábek P. The least eigenvalue of nonhomogeneous degenerated quasilinear eigenvalue problems. Math Bohem 1995;120(2):169–95.
[76] Dunford N, Schwartz JT. Linear operators. I. General theory. London: Interscience Publishers, Ltd.; 1958.
[77] El Manouni S, Perera K, Shivaji R. On singular quasi-monotone (p, q)-Laplacian systems. Proc Roy Soc Edinburgh Sect A 2012;142(3):585–94.
[78] Evans LC, Gariepy RF. Measure theory and fine properties of functions. Stud Adv Math. Boca Raton (FL): CRC Press; 1992.
[79] Fan K. Some properties of convex sets related to fixed point theorems. Math Ann 1984;266(4):519–37.
[80] Faraci F, Motreanu D, Puglisi D. Positive solutions of quasi-linear elliptic equations with dependence on the gradient. Calc Var Partial Differential Equations 2015;54(1):525–38.
[81] Faria LFO, Miyagaki O, Motreanu D. Comparison and positive solutions for problems with the (p, q)-Laplacian and a convection term. Proc Edinb Math Soc (2) 2014;57(3):687–98.
[82] Faria LFO, Miyagaki O, Motreanu D, Tanaka M. Existence results for nonlinear elliptic equations with Leray–Lions operator and dependence on the gradient. Nonlinear Anal 2014;96:154–66.
[83] Felmer P, Montenegro M, Quass A. A note on the strong maximum principle and the compact support principle. J Differential Equations 2009;246(1):39–49.
[84] Fernández Bonder J, Rossi JD. Existence results for the p-Laplacian with nonlinear boundary conditions. J Math Anal Appl 2001;263(1):195–223.
[85] Filippov AF. On some problems of the theory of optimal control. Vestn Mosk Univ 1959;2:25–32.
[86] Fitzgibbon WE. Semilinear functional differential equations in Banach spaces. J Differential Equations 1978;29(1):1–14.
[87] Fu XL, Zhang WE. Exact null controllability of non-autonomous functional evolution system with nonlocal conditions. Acta Math Sci Ser B 2013;3(3):747–57.
[88] Fučík S. Boundary value problems with jumping nonlinearities. Čas Pěst Mat 1976;101(1):69–87.
[89] Fučík S. Solvability of nonlinear equations and boundary value problems. Dordrecht: D. Reidel Publishing Co.; 1980.
[90] Fukagai N, Narukawa K. On the existence of multiple positive solutions of quasilinear elliptic eigenvalue problems. Ann Mat Pura Appl (4) 2007;186(3):539–64.
[91] García Azorero JP, Manfredi J, Peral Alonso I. Sobolev versus Hölder local minimizers and global multiplicity for some quasilinear elliptic equations. Commun Contemp Math 2000;2(3):385–404.
[92] García-Huidobro M, Manásevich R, Ubilla P. Existence of positive solutions for some Dirichlet problems with an asymptotically homogeneous operator. Electron J Differential Equations 1995;1995(10). 22 pp.
[93] Gasiński L, Papageorgiou NS. Positive solutions for nonlinear elliptic problems with dependence on the gradient. J Differential Equations 2017;263(2):1451–76.
[94] Gasiński L, Papageorgiou NS. Nonsmooth critical point theory and nonlinear boundary value problems. Ser Math Anal Appl, vol. 8. Boca Raton (FL): Chapman & Hall; 2005.
[95] Gasiński L, Papageorgiou NS. Nonlinear analysis. Ser Math Anal Appl, vol. 9. Boca Raton (FL): Chapman & Hall; 2006.
[96] Ghergu M. Lane–Emden systems with negative exponents. J Funct Anal 2010;258(10):3295–318.
[97] Ghergu M. Lane–Emden systems with singular data. Proc Roy Soc Edinburgh Sect A 2011;141(6):1279–94.

[98] Ghoussoub N. Duality and perturbation methods in critical point theory. Cambridge Tracts in Math, vol. 107. Cambridge: Cambridge University Press; 1993.
[99] Giacomoni J, Hernández J, Moussaoui A. Quasilinear and singular systems: the cooperative case. In: Nonlinear elliptic partial differential equations. Contemp Math, vol. 540. Providence (RI): Amer. Math. Soc.; 2011. p. 79–94.
[100] Giacomoni J, Hernández J, Sauvy P. Quasilinear and singular elliptic systems. Adv Nonlinear Anal 2013;2(1):1–41.
[101] Giacomoni J, Schindler I, Takáč P. Sobolev versus Hölder local minimizers and existence of multiple solutions for a singular quasilinear equation. Ann Sc Norm Super Pisa Cl Sci (5) 2007;6(1):117–58.
[102] Giannessi F, Khan A. Regularization of non-coercive quasi variational inequalities. Control Cybernet 2000;29(1):91–110.
[103] Gilbarg D, Trudinger N. Elliptic partial differential equations of second order. Classics Math. Berlin: Springer-Verlag; 2001.
[104] Goeleven D, Motreanu D, Dumont Y, Rochdi M. Variational and hemivariational inequalities: theory, methods and applications, vol. I. Unilateral analysis and unilateral mechanics. Nonconvex Optim Appl, vol. 69. Boston: Kluwer Academic Publishers; 2003.
[105] Goeleven D, Motreanu D. Variational and hemivariational inequalities: theory, methods and applications, vol. II. Unilateral problems. Nonconvex Optim Appl, vol. 70. Boston: Kluwer Academic Publishers; 2003.
[106] Gol'dshtein V, Motreanu D, Motreanu VV. Non-homogeneous Dirichlet boundary value problems in weighted Sobolev spaces. Complex Var Elliptic Equ 2015;60(3):372–91.
[107] Granas A, Dugundji J. Fixed point theory. Springer Monogr Math. New York: Springer-Verlag; 2003.
[108] Gwinner J. On a new class of differential variational inequalities and a stability result. Math Program Ser B 2013;139(1–2):205–21.
[109] Hai DD. On a class of singular p-Laplacian boundary value problems. J Math Anal Appl 2011;383(2):619–26.
[110] Haslinger J, Panagiotopoulos PD. Optimal control of systems governed by hemivariational inequalities. Existence and approximation results. Nonlinear Anal 1995;24(1):105–19.
[111] Hernández J, Mancebo FJ, Vega JM. Positive solutions for singular semilinear elliptic systems. Adv Differential Equations 2008;13(9–10):857–80.
[112] Hu S, Papageorgiou NS. Handbook of multivalued analysis, vol. I. Theory. Math Appl. Dordrecht: Kluwer Academic Publishers; 1997.
[113] Iannizzotto A, Papageorgiou NS. Existence of three nontrivial solutions for nonlinear Neumann hemivariational inequalities. Nonlinear Anal 2009;70(9):3285–97.
[114] Jin Z. Multiple solutions for a class of semilinear elliptic equations. Proc Amer Math Soc 1997;125(12):3659–67.
[115] Kamenskii M, Obukhovskii V, Zecca P. Condensing multivalued maps and semilinear differential inclusions in Banach spaces. De Gruyter Ser Nonlinear Anal Appl, vol. 7. Berlin: Walter de Gruyter & Co.; 2001.
[116] Kavian O. Introduction à la théorie des points critiques et applications aux problèmes elliptiques. Math Appl. Paris: Springer-Verlag; 1993.
[117] Kenmochi N. Nonlinear operators of monotone type in reflexive Banach spaces and nonlinear perturbations. Hiroshima Math J 1974;4:229–63.
[118] Khan AA, Motreanu D. Existence theorems for elliptic and evolutionary variational and quasi-variational inequalities. J Optim Theory Appl 2015;67(3):1136–61.
[119] Khan AA, Tammer C, Zălinescu C. Regularization of quasi variational inequalities. Optimization 2015;64(8):1703–24.
[120] Kim Y-H. A global bifurcation for nonlinear equations with nonhomogeneous part. Nonlinear Anal 2009;71(12):e738–43.
[121] Kim I-S, Kim Y-H. Global bifurcation for equations involving nonhomogeneous operators in an unbounded domain. Nonlinear Anal 2010;73(4):1057–64.

[122] Kinderlehrer D, Stampacchia G. An introduction to variational inequalities and their applications. Philadelphia (PA): SIAM; 1987.
[123] Kluge R. On some parameter determination problems and quasi-variational inequalities. In: Theory of nonlinear operators, Proc. fifth internat. summer school. Berlin: Akademie-Verlag; 1978. p. 129–39.
[124] Kravchuk AS, Neittaanmäki PJ. Variational and quasi-variational inequalities in mechanics. Dordrecht: Springer; 2007.
[125] Kyritsi ST, O'Regan D, Papageorgiou NS. Existence of multiple solutions for nonlinear Dirichlet problems with a nonhomogeneous differential operator. Adv Nonlinear Stud 2010;10(3):631–57.
[126] Ladyzhenskaya OA, Uraltseva NN. Linear and quasilinear elliptic equations. New York: Academic Press; 1968.
[127] Lê A. Eigenvalue problems for the p-Laplacian. Nonlinear Anal 2006;64(5):1057–99.
[128] Li X, Huang N, O'Regan D. Differential mixed variational inequalities in finite dimensional spaces. Nonlinear Anal 2010;72(9–10):3875–86.
[129] Lieberman GM. Boundary regularity for solutions of degenerate elliptic equations. Nonlinear Anal 1988;12(11):1203–19.
[130] Lieberman GM. The natural generalization of the natural conditions of Ladyzhenskaya and Ural'tseva for elliptic equations. Comm Partial Differential Equations 1991;16(2–3):311–61.
[131] Lindqvist P. On the equation $\operatorname{div}(|\nabla u|^{p-2}\nabla u) + \lambda|u|^{p-2}u = 0$. Proc Amer Math Soc 1990;109(1):157–64.
[132] Liu Z. Existence results for evolution noncoercive hemivariational inequalities. J Math Anal Appl 2004;120(2):417–27.
[133] Liu Z. Existence results for quasilinear parabolic hemivariational inequalities. J Differential Equations 2008;244(6):1395–409.
[134] Liu Z, Li X. Approximate controllability for a class of hemivariational inequalities. Nonlinear Anal Real World Appl 2015;22:581–91.
[135] Liu Z, Li X, Motreanu D. Approximate controllability for nonlinear evolution hemivariational inequalities in Hilbert spaces. SIAM J Control Optim 2015;531(5):3228–44.
[136] Liu Z, Loi NV, Obukhovskii V. Existence and global bifurcation of periodic solutions to a class of differential variational inequalities. Internat J Bifur Chaos Appl Sci Engrg 2013;23:1350125.
[137] Liu Z, Lv J, Sakthivel R. Approximate controllability of fractional functional evolution inclusions with delay in Hilbert spaces. IMA J Math Control Inform 2014;31(3):363–83.
[138] Liu Z, Zeng S. Existence results for a class of hemivariational inequalities involving the stable (g, f, a)-quasimonotonicity. Topol Methods Nonlinear Anal 2016;47(1):195–217.
[139] Liu Z, Zeng S, Motreanu D. Evolutionary problems driven by variational inequalities. J Differential Equations 2016;260(9):6787–99.
[140] Lunsford ML. Generalized variational and quasi-variational inequalities with discontinuous operators. J Optim Theory Appl 1997;214(1):245–63.
[141] Lloyd NG. Degree theory. Cambridge Tracts in Math, vol. 73. Paris: Cambridge University Press; 1978.
[142] Mahmudov NI. Approximate controllability of evolution systems with nonlocal conditions. Nonlinear Anal 2008;68(3):536–46.
[143] Marano SA, Mosconi SJN, Papageorgiou NS. Multiple solutions to (p, q)-Laplacian problems with resonant concave nonlinearity. Adv Nonlinear Stud 2016;16(1):51–65.
[144] Marano SA, Motreanu D, Puglisi D. Multiple solutions to a Dirichlet eigenvalue problem with p-Laplacian. Topol Methods Nonlinear Anal 2013;42(2):277–91.
[145] Marano SA, Papageorgiou NS. Constant-sign and nodal solutions of coercive (p, q)-Laplacian problems. Nonlinear Anal 2013;77:118–29.
[146] Marcus M, Mizel VJ. Nemitsky operators on Sobolev spaces. Arch Ration Mech Anal 1973;15:347–70.

[147] Marino A, Micheletti AM, Pistoia A. A nonsymmetric asymptotically linear elliptic problem. Topol Methods Nonlinear Anal 1994;4(2):289–339.

[148] Marino A, Prodi G. Metodi perturbativi nella teoria di Morse. Boll Unione Mat Ital (4) 1975;11:1–32.

[149] Martínez SR, Rossi JD. Isolation and simplicity for the first eigenvalue of the p-Laplacian with a nonlinear boundary condition. Abstr Appl Anal 2002;7(5):287–93.

[150] Martínez SR, Rossi JD. On the Fučik spectrum and a resonance problem for the p-Laplacian with a nonlinear boundary condition. Nonlinear Anal 2004;59(6):813–48.

[151] Mawhin J. Generic properties of nonlinear boundary value problems. In: Differential equations and mathematical physics. Boston (MA): Academic Press; 1992. p. 217–34.

[152] Mawhin J, Willem M. Critical point theory and Hamiltonian systems. Appl Math Sci, vol. 74. New York: Springer-Verlag; 1989.

[153] Migórski S, Ochal A. Optimal control of parabolic hemivariational inequalities. J Global Optim 2000;17(1–4):285–300.

[154] Migórski S, Ochal A. Quasi-static hemivariational inequality via vanishing acceleration approach. SIAM J Math Anal 2009;41(4):1415–35.

[155] Migórski S, Ochal A, Sofonea M. Nonlinear inclusions and hemivariational inequalities. Models and analysis of contact problems. Adv Mech Math, vol. 26. New York: Springer; 2013.

[156] Miyajima S, Motreanu D, Tanaka M. Multiple existence results of solutions for the Neumann problems via super- and sub-solutions. J Funct Anal 2012;262(5):1921–53.

[157] Mihăilescu M. An eigenvalue problem possessing a continuous family of eigenvalues plus an isolated eigenvalue. Commun Pure Appl Anal 2011;10(2):701–8.

[158] Montenegro M, Suarez A. Existence of a positive solution for a singular system. Proc Roy Soc Edinburgh Sect A 2010;140(2):435–47.

[159] Mosco U. Convergence of convex sets and of solutions of variational inequalities. Adv Math 1969;3:512–85.

[160] Mosco U. Implicit variational problems and quasi variational inequalities. In: Nonlinear operators and the calculus of variations. Lecture Notes in Math. Berlin: Springer; 1976. p. 83–156.

[161] Motreanu D. Generic existence of Morse functions on infinite-dimensional Riemannian manifolds and applications. In: Global differential geometry and global analysis. Lecture Notes in Math, vol. 1481. Berlin: Springer; 1991. p. 175–84.

[162] Motreanu D, Motreanu VV. Transversality theory with applications to differential equations. In: Essays in mathematics and its applications. Cham: Springer; 2016. p. 133–57.

[163] Motreanu D, Motreanu VV. Generic existence of nondegenerate homoclinic solutions. Lobachevskii J Math 2017;38(2):322–9.

[164] Motreanu D, Motreanu VV, Moussaoui A. Location of nodal solutions for quasilinear elliptic equations with gradient dependence. Discrete Contin Dyn Syst Ser S 2018;11(2):293–307.

[165] Motreanu D, Motreanu VV, Papageorgiou NS. A unified approach for multiple constant sign and nodal solutions. Adv Differential Equations 2007;12(12):1363–92.

[166] Motreanu D, Motreanu VV, Papageorgiou NS. Multiple nontrivial solutions for nonlinear eigenvalue problems. Proc Amer Math Soc 2007;135(1):3649–58.

[167] Motreanu D, Motreanu VV, Papageorgiou NS. Positive solutions and multiple solutions at non-resonance, resonance and near resonance for hemivariational inequalities with p-Laplacian. Trans Amer Math Soc 2008;360(5):2527–45.

[168] Motreanu D, Motreanu VV, Papageorgiou NS. Multiple constant sign and nodal solutions for nonlinear Neumann eigenvalue problems. Ann Sc Norm Super Pisa Cl Sci (5) 2011;10(2):729–55.

[169] Motreanu D, Motreanu VV, Papageorgiou NS. Topological and variational methods with applications to nonlinear boundary value problems. New York: Springer; 2014.

[170] Motreanu D, Moussaoui A. Existence and boundedness of solutions for a singular cooperative quasilinear elliptic system. Complex Var Elliptic Equ 2014;59(2):285–96.

[171] Motreanu D, Moussaoui A. A quasilinear singular elliptic system without cooperative structure. Acta Math Sci Ser B Engl Ed 2014;34(3):905–16.

[172] Motreanu D, Moussaoui A. An existence result for a class of quasilinear singular competitive elliptic systems. Appl Math Lett 2014;38:33–7.

[173] Motreanu D, Moussaoui A, Zhang Z. Positive solutions for singular elliptic systems with convection term. J Fixed Point Theory Appl 2014;19(3):2165–75.

[174] Motreanu D, Panagiotopoulos PD. Minimax theorems and qualitative properties of the solutions of hemivariational inequalities. Optim Appl, vol. 29. Dordrecht: Kluwer Academic Publishers; 1999.

[175] Motreanu D, Papageorgiou NS. Multiple solutions for nonlinear elliptic equations at resonance with a nonsmooth potential. Nonlinear Anal 2004;56(8):1211–34.

[176] Motreanu D, Papageorgiou NS. Multiple solutions for nonlinear Neumann problems driven by a nonhomogeneous differential operator. Nonlinear Anal 2011;139(10):3527–35.

[177] Motreanu D, Rădulescu V. Variational and non-variational methods in nonlinear analysis and boundary value problems. Optim Appl, vol. 67. Dordrecht: Kluwer Academic Publishers; 2003.

[178] Motreanu D, Tanaka M. Sign-changing and constant-sign solutions for p-Laplacian problems with jumping nonlinearities. J Differential Equations 2010;249(12):3352–76.

[179] Motreanu D, Tanaka M. Existence of solutions for quasilinear elliptic equations with jumping nonlinearities under the Neumann boundary condition. Calc Var Partial Differential Equations 2012;43(1–2):231–64.

[180] Motreanu D, Tanaka M. Generalized eigenvalue problems of nonhomogeneous elliptic operators and their application. Pacific J Math 2013;265(1):151–84.

[181] Motreanu D, Tanaka M. Multiple existence results of solutions for quasilinear elliptic equations with a nonlinearity depending on a parameter. Ann Mat Pura Appl 2014;193(4):1255–82.

[182] Motreanu D, Tanaka M. On a positive solution for (p,q)-Laplace equation with indefinite weight. Minimax Theory Appl 2016;1(1):1–20.

[183] Motreanu D, Tanaka M. Existence of positive solutions for nonlinear elliptic equations with convection terms. Nonlinear Anal 2017;152:38–60.

[184] Motreanu D, Vetro C, Vetro F. A parametric Dirichlet problem for systems of quasilinear elliptic equations with gradient dependence. Numer Funct Anal Optim 2016;37(12):1551–61.

[185] Motreanu D, Vetro C, Vetro F. Systems of quasilinear elliptic equations with dependence on the gradient via subsolution–supersolution method. Discrete Contin Dyn Syst Ser S 2018;11(2):309–21.

[186] Motreanu D, Winkert P. On the Fučik spectrum for the p-Laplacian with a Robin boundary condition. In: Nonlinear analysis. Springer Optim Appl, vol. 68. New York: Springer; 2012. p. 471–85.

[187] Motreanu D, Winkert P. On the Fučik spectrum for the p-Laplacian with a Robin boundary condition. Nonlinear Anal 2011;74(14):4671–81.

[188] Motreanu VV. Uniqueness results for a Dirichlet problem with variable exponent. Commun Pure Appl Anal 2010;9(5):1399–410.

[189] Motreanu VV. Existence results for constrained quasivariational inequalities. Abstr Appl Anal 2013:427908.

[190] Moussaoui A, Khodja B, Tas S. A singular Gierer–Meinhardt system of elliptic equations in $\mathbb{R}^N$. Nonlinear Anal 2009;71(3–4):708–16.

[191] Mugnai D, Papageorgiou NS. Wang's multiplicity result for superlinear (p,q)-equations without the Ambrosetti–Rabinowitz condition. Trans Amer Math Soc 2014;366(9):4919–37.

[192] Naniewicz Z, Panagiotopoulos PD. Mathematical theory of hemivariational inequalities and applications. Monogr Textb Pure Appl Math, vol. 188. New York: Marcel Dekker; 1995.

[193] Noor MA, Rassias TM. A class of projection methods for general variational inequalities. J Math Anal Appl 2002;268(1):334–43.

[194] Panagiotopoulos PD. Hemivariational inequalities. Applications in mechanics and engineering. Appl Math Sci, vol. 44. Berlin: Springer; 1993.
[195] Pang J. Frictional contact models with local compliance: semismooth formulation. Z Angew Math Mech 2008;88(6):454–71.
[196] Pang J, Stewart DE. Differential variational inequalities. Math Program Ser A 2007;113(2):345–424.
[197] Park JY, Park SH. Optimal control problems for anti-periodic quasi-linear hemivariational inequalities. Optimal Control Appl Methods 2007;28(4):275–87.
[198] Pazy A. Semigroups of linear operators and applications to partial differential equations. Appl Math Sci, vol. 44. New York: Springer-Verlag; 1983.
[199] Peng ZJ, Liu ZH, Liu XY. Boundary hemivariational inequality problems with doubly nonlinear operators. Math Ann 2013;356(4):1339–58.
[200] Perera K, Agarwal RP, O'Regan D. Morse theoretic aspects of p-Laplacian type operators. Math Surveys Monogr, vol. 161. Providence (RI): American Mathematical Society; 2010.
[201] Pohožaev SI. Equations of the type $\Delta u = f(x, u, Du)$. Mat Sb (NS) 1980;113(155)(2(10)):324–38.
[202] Ponstein J. Approaches to the theory of optimization. Cambridge Tracts in Math, vol. 77. Cambridge: Cambridge University Press; 1980.
[203] Prado H, Ubilla P. Existence of nonnegative solutions for generalized p-Laplacians. In: Reaction diffusion systems. Lect Notes Pure Appl Math, vol. 194. New York: Dekker; 1998. p. 289–98.
[204] Precup R. Linear and semilinear partial differential equations. An introduction. de Gruyter Textbook. Basel: Walter de Gruyter & Co.; 2013.
[205] Pucci P, Serrin J. The maximum principle. Progr Nonlinear Differential Equations Appl, vol. 73. Basel: Birkhäuser Verlag; 2007.
[206] Pucci P, Serrin J. The strong maximum principle revisited. J Differential Equations 2004;196(1):1–66.
[207] Rabinowitz PH. Minimax methods in critical point theory with applications to differential equations. Providence (RI): American Mathematical Society; 1986. Published for the Conference Board of the Mathematical Sciences, Washington, DC.
[208] Rădulescu V, Repovš D. Existence results for variational–hemivariational problems with lack of convexity. Nonlinear Anal 2010;73(1):99–104.
[209] Ricceri B. Infinitely many solutions of the Neumann problem for elliptic equations involving the p-Laplacian. Bull Lond Math Soc 2001;33(3):331–40.
[210] Ricceri B. Nonlinear eigenvalue problems. In: Handbook of nonconvex analysis and applications. Somerville (MA): Int. Press; 2010. p. 543–95.
[211] Robinson SB. On the second eigenvalue for nonhomogeneous quasi-linear operators. SIAM J Math Anal 2004;35(5):1241–9.
[212] Ruiz D. A priori estimates and existence of positive solutions for strongly nonlinear problems. J Differential Equations 2004;199(1):96–114.
[213] Rykaczewski K. Approximate controllability of differential inclusions in Hilbert spaces. Nonlinear Anal 2012;75(5):2701–12.
[214] Saut J-C. Generic properties of nonlinear boundary value problems. In: Partial differential equations. Warsaw: Banach Center Publ., PWN; 1983. p. 331–51.
[215] Saut J-C, Temam R. Generic properties of nonlinear boundary value problems. Comm Partial Differential Equations 1979;4:293–319.
[216] Schechter M. The Fučík spectrum. Indiana Univ Math J 1994;43(4):1139–57.
[217] Showalter RE. Monotone operators in Banach space and nonlinear partial differential equations. Math Surveys Monogr, vol. 49. Providence (RI): American Mathematical Society; 1997.
[218] Sidiropoulos NE. Existence of solutions to indefinite quasilinear elliptic problems of p–q-Laplacian type. Electron J Differential Equations 2010;(162). 23 pp.
[219] Smale S. An infinite dimensional version of Sard's theorem. Amer J Math 1965;87:861–6.

[220] Struwe M. A note on a result of Ambrosetti and Mancini. Ann Mat Pura Appl 1982;131(4):107–15.

[221] Szulkin A. Minimax principles for lower semicontinuous functions and applications to nonlinear boundary value problems. Ann Inst H Poincaré Anal Non Linéaire 2009;3(2):77–109.

[222] Tanaka M. Existence of constant sign solutions for the p-Laplacian problems in the resonant case with respect to Fučik spectrum. SUT J Math 2009;45(2):149–66.

[223] Tanaka M. The antimaximum principle and the existence of a solution for the generalized p-Laplace equations with indefinite weight. Differ Equ Appl 2012;4:581–613.

[224] Tanaka M. Generalized eigenvalue problems for (p, q)-Laplacian with indefinite weight. J Math Anal Appl 2014;419(2):1181–92.

[225] Tanaka M. Existence of the Fucík type spectrums for the generalized p-Laplace operators. Nonlinear Anal 2012;7:3407–35.

[226] Tanaka M. Existence of a positive solution for quasilinear elliptic equations with nonlinearity including the gradient. Bound Value Probl 2013;2013(173). 11 pp.

[227] Tolksdorf P. Regularity for a more general class of quasilinear elliptic equations. J Differential Equations 1984;51(1):126–50.

[228] Troianiello GM. Elliptic differential equations and obstacle problems. New York: Plenum Press; 1987.

[229] Vázquez JL. A strong maximum principle for some quasilinear elliptic equations. Appl Math Optim 1984;12(3):191–202.

[230] Wei J, Wang J. Infinitely many homoclinic orbits for the second order Hamiltonian systems with general potentials. J Math Anal Appl 2010;366(2):694–9.

[231] Winkert P. Constant-sign and sign-changing solutions for nonlinear elliptic equations with Neumann boundary values. Adv Differential Equations 2010;18(5–6):561–99.

[232] Winkert P. Multiple solution results for elliptic Neumann problems involving set-valued nonlinearities. J Math Anal Appl 2011;377(1):121–34.

[233] Wu M, Yang Z. A class of p–q-Laplacian type equation with potentials eigenvalue problem in $\mathbb{R}^N$. Bound Value Probl 2009:185319.

[234] Yin H, Yang Z. A class of p–q-Laplacian type equation with concave–convex nonlinearities in bounded domain. J Math Anal Appl 2011;382(2):843–55.

[235] Yin H, Yang Z. Multiplicity of positive solutions to a p–q-Laplacian equation involving critical nonlinearity. Nonlinear Anal 2012;75(6):3021–35.

[236] Yosida K. Functional analysis. Classics Math. Berlin: Springer-Verlag; 1995.

[237] Zeidler E. Nonlinear functional analysis and its applications. I. Fixed-point theorems. New York: Springer-Verlag; 1986.

[238] Zeidler E. Nonlinear functional analysis and its applications, vol. II B. Berlin: Springer; 1990.

[239] Zeidler E. Nonlinear functional analysis and its applications. IV. Applications to mathematical physics. New York: Springer-Verlag; 1988.

[240] Zhang J, Li S, Whang Y, Xue X. Multiple solutions for semilinear elliptic equations with Neumann boundary condition and jumping nonlinearities. J Math Anal Appl 2010;371(2):682–90.

[241] Zhang Q, Motreanu D. Existence and blow-up rate of large solutions of $p(x)$-Laplacian equations with large perturbation and gradient terms. Adv Differential Equations 2016;21(7–8):699–734.

[242] Zou HH. A priori estimates and existence for quasi-linear elliptic equations. Calc Var Partial Differential Equations 2008;33(4):417–37.

Index

T

U

V

W

Y

Kolibri 360 GmbH
Ferdinand-Jühlke-Straße 7
99095 Erfurt, Deutschland
produktsicherheit@kolibri360.de